Sustainable Agriculture Reviews

Volume 53

Other Publications by Dr. Eric Lichtfouse

Books

Scientific Writing for Impact Factor Journals
https://www.novapublishers.com/catalog/product_info.php?products_id=42242

Environmental Chemistry
http://www.springer.com/978-3-540-22860-8

Sustainable Agriculture
Volume 1: http://www.springer.com/978-90-481-2665-1
Volume 2: http://www.springer.com/978-94-007-0393-3

Book series

Environmental Chemistry for a Sustainable World
http://www.springer.com/series/11480

Sustainable Agriculture Reviews
http://www.springer.com/series/8380

Journal

Environmental Chemistry Letters
http://www.springer.com/10311

Sustainable agriculture is a rapidly growing field aiming at producing food and energy in a sustainable way for humans and their children. Sustainable agriculture is a discipline that addresses current issues such as climate change, increasing food and fuel prices, poor-nation starvation, rich-nation obesity, water pollution, soil erosion, fertility loss, pest control, and biodiversity depletion.

Novel, environmentally-friendly solutions are proposed based on integrated knowledge from sciences as diverse as agronomy, soil science, molecular biology, chemistry, toxicology, ecology, economy, and social sciences. Indeed, sustainable agriculture decipher mechanisms of processes that occur from the molecular level to the farming system to the global level at time scales ranging from seconds to centuries. For that, scientists use the system approach that involves studying components and interactions of a whole system to address scientific, economic and social issues. In that respect, sustainable agriculture is not a classical, narrow science. Instead of solving problems using the classical painkiller approach that treats only negative impacts, sustainable agriculture treats problem sources.

Because most actual society issues are now intertwined, global, and fast-developing, sustainable agriculture will bring solutions to build a safer world. This book series gathers review articles that analyze current agricultural issues and knowledge, then propose alternative solutions. It will therefore help all scientists, decision-makers, professors, farmers and politicians who wish to build a safe agriculture, energy and food system for future generations.

More information about this series at https://link.springer.com/bookseries/8380

Mohammad Faizan • Shamsul Hayat
Fangyuan Yu
Editors

Sustainable Agriculture Reviews 53

Nanoparticles: A New Tool to Enhance Stress Tolerance

Springer

Editors
Mohammad Faizan
Department of Silviculture
Nanjing Forestry University
Nanjing, Jiangsu, China

Fangyuan Yu
Department of Silviculture
Nanjing Forestry University
Nanjing, Jiangsu, China

Shamsul Hayat
Department of Botany, Faculty of Life Sciences, Plant Physiology & Biochemistry Section
Aligarh Muslim University
Aligarh, Uttar Pradesh, India

ISSN 2210-4410 ISSN 2210-4429 (electronic)
Sustainable Agriculture Reviews
ISBN 978-3-030-86875-8 ISBN 978-3-030-86876-5 (eBook)
https://doi.org/10.1007/978-3-030-86876-5

This Springer imprint is published by the registered company Springer Nature Switzerland AG
The registered company address is: Gewerbestrasse 11, 6330 Cham, Switzerland

Preface

Nanotechnology is an interdisciplinary science, growing rapidly due to its incorporation into various scientific sectors, such as medicine, engineering, pharmacy, industry, wastewater treatment, paper preservation, and agriculture. Nanoparticles (NPs) are aggregates of atoms or molecules at a nano-size, ranging from 1 to 100 nm. Nanomaterials have unique physical and chemical properties that differ from their bulk forms. These properties include a larger surface-to-volume ratio in comparison with bulk materials, size, shape, crystal structure, thermal stability, charge, and zeta potential. This leads to increased sustainability which is an essential requirement for enhanced agricultural production and resistance against abiotic stress. In countries with high population such as India and China, the use of NPs might have a remarkable influence on mitigating poverty, hunger, malnutrition, and child mortality. Metal and metal oxides NPs have been reported to promote plant nutrition, increase seed germination, boost plant defense system, and enhance soil conditions. This book is dedicated to presenting the latest developments of the role of NPs to enhance stress tolerance in plants.

The book is comprised of 16 chapters. Chapter 1 discusses the role of quantum dots, polymeric NPs, and dendrimers in emphasizing crops tolerate biotic and abiotic stresses. Chapter 2 deals with climate change mitigation and nanotechnology. Nanoparticles as a new promising tool to increase plant immunity against abiotic stress is covered in Chap. 3. In Chap. 4, mechanism of nanomaterial-mediated alleviation is discussed. Chapter 5 highlights the alleviation mechanism of drought stress in plants using metal NPs. Chapter 6 addresses the role of various NPs in countering heavy metal, salt, and drought stress in plants. Chapter 7 deals with mode of action and signaling of NPs to alleviate abiotic stress in crop plants. In Chap. 8, impact of NPs and nanoparticle-coated biomolecules to ameliorate salinity stress in plants, with special reference to physiological, biochemical, and molecular mechanism of action, is discussed briefly. Chapter 9 describes the impact of carbon nanotubes on abiotic stress response in plants. In Chap. 10, responses of crop plants under nanoparticles supply in alleviating biotic and abiotic stresses are discussed. This chapter also addresses the modulation of gene expression by nanoparticle supply. Chapter 11 deals with the nanotechnological approaches for efficient delivery of plant ingredients. In Chap. 12, enhancement of stress tolerance of crop plants by ZnO nanoparticles is presented. Chapter 13 demonstrates the effects of nanoparticles on alleviating phytotoxicity of soil heavy metals: potential for enhancing phytoremediation. In Chap. 14, bio-fabricated silver nanoparticles are discussed. Chapter 15 covers nano-proteomics of stress tolerance in crop plants. This chapter also discusses the proteomic technology adapted by plant sciences, plant proteomics under nanoparticles stress, and nanoparticles uptake and mode of action under stressed conditions. Chapter 16 covers the role of chitosan nanoparticles in regulation of plant physiology under abiotic stress.

This book is not an encyclopedia of reviews but rather a compendium of newly composed, integrated, and illustrated chapters describing our knowledge of nanoparticles and their role in minimizing biotic and abiotic stress. The chapters incorporate both theoretical and practical aspects and may serve as baseline information for future research through which significant developments are possible. It is intended that this book will be useful to the students, researchers, and instructors, both in universities and research institutes, especially in relation to biological and agricultural sciences.

With great pleasure, we extend our sincere thanks to all the contributors for their timely response, their excellent and up-to-date contributions, and their consistent support and cooperation. We are thankful to all who have helped us in any capacity during the preparation of this volume. We are extremely thankful to Springer for their expeditious acceptance of our proposal and completion of the review process. The subsequent cooperation and understanding by their staff are gratefully acknowledged. We express our sincere thanks to our family members for all the support they provided and the neglect and loss they suffered during the preparation of this book.

Finally, we are thankful to the Almighty who provided and guided all the channels to work in cohesion on the concept to the development of the final version of this treatise, *Nanoparticles: A New Tool to Enhance Stress Tolerance.*

Nanjing, China — Mohammad Faizan
Aligarh, India — Shamsul Hayat
Nanjing, China — Fangyuan Yu

Contents

About the Editors

Mohammad Faizan is a postdoctoral scientist/assistant teacher in the Department of Silviculture, College of Forestry, Nanjing Forestry University, Nanjing, China. In 2014, he obtained his M.Sc. degree in botany from Chhatrapati Shahu Ji Maharaj University, Kanpur, India, and his Ph.D. in 2018 from Aligarh Muslim University, Aligarh, India. The findings of his work have revealed that exogenous application of brassinosteroids and nanoparticles improved the yield and quality of plants under natural as well as stressful conditions. He has written one book published by Springer addressing nanotechnology for plant growth and development and has authored or co-authored six book chapters. Findings of his research works have been published in *Ecotoxicology and Environmental Safety*, *Nitric Oxide*, *Plant Physiology and Biochemistry*, the *Journal of Plant Growth Regulation*, *Current Protein & Peptide Science*, *Photosynthetica*, the *South African Journal of Botany*, *IET Nanotechnology*, the *Russian Journal of Plant Physiology*, *Biocell*, and many more, with more than 540 google scholar citation and 10 h-index. He is also member of the editorial board of the *Journal of Plant Sciences* and is also member of various societies.

Shamsul Hayat is a professor in the Department of Botany, Aligarh Muslim University, Aligarh, India. He received his Ph.D. degree in botany from Aligarh Muslim University, Aligarh, India. Before joining the department as faculty, he worked there as research associate and young scientist. He has also worked as associate professor in King Saud University, Riyadh, Saudi Arabia; as a BOYSCAST Fellow at the National Institute of Agrobiological Sciences, Tsukuba, Japan; and as visiting scientist through INSA-Bilateral exchange program in the Faculty of Biology and Chemistry, Institute of Biology, Department of Plant Biochemistry and Toxicology, University of Bialystok, Poland. His major area of research includes plant hormone, nanoscience, and abiotic stress in plants. It has been reported from his group that phytohormones such as brassinosteroids and salicylic acid play an important role in increasing the photosynthetic efficiency of the plant and regulate the antioxidant system even under abiotic stress. He is also studying the protein profiling in hormone-treated plants under abiotic stress. Dr. Hayat has been awarded

Prof. Hira Lal Chakravarty Award by the Indian Science Congress Association, Kolkata, India; Associate of National Academy of Agricultural Sciences, New Delhi, India; BOYSCAST Fellow by Department of Science & Technology, Government of India, New Delhi; and Young Scientist by the Association of the Advancement of Science, Aligarh, India. Dr. P.S. Khankhoje gold medal has also been awarded to him by Dr. Punjab Rao Deshmukh Krishi Vidyapeeth, Akola. He has been the principal investigator of various projects sanctioned by different agencies, and has guided seven Ph.D. students and two M.Phil. students besides a number of M.Sc. students. Dr. Hayat has published more than 160 research papers in leading journals of the world such as *Ecotoxicology and Environmental Safety*, *Environmental & Experimental Botany*, *Plant Physiology & Biochemistry*, *Environmental Pollution*, *Nitric Oxide*, *Protoplasma*, *Plant Signalling & Behaviour*, *Photosynthetica*, and *Acta Physiologiae Plantarum* with high impact factor, and he has also published 12 books by Kluwer Academic, Springer, Wiley-VCH, Science Publisher, and Narosa Publishing House. Besides this, he has also published 25 book chapters. Dr. Hayat has presented his work at several national and international conferences in Japan, Brazil, Spain, China, Poland, and Saudi Arabia. He is a regular reviewer and on the panel of editorial boards of National and International journals. Dr. Hayat is also member of important national and international scientific societies. Recently, he has been appointed as Indian representative in Asia Association of Plant Scientists.

Fangyuan Yu is a professor in the Department of Silviculture, College of Forestry, Nanjing Forestry University, Nanjing, China. He received his Ph.D. degree in botany from Nanjing Forestry University, Nanjing, China. Dr. Yu was trained in the Department of Forest Science, The University of British Columbia, for 1 year. His major areas of research are tree seed physiology, nursery practices, seedling ecophysiology, mechanism of seed oil, and flower color and fragrance development. He has been the principal investigator of various projects sanctioned by different agencies of China and guided several Ph.D. students. Findings of his research works have been published in international journals of high impact factor such as *BMC Plant Biology*, *Tree*, the *International Journal of Agriculture & Biology*, *Metabolites*, *Forests*, the *Journal of Plant Growth Regulation*, *New Forests*, and the *Journal of Forestry Research*, and he has also published three books.

Contributors

Heba S. Abbas National Organization for Drug Control and Research, Egyptian Drug Authority, Cairo, Egypt

Magdi T. Abdelhamid Botany Department, National Research Centre, Cairo, Egypt

Awatif M. Abdulmajeed Biology Department, Faculty of Science, University of Tabuk, Umluj, Kingdom of Saudi Arabia

Farrag F. B. Abu-Ellail Department of Breeding and Genetics, Sugar Crops Research Institute, Agricultural Research Center (ARC), Giza, Egypt

Muhammad Ansar Ahmad Department of Plant Sciences, Quaid-i-Azam University, Islamabad, Pakistan

Suliman Mohammed Alghanem Biology Department, Faculty of Science, Tabuk University, Tabuk, Kingdom of Saudi Arabia

Haifa Abdulaziz S. Alhaithloul Biology Department, College of Science, Jouf University, Sakaka, Kingdom of Saudi Arabia

Baber Ali Department of Plant Sciences, Quaid-i-Azam University, Islamabad, Pakistan

Taghreed S. Alnusaire Biology Department, College of Science, Jouf University, Sakaka, Saudi Arabia

Amna Department of Plant Sciences, Quaid-i-Azam University, Islamabad, Pakistan

Yamshi Arif Plant Physiology Section, Department of Botany, Faculty of Life Sciences, Aligarh Muslim University, Aligarh, India

Muhammad Atif Azeem Department of Plant Sciences, Quaid-i-Azam University, Islamabad, Pakistan

Babli Genetics Lab Department of Plant Science, M.J.P.R.U, Bareilly, Uttar Pradesh, India

Mahendar Singh Bhinda ICAR-Vivekananda Parvatiya Krishi Anusandhan Sansthan, Almora, Uttarakhand, India

Hassan Javed Chaudhary Department of Plant Sciences, Quaid-i-Azam University, Islamabad, Pakistan

Padmanabh Dwivedi Department of Plant Physiology, Institute of Agricultural Sciences, Banaras Hindu University, Varanasi, India

Nesma ElSayed Botany Department, Faculty of Science, Benha University, Benha, Egypt

Mohammad Faizan Collaborative Innovation Center of Sustainable Forestry in Southern China, College of Forest Science, Nanjing Forestry University, Nanjing, China

Atikah Farooq Department of Plant Sciences, Quaid-i-Azam University, Islamabad, Pakistan

Akankhya Guru Department of Plant Physiology, Institute of Agricultural Sciences, Banaras Hindu University, Varanasi, India

Murtaza Hasan Department of Biotechnology, The Institute of Biochemistry, Biotechnology and Bioinformatics, The Islamia University of Bahawalpur, Bahawalpur, Pakistan

Hanan A. Hashem Botany Department, Faculty of Science, Ain Shams University, Cairo, Egypt

Shamsul Hayat Plant Physiology Lab Department of Botany, Aligarh Muslim University, Aligarh, India

Anjuman Hussain Plant Physiology Lab Department of Botany, Aligarh Muslim University, Aligarh, India

Ramakanth Illa Department of Chemistry, Rajiv Gandhi University of Knowledge Technologies, Nuzvid, India

Sumera Iqbal Department of Botany, Lahore College for Women University, Lahore, Pakistan

Khajista Jabeen Department of Botany, Lahore College for Women University, Lahore, Pakistan

Ajai Kumar Jaitly Microbiology Lab Department of Plant Science, M.J.P.R.U, Bareilly, Uttar Pradesh, India

Sumera Javad Department of Botany, Lahore College for Women University, Lahore, Pakistan

Muhammad Tariq Javed Department of Botany, Government College University, Faisalabad, Pakistan

Radwan Khalil Faculty of Science, Botany Department, Benha University, Benha, Egypt

Marek Kolenčík Department of Soil Science and Geology, Faculty of Agrobiology and Food Resources, Slovak University of Agriculture in Nitra, Nitra, Slovakia
Nanotechnology Centre, VŠB Technical University of Ostrava, Ostrava, Czech Republic

Muddukrishnaiah Kotakonda Department of Technology, Anna University, Chennai, India
Centre for Biotechnology and Phyto Pharmacognosy Research, Coimbatore, India

Asha Kumari ICAR-Vivekananda Parvatiya Krishi Anusandhan Sansthan, Almora, Uttarakhand, India

Sindy Kurtinová Institute of Laboratory Research on Geomaterials, Faculty of Natural Sciences, Comenius University in Bratislava, Bratislava, Slovakia

Weitao Liu MOE Key Laboratory of Pollution Processes and Environmental Criteria, Tianjin Engineering Center of Environmental Diagnosis and Contamination Remediation College of Environmental Science and Engineering, Nankai University, Tianjin, China

Amira S. Mahmoud Department of Environmental Studies, Institute of Graduate Studies and Research, Alexandria University, Alexandria, Egypt

Saba Maqsood Department of Botany, Lahore College for Women University, Lahore, Pakistan

Saba Mehreen Department of Chemistry, Quaid-i-Azam University, Islamabad, Pakistan

Tatiana Minkina Academy of Biology and Biotechnology, Southern Federal University, Rostov-on-Don, Russia

Anayat Rasool Mir Plant Physiology Lab Department of Botany, Aligarh Muslim University, Aligarh, India

Ghazala Mustafa Department of Plant Sciences, Quaid-i-Azam University, Islamabad, Pakistan

Iqra Naseer Department of Botany, Lahore College for Women University, Lahore, Pakistan

Nazish Microbiology Lab Department of Plant Science, M.J.P.R.U, Bareilly, Uttar Pradesh, India

Manoj Parihar ICAR-Vivekananda Parvatiya Krishi Anusandhan Sansthan, Almora, Uttarakhand, India

Sameer H. Qari Biology Department, Aljumum University College, Umm Al – Qura University, Makkah, Saudi Arabia

Ayesha Qayyum Department of Plant Sciences, Quaid-i-Azam University, Islamabad, Pakistan

Vishnu D. Rajput Academy of Biology and Biotechnology, Southern Federal University, Rostov-on-Don, Russia

Zarqa Riaz Department of Plant Sciences, Quaid-i-Azam University, Islamabad, Pakistan

Payel Saha Department of Plant Physiology, Bidhan Chandra Krishi Vishwavidyalaya, Mohanpur, West Bengal, India

Soumya Kumar Sahoo Crop Physiology and Biochemistry Division, ICAR-NRRI, Cuttack, India

Maysoun M. Saleh Genetic Resources Department, General Commission for Scientific Agricultural Research (GCSAR), Damascus, Syria

Khaled F. M. Salem Department of Plant Biotechnology, Genetic Engineering and Biotechnology Research Institute (GEBRI), University of Sadat City, Sadat City, Egypt
Department of Biology, College of Science and Humanitarian Studies, Shaqra University, Qwaieah, Saudi Arabia

Martin Šebesta Institute of Laboratory Research on Geomaterials, Faculty of Natural Sciences, Comenius University in Bratislava, Bratislava, Slovakia

Shareen Department of Agricultural Microbiology, Faculty of Natural Sciences, Aligarh Muslim University, Aligarh, India

Binny Sharma Department of Plant Physiology, Institute of Agricultural Sciences, Banaras Hindu University, Varanasi, Uttar Pradesh, India

Husna Siddiqui Plant Physiology Section, Department of Botany, Faculty of Natural Sciences, Aligarh Muslim University, Aligarh, India

Ajit Singh Malaysia School of Biosciences, University of Nottingham, Semenyih, Malaysia

Mona H. Soliman Botany and Microbiology Department, Faculty of Science, Cairo University, Giza, Egypt
Biology Department, Faculty of Science, Taibah University, Yanbu, Kingdom of Saudi Arabia

Sidra tul Muntha Laboratory of Germplasm Innovation and Molecular Breeding, College of Agriculture and Biotechnology, Zhejiang University, Hangzhou, China

Aurang Zeb MOE Key Laboratory of Pollution Processes and Environmental Criteria, Tianjin Engineering Center of Environmental Diagnosis and Contamination Remediation College of Environmental Science and Engineering, Nankai University, Tianjin, China

Yinlong Zhang Co-Innovation Center for Sustainable Forest in Southern China, Nanjing Forestry University, Nanjing, China

Chapter 1
Role of Quantum Dots, Polymeric NPs and Dendrimers in Emphasizing Crops Tolerate Biotic and Abiotic Stresses

Khaled F. M. Salem, Maysoun M. Saleh, Farrag F. B. Abu-Ellail, Heba S. Abbas, and Amira S. Mahmoud

Abstract Quantum dots made of carbon (CQDs), carbon nanoparticles with fluorescent properties, are a modern form of carbon nanomaterials that have attracted attention as potential competitors to traditional semiconductor quantum dots. Colloidal quantum droplets have the desirable properties of low toxicity, environmental friendliness, low cost, and clear synthetic methods, in addition to their related optical properties. Furthermore, surface passivation and activation of colloid quantum droplets allow physical and chemical properties to be controlled. Colloidal quantum dots have a diverse set of uses since their discovery, including biosensing, bioimaging, nanomedicine, photocatalysis, chemical identifying and electrocatalysis. This chapter provides a past and present overview of the role of quantum dots, polymeric nanoparticles (NPs) and dendrimers in emphasizing crops tolerate biotic and abiotic stresses, properties, with a focus on the synthesis, characterizations and

K. F. M. Salem (✉)
Department of Plant Biotechnology, Genetic Engineering and Biotechnology Research Institute (GEBRI), University of Sadat City, Sadat City, Egypt

Department of Biology, College of Science and Humanitarian Studies, Shaqra University, Qwaieah, Saudi Arabia
e-mail: khaled.salem@gebri.usc.edu.eg; khaledfathi@su.edu.sa

M. M. Saleh
Genetic Resources Department, General Commission for Scientific Agricultural Research (GCSAR), Damascus, Syria
e-mail: mzainsamasaleh@gcsar.gov.sy

F. F. B. Abu-Ellail
Department of Breeding and Genetics, Sugar Crops Research Institute, Agricultural Research Center (ARC), Giza, Egypt

H. S. Abbas
National Organization for Drug Control and research, Egyptian Drug Authority, Giza, Egypt

A. S. Mahmoud
Department of Environmental studies, Institute of Graduate Studies and Research, Alexandria University, Alexandria, Egypt

M. Faizan et al. (eds.), *Sustainable Agriculture Reviews 53*, Sustainable Agriculture Reviews 53, https://doi.org/10.1007/978-3-030-86876-5_1

toxicity of quantum dots, as well as some discussion of the challenges and opportunities in this exciting and promising field.

Keywords Biotic and abiotic stresses · Dendrimers · Polymeric nanoparticles · Sustainability development · Toxicity · Quantum dots

Abbreviations

QDs	Quantum dots
CdS	Cadmium Sulphide
CdTe	Cadmium Tellurium
H	Hydrogen
O_2	Oxygen
Ni	Nickel
Al	Aluminum

1.1 Introduction

1.1.1 Background

Several challenges concerning climatic changes and food security are emerging as a result of the unprecedented increase in world population (Ebert 2014). The main goal to achieve in the agriculture sector is to improve crop production, for that researchers worked in the last decades to explore more methods to apply such as nanomaterials in agriculture (Fig. 1.1) (Sharma et al. 2019). Recently, nanomaterials are playing a critical function in most appliance constructions in agronomy science, as well as enhancing plant tolerance to stresses (Marx et al. 2004). NMs are applied in agriculture for many purposes such as improving crop yield and protection (Abd-Elsalam and Prasad 2018; Chen et al. 2015; Gupta et al. 2018; Ismail et al. 2017; Karuppanapandian et al. 2011). The nano-scale size considers as the main base of nanomaterials categorization, that diverges in classical concepts from 1–100 nanometer (Ashkavand et al. 2015). Prajitha et al. (2019) showed the dimension variations between nano-scale and both micro and macro-scale forms (Fig. 1.2). The exceptional characters of size, high level of activity and large surface area which were resulted from improving nanomaterials (Prasad et al. 2016), make them proper for handling in several fields of sciences (Sharma et al. 2012) such as fertilizers, pesticides and herbicides in agriculture (Prasad et al. 2014). Scientists studied different effects of nanomaterials on crop defense systems and growth under different environments (Drew and Armstrong et al. 2002). Different kinds of NMs like quantum dots and polymeric NMs were produced and used in many appliances for different fields of sciences (Chauhan et al. 2016; Joshi et al. 2018; Talreja et al. 2016), for

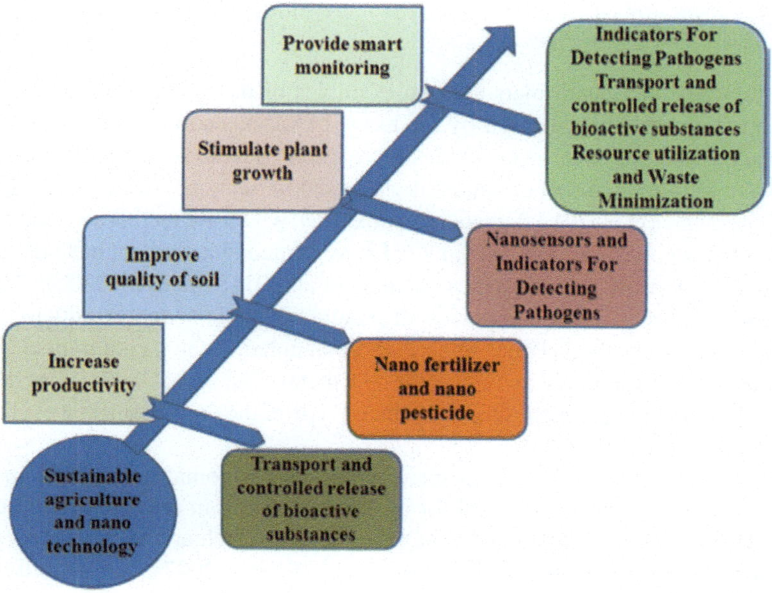

Fig. 1.1 Some aspects of using nanotechnology in the agriculture sector. (Source: Sharma et al. 2019)

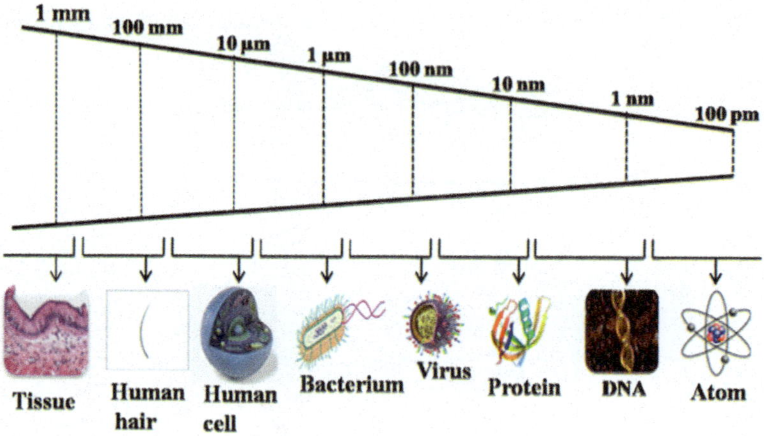

Fig. 1.2 Illustrative demonstration of nano-scale comparing to both macro& micro-scale (Source: Prajitha et al. 2019)

instance in agriculture (Arora et al. 2012). Scientists confirmed the NMs ability to strengthen plant growth under water shortage (Khan et al. 2017) and other abiotic stress (Prasad et al. 2017), in addition to biotic stress since NMs act as an antioxidant material (Aziz et al. 2019). Scientists classified the nanomaterials into many types according to their structure to carbon-based NMs, Metal-based NMs, nano-composites (include polymeric NMs) and dendrimers (Fig. 1.3) (Prajitha et al. 2019).

1.1.2 Definition

Quantum dots and dendrimers are two examples of the engineered nanomaterials which are commonly improved by the usage of various sorts of materials such as metals, organic polymer, metal-oxides and carbon (Bandala et al. 2019; Rana et al. 2013). A well-defined sort of nanomaterials used in many applications is the quantum dots (Matea et al. 2017; Valizadeh et al. 2012), which are metalloid-nanocrystals considered according to Rocha et al. (2017) as semiconductors that have distinctive electrical and visual characteristics (Mansur et al. 2011; Hardman 2006). The core of QDs contains different composites of (magnetic, noble or semiconductors) metals, with an outlier shell (Pelley et al. 2009) that enhance the quantum dots characteristics and diminishes the toxicity of the core metal (Chang et al. 2006; King-Heiden et al. 2009) that differs according to the metal type in the core (Jamieson et al. 2007; Libralato et al. 2017).

In the previous ten years, fluorescent synthesized nanomaterials called quantum dots appeared as a promising tool for molecule tracking in internal cells of organisms (Pierobon and Cappello 2012). Quantum dots which are nano-crystals (Chan et al. 2002) demonstrate unique levels of energy (Klimov 2007), high stability under light (Wu et al. 2006), in which they release definite color which differs according to the sort of material (Bakalova et al. 2004; Deb et al. 2011). The first types of nanomaterials that were applied in sciences contain quantum dots (Azzazy et al. 2007; Chan and Nie 2018) such as the quantum dots whose core consists of CdSe-ZnS (Deerinck 2008). The diameter of classical-quantum dots is about 2–10

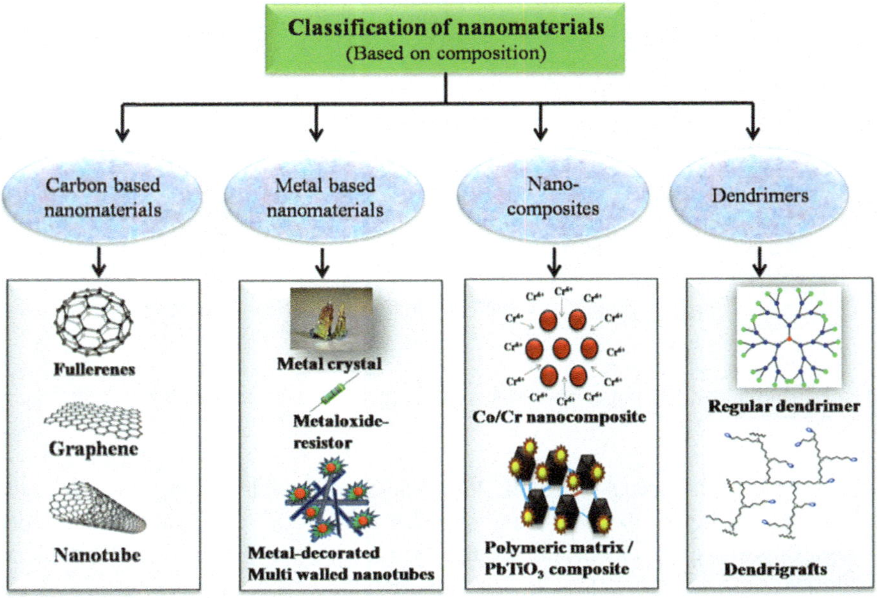

Fig. 1.3 Nanomaterials classification according to their structure. (Source: Prajitha et al. 2019)

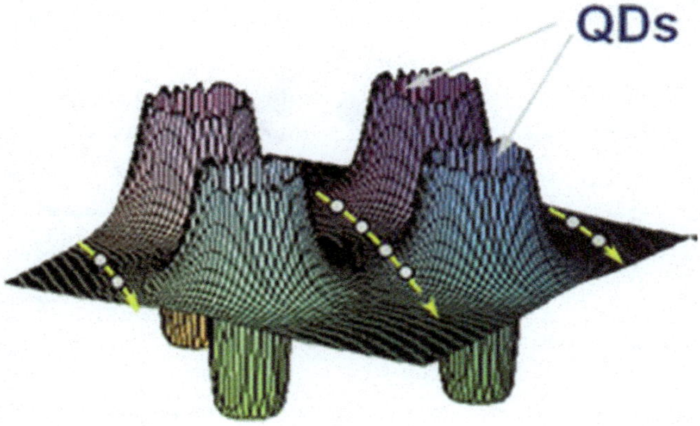

Fig. 1.4 Regular distribution of quantum dots construction with barriers. (Source: Sablon et al. 2012)

nanometer (Ghasemi et al. 2009). The usage of quantum dots as markers raised rapidly in the last few years (Corredor et al. 2009) because they can manage the difficulties in photo-bleaching because of their high stability to light (Fig. 1.4) (Sablon et al. 2012).

Polymeric nanomaterials are used especially to both nano-capsules and nanospheres and other polymeric NMs. They were rapidly improved to act as a key elements in several scientific sectors, for instance pollution, as well as ecological management concerning releasing targeted ingredients (Chandra et al. 2015) (Fig. 1.5). It is expected that polymeric nanomaterials will show the way to more innovations in the next years or decades (Pradhan et al. 2013).

Dendrimers consider as a famous example of some nanomaterials that consist of polymeric organic molecules displayed in a special structure that is branched and symmetrical, this structure enables the polymeric nanomaterials to achieve the insertion or conjunction of the needed remedial agent on their cap (Weir et al. 2012). This sort of NMs which are distinguished from the three-dimension structure is called dendrimers as shown in Fig. 1.6. The dendrimers consist of the core which is located in the middle and the dendrons (the branched composition), while the internal space is identified as the central cavity (Baan et al. 2006).

1.2 Properties of Quantum Dots (QDs)

QDs consist of a central semiconductor core surrounded by a stable inorganic salt shell. These semiconductor nanocrystals have a size range between 1 and 20 nm (Mansur 2010). There are several examples of core/shell structured QDs like cadmium telluride, ZnS, cadmium telluride/ cadmium sulfide or core/shell/shell such as Cadmium Telluride/ Cadmium Sulfide/ Zinc Sulfide. Because of their small scale,

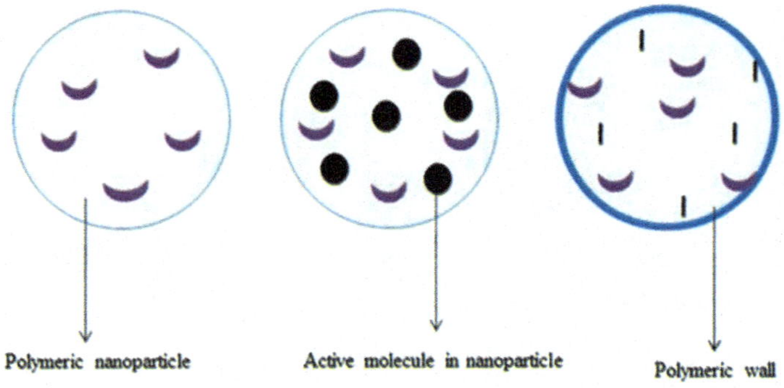

Fig. 1.5 Formulation of the release- management of polymeric nanoparticles. (Source: Campos et al. 2018)

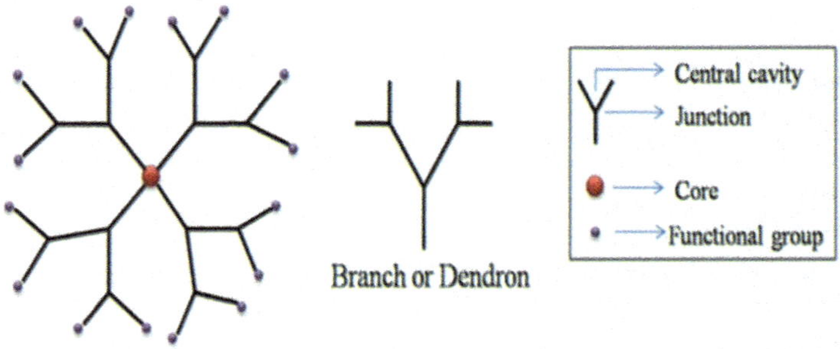

Fig. 1.6 The dendrimer regular composition. (Source: Prajitha et al. 2019)

QDs emit physically confined electrons, also known as quantum confinements. They are subject to several unique optical, mechanical and chemical features as a result of their spatial features, which are not present in other materials. QDs properties can be simply under control by adjusting their size, which has given their potential for several medical and industrial applications (Hong 2019).

1.2.1 *Optical Characteristics of Quantum Dots (QDs)*

QDs have distinct optical characteristics due to their physical phenomenon called quantum confinements. When a photon with enough energy hits QDs, the electron will be excited from an electron-filled band to an electron-deficient band and a positive hole will appear in its position, which is known as the exciton phenomenon. The exciton size in QDs is smaller than QDs size, resulting in quantum confinement as compressing the exciton into QDs material (Hong 2019; Simon et al. 2010). Recently, the use of QD has gained a lot of popularity. It has been investigated in several commercial fields, including biomedical applications (Field et al. 2020).

Drug delivery is the most common application of QDs. The ability to detect fluorescent-carrying drug dots makes them ideal for imaging in the diagnosis of cancer diseases (Badıllı et al. 2020). The wide absorption and narrow radiation spectra of QDs can be adjusted. Photochemical oxidation and photobleaching are no problems for QDs. A noticeable stimulated shift is a property that enables the recognition of low signal fluorescence. Aside from their distinctive optical features, QDs have favorable electrical features (Badıllı et al. 2020).

The size of QDs is largely responsible for their unique features. They have an inherent bandgap through which electrons can be promoted by incident light excitation as semiconducting materials. Dissimilar bulk semiconducting substances, QDs are too scarce to form incessant electron-filled and electron-deficient bands. QDs, on the other hand, create an electronic structure more akin to the distinct electronic states, which exist in the single atoms, the greater the QD, the smaller distance between energy bands. As a result, as a QD grows larger, its electron-filled and electron-deficient bands formed continuously and the bandgap approaches that of the bulk materials. So, changing the size of QDs allows the variation in the bandgap, which is never happened in a single atom (Hong 2019; Simon et al. 2010). As a result of releasing UV light to the small-scaled QDs, electrons that migrate to greater energy states travel a greater energy distance until they miss further energy and coming back to a balanced level. Consequently, QDs emit more energy from visible light and appear to be blue. However, by releasing UV to greater QDs, the electrons that migrate to the greater energy state have a narrower energy difference as they miss further energy and coming back to the balanced level, so they radiate visible light of a reduced amount of energy, which appears to be red. Light emission differs according to the elements that make up the center of QDs. CDs emit light in the UV blue region, but CdTe emits light in the far red/near-infrared region (Krishnaswamy and Orsat 2017).

1.2.2 *Effect of Core-Shell Materials on Quantum Dots (QDs) Bioactivity*

Several biologically active molecules may bind to the QDs shell surface, including proteins, peptides and lipids. For example, the thiol group improves the water solubility, when it is attached to the quantum dots surface (Idowu et al. 2008). Furthermore, polymer biomolecules are allowing QDs semiconductors to be directed to a certain organ inside the body for disease diagnosis or treatment (Wang et al. 2012). Polyethylene glycol is the lion's share commonly used polymer shell, particularly in nanomedicine. A QD external shell can be functionalized by various techniques like electrostatic interactions, physical adsorption, and multivalent chelating. Surface attachments can influence the dimensions of the QDs significantly (Clift and Stone 2012). To summarize, numerous surfaces may be added to the surface of QDs, resulting in variations in their physiochemical characteristics

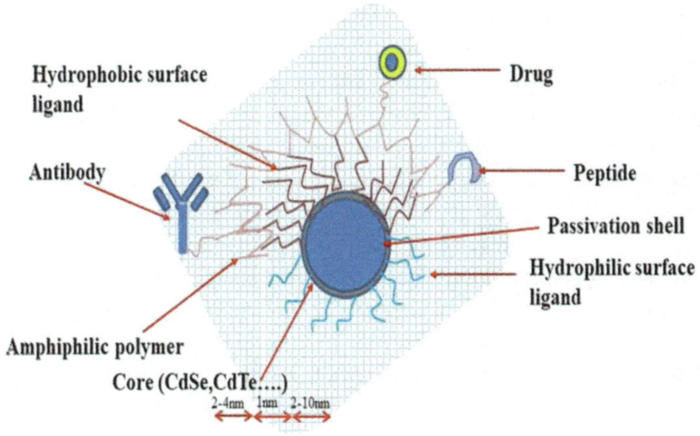

Fig. 1.7 Quantum dot coated with surface bioactive molecules. (Source: Adapted from Maysinger et al. 2007)

(Maysinger et al. 2007). The configuration and various core regions of the quantum dot, as well as the typical superficial capping agents, are depicted schematically in Fig. 1.7.

1.2.3 Electrical Characteristics of Quantum Dots (QDs)

Besides the optical features of QDs, electrical characteristics are examined by connecting metal particles to the metal electrodes using tunnel contacts as a kind of QDs. Another kind of QDs is a 3D potential well for electrons that are created by applying an electric power between the anode and cathode of a nano-varied configuration on the solid surface. In such devices, when analyzing the conductivity, the influence of the Coulomb blockade occurrence and electron's resonance tunneling are noticed (Cartoixà and Wang 2005; Pool and Owens 2003). Kharlamov et al. (2013) documented the electrical properties of CuO and NiO QDs, as well as the influence of altering the QD surface state during the chemisorption of the gas molecule on variance in electrical conductivity. They explained the electrical characteristics of a nano-varied configuration composed of two electrodes and spherical Ni and Cu nanoparticles upon the existence of CuO and NiO QDs. The results showed that electrical conductivity varied due to the chemical reactions on the surface of QDs and the chemisorption of gas molecules. The nano-varied configuration before its treatment with hydrogen, the intensity of electric current through Al/Cu layer-value and voltage polarity across Al electrode (I-V) distinctive shape is superlinear (Fig. 1.8a). Thus, by growing the voltage, the differential resistance of the nano-varied configuration decreases. The (I–V) distinctive form is independent of the existence (air, H atom) and pressure of the gas molecule = 0.15–10^5 Pascal in this situation. With voltage, the current rises linearly at first; then, once the voltage reaches an acute value, the current amplitude declines in hops of up to 10^4 times (in less than 1 s). For instance, the nano-varied configuration with copper particles, a resulting drop in voltage outcomes in the incomplete restoration of the electrical conductivity for this configuration, and its postponement is noticed (Fig. 1.8b).

However, in the nano-varied configuration with Ni particles, a reduction in the voltage does not happen. The transformation from non-conducting to conducting level is given only by the shift in the polarity of the voltage. Because of the adsorption of H-atom on its surface and following desorption, the turn-on and subsequent turn-off of the hydrogen atom flux are followed by a flat rise and accompanying decline in the nano-varied configurations electrical conductivity (Fig. 1.9).

The adsorption of oxygen molecules alters the nano-varied configuration electrical conductivity and it is I–V features significantly. During the nano-varied

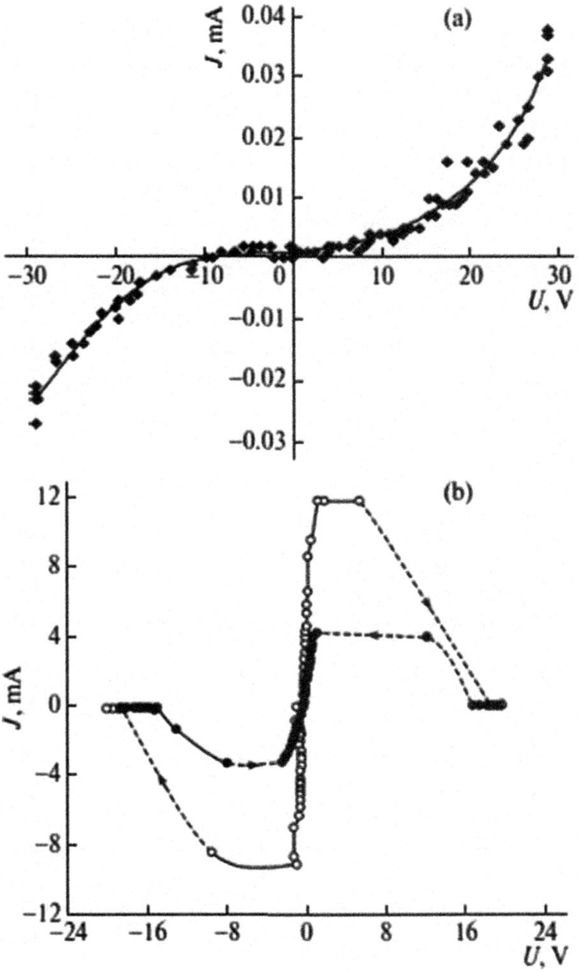

Fig. 1.8 Dependence of the I-V characteristics. (a) without and (b) with the exposure of Al electrode with the placed Cu particles to H-plasma for 0.5 h. The white and black circles correspond to an elevation and reduction in voltage, respectively. The spotted line defines the jump-like differences in the electric current and voltage. θ = 0.9, T = 295 K. (Source: Kharlamov et al. 2013)

configuration sensitivity to the O_2 molecule environment, the electrical conductivity decreases non-monotonically (Fig. 1.10). In the case of curve 6, a share of a negatively differential resistance in the intensity of electric current–polarity of voltage feature was detected after the nano-varied configuration exposure to the O_2 atom environment, with its reversible value reaching zero with raising the intensity of electric current (Fig. 1.10). Also, a present saturation is found in the intensity of electric current–polarity of voltage feature after a decline in the resistance of the nano-varied configuration to 5–8 Ω under the exposure of hydrogen atoms. Initially, the present amplitude of the current rises with voltage as stated by linear law but becomes independent of it afterward the break in the intensity of electric current–polarity of voltage feature (Kharlamov et al. 2013).

1.3 Synthesis and Characterization

QDs are an archetype nanomaterial that has piqued the interest of a wide range of researchers. Because of the quantum size effect, QDs have unique electronic and optical features, for example, narrow and symmetrical emission, admirable photostability, and a substantial Stokes change. (Alivisatos 1996; He and Ma 2014; Howes et al. 2014; Zrazhevskiy et al. 2010). Advanced QD synthesis methods for various routes include living organism synthesis, biomolecule-templated synthesis, polymer-templated synthesis, top-down synthesis and bottom-up colloidal synthesis. The most popular method for making highly dispersed QDs for bioapplications is bottom-up colloidal synthesis (Goryacheva et al. 2015). Pang and Gong (2019) studied QDs effects on plants and their future applications in botany (Fig. 1.11a). Figure 1.11b shows how surface modification techniques including ligand exchange, silanization of surface, and coating of the cross-linked shell can improve QD

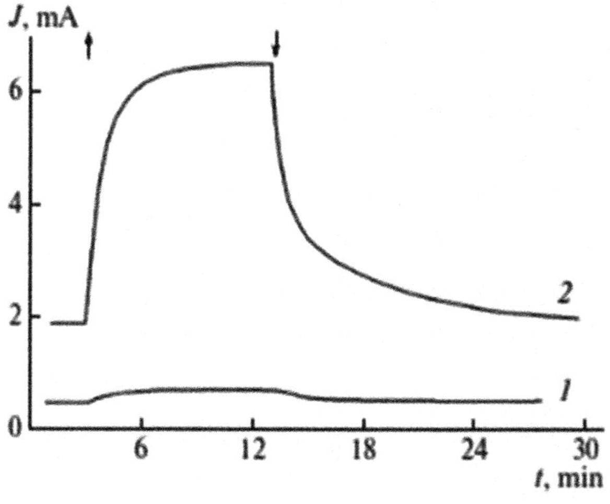

Fig. 1.9 Dependence of the I-V characteristics on the time next to (↑) turn on and (↓) turn off of the H-atom flux. U: (1) 1; (2) 3 V. θ = 0.8, T = 295 K. (Source: Kharlamov et al. 2013)

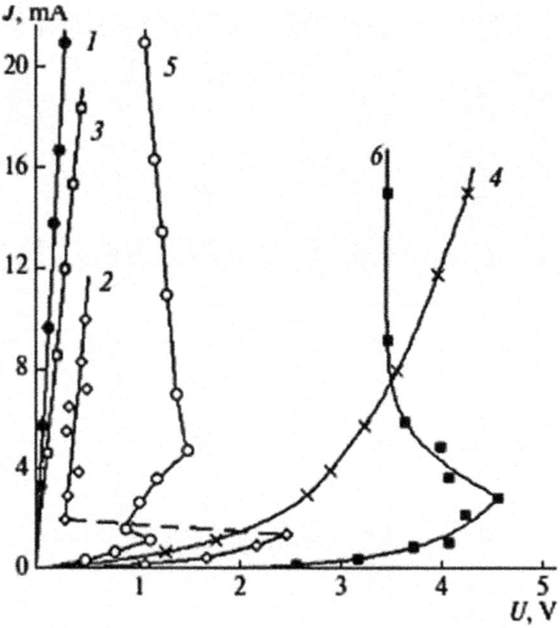

Fig. 1.10 I–V features of the Al–Ni particle layer–Al structure after exposure of Al electrode with the placed Ni nanoparticles to the H atom environment for (1) 3; consequent exposure to O_2 molecule: (2) 0.5, (3) 3, (4) 16 h; to H- atom: (5) 0.17 h; to O- atom: (6) 0.17 h; θ = 0.9, T = 295 K. (Source: Kharlamov et al. 2013)

attributes including solubility, electrophoretic mobility, and targeted activities (Pang and Gong 2019).

For applications such as bioimaging or drug delivery, QDs are frequently attached to function biomolecules especially enzymes, proteins or nucleic acids by adsorption or covalently coupling (Karakoti et al. 2015). The detection of new forms of QDs has resulted in their use in biological and medical fields (Pang and Gong 2019) (Fig. 1.12). Moreover, the absorption, transport and transformation processes of QDs affect plants and ecosystems, according to meta-analysis studies (Oh et al. 2016).

Finally, it was discovered that the size and surface properties of QDs are the most important correlated attributes for QD toxicity (Nair et al. 2010). Understanding trophic transmission and ecological consequences of QDs, particularly in various matrices, requires effectively describing QD features including distribution of hydrated particle size, the rate dissolving, surface charge, the chemical and physical properties. (Schwab et al. 2016). To ensure the safety of newly emerging nanotechnological applications, it is critical to determine the environmental fate of QDs (Miralles et al. 2012) and to define relevant standards and protocols for the evaluation of QDs phytotoxicity (Farre et al. 2011; Montes et al. 2017; Nel et al. 2006).

Another study was conducted one of the most important human efforts in the history of humanity has been the cultivation of plants for food. So, carbon quantum dot CDs have sparked a lot of interest in agriculture as a way to enhance plant

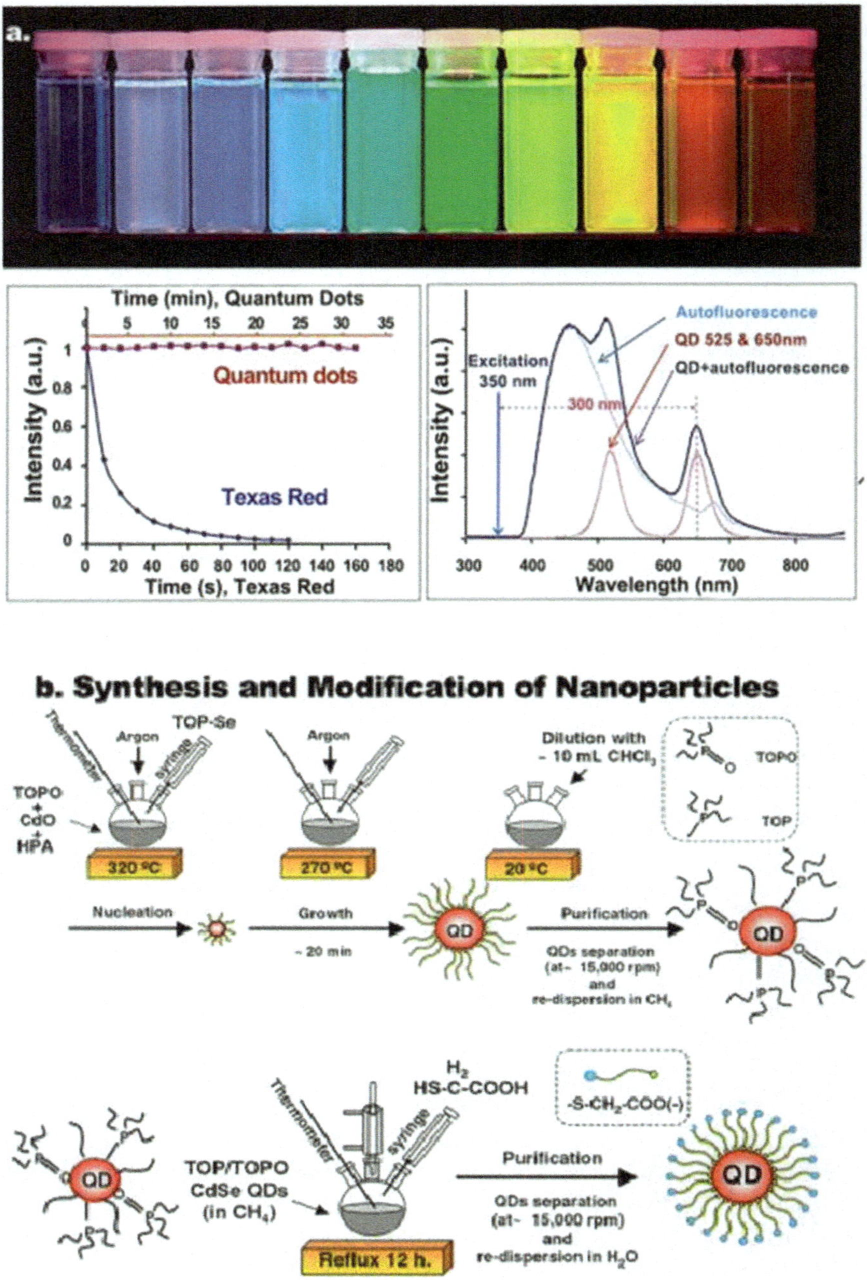

Fig. 1.11 (**a**) QD probes photophysical features. (Source: Zrazhevskiy et al. 2010) and (**b**) A typical luminous QD production method and surface modification using CdO as a precursor. (Source: Costa-Fernandez et al. 2006)

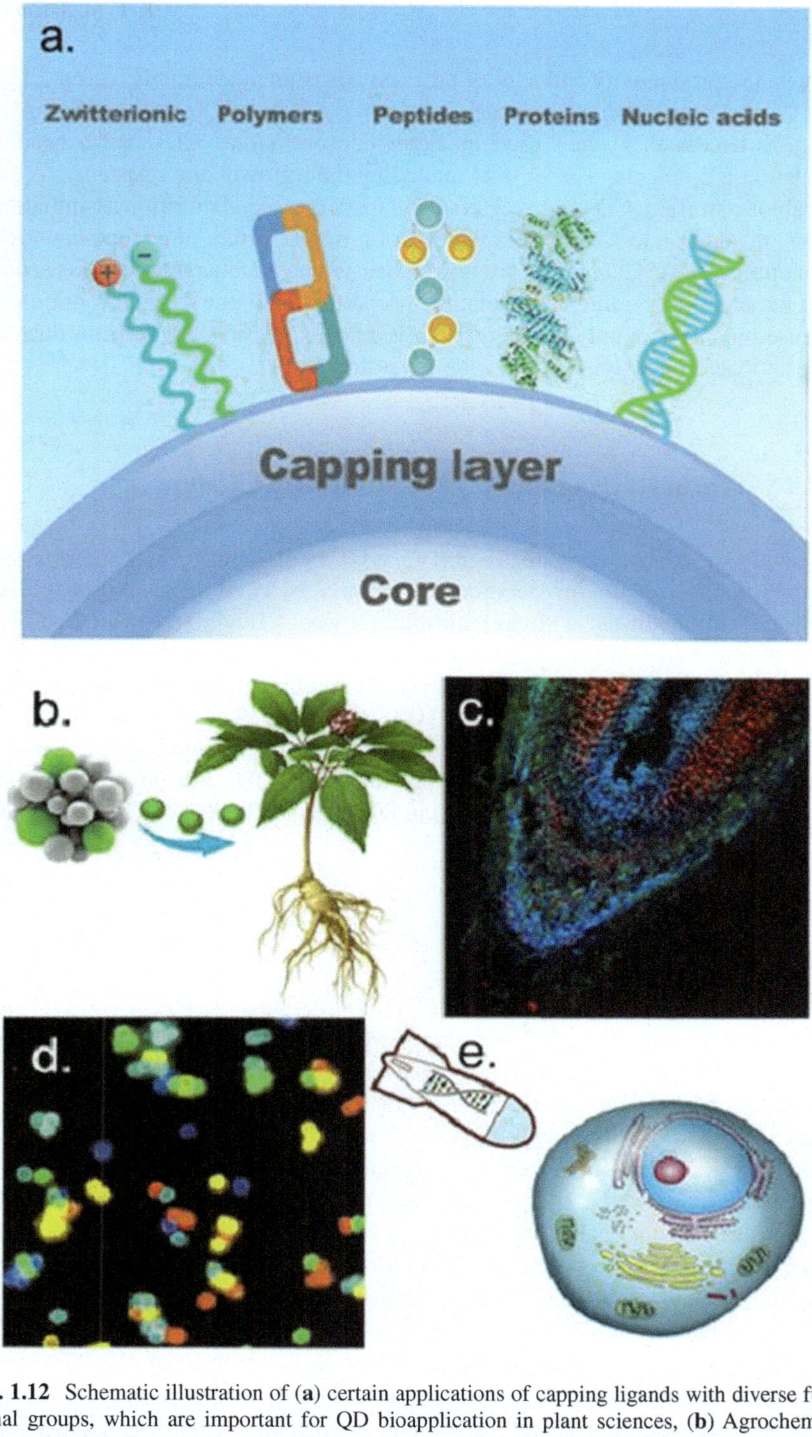

Fig. 1.12 Schematic illustration of (**a**) certain applications of capping ligands with diverse functional groups, which are important for QD bioapplication in plant sciences, (**b**) Agrochemical nanocarriers based on quantum dots (Source: Pang and Gong 2019), (**c**) Cell imaging and labeling with QDs. (Source: Wang et al. 2014), (**d**) Biomolecular multicolor coding based on QDs. and (**e**) Genes or bioactive compounds in QD-based nanocarriers for plant cells. (Source: Han et al. 2001)

growth and productivity and a lot of progress has been made in this region. Li et al. (2019a, b) examined current advancements in the field of CD applications in agriculture., Unless they were used in larger concentrations, CDs had a beneficial impact on crop plants in most tests, including their growth and resistance to biotic and abiotic stresses. CDs can accelerate photosynthesis and nutrition assimilation in plants, in addition to Azotobacter fixation of nitrogen, promoting crop plant growth at an optimal dose. CDs can improve crop resistance to biotic/abiotic stresses due to their inherent free radical scavenging properties and effect on the antioxidant defense mechanism and disease resistance gene expression. Aside from that, CDs have antibacterial activity over a wide spectrum.

1.3.1 Synthesis of Carbon Quantum Dots (CQDs)

Mahat et al. (2020) have synthesized (CQDs) from oil palm by hydrothermal treatment carbonization at a low-temperature process. by combining 12 g of oil palm-activated carbon powder with 300 milliliter DI water. The mixture was then placed in a 500-milliliter Teflon-lined autoclave. Set the autoclave for 3 hours at 200 °C. Allow the reaction container to cool to ambient temperature before centrifuging the solution for 30 minutes at 1000 rpm. After that, the solution is filtered through a 0.2-micrometer filter. Then, the solution is filtered through 0.2 m filters to eliminate micron-sized particles. To obtain solid CQDs, the final solution was lyophilized. The CQDs were dispersed in DI water and kept at 4 °C at a concentration of 7.6 mg/mL.

1.3.2 Characterization

Mahat and Shamsudin (2020) used SEM to measure CQDs characteristics. Fig. 1.13a illustrates the morphology of The oil palm empty fruit bunches (OPEFBs) as determined by scanning electron microscopy. The matrix material covers the entire surface of the fiber in OPEFBs, which have a hard surface and a composites layer with a matrix (lignin) and reinforcement or fiber Fig. 1.13b shows the AC generated after the carbonization–activation phase. Due to the activation of carbon, the AC surfaces are coarse and have thick ridges. During the manufacture of AC (Joseph et al. 2017), the structure of hydrothermally synthesized Transmission electron microscopy was used to confirm CQDs from AC–OPEFBs (Fig. 1.13c). As seen in the histogram, The image depicts a homogeneous dispersion of spherical CQDs with a diameter of 5–7 nm (inset), according to a high-resolution TEM image (Fig. 1.13d) (Kwon et al. 2014). In comparison, CQDs have an approximate size range of 5–10 nm, with a peak of 7 1 nm, according to dynamic light scattering

Fig. 1.13 (**a**, **b**) SEM of biomass of OPEFBs, AC-OPEFBs, respectively, (**c**) TEM of histogram distribution and (**d**) Hrtem of CQDs synthesis. (Source: Mahat and Shamsudin 2020).

(DLS). (Fig. 1.14a). This discrepancy arises since DLS calculates the particle hydrodynamic scale, which is greater than the CQDs real size as calculated by TEM. The CQDs have an amorphous structure, as shown by the peaks (100) and (001) at bands 2 = 18.5° and 2 = 21.92°, respectively, in the X-Ray Diffraction (XRD) pattern (Fig. 1.14b). CQDs may have amorphous or nanocrystalline structures, and their cores are made up of sp2 graphitic (Chan et al. 2018; Wang et al. 2017).

The development of amorphous, porous graphitic carbon structures is aided by high-temperature treatment during carbonization (Sharma et al. 2017; Sudhan et al. 2017). In the XRD spectrum, the main diffraction peak occurs at 21.92°, with a complete range at a partial limit of about 5.1°, which is related to a 0.35 nm interlayer gap between the two layers. The graphite structure area planar carbon-based sheets (Gomez et al. 2018; Wang et al. 2012). Also, Rani et al. (2014) said that the activated carbon from OPEFBs. The existence of an amorphous structure was revealed by two broad diffraction peaks at 2 = 20–30° and 40–50°. Carbon rings hold the adsorbent together, making it easier to produce well-defined adsorbents.

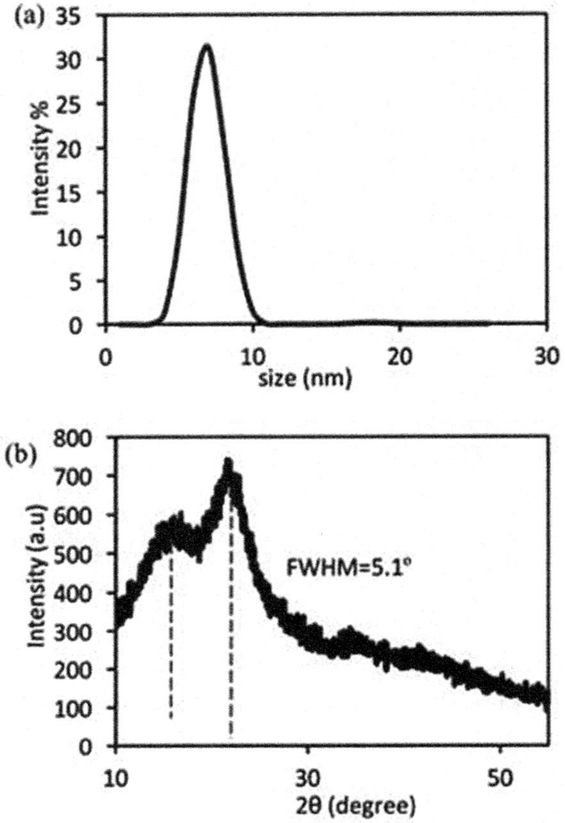

Fig. 1.14 (**a**) CQDs size distribution and (**b**) XRD pattern at CQDs at $2\theta = 18.5°$ and $2\theta = 21.92°$. (Source: Mahat and Shamsudin 2020)

1.3.3 Cadmium Selenide Quantum Dots Synthesis and Characterization (CdSe QDs)

Banerjee et al. (2021) were synthesized CdSe quantum dots for biomedical applications. They studied the CdSe QDs cytotoxicity and genotoxicity in onion as a model plant. They prepared the CdSe QDs by using the method of Nordell et al. (2005). The results illustrated that the existence of distinct cadmium and selenium peaks was revealed by EDX spectrum analysis. The presence of impurities in the copper grid may explain the occurrence of copper peaks (Fig. 1.15a). CdSe QDs are mainly monodispersed and spherical, with an average diameter of 7.69 ± 0.22 nm, according to TEM images (Fig. 1.15a). The colloidal sample of CdSe QDs is orange-red when viewed under ambient light (Fig. 1.15b). The CdSe QDs colloidal sample appears orange-red and when viewed under UV illumination, it appears yellow-green (Fig. 1.15c). The optical characterization of the QDs shows that the absorption energy is higher than the emission energy. A narrow fluorescence emission peak appeared at 558 nm and an absorption peak appeared at 442 nm in QD nanocrystals (Fig. 1.15d,e).

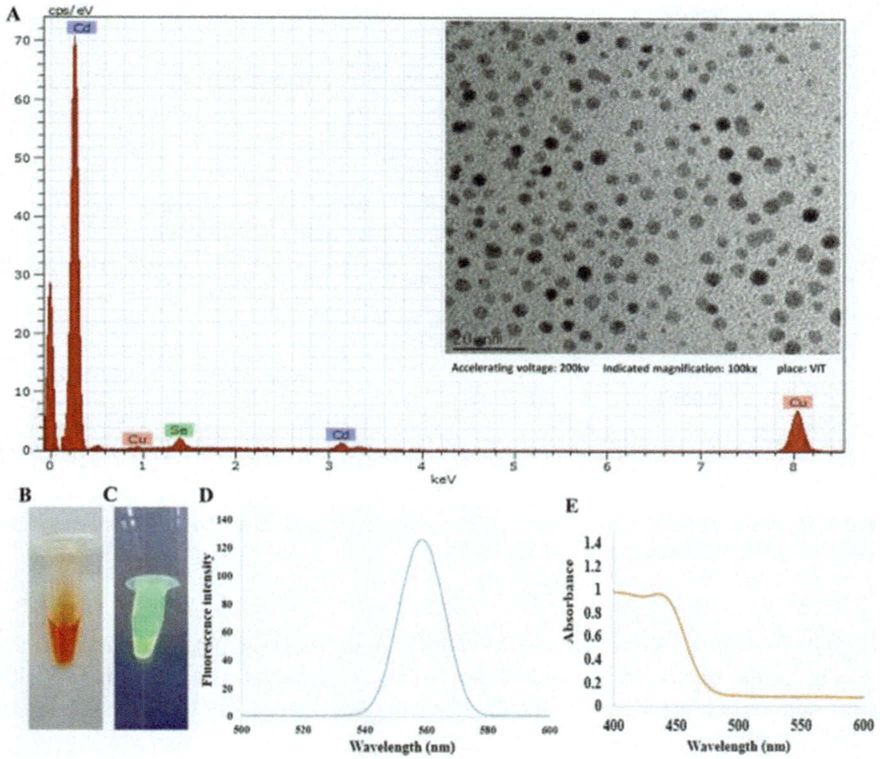

Fig. 1.15 CdSe QD Primary characterization. (**a**) CdSe QD EDX spectra, CdSe QD photomicrographs illustrate typical primary diameters, (**b**) sample of QD as pictured below white light, (**c**) QD sample as pictured below UV light, (**d**) Fluorescence spectra and (**e**) CdSe QD spectra. (Source: Banerjee et al. 2021)

1.4 Application for Plant Stress Tolerance

Nowadays, it is very important to assess the positive effects of quantum dots, polymer NPs and dendrimers on crops cultivated under a variety of stresses (biotic & abiotic) (Fig. 1.16), since more than ninety percent of arable lands are suffering from various ecological stresses (Rolli et al. 2015; Verslues et al. 2006).

1.4.1 A Biotic Stress

Harmful damages on crops development parameters, in addition to production are resulted because of salt stress (Shu et al. 2015; Wani et al. 2020), which affect negatively the metabolism process in plants (Parihar et al. 2015; Xiong and Zhu 2002) as a result of decreasing potassium and increasing sodium ions in crops (Mozafari

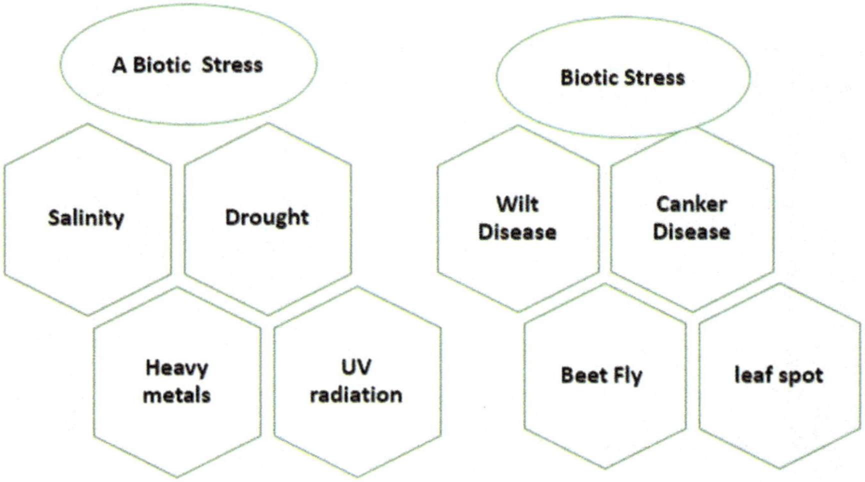

Fig. 1.16 Types of biotic and abiotic stresses positively affected by quantum dots, polymer and dendrimers NPs. (Constructed by: M.M. Saleh)

and Ghaderi 2018; Khoshbakht et al. 2017; Quinet et al. 2010), in addition to the decline in the weight of leaves (Zhang 2014). The mixture of some nanoparticles with some chemicals can diminish the damaging effects of salt stress (Gohari et al. 2020a).

It was noticed that the usage of carbon quantum dots together with the chemical material putrescine via priming reduced salinity damages impacts on plants because this combination could enhance leaves weight, as well as maintain the percentage of each of potassium, phenolic compounds and proline in addition to emphasize the activity of antioxidant enzymes (Gohari et al. 2020b; Gohari et al. 2020c).

Few kinds of research were done about the effects of the combination of carbon quantum dots with putrescine. Salt stress delays the metabolism in general (Borghesi et al. 2011), but the results of some researches confirmed the positive impact of carbon quantum dots in the enhancement of biosynthesis and increasing the carotenoids accumulation which considers as antioxidants that support plant tolerance to salinity. Furthermore, the lower concentration of graphene quantum dots improves plant tolerance to salinity (Feng et al. 2019).

According to Wani et al. (2020), salinity leads to several modifications in different crops traits on the nutrition, physiology and morphology levels, furthermore, it increases the toxic level of sodium ions and encourages oxidative stress ending with reduction of the final yield.

Tang et al. (2015) explained that the reaction toward salinity differs from one crop to another comprising defense system, electrons transportation of photosynthesis, absorption of sodium ions, assimilation of dioxide carbon and osmotic adjusting (Yan et al. 2013). Some researchers confirmed the ability to reduce the harmful effects of salt stress by using the chemical priming methods in which their efficiency can easily be emphasized by applying nanotechnology (Pollastri et al. 2018), via the advanced formation of nanoparticles (Ioannou et al. 2020; Khan et al. 2017). Furthermore, some nano-particles, for example QDs could be invested as carriers (Bai et al. 2019) since they can raise crop tolerance to salinity and several abiotic stresses by assuring the defense system and reducing the oxidative process caused by environmental stresses (Khan et al. 2020).

Lim et al. (2015) mentioned that carbon quantum dots consider as an extra model of quantum dots consist of oxygen which displays several traits of others carbon nanoparticles, like very the small percent of toxicity (Zhang et al. 2015). Many studies referred to the valuable role of many quantum dots in crops such as using graphene quantum dots with coriander and garlic seeds which revealed a rise in plant growth relating to vegetative and productive parts and roots (Chakravarty et al. 2015). Also, Li et al. (2018) found that carbon quantum dots emphasized the production of carbohydrates, germination of seeds, elongation of roots, the activity of some enzymes. The combination of carbon quantum dots with proline sounds to be an effective tool to diminish the negative impacts of salinity on plants like grapevine plants (Gohari et al. 2021).

According to (Behboudi et al. 2019), chitosan nanoparticles which are a polymer can reduce the different negative effects of water shortage stress on several plants. It was found by Su et al. (2018) that an efficient concentration of carbon dots can control positively peanut crop resistance to drought imitated by polyethylene glycol (PEG), they mentioned that the growth parameters of peanut were hindered but were alleviated when using the carbon dots which encouraged the action of antioxidant-enzymes. Many researchers clarified the role of carbon dots in protecting crops from several abiotic stresses in combination with the enhancement of the antioxidant system (Li et al. 2017; Qian et al. 2018; Wang et al. 2019). Also, Zhang et al. (2018) concluded that carbon dots can protect *Chlorella vulgaris* plants from UV radiation stress by playing an antioxidant function in improving the rate of crops development and growth parameters.

In addition, carbon dots can bind the dissolved cadmium ions and diminish its up-taken from wheat or citrus roots under heavy metal stress, furthermore, carbon dots improve each antioxidant system and the content of anthocyanin and at the same time reduces the damages caused by most of the abiotic stresses in the cell membrane (Li et al. 2019a, b; Xiao et al. 2019) (Fig. 1.17).

1.4.2 Biotic Stress

Applying chitosan nanoparticles in the seedlings of date palm emphasizes the resistance of biotic stress (Mohamed 2019). Shahryari et al. (2020) clarified that the way to decrease canker disease in stone fruit trees is to use silver-chitosan nanoparticles. In addition, Sabbour and Soleiman (2019) mentioned that the usage of nano-chitosan can manage the infection of (*Pegomya hyoscyami*) beet fly in a sugar beet crop.

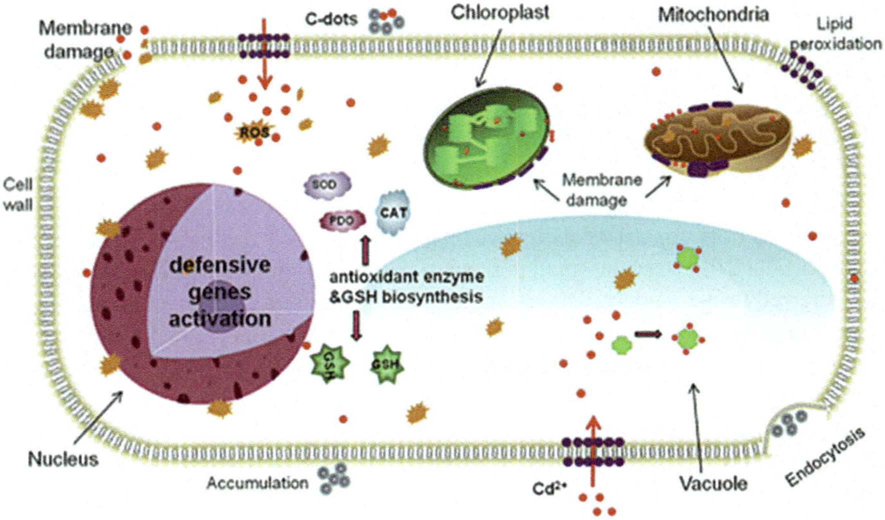

Fig. 1.17 The internal damages in Citrus plantlets due to several concentrations of Cd^{2+}. (Source: Li et al. 2019a)

One of the common biotic stresses is the contamination with phytopathogens that ending with a harsh shrink in crop yield. It was noticed that the rice leaves which were pre-cultured with carbon dots revealed good resistance toward this disease (Li et al. 2018). Regarding the control of crop disease, Kashyap et al. (2015) clarified that the usage of biopolymer nanoparticles could be a promising technique. The vegetative treatment of chitosan nanoparticles hinders the appearance of wilt disease and protect tomato crop from it, ending with yield raise (Sathiyabama and Charles 2015). Furthermore, many researchers such as Saharan et al. (2015) and Kheiri et al. (2016) confirmed the efficient task of chitosan nanoparticles in the management of phytopathogens, and this role was supported when Chandra et al. (2015) confirmed the efficacy of chitosan nanoparticles in the management of crops diseases.

Choudhary et al. (2017, 2019) improved the copper-chitosan nanoparticles and zinc-chitosan nanoparticles, then used these two combinations successfully in controlling the disease leaf spot in maize. Similar findings were confirmed by Sathiyabama and Manikandan (2018) for the management of finger millet blast disease. In addition, the same combination was successful when applied to control some pathogens. A dual role of copper-chitosan nanoparticles and zinc-chitosan nanoparticles was confirmed previously by Saharan and Pal (2016) in enhancing crop growth, inhibiting crop infection of phytopathogens and defending crops from various diseases.

1.5 Toxicity

Apart from their many industrial and medical applications, NPs and other nanomaterials have some toxicities (Bahadar et al. 2016; Ibrahim 2013; Khlebtsov and Dykman 2011; Khlebtsov et al. 2010) and fundamental information is needed to properly fight these toxic effects. During various human activities, NPs infiltrate the atmosphere through water, soil and air. Using NPs for environmental remediation, on the other hand, involves injecting or vacuuming designed NPs into the soil or aqueous techniques. As a result, all stakeholders are more interested in the project. Magnetic nanomaterial advantages, for example, small size and superior reactivity and large capability, can make them potentially lethal by causing cytotoxic and adverse effects, which are uncommon in micron-sized meter sections. In addition, studies have shown that NPs be able to reach living organisms through consumption or breathing and that they can spread throughout the body to numerous organs and tissues, anywhere they can respond to the effects of toxins. While several studies have looked at the toxic effects of NPs on animal and plant cells, there have been few toxicological studies on plants using magnetic NPs to date. Many consumer goods contain Ag NPs, which are released into the aquatic environment as dissolved microscopic particles, causing harmful effects on aquatic species, for example, algae, bacteria, fish and daphnia (Navarro et al. 2008). Since it absorbs the entire cardiac output as well as being an entrance gate for inhaled particles, the breathing system provides a specific goal for the potential poisonousness of NPs (Ferreira et al. 2013).

While nanoparticles are commonly utilized in biological treatments, the capability for harmful physical conditions impacts from extended contact at numerous strength amounts in human beings needs yet to be demonstrated in the environment, considering nanotechnology‘s rapid advancement and early adoption. Nuclear power plants, on the other hand, are projected to have a greater environmental effect in the future. The ability of NPs to control protein concentration, which is dependent on particle size, curvature, form and surface properties, functional groups, and free energy, is one of their toxicities. As a result of this interaction, some particles cause biological consequences such as protein shivering, thiol crosslinking, and enzymatic activity loss. Another example is when the thermodynamic qualities of raw materials support dissolving particles in a suspension or organic system, resulting in the release of toxic ions (Xia et al. 2008).

NPs appear toward congregate in abrasive water and saltwater and some forms of biological material or other biological particles (colloids) found in freshwater have a big impact on them. Although the state of dispersion affects environmental toxicity, several abiotic elements that influence it, for example, organic matter, acidity and salinity, are still being studied as part of environmental toxicity studies (Handy et al. 2008).

The toxicological data for recorded nanoparticles currently contains a lot of variables and is not comparable, making it hard to build a model or draw broad conclusions (Santos et al. 2010). For the study and treatment of QDs in botany, a better understanding of their toxicity mechanism is needed (Ravindran et al. 2005). Physical and chemical characteristics of QDs (substance for the coating, volume, core/coat) as well as the environmental media in which they are used all contribute to their toxicity (Minnaar and Anderson 2018). However, the essence of the individual subject matter, on the other hand, is significant.

There are currently few reports on the study of environmental toxicity of nanomaterials from the level of biological morphology and physiology, and there no analysis of the relationship between environmental toxicity results and environmental properties of nanomaterials, limiting the promotion of this method in assessing nanomaterials environmental toxicity. The majority of investigations on the mechanism of QD harmfulness in lower plants are based on the release of basic mineral substances. QDs, for example, *Lemna minor* L. and *Thalassiosira pseudonana* show cadmium ion-related toxicity, as well as toxicity-dependent toxicity to *Lemna minor* L. and *Thalassiosira pseudonana* and toxicity dependent on pH and dosage (Modlitbova et al. 2018a; Zhang et al. 2013). QD strain *Chlamydomonas reinhardtii* influence has already been clarified at the molecular level, and RNA sequencing and genetic ontology review of the entire transcriptome revealed that free cadmium and QDs have different biological effects (Domingos et al. 2011). In the real environment, a variety of nanomaterials can coexist, and the toxic effect is unaffected. Ion release due to antagonism or synergy between the different nanomaterials should simply be attributed to QDs (Yu et al. 2018). A variety of molecular mechanisms and extrinsic factors are involved in the effect, according to experiments, especially those that used forest regression model analysis (Oh et al. 2016; Winnik and Maysinger 2013). Unfortunately, studies documenting QD biotransformation in individuals and plant ecosystems (Modlitbova et al. 2018b), QD transfer to the next generation of plants is a possibility, as well as the potential for QD biomagnification within the food chain, are lacking (Remedios et al. 2012). Several proposed pathways of QD-induced phytotoxicity are summarized in Fig. 1.18.

1.6 Conclusion and Prospects

Agriculture could be revolutionized by nanotechnology technologies that allow for improved input management and conservation. Through applications in agriculture and food systems, nanotechnology researchers have a lot of potentials to help society. Nanotechnology has the ability to transform the agricultural industry by contributing to the resolution of critical issues. Such as food shortage, crop productivity, abiotic stresses, and sustainability, as demonstrated by the benefits and applications of nano-delivery systems listed above. The researchers presented a number of natural polymers that can be turned into nano-delivery systems for agricultural uses. It's also worth noting that the cost of the material and its processing into nano-delivery systems would have a big impact on the materials chosen and their use in each of the above-mentioned applications.

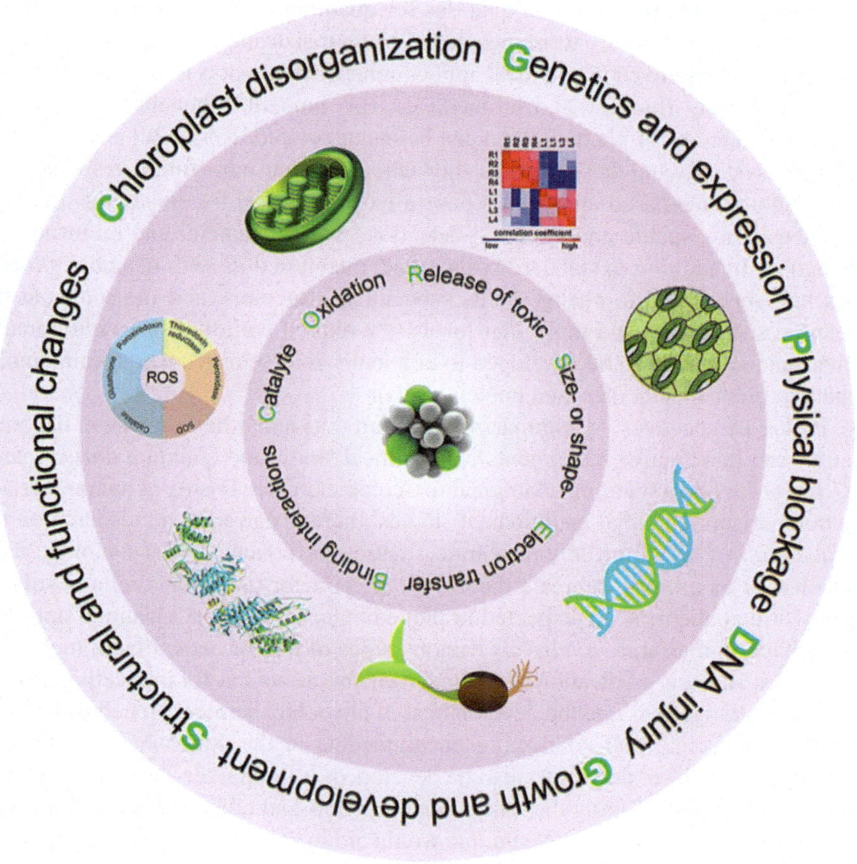

Fig. 1.18 Several proposed pathways for QD-induced phytotoxicity are depicted in this diagram. (Source: Pang and Gong 2019)

Nanomaterials for botanical applications have created a new generation of technology with great promise to change botanical science. Because of their distinctness and adaptable optical properties, semiconductor quantum dots (QDs) have piqued the interest of a wide range of researchers. We focus on the most current advances in quantum dots-based botanical science and abiotic stresses, as well as quantum dots absorption, translocation, and effects on plants, as well as possible quantum dots applications in agriculture. The existing limits of quantum dots technology are explored, as well as possible applications in botanical research using quantum dots. It is crucial to understand abiotic stresses and the environmental fate of quantum dots and to create applicable criteria and methodologies for the evaluation of quantum dots, phytotoxicity to ensure the safety of rapidly developing nanotechnological applications. According to the meta-analysis, the toxicity of Cd-based quantum dots is affected by the exposure period, exposure amount, cell type, and exposure environment, as well as quantum dots, features (size, surface ligand, and shell).

Finally, the most critical linked qualities for quantum dots on tolerance to abiotic stresses, and their toxicity were revealed to be their size and surface characteristics. To be widely employed in botanical applications, quantum dots must also have great biocompatibility (biology and biomedical). Biocompatible quantum dots for in-depth applications of plants in vivo can be manufactured using biological and biomimetic systems, and these systems show a lot of promise for future growth.

Quantum dots-based approaches have a lot of potential for enhancing accuracy and sensitivity, spatial and temporal detection, multi-parameter and multifunction detection. In addition to standard bioimaging, quantum dots with magnetic properties have been used to change cells, subcellular structures, and molecules at the cellular, subcellular, and molecular levels (the axis of cell division). Additionally, smart nanocarriers can be developed to efficiently release or recover specific molecules (proteins, medicines, and nucleic acids).

Future perspectives, to summarize, quantum dots and other innovative nanomaterials can be effective instruments for botanical study, and quantum dots research is crucial for understanding ecological toxicological mechanisms of nanomaterials. In comparison to animal and medical studies, there is currently no research on the application of quantum dots in abiotic stresses on botany. Novel nanomaterials, which will be utilized to create theoretical support for understanding and solving real-world challenges, are expected to pique botanists' interest. Quantum dots will be utilized in the future to classify various types of cancer cells, illness molecular pathways, and new medication action mechanisms, as well as for intracellular/extracellular investigations and the development of novel biochemical procedures. Novel ways for detecting and controlling quantum dots in the atmosphere, as well as reducing their dispersion, must also be developed. A biomarker strategy could be a useful tool for determining the risk of quantum dots and other engineered nanomaterials. This novel integrated technique would aid in precise risk categorization and ensure quantum dot long-term use in a variety of applications.

References

Abd-Elsalam KA, Prasad R (2018) Nanobiotechnology applications in plant protection. Springer International Publishing, Cham

Alivisatos AP (1996) Semiconductor clusters, nanocrystals and quantum dots. Science 271(5251):933

Arora S, Sharma P, Kumar S et al (2012) Gold-nanoparticle induced enhancement in growth and seed yield of Brassica juncea. Plant Growth Regul 66(3):303–310

Ashkavand P, Tabari M, Zarafshar M et al (2015) Effect of SiO_2 nanoparticles on drought resistance in hawthorn seedlings. For Res Paper 76(4):350–359

Aziz N, Faraz M, Sherwani MA et al (2019) Illuminating the anticancerous efficacy of a new fungal chassis for silver nanoparticle synthesis. Front Chem 7:65. https://doi.org/10.3389/fchem.2019.00065

Azzazy HM, Mansour MM, Kazmierczak SC (2007) From diagnostics to therapy: prospects of quantum dots. Clin Biochem 40(13–14):917–927

Baan R, Straif K, Grosse Y et al (2006) Carcinogenicity of carbon black, titanium dioxide and talc. Lancet Oncol 7(4):295–296

Badıllı U, Mollarasouli F, Bakirhan NK et al (2020) Role of quantum dots in pharmaceutical and biomedical analysis, and its application in drug delivery. TrAC Trends Anal Chem 131:116013. https://doi.org/10.1016/j.trac.2020.116013

Bahadar H, Maqbool F, Niaz K, Abdollahi M (2016) Toxicity of nanoparticles and an overview of current experimental models. Iran Biomed J 20:1–11. https://doi.org/10.7508/ibj.2016.01.001

Bai X, Purcell-Milton F, Gunko YK (2019) Optical properties, synthesis, and potential applications of Cu-based ternary or quaternary anisotropic quantum dots, polytypic nanocrystals, and core/shell heterostructures. Nanomaterials 9(1):85

Bakalova R, Ohba H, Zhelev Z (2004) Quantum dots as photosensitizers? Nat Biotechnol 22(11):1360–1361

Bandala ER, Berli M (2019) Engineered nanomaterials (ENMs) and their role at the nexus of food, energy and water. Mater Sci Energy Technol 2:29–40

Banerjee R, Goswami P, Chakrabarti M et al (2021) Cadmium selenide (CdSe) quantum dots cause genotoxicity and oxidative stress in Allium cepa plants. Mutation Res Genet Toxicol Env Mutagenesis 865:503338. https://doi.org/10.1016/j.mrgentox.2021.503338

Behboudi F, Tahmasebi-Sarvestani Z, Kassaee MZ (2019) Evaluation of chitosan nanoparticles effects with two application methods on wheat under drought stress. J Plant Nutr 42:1439–1451

Borghesi E, González-Miret ML, Escudero-Gilete ML et al (2011) Effects of salinity stress on carotenoids, anthocyanins, and color of diverse tomato genotypes. J Agric Food Chem 59(21):11676–11682

Campos EVR, Proença PLF, Oliveira JL et al (2018) Chitosan nanoparticles functionalized with β-cyclodextrin: a promising carrier for botanical pesticides. Sci Rep 8:7964

Cartoixà X, Wang LW (2005) Microscopic dielectric response functions in semiconductor quantum dots. Phys Rev Lett 94(23):236804. https://doi.org/10.1103/PhysRevLett.94.236804

Chakravarty D, Erande MB, Late DJ (2015) Graphene quantum dots as enhanced plant growth regulators: effects on coriander and garlic plants. J Sci Food Agric 95(13):2772–2778

Chan KK, Yap SHK, Yong KT (2018) Biogreen synthesis of carbon dots for biotechnology and nanomedicine applications. Springer Berlin Heidelberg. https://doi.org/10.1007/s40820-018-0223-3

Chan WC, Maxwell DJ, Gao X et al (2002) Luminescent quantum dots for multiplexed biological detection and imaging. Curr Opin Biotechnol 13(1):40–46

Chandra S, Chakraborty N, Dasgupta A et al (2015) Chitosan nanoparticles: a positive modulator of innate immune responses in plants. Sci Rep 5:15195

Chang E, Thekkek N, Yu WW et al (2006) Evaluation of quantum dot cytotoxicity based on intracellular uptake. Small 2:1412–1417

Chauhan D, Afreen S, Mishra S et al (2016) Synthesis, characterization and application of zinc augmented aminated PAN nanofibers towards decontamination of chemical and biological contaminants. J Ind Eng Chem 55:50–64

Chen J, Liu X, Wang C et al (2015) Nitric oxide ameliorates zinc oxide nanoparticles-induced phytotoxicity in rice seedlings. J Hazard Mater 297:173–182

Choudhary RC, Kumaraswamy RV, Kumari S et al (2017) Cu-chitosan nanoparticle boost defense responses and plant growth in maize (Zea mays L.). Sci Rep 7:9754

Choudhary RC, Kumaraswamy RV, Kumari S et al (2019) Zinc encapsulated chitosan nanoparticle to promote maize crop yield. Int J Biol Macromol 127:126–135

Clift MJ, Stone V (2012) Quantum dots: an insight and perspective of their biological interaction and how this relates to their relevance for clinical use. Theranostics 2:668

Corredor E, Testillano PS, Coronado MJ et al (2009) Nanoparticle penetration and transport in living pumpkin plants: in situ subcellular identification. BMC Plant Biol 9:45

Costa-Fernandez JM, Pereiro R, Sanz-Medel A (2006) The use of luminescent quantum dots for optical sensing. Trends Anal Chem 25(3):207–218

Deb P, Bhattacharyya A, Ghosh SK et al (2011) Excellent biocompatibility of semiconductor quantum dots encased in multifunctional poly (N-isopropylacrylamide) nanoreservoirs and nuclear specific labeling of growing neurons. Appl Phys Lett 98(10):103702–103703

Deerinck TJ (2008) The application of fluorescent quantum dots to confocal, multiphoton and electron microscopic imaging. Toxicol Pathol 36(1):112–116

Domingos RF, Simon DF, Hauser C, Wilkinson KJ (2011) Bioaccumulation and effects of CdTe/CdS quantum dots on Chlamydomonas reinhardtii nanoparticles or the free ions? Environ Sci Technol 45(45):7664–7669

Drew MC, Armstrong W (2002) Root growth and metabolism under oxygen deficiency. In: Waisel Y, Eshel A, Kafkafi U (eds) Plant roots: the hidden half, 3rd edn. Marcel Dekker, New York, pp 729–761

Ebert AW (2014) Potential of underutilized traditional vegetables and legume crops to contribute to food and nutritional security, income and more sustainable production systems. Sustainability 6(1):319–335

Farre M, Sanchís J, Barcelo D (2011) Analysis and assessment of the occurrence, the fate and the behavior of nanomaterials in the environment. Trends Anal Chem 30(3):517–527

Feng P, Geng B, Cheng Z et al (2019) Graphene quantum dots-induced physiological and biochemical responses in mung bean and tomato seedlings. Braz J Bot 42(1):29–41

Ferreira AJ, Cemlyn-Jones J, Robalo Cordeiro C (2013) Nanoparticles, nanotechnology and pulmonary nanotoxicology. Rev Port Pneumol 19:28–37. https://doi.org/10.1016/j.rppneu.2012.09.003

Field LD, Chen YC, Delehanty JB (2020) Semiconductor quantum dots for visualization and sensing in neuronal cell systems. In: Wright N (ed) Basic neurobiology techniques, neuro methods, vol 152. Humana, New York. https://doi.org/10.1007/978-1-4939-9944-6_1

Ghasemi Y, Peymani P, Afifi S (2009) Quantum dot: magic nanoparticle for imaging, detection and targeting. Acta Biomed 80(2):156–165

Gohari G, Alavi Z, Esfandiari E et al (2020a) Interaction between hydrogen peroxide and sodium nitroprusside following chemical priming of Ocimum basilicum L. against salt stress. Physiol Plant 168(2):361–373

Gohari G, Mohammadi A, Akbari A et al (2020b) Titanium dioxide nanoparticles (TiO_2 NPs) promote growth and ameliorate salinity stress effects on essential oil profile and biochemical attributes of Dracocephalum moldavica. Sci Rep 10(1):1–14

Gohari G, Panahirad S, Sepehri N et al (2021) Enhanced tolerance to salinity stress in grapevine plants through application of carbon quantum dots functionalized by proline. Environ Sci Pollut Res. https://doi.org/10.1007/s11356-021-13794-w

Gohari G, Safai F, Panahirad S et al (2020c) Modified multiwall carbon nanotubes display either phytotoxic or growth-promoting and stress protecting activity in Ocimum basilicum L in a concentration-depended manner. Chemosphere:126171

Gomez IJ, Arnaiz B, Cacioppo M (2018) Nitrogen-doped carbon nanodots for bioimaging and delivery of paclitaxel. J Mater Chem B 6:1–6. https://doi.org/10.1039/x0xx00000x

Goryacheva IY, Speranskaya ES, Goftman VV et al (2015) Synthesis and bioanalytical applications of nanostructures multiloaded with quantum dots. Trends Anal Chem 66:53–62

Gupta N, Upadhyaya CP, Singh A et al (2018) Applications of silver nanoparticles in plant protection. In: Abd-Elsalam K, Prasad R (eds) Nanobiotechnology applications in plant protection. Springer International Publishing AG, Cham, pp 247–266

Han M, Gao X, Su JZ, Nie S (2001) Quantum-dot-tagged microbeads for multiplexed optical coding of biomolecules. Nat Biotechnol 19(7):631–635

Handy RD, von der Kammer F, Lead JR et al (2008) The ecotoxicology and chemistry of manufactured nanoparticles. Ecotoxicology 17:287–314. https://doi.org/10.1007/s10646-008-0199-8

Hardman RA (2006) Toxicologic review of quantum dots: toxicity depends on physicochemical and environmental factors. Environ Health Perspect 114:65–172

He X, Ma N (2014) An overview of recent advances in quantum dots for biomedical applications. Colloids Surf B Biointerfaces 124:118–131

Hong NH (2019) Introduction to nanomaterials: basic properties, synthesis and characterization. In: Hong NH (ed) Nano-sized multifunctional materials. Elsevier, pp 1–19. https://doi.org/10.1016/B978-0-12-813934-9.00001-3
Howes PD, Chandrawati R, Stevens MM (2014) Colloidal nanoparticles as advanced biological sensors. Science 346(6205):1247390
Ibrahim KS (2013) Carbon nanotubes-properties and applications: a review. Carbon Lett 14:131–144. https://doi.org/10.5714/CL.2013.14.3.131
Idowu M, Lamprecht E, Nyokong T (2008) Interaction of water-soluble thiol capped CdTe quantum dots and bovine serum albumin. J Photochem Photobiol A Chem 198:7–12
Ioannou A, Gohari G, Papaphilippou P et al (2020) Advanced nanomaterials in agriculture under a changing climate: the way to the future? Environ Exp Bot 176:104048
Ismail M, Prasad R, Ibrahim AIM et al (2017) Modern prospects of nanotechnology in plant pathology. In: Prasad R, Kumar M, Kumar V (eds) Nanotechnology. Springer Nature Singapore Pte Ltd, Singapore, pp 305–317
Jamieson T, Bakhshi R, Petrova D et al (2007) Biological applications of quantum dots. Biomaterials 28:4717–4732
Joseph CG, Quek KS, Daud WMAW, Moh PY (2017) Physical activation of oil palm empty fruit bunch via CO_2 activation gas for CO_2 adsorption. IOP Conf Ser Mater Sci Eng. https://doi.org/10.1088/1757-899X/206/1/012003
Joshi N, Jain N, Pathak A et al (2018) Biosynthesis of silver nanoparticles using Carissa carandas berries and its potential antibacterial activities. J Sol-Gel Sci Techn 86(3):682–689. https://doi.org/10.1007/s10971-018-4666-2
Karakoti AS, Shukla R, Shanker R, Singh S (2015) Surface functionalization of quantum dots for biological applications. Adv Colloid Interface Sci 215:28–45
Karuppanapandian T, Wang HW, Prabakaran N et al (2011) 2,4-Dichlorophenoxyacetic acid-induced leaf senescence in mung bean (Vigna radiate L. Wilczek) and senescence inhibition by co-treatment with silver nanoparticles. Plant Physiol Biochem 49(2):168–177
Kashyap PL, Xiang X, Heiden P (2015) Chitosan nanoparticle-based delivery systems for sustainable agriculture. Int J Biol Macromol 77:36–51
Khan I, Raza MA, Awan SA et al (2020) Amelioration of salt-induced toxicity in pearl millet by seed priming with silver nanoparticles (AgNPs): the oxidative damage, antioxidant enzymes and ions uptake are major determinants of salt-tolerant capacity. Plant Physiol Biochem 156:221–232
Khan MN, Mobin M, Abbas ZK et al (2017) Role of nanomaterials in plants under challenging environments. Plant Physiol Biochem 110:194–209
Kharlamov VF, Korostelev DA, Bogoraz IG et al (2013) Electrical properties of semiconductor quantum dots. Semiconductors 47:494–500. https://doi.org/10.1134/S1063782613040131
Kheiri A, Moosawi Jorf SA, Malihipour A et al (2016) Application of chitosan and chitosan nanoparticles for the control of Fusarium head blight of wheat (Fusarium graminearum) in vitro and greenhouse. Int J Biol Macromol 93:1261–1272
Khlebtsov N, Dykman L (2011) Biodistribution and toxicity of engineered gold nanoparticles: a review of in vitro and in vivo studies. Chem Soc Rev 40:1647–1671. https://doi.org/10.1039/C0CS00018C
Khlebtsov NG, Dykman LA (2010) Optical properties and biomedical applications of plasmonic nanoparticles. J Quant Spectrosc Radiat Transf 111:1–35. https://doi.org/10.1016/j.jqsrt.2009.07.012
Khoshbakht D, Asghari MR, Haghighi M (2017) Influence of foliar application of polyamines on growth, gas-exchange characteristics and chlorophyll fluorescence in Bakraii citrus under saline conditions. Photosynthetica 56(2):731–742
King-Heiden TC, Wiecinski PN, Mangham AN et al (2009) Quantum dot nanotoxicity assessment using the zebrafish embryo. Environ Sci Technol 43:605–1611
Klimov VI (2007) Spectral and dynamical properties of multiexcitons in semiconductor nanocrystals. Annu Rev Phys Chem 58:635–673

Krishnaswamy K, Orsat V (2017) Chapter 2: sustainable delivery systems through green nanotechnology in nano and microscale drug delivery systems. In: Design and fabrication. Elsevier, pp 17–32
Kwon W, Lee G, Do S et al (2014) Size-controlled soft-template synthesis of carbon nanodots toward versatile photoactive materials. Small 10:506–513. https://doi.org/10.1002/smll.201301770
Li F, Li T, Sun C et al (2017) Selenium-doped quantum dots for free-radical scavenging. Chem Int Ed 56:9910–9914
Li H, Huang J, Lu F et al (2018) Impacts of carbon dots on rice plants: boosting the growth and improving the disease resistance. Appl Bio Mater 1(3):663–672
Li J, Xiao L, Cheng Y et al (2019a) Applications of carbon quantum dots to alleviate Cd_2 phytotoxicity in Citrus maxima seedlings. Chemosphere 236:124385
Li Y, Xu X, Wu Y et al (2019b) A review on the effects of carbon dots in the plant system. Mat Chem Front. https://doi.org/10.1039/C9QM00614A
Libralato G, Galdiero E, Falanga A et al (2017) Toxicity effects of functionalized quantum dots, gold and polystyrene nanoparticles on target aquatic biological models: a review. Molecules 22:1439
Lim SY, Shen W, Gao Z (2015) Carbon quantum dots and their applications. Chem Soc Rev 44(1):362–381
Mahat A, Shamsudin S (2020) Transformation of oil palm biomass to optical carbon quantum dots by carbonization-activation and low-temperature hydrothermal processes. Diamond Related Mat 102:107660
Mansur HS (2010) Quantum dots and nanocomposites. Wiley Interdiscip Rev Nanomed Nanobiotechnol 2:113–129
Mansur HS, Mansur AAP, González JC (2011) Synthesis and characterization of CdS quantum dots with carboxylic-functionalized poly (vinyl alcohol) for bioconjugation. Polym Guildf 52:1045–1054
Marx S, Zaltsman A, Turyan I et al (2004) Parathion sensor based on molecularly imprinted sol-gel films. Ann Chem 76:120–126
Matea CT, Mocan T, Tabaran F et al (2017) Quantum dots in imaging, drug delivery and sensor applications. Int J Nanomedicine 12:5421–5431
Maysinger D, Lovrić J, Eisenberg A, Savić R (2007) Fate of micelles and quantum dots in cells. Eur J Pharm Biopharm 65(3):270–281
Minnaar C, Anderson B (2018) A novel pollen-tracking method: using quantum dots as pollen labels. BioRxiv 286047
Miralles P, Church TL, Harris AT (2012) Toxicity, uptake and translocation of engineered nanomaterials in vascular plants. Environ Sci Technol 46(17):9224
Modlitbova P, Novotny K, Porízka P et al (2018a) Comparative investigation of toxicity and bioaccumulation of Cd-based quantum dots and Cd salt in freshwater plant Lemna minor L. Ecotoxicol Environ Saf 147:334–341
Modlitbova P, Porízka P, Novotny K et al (2018b) Short-term assessment of cadmium toxicity and uptake from different types of Cd-based quantum dots in the model plant Allium cepa L. Ecotoxicol Environ Saf 153:23–31
Mohamed EA (2019) Copper and chitosan nanoparticles as potential elicitors of innate immune response in date palm: a comparative study. Arch Phytopathol Plant Prot 52:1276–1288
Montes A, Bisson MA, Gardella JA Jr, Aga DS (2017) Uptake and transformations of engineered nanomaterials: critical responses observed in terrestrial plants and the model plant Arabidopsis thaliana. Sci Total Environ 607:1497–1516
Mozafari AA, Ghaderi N (2018) Grape response to salinity stress and role of iron nanoparticle and potassium silicate to mitigate salt-induced damage under in vitro conditions. Physiol Mol Biol Plants 24(1):25–35
Nair R, Varghese SH, Nair BG (2010) Nanoparticulate material delivery to plants. Plant Sci 179(3):154–163

Navarro E, Piccapietra F, Wagner B et al (2008) Toxicity of silver nanoparticles to Chlamydomonas reinhardtii. Environ Sci Technol 42:8959–8964. https://doi.org/10.1021/es801785m

Nel A, Xia T, Madler L, Li N (2006) Toxic potential of materials at the nanolevel. Science 311(5761):622

Nordell KJ, Boatman EM, Lisensky GC (2005) A safer, easier, faster synthesis for CdSe quantum dot nanocrystals. J Chem Educ 82(11):1697. https://doi.org/10.1021/ed082p1697

Oh E, Liu R, Nel A et al (2016) Meta-analysis of cellular toxicity for cadmium containing quantum dots. Nat Nanotechnol 11:479–486

Pang C, Gong Y (2019) Current status and future prospects of semiconductor quantum dots in botany. J Agric Food Chem 67:7561–7568

Parihar P, Singh S, Singh R (2015) Effect of salinity stress on plants and its tolerance strategies: a review. Environ Sci Pollut Res 2:4056–4075

Pelley JL, Daar AS, Saner MA (2009) State of academic knowledge on toxicity and biological fate of quantum dots. Toxicol Sci 112:276–296

Pierobon P, Cappello G (2012) Quantum dots to tail single bio-molecules inside living cells. Adv Drug Deliv Rev 64(2):167–178

Pollastri S, Savvides A, Pesando M et al (2018) Impact of two arbuscular mycorrhizal fungi on Arundo donax L. response to salt stress. Planta 247(3):573–585

Pool CP Jr, Owens FJ (2003) Introduction to nanotechnology. Wiley Interscience, London

Pradhan S, Patra P, Das S et al (2013) Photochemical modulation of bio-safe manganese nanoparticles on Vigna radiata: a detailed molecular, biochemical and biophysical study. Environ Sci Technol 47:13122–13131

Prajitha N, Athira SS, Mohanan PV (2019) Bio-interactions and risks of engineered nanoparticles. Environ Res 172:98–108. https://doi.org/10.1016/j.envres.2019.02.003

Prasad R, Gupta N, Kumar M et al (2017) Nanomaterials act as plant defense mechanism. In: Prasad R, Kumar M, Kumar V (eds) Nanotechnology. Springer Nature Singapore Pte Ltd, Singapore, pp 253–269

Prasad R, Kumar V, Prasad KS (2014) Nanotechnology in sustainable agriculture: present concerns and future aspects. Afr J Biotechnol 13(6):705–713

Prasad R, Pandey R, Barman I (2016) Engineering tailored nanoparticles with microbes: quo vadis. WIREs Nanomed Nanobiotechnol 8:316–330

Qian K, Guo H, Chen G et al (2018) Distribution of different surface-modified carbon dots in pumpkin seedlings. Sci Rep 8:7991

Quinet M, Ndayiragije A, Lefevre I (2010) Putrescine differently influences the effect of salt stress on polyamine metabolism and ethylene synthesis in rice cultivars differing in salt resistance. J Exp Bot 61(10):2719–2733

Rana S, Kalaichelvan PT (2013) Ecotoxicity of nanoparticles. ISRN Toxicol 2013:1–11. https://doi.org/10.1155/2013/574648

Rani NHA, Mohamad NF, Matali S, Kadir SASA (2014) Preparation and characterization of activated carbon made from oil palm empty fruit bunch. Key Eng Mater 594–595:44–48

Ravindran S, Kim S, Martin R et al (2005) Quantum dots as bio-labels for the localization of a small plant adhesion protein. Nanotechnology 16(1):1

Remedios C, Rosario F, Bastos V (2012) Environmental nanoparticles interactions with plants: morphological, physiological and genotoxic aspects. J Bot 2012:751686

Rocha TL, Mestre NC, Sabóia-Morais SMT et al (2017) Environmental behaviour and ecotoxicity of quantum dots at various trophic levels: a review. Environ Int 98:1–17

Rolli E, Marasco R, Vigani G et al (2015) Improved plant resistance to drought is promoted by the root-associated microbiome as a water stress-dependent trait. Environ Microbiol 17:316–331

Sabbour MM, Soleiman NY (2019) Control of beet fly (Pegomya hyoscyami) (Diptera: Anthomyidae) using chitosan and nanochitosan. Plant Arch 19:462–465

Sablon KA, Mitin V, Little JW et al (2012) Quantum dots with built-in charge for enhancing quantum dot solar cells and infrared photodetectors. In: Wang ZM (ed) Quantum dot devices, lecture notes in nanoscale science and technology 13. Springer Science Business Media, New York

Saharan V, Pal A (2016) Chitosan-based nanomaterials in plant growth and protection. Springer, New Delhi

Saharan V, Sharma G, Yadav M et al (2015) Synthesis and in vitro antifungal efficacy of Cu-chitosan nanoparticles against pathogenic fungi of tomato. Int J Biol Macromol 75:346–353

Santos AR, Miguel AS, Tomaz L et al (2010) The impact of CdSe/ZnS quantum dots in cells of Medicago sativa in suspension culture. J Nanobiotechnol 8(1):24

Sathiyabama M, Charles RE (2015) Fungal cell wall polymer-based nanoparticles in protection of tomato plants from wilt disease caused by Fusarium oxysporum f. sp. lycopersici. Carbohydr Polym 133:400–407

Sathiyabama M, Manikandan A (2018) Application of copper-chitosan nanoparticles stimulate growth and induce resistance in finger millet (Eleusine coracana Gaertn.) plants against blast disease. J Agric Food Chem 66:1784–1790

Schwab F, Zhai G, Kern M et al (2016) Barriers, pathways and processes for uptake, translocation and accumulation of nanomaterials in plants-Critical review. Nanotoxicology 10(3):257–278

Shahryari F, Rabiei Z, Sadighian S (2020) Antibacterial activity of synthesized silver nanoparticles by sumac aqueous extract and silver-chitosan nanocomposite against Pseudomonas syringae pv. Syringae. J Plant Pathol 102:469–475. https://doi.org/10.1007/s42161-019-00478-1

Sharma D, Sharma, Dhuriya YK (2019) Nanotechnology: A novel strategy against plant pathogens. In: Panpatte DG, Jhala YK (eds) Nanotechnology for Agriculture: Crop production & protection, pp 153–170. https://doi.org/10.1007/978-981-32-9374-8_9

Sharma P, Jha AB, Dubey RS et al (2012) Reactive oxygen species, oxidative damage and antioxidative defense mechanism in plants under stressful conditions. Aust J Bot 2012:1–26

Sharma V, Tiwari P, Mobin SM (2017) Sustainable carbon-dots: recent advances in green carbon dots for sensing and bioimaging. J Mater Chem B 5:8904–8924. https://doi.org/10.1039/C7TB02484C

Shu S, Yuan Y, Chen J et al (2015) The role of putrescine in the regulation of proteins and fatty acids of thylakoid membranes under salt stress. Sci Rep 5:14390

Simon J, Protasenko V, Lian C et al (2010) Polarizationinduced hole doping in wide-band-gap uniaxial semiconductor heterostructures. Science 327(5961):60–64

Su L, Ma XL, Zhao KK et al (2018) Carbon nanodots for enhancing the stress resistance of peanut plants. ACS Omega 3:17770–17777

Sudhan N, Subramani K, Karnan M et al (2017) Biomass-derived activated porous carbon from rice straw for a high-energy symmetric supercapacitor in aqueous and nonaqueous electrolytes. Energy Fuel 31:977–985. https://doi.org/10.1021/acs.energyfuels.6b01829

Talreja N, Verma N, Kumar D (2016) Carbon bead-supported ethylene diamine functionalized carbon nanofibers: an excellent adsorbent for salicylic acid. CLEAN Soil Air Water 44(11):1461–1470

Tang X, Mu X, Shao H (2015) Global plant-responding mechanisms to salt stress: physiological and molecular levels and implications in biotechnology. Crit Rev Biotechnol 35:425–437

Valizadeh A, Mikaeili H, Samiei M et al (2012) Quantum dots: synthesis, bioapplications and toxicity. Nanoscale Res Lett 7:480

Verslues PE, Agarwal M, Katiyar-Agarwal S et al (2006) Methods and concepts in quantifying resistance to drought, salt and freezing, abiotic stresses that affect plant water status. Plant J 45:523–539

Wang H, Xie Y, Na X et al (2019) Fluorescent carbon dots in baked lamb: formation, cytotoxicity and scavenging capability to free radicals. Food Chem 286:405–412

Wang J, Han S, Ke D, Wang R (2012) Semiconductor quantum dots surface modification for potential cancer diagnostic and therapeutic applications. J Nanomater 2012:1–8

Wang J, Yang Y, Zhu H et al (2014) Uptake, translocation, and transformation of quantum dots with cationic versus anionic coatings by Populus deltoides × nigra cuttings. Environ Sci Technol 48(12):6754–6762

Wang Y, Zhu Y, Yu S, Jiang C (2017) Fluorescent carbon dots: rational synthesis, tunable optical properties and analytical applications. RSC Adv 7:40973–40989. https://doi.org/10.1039/c7ra07573a

Wani SH, Kumar V, Khare T et al (2020) Engineering salinity tolerance in plants: progress and prospects. Planta 251:76

Weir A, Westerhoff P, Fabricius L et al (2012) Titanium dioxide nanoparticles in food and personal care products. Environ Sci Technol 46(4):2242–2250

Winnik FM, Maysinger D (2013) Quantum dot cytotoxicity and ways to reduce it. Acc Chem Res 46(3):672–680

Wu YL, Lim CS, Fu S et al (2006) Water-soluble quantum dots for biomedical applications. Biochem Biophys Res Commun 348(3):781–786

Xia T, Kovochich M, Liong M et al (2008) Comparison of the mechanism of toxicity of zinc oxide and cerium oxide nanoparticles based on dissolution and oxidative stress properties. ACS Nano 2:2121–2134. https://doi.org/10.1021/nn800511k

Xiao L, Guo H, Wang S et al (2019) Carbon dots alleviate the toxicity of cadmium ions (Cd^{2+}) toward wheat seedlings. Environ Sci Nano 6:1493–1506

Xiong L, Zhu JK (2002) Molecular and genetic aspects of plant responses to osmotic stress. Plant Cell Environ 25(2):131–139

Yan K, Shao H, Shao C et al (2013) Physiological adaptive mechanisms of plants grown in saline soil and implications for sustainable saline agriculture in coastal zone. Acta Physiol Plant 35(10):2867–2878

Yu Z, Hao R, Zhang L, Zhu Y (2018) Effects of TiO_2, SiO_2, Ag and CdTe/CdS quantum dots nanoparticles on toxicity of cadmium towards Chlamydomonas reinhardtii. Ecotoxicol Environ Saf 156:75–86

Zhang G, Xu S, Hu Q et al (2014) Putrescine plays a positive role in salt-tolerance mechanisms by reducing oxidative damage in roots of vegetable soybean. J Integr Agric 13:349–357

Zhang M, Gao B, Chen J, Li YC (2015) Effects of graphene on seed germination and seedling growth. J Nanopart Res 17(2):78

Zhang M, Wang H, Song Y et al (2018) Pristine carbon dots boost the growth of Chlorella vulgaris by enhancing photosynthesis. Appl Bio Mater 1:894–902

Zhang S, Jiang Y, Chen CS (2013) Ameliorating effects of extracellular polymeric substances excreted by Thalassiosira pseudonana on algal toxicity of CdSe quantum dots. Aquat Toxicol 126:214–223

Zrazhevskiy P, Sena M, Gao X (2010) Designing multifunctional quantum dots for bioimaging, detection, and drug delivery. Chem Soc Rev 39(11):4326–4354

Chapter 2
Climate Change Mitigation and Nanotechnology: An Overview

Asha Kumari, Mahendar Singh Bhinda, Binny Sharma, and Manoj Parihar

Abstract Global climate change is one of the most pressing issues confronting the world today, resulting in decreased crop productivity and soil fertility. Plant physiology, morphology, and metabolism all suffer as a result of climate change. Several strategies for limiting environmental stress have been adopted to improve stress tolerance in plants through transgenic approaches and changes in overall genetic composition. Nanotechnology, a fast-growing science which is emerging, plays a vital role in agriculture in promoting positive changes in yield, chlorophyll content and gene expression in plants. Drought tolerance is boosted by nanoparticles, which increase root hydraulic conductance and water uptake while also showing a differential assortment of proteins involved in oxidation-reduction, ROS(Reactive oxygen sp.) removal, stress signalling, and hormonal mechanisms. Since nanoparticles, in their nature, are highly mobile, the nutrients can be easily transported to all the plant parts. Thus, the main idea behind writing this chapter is to find ways to increase the potential of plants in stressful conditions by using nano-preparations.

Keywords Nano-preparations · Climate change · Environmental stresses · ROS (reactive oxygen species)

2.1 Introduction

As global climate change progresses, abiotic and biotic stress is a serious constraint on the production of crops. As estimated, about 70 percent of the crop yield reduction is stimulated through means of different abiotic stresses (Acquaah 2007).

A. Kumari (✉) · M. S. Bhinda · M. Parihar
ICAR-Vivekanada Parvatiya Krishi Anusandhan Sansthan, Almora, Uttarakhand, India
e-mail: asha.kumari@icar.gov.in

B. Sharma
Department of Plant Physiology, Institute of Agricultural Sciences, Banaras Hindu University, Varanasi, Uttar Pradesh, India

M. Faizan et al. (eds.), *Sustainable Agriculture Reviews 53*, Sustainable Agriculture Reviews 53, https://doi.org/10.1007/978-3-030-86876-5_2

Climate changes and other interferences caused by humans are also causing a serious decrease in fertile soils and apparent food cultivation environments (Khan and Akhtar 2015; Chen et al. 2017). Plants that cause enormous losses in the crop harvest worldwide comprise drought, salinity, high temperatures, flooding, cold, freezing, UV radiation, etc. (Wani et al. 2016; Li et al. 2017a, b).

Drought, salinity and extreme temperatures are therefore among the various abiotic stresses most commonly reported and pose a threat to global food safety. Abiotic and biotic stresses lead to a sequence of morphological, physiological, biochemical and molecular changes that negatively impact the development and productivity of crop plants. Developing stress-tolerant plants is often a worthwhile way of conquering the issue of decreasing global food production. Limited success in refining the stress tolerances of crop plants, including inter-specific or inter-generic hybridization, has been achieved with traditional plant breeding initiatives. Traditional breeding processes are limited by the means of prolonging stress-tolerant attributes, in addition to the low genetic variability of yield components under stress conditions and the lack of efficient selections. Consequently, further techniques to promote stress-tolerant crops should be developed.

2.1.1 What Is the Definition of Nano-Technology?

Global agriculture production and changes in the climate, the exhaustion of water and soil resources, energy issues, abiotic stresses, biotic stresses, etc. face great challenges. These challenges should be addressed in an economically viable and ecologically sustainable way. Today, there is a global demand for massive food production using limited available resources, but with a minimum use of fertilizers and pesticides that can minimize environmental damage in the surrounding areas, as well as the redesign of existing agriculture into high-tech agricultural innovations.

One of the solutions to this problem is to use new technologies in irrigation and fertilisation, as well as to look for less toxic forms of plant nutrition. On the other hand, it is important to promote resistant varieties which will be able to resist abiotic stress. In the development stage of agriculture and plant biotechnology, nanotechnology is a highly promising technology (Scrinis et al. 2007). Nanotechnology has been broadly defined as materials, systems, and processes that operate at a scale of 100 nanometers (nm) or less. The term 'Nano' generally refers to a size scale ranging from 1 to 100 nm. These nano-materials are made up of very small components, and these constituents have an impact on the properties of materials at a macro level. When compared to the same mass of fibres produced in a larger form, nano-materials have a comparatively higher surface area. Nanoparticles may make substances more chemically reactive, affecting their strength or electrical assets. These particles have a higher surface-to-volume ratio, which increases their excitability and the potential for regenerative capacity (Dubchak et al. 2010). In this regard, agri-nanotechnology is regarded as one of the most significant and promising techniques. Various researchers have discovered that nano-materials can reduce the damage caused by various abiotic stresses by activating the plant defense system. It is also well known that, due to their small size, these nano-particles have the ability to penetrate plant tissues. These nanoparticles can also increase surface area, making them more effective in the adsorption process and in the delivery of targeted substances. Because of the nanoparticles, they can control water uptake by plant tissues, stimulating seed germination and plant growth (Elsakhawy et al. 2018). Natural nanoparticles are a native constituent of biological systems with diverse structures and broad biological roles (nano-clay, many chemicals derived from soil organic matter, lipoproteins, exosomes, magnetosomes, viruses, ferritin) (Li et al. 2013; Hedayati et al. 2016). As a result, nanotechnology, a new evolving and rapidly growing technology, can help to alleviate many of these stress factors through various mechanisms such as antioxidant defence mechanisms and the application of less toxic and more efficient fertilizers (Zuverza-Mena et al. 2017).

2.2 Application of Nanotechnology in Major Abiotic Stresses

Because of more than one abiotic stresses and continuous increase in worldwide climate change, apprising want is demand for abiotic stress management and adaptability in plants (Grover et al. 2016; Wani et al. 2016). Consequently, the study of various consequences of abiotic stresses on plant growth and development is vital issue at distinctive stages which includes biochemical, physiological and molecular levels (Wani et al. 2016). Different plant mechanisms in counteracting the abiotic stresses and sustaining their growth need to be a superb implication. The usage of nano-materials is an emergent resolution amid the recognized clarifications of plants towards abiotic stresses (Hatami et al. 2016; Reddy et al. 2016; de la Rosa et al. 2017; Khan et al. 2017).

The applications of nanoparticles or nano-devices have an effect on different growth phases both encouraging as well as deleterious influence on plant development. Nanotechnology encompasses unique nano-material features that facilitate for agricultural research in crop enhancement in addition to mitigation to stresses (Carmen et al. 2003).The speedy improvement of nanotechnology has endorsed metal-based nanoparticles of 1–100 nm in size and in enormously small amounts have been to be confirmed as alternative plant mineral nutrients and stimulating substance. Khan et al. (2017) postulated by analyzing their own work, and related studies that nanoparticles can also additionally facilitate adaptation and tolerance to stress in plant via the stimulation of defense mechanisms and stress-associated gene manifestation. Plants react to surrounding stress elements through means of activating numerous transcellular membrane sensors, mainly by employing Ca^{2+} channels and Ca^{2+} binding proteins (Thapa et al. 2011).Consequent downstream processes result in alteration in levels of gene manifestation and eventually to plant adaptability for stress. When compared to traditional bulk fertilizer, the use of nanoparticles augments plant responses towards drought pressure (Saxena et al. 2016). For example, the application of ZnO nanoparticles improved germination percentage under water stress in soybean seeds (Sedghi et al. 2013). Likewise, with ZnO nanoparticles application in soil, the sorghum productiveness and nitrogen procurement upgraded remarkably (Dimkpa et al. 2019). Moreover, ZnO nanoparticles added to the culture media boost somatic embryo development and plant rejuvenation, as well as raised stress tolerance in banana plants grown up under in-vitro (Helaly et al. 2014). Cu and Zn nano-particles, On the other hand, efficiently lessened drought consequences on wheat plants via means of enhancing antioxidant enzyme accomplishments and relative water content, dropping thiobarbituric acid reactive substance (TBARS) accretion, and photosynthetic process stabilization (Taran et al. 2017).The application of zero-valent Fe nanoparticles become observed to aid normal drought sensitivity upkeep, plasma membrane Hþ-ATPase stimulation, and opening of stomata, in addition to chlorophyll and plant biomass matter, even as CO_2 absorption in *Arabidopsis thaliana* elevated throughout water deficit condition (Kim et al. 2015). While the potential function of nanoparticles in advancing plant protection responses towards abiotic stress, counting water stress, had been comprehensively considered, but several more facets endure to be exposed. The impact of metal-based nanoparticles for enhancement of adaptability towards various abiotic stresses in many plant species has been comprehensively confirmed by various researchers (Da Costa and Sharma 2016; Askary et al. 2017; Choudhary et al. 2017; Jalali et al. 2017; Latef et al. 2017; Iqbal et al. 2019; Pérez-Labrada et al. 2019), although, their involvement in convincing gene expression for stress conditions, has not been determined yet.

2.2.1 Nanoparticles Impact on Abiotic Stresses in the Plants

Drought, heat, salinity, alkalinity, flooding or water submergence, and heavy metal toxicity/deficiencies are some of the most common abiotic stress factors which have a negative influence on development and yield of most of crop species (Boyer 1982). Plants are exposed to a variety of environmental challenges during their life span, and as a result, they develop their defence system to counter these pressures at multiple levels with the aids of employing molecular, biochemical, and physiological regulating paths. Plants undertake divergent molecular paths to regulate proper adjustment of gene expressions in order to cope with these strains. It's been shown by several research investigations have revealed that nanoparticles influence on plants growth and development dependent on the amount of these novel particles.

Nanoparticles are concerned in upgrading the actions of antioxidant enzymes such as SOD, CAT and POD (Laware and Raskar 2014).There is complexity within side the reaction of plants to abiotic stress involving changes of their biology, morphology and metabolism. A collection of techniques had been followed with a purpose to raise the tolerance against abiotic stress in plants that comprehend engineering genetic varieties with distinctive genetic organizations which make able to perform better under stress conditions. The application of nanoparticles has brought about the improved germination and seedling growth rate besides some physiological activities like nitrogen metabolism, carbohydrate contents, protein and variety of positive alterations in gene expression demonstrating the fact that their applications in plants leads to crop Improvement. It is now evident that nano-structured substances are capable of enhancing the water stress tolerance via water uptake and increasing in the root hydraulic conductance in addition to differential mobility resulting in a quick transportation of nutrients to almost all parts of plants. The mitigating effects of various nano-materials on major abiotic stresses (drought, salinity, flooding, temperature and heavy metal stresses) in crop plants are given in the Table 2.1.

Among the various abiotic stresses, the drought is a major stress that depressingly influences plant development as well as crop yield. To cope up with this drought stress various techniques are employed in plants, among those, application of nanoparticles is thought to be most efficient and capable. An investigation regarding the performances of soybean crop plants that were exposed with iron, copper, cobalt, and zinc oxide nanoparticles under drought-induced conditions was conducted by Linh et al. (2020). The experimental results revealed that these metal-based nanoparticles support the drought tolerance in exposed plants. Under this study the desirable physiological characteristics especially, in iron nanoparticles treated plants were considerably enhanced such as relative water content, drought tolerance index, and biomass reduction rate.

Likewise, the use of different concentrations of silica nanoparticles in two sorghum cultivars has led to improvement in drought tolerance through the positive effects on malondialdehyde (MDA), photosynthesis parameters, membrane electrolyte leakage, chlorophyll, proline contents, carbohydrate and carotenoid contents. Various researchers discovered that the expression of the drought tolerance marker genes GmRD20A, GmDREB2, GmERD1, GmFDL19, GmNAC11, GmWRKY27, GmMYB118, and GmMYB174, were up-regulated in the roots or shoots (or both) of nanoparticles exposed plants under drought conditions. In various *Arabidopsis* investigations, the genes listed above have been used as water deficiency tolerance indicators (Huang et al. 2012) and other species, including soybean (Neves-Borges et al. 2012; Stolf-Moreira et al. 2011). A gene-, tissue-, and nanoparticles-dependent amplification of gene expression were revealed through quantitative PCR investigation at the molecular level of many drought-responsive genes.

Among these, Fe nanoparticles treatment resulted in the expression of all tested genes in roots; as well as the expression of three drought responsive genes amplified in leaves of all nanoparticles treated plants, whereas the expression of GmERD1 (Early Responsive to Dehydration 1) was prompted in both roots and shoots under the four nanoparticles treatments tested. These conclusions revealed that using nanoparticles can advance drought tolerance capacity in soybean plants by stimulating drought-related gene expression.

Another prominent abiotic stress is salinity stress which has caused concern about crop production sustainability. It is expected based on studies that approximately more than 20% of cultivated land all around the globe is distress from salinity stress and also the trouble is day by day expanding. Salinity stress trigger favorable effects on numerous physiological and biochemical pathways that are

Table 2.1 Mitigating effects of various nanomaterials on major abiotic stresses in crop plants (Taken from – Tamer Elsakhawy et al. 2020) (Source: from Khan et al. 2017)

Nano-materials	Crop plants	Plant characters improved	References
Drought stress			
Nano TiO_2	Wheat (*Triticum aestivum* L.)	Improving wheat growth, production, as well as gluten and starch quantity	Jaberzadeh et al. (2013)
Nano TiO_2	Flax (*Linum usitatissimum* L.)	Enhancing chlorophyll and carotenoids, improving development and yield characteristics, decreasing H_2O_2 and malondialdehyde (MDA) levels	Aghdam et al. (2015)
Nano TiO_2	Basil (*Ocimum basilicum* L.)	Improving the toxic consequences of drought stress on basil plants	Kiapour et al. (2015)
Nano Fe	*Carthamus tinctorius*	Reducing drought effect and improving safflower productivity	Yousefi et al., (2014)
Nano zero valent Fe	*Arabidopsis thaliana* L.	In *Arabidopsis* plants leads to plasma membrane activation, H^+ -ATPase, stomata opening, raised chlorophyll content and plant biomass, as well as upheld normal drought sensitivity, and amplified CO_2 absorption	Kim et al. (2015)
Nano SiO_2	*Crataegus sp.*	Photosynthetic rate, stomatal conductance, and plant biomass all showed a positive significant influence. However, chlorophyll and carotenoid amount shows non-significant result	Ashkavand et al. (2015)
Nano ZnO	Soybean (*Glycine max* L.)	Germination percent and rate improved whereas seed residual fresh and dry weight decreased in soybean	Sedghi et al. (2013)
Salinity stress			
Nano SiO_2	Tomato (*Lycopersicum esculentum* L.)	Germination capability, root length, and dry weight were improved at low nano-SiO_2 application. But, at higher levels seed germination properties were decreased	Haghighi et al. (2012)
Nano SiO_2	Cherry tomatoes (*Solanum lycopersicum* L.)	Relieving the salinity's influence on fresh quantity, chlorophyll level, photosynthetic rate, and leaf water amount	Haghighi and Pessarakli (2013)
Nano SiO_2	Basil (*Ocimum basilicum* L.)	Fresh and dry weight, chlorophyll as well as proline levels were raised	Kalteh et al., (2014)
Nano SiO_2	Lens (*Lens culinaris* Medik.)	Enhancing emergence and seedling growth rate	Sabaghnia and Janmohammadi (2014)
Nano SiO_2 (10 nm)	Squash (*Cucurbita pepo* L.)	Seed germination and growth features improved, MDA, H_2O_2 and electrolyte leakage was decreased, chlorophyll breakdown and oxidative damage was minimized, and also photosynthetic variables and antioxidant enzymes levels were enhanced	Siddiqui et al. (2014)

(continued)

Table 2.1 (continued)

Nano-materials	Crop plants	Plant characters improved	References
Nano SiO_2	Faba bean (*Vicia faba* L.)	Enhancing seed germination, growth parameters, antioxidant enzymes activity, relative water content and overall yield	Qados and Moftah (2015)
Nano SiO_2 (20 nm)	Tomato (*Solanum lycopersicum* L.)	Influence of salinity stress on germination rate, root length, and fresh weight was repressed by up-regulation of expression level of four salt stress genes whereas down-regulation the expression pattern of six genes.	Almutairi (2016)
Nano ZnO and Fe_3O_4	*Moringa peregrina*	Increased N, P, K^+, Ca^{2+}, Mg^{2+}, Fe, Zn, total chlorophyll, carotenoids, proline, carbohydrates, crude protein amounts, and enzymatic and non-enzymatic antioxidants while decreasing Na + and Cl- levels.	Amira et al. (2015)
Nano ZnO	Sunflower (*Helianthus annuus* L.)	Increased growth, net CO_2 absorption rate, sub-stomatal CO_2quantity, chlorophyll content, Fv/Fm and Zn levels and lessening Na^+level in leaves	Torabian et al. (2016)
Flooding stress			
Nano Ag	Saffron (*Crocus sativus)*	Inhibiting ethylene signals, boost root development	Rezvani and Sorooshzadeh (2012)
Nano Al_2O_3	Soybean (*Glycine max* L.)	Accelerated growth, modulation of energy metabolism and cell death	Mustafa et al. (2015)
Nano Ag	Soybean (*Glycine max* L.)	Reduced production of cytotoxic glycolysis by products, increased quantity of stress-associated proteins, and enhanced seedling growth	Mustafa et al. (2015)
Heavy metal toxicity			
FeNPs	Wheat	Increases growth, photosynthesis, Fe content in wheat seedlings, reduced oxidative stress	Adrees et al. (2019)
Nano SiO_2	Maize	Mitigates Al phytotoxicity and improves ROS scavenging system	de Sousa et al. (2019)
TiO_2 NPs	Maize	Reduced Cd content	Lian et al. (2019)
ZnO NPs	*Leucaena leucocephala*	Decreased Cd & Pb genotoxicity, increased antioxidant defense system	Venkatachalam et al. (2017)
Mel-AuNPs	Rice	Enhanced melatonin absorption and decreased Cd uptake	Jiang et al. (2021)
Fe_3O_4 NPs	Tomato	Decreased Cd accumulation, increased nutrient uptake and defense mechanism of plants	Rahmatizadeh et al. (2019)
ZnO NPs	Spinach, parsley, cilantro	Reduced Pb and Cd toxicity	Sharifan et al. (2020).
ZnO & Fe NPs	Wheat	Increased growth, reduced Cd uptake and oxidative stress	Rizwan et al. (2019).

(continued)

Table 2.1 (continued)

Nano-materials	Crop plants	Plant characters improved	References
ZnO NPs	Soybean	Reduced As uptake, improved biochemical attributes, antioxidant enzymes, ascorbate-glutathione cycle and glyoxalase system	Ahmad et al. (2020)
Temperature stress			
Se NPs	Sorghum	Increased fatty acid unsaturation, pollen germination and reduced oxidative stress	Djanaguiraman et al. (2018)
Ag NPs	Wheat	Reduced heat stress	Iqbal et al. (2018)
CeO_2 NPs	Maize	Up-regulation of heat shock proteins, reduced heat stress	Zhao et al. (2012)
TiO_2 NPs	Tomato	Improved photosynthesis, stomatal opening, reduced heat stress	Qi et al. (2013)
ZnO NPs	Rice	Enhanced chilling tolerance	Song et al. (2021)
TiO_2	Chickpea	Reduced ion leakage, enhanced cold stress tolerance	Mohammadi et al. (2013)
Ag NPs	Green beans	Enhanced germination and plant morphological traits under chilling stress	Prażak et al. (2020)

directly related in the course of plant growth and yield. Under salinity stress condition the application of silicon nanoparticles and silicon based fertilizers in basil have desirable effects on morphological and physiological traits as well as vegetative features. The results revealed a significant enhances on plants growth, chlorophyll content and proline level in basil plants when treated with silicon nanoparticles that are competent to interact with plants biological systems mechanically and chemically mainly owing to their intrinsic catalytic reactivity, smaller size and huge surface area.

Similarly, Jiang et al., (2012) reported that application of silver nanoparticles of particle size of about 6 nm when treated at a 5 mg/l concentration, in the *Spirodela polyrhiza* plant exhibited activation of antioxidant system and also an increase in the activity of superoxide, dismutase, catalase and peroxidase. In *Brassica juncea* seedlings the use of gold nanoparticles resulted into outstanding enhancement of antioxidant enzymes activity, guaiacol peroxidase, catalase and glutathione reductase and peroxidase along with the accrual of hydrogen peroxide and proline content in plants. (Das and Das 2019).Likewise, Sharma et al. (2012) found that the treatment of *Brassica juncea* with silver nanoparticles resulted in improved the performance of antioxidant enzymes (ascorbate peroxidase, guaiacol peroxidase and catalase) that leads to reduced level of reactive oxygen species (ROS).

Heavy metals such as Cd, Pb, Hg, As, Fe etc. into the soil which is major environmental concern accumulate as a result of speedy industrialization, modern farming systems, mining, and other human activities. Heavy metal toxicity not only deteriorates the soil quality but also uptake by plants and hampers plant metabolism. Heavy metals present in crop species ultimately become part of food chain which directly or indirectly affects human health also. Exposure of heavy metals to plant may stimulates aberrations in morpho- physiological processes like decrease in water potential, oxidative burst, growth inhibitions, cell death. Heavy metal toxicity also alters molecular and structural deformities. The effect of heavy metal on plant metabolism and sequestration has been thoroughly depicted in several studies (Yadav et al. 2010, Maleki et al. 2017, Baig et al. 2020, Goyal et al. 2020). Plants develop several approaches to lessen the negative consequences of heavy metal toxicity within its system. However, the use of nanoparticles can be a possible effective approach to clean the contaminated environment with heavy metal. The magnetic Fe_3O_4 nanoparticles play important role in mitigating toxicity of Pb, Zn, Cd, Cu in wheat seedlings (Konate et al. 2017). Venkatachalam et al. 2017 studied that ZnO NPs plays potential role in mitigating Cd & Pb toxicity in *Leucaena leucocephala* (Lam.) seedlings by reducing oxidative stress, MDA content and increasing plant physiological traits. Foliar application of TiO_2 NPs reduced Cd toxicity in maize (Lian et al. 2020). Similarly, silicon dioxide NPs ameliorates Al toxicity in maize (De Sousa et al. 2019).

Temperature is a fundamental environmental factor that controls plant growth and metabolism to a great extent. The fluctuating average global temperatures possess major threat to agricultural productivity. The increase or decrease in universal temperature disturbs plant growth and development adversely and causes injuries to plants. Mitigating the harmful effects of fluctuating temperature is a matter of global concern for researchers worldwide. Selenium NPs plays important role in alleviating elevated temperature by fatty acid unsaturation, pollen germination, photosynthetic rate and decrease in oxidative stress in Sorghum bicolour (Djanaguiraman et al. 2018). Similarly, Se NPs also mitigated heat stress by increase hydration ability, chlorophyll content and plant development in tomato (Haghighi et al. 2014). Silver nanoparticles alleviate harmful effects on heat stress by enhancing morphological attributes in wheat (Iqbal et al. 2018).The potential role of nano-materials in combating heat stress tolerance had been studied in various crop species (Khodakovkazya et al. 2011; Zhao et al. 2012; Qi et al. 2013, Borai et al. 2017; Singh et al. 2019).Seasonal changes and alterations in climatic patterns cause chilling stress in plants. They occur at very low temperature and causes adverse effects on plant developmental patterns like germination, growth, yield etc. Chilling stress stimulates membrane disintegration and ion leakage and ultimately death to plant tissues (Jalil et al. 2019).Application of nanoparticles has positive effects on mitigating chilling stress in plants. ZnO NPs alleviates chilling tolerance by stimulating antioxidant defense system and transcription factors in rice crop (Song et al. 2021). Similarly, the TiO_2NPs also reduces the chilling stress by reducing ion leakage in chickpea (Mohammadi et al. 2013, 2014).Numerous studies have been conducted to implicit the role of nanoparticles in mitigating cold stress in the plants (Elsheery et al. 2020, Prażak et al. 2020, Gao et al. 2006a, b, Ze et al. 2011, Giraldo et al. 2014, Hasanpour et al. 2015, Xu et al. 2014). Thus, nanoparticles are small size molecules that perform vital role in promoting plant growth and developmental patterns and also combats stress responses in plants by inducing defense mechanism, reducing lipid peroxidation, lessening oxidative stress, and stimulating defense related genes and transcription factors.

2.3 Potential Role of Nanotechnology to Confer Biotic Stress Tolerance in the Plants

Biotic stresses such as pathogen infection, insect and pest incidence, herbivore attack are major constraints of agriculture which not only impart negative effects on plant morphology but also affect crop production adversely. Plants thrive under variable existing environments as physiological, ecological and genetic unit with various microorganisms which can be beneficial, neutral or pathogenic to them. Biotic stresses arise due to incidence of such harmful microorganisms, insects and pests and herbivores respectively. They are extremely detrimental to plant health and functional status. The recurrent occurring biotic stresses due to changing environmental patterns negatively affect food quality and also cause pre-harvest and post harvest losses of crops as well. According to Food and Agriculture Organization (FAO) of the United Nations loss of crop production occurred from 20 to 40% due to pest incidence, phytopathogenic organisms and weeds. (Mitra 2021; Khan et al. 2021). The onset of biotic stresses like pathogen invasion, weeds, insects and pests has diverse effects on the plants. The interaction of plants with pathogenic organisms may alter physiological responses and manipulates carbohydrate metabolism. Besides this, plants develop chlorotic and necrotic patches in different plant parts along with diminished photosynthetic rate (Berger et al. 2007). They influence plant metabolic functions and yield attributes accordingly. The frequently changing climatic conditions stimulate the increased incidence of biotic stresses in either way. However, our knowledge regarding production losses due to pathogen infection are limited. The researchers and agriculturalists globally aim to increase the production of agricultural commodities in order to sustain the rising population. The climatic abnormalities, limited resources and water scarcity worsen the scenario, thereby decreasing production of crops and food grains. With progression of pathogen infestations, insect and pest attack the demand of agrochemicals have been potentially increased. The limitless use of pesticides and fungicides adhere the risk of accumulation of toxic chemicals in the produce thereby affecting food chain and deteriorating plant and human health. The conventional methods for biotic stress mitigation are also not worth promising in many aspects, breeding techniques consumes more time and is slow process. Also the transgenic crops also face ethical issues and safety conflicts. Therefore, in order to cope with biotic stresses, to increase crop production and sustainability modern technologies are needed to be intricated in the agricultural system. Nanotechnology is one of the promising tools to maintain plant health and ensures food security. Nanotechnology is a multidisciplinary and interdisciplinary technology which deals with particles ranging of size 0.1 to 100 nanometers (Khan et al. 2019). Nanotechnology has wide applications in agriculture. Nanotechnology is a cutting edge method that provides more advantages over conventional methods to increase crop production and productivity. The nanomaterials enhance seed germination, soil health status, modulate crop growth, enhance stress tolerance, affect gene expression and bring ecological sustainability (Khan et al. 2021, Banerjee et al. 2016). Plant nano biotechnology is also used to confer disease resistance and mitigate biotic stress responses in the plants. Nanosensors/ nanobiosensors, nano formulations of agrochemicals have been used as nanopesticides, nanofungicides, nanofertilizers in agricultural system, crop improvement techniques, disease identification and control, genetic manipulation of agricultural crops, postharvest management with smarter, stronger and cost-effective techniques (Kerry et al. 2017).

2.3.1 Uptake, Synthesis and Characterization of Nanoparticles

The uptake of NPs is being progressed from soil, atmosphere, water etc. It undergoes a series of bio geo-transformation which determines its availability within plant system. The NPs are absorbed by the roots and then translocated to different cellular compartments. However, the mechanism of NPs uptake and absorption inside plants is not well established yet but studies conducted reveals this might often progressed along with nutrients and water uptake by cellular penetration (Mukherjee et al. 2016). The bioaccumulation and adsorption is greatly dependent on its size, surface charge, segregation, plant species, growth condition/ environment which is crucial for its transportation across various cell organelles and tissues. The uptake of ZnO NPs is being studied in ryegrass (Lin & Xing 2007). Koemel et al. 2013 studied the uptake of Gold NPs in rice particles. NPs small in size enters plant roots through osmotic pressure, capillary forces or directly to epidermal tissues. The entry of NPs in cell membrane depends on size of NPs. They further move apoplastically after entering cell wall to vascular system. NPs when gets transported through Casparian strip to vascular system shows symplastically movement. Further movement consists of binding to carrier proteins present in endodermal cell membrane, pore formation and transport. NPs also transport from one cell to another through plasmodesmata (Tripathi et al. 2017, Perez-de-Luque et al. 2017). The absorbed NPs by the plants occur in cytoplasm, nucleus, epidermis and other cell organelles. NPs absorbed by seed coat enter seeds through parenchymous spaces, diffusion in cotyledon (Ali et al. 2021). Foliar application of nanomaterials stimulates their entry via stomata or cuticular surfaces. The nanomaterials possessing size greater than 10 nm follows both apoplastic and symplastic pathways while size 10–50 nm follows symplastic route. Similarly, NPs of size 50–200 nm follows apoplastic path. Furthermore, NPs which are present internally moves along sugars in phloem sieve tubes and gets accumulated in different plant parts like roots, stems, fruits etc. After being absorbed by root system, NPs gets translocated in different plant parts as metallic or non- metallic oxides. Plant physiological and morphological structure plays very crucial role in its translocation. AuNPs which gets accumulated in shoot of *Oryza sativa* cannot do so in shoot of *Cucurbita pepo* and *Raphanus raphanistrum* (Zhu et al. 2012). The accumulation of metallic and non-metallic oxide NPs has been thoroughly investigated in various crop species (Ali et al. 2021). The coated nano-anatase titanium oxide has been well absorbed through roots but not gets translocated in rice plants (Kurepa et al. 2010). The translocation of NPs has been broadly studied in various crops through various pathways in corn, cucumber, tomato and further concluded that it varies considerably different with different plant species (Mukherjee et al. 2016). NPs can be categorised as natural, incidental and engineered as per their origin (Monica and Cremonini 2009). Various forms of nanomaterials used in plant system includes metalloids, metallic oxides, non-metallic NPs, carbon nanomaterials, quantum dots, liposomes, dendrimers etc. NPs are extremely small size molecules synthesized by physical, chemical and biological methods with approaches viz., top-down and bottom-up processes. (Fig. 2.1).

2.4 Role of Nanotechnology in Mitigating Biotic Stress

Our agricultural system is based on several management practices. In the awake of progressing plant diseases, insect-pests and weeds attack the crop management required potential approaches to minimize biotic stresses thereby minimizing crop losses. The extensive use of agrochemicals to control phytopathogen is well established in the agricultural system. The use of nanotechnology in disease management is beneficial to crop production as synthetic chemicals are harmful for plants, environment and human health. The nanoformulations like nanopesticides, nanofungicides and nanofertilizers have been widely used in crop production and protection in recent times. The optimum dosage of these NPs not only minimizes plant diseases but also ensures sustainability to the environment. They possess antimicrobial activity. Silver NPs consists of antibacterial and fungicidal properties used in various crop species (Kim et al. 2012). AgNPs contain antifungal properties against *Alternaria alternata, A. citri, Rhizoctonia solani, Sclerotinia sp., Penicillium digitatum, Botrytis cinerea* (Abdelmale and Salaheldin 2016., Krishnaraj et al. 2012). Cui et al. 2009 reported that TiO_2 NPs possess antibacterial properties against *P. synringae pv. lachrymans* and *P. cubensis*. Chun et al. 2018 reported that chitosan NPs enhance biotic stress tolerance against *Fusarium andiyazi* in tomato by inducing up-regulation of pathogenesis- related proteins and antioxidant genes. Danish et al. 2021 reported that silver NPs synthesized from *Senna siamea* mitigate biotic stress induced by *Meloidogyne incognita* enhance plant growth and biochemical attributes in *Trachyspermum ammi*. In wheat, titanium dioxide nanoparticles (TiO_2 NPs) synthesized using *Moringa oleifera* Lam. aqueous leaf extract showed antifungal properties against *Bipolaris sorokiniana* (Satti et al. 2021).

2.4.1 Concept of Green Nanotechnology in Biotic stress Management

Nanotechnology is most prominent approach for sustainable agricultural production. Nanotechnology deals with effective utilization of nano particles ranging from 1 to 100 nm which is crucial for plant growth and defense mechanism. The

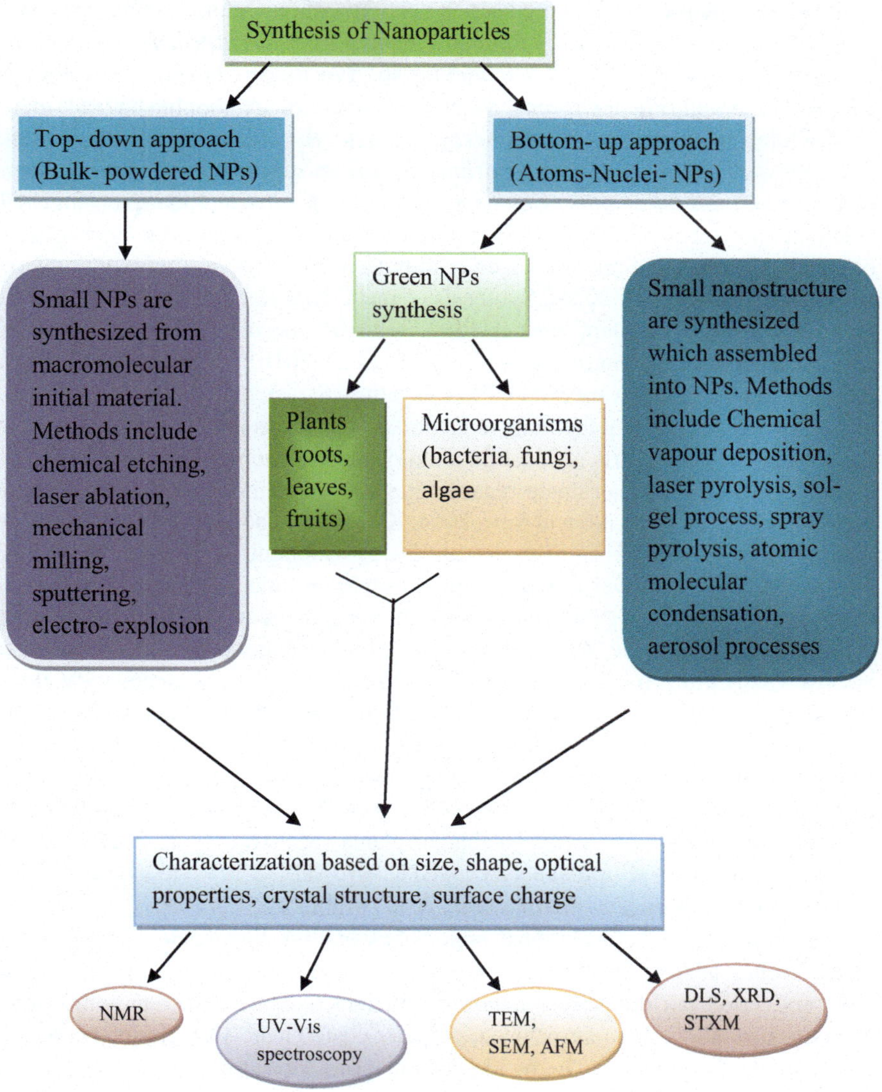

Fig. 2.1 Synthesis and characterization of nanoparticles

synthesis of nano particles may involves physical and chemical processes, but the utilization of plants and microorganisms for synthesis of nano particles are effectively drawing greater attention to researchers and agriculturalists as they are eco-friendly, relatively inexpensive, easy & maintains environmental sustainability. The green nanotechnology is gaining rounds in the modern era of agriculture. Plants parts like roots, stems, fruits, flowers etc. are effectively used in synthesis of nanoparticles. However, various microorganisms like bacteria, fungi, algae are also used in this process for synthesis of metal nanoparticles. Many unicellular and multicellular microorganisms are used in synthesis of nanoparticles. Bacterial species like *Streptomyces* sp., *Rhodopseudomonas capsulate, Enterobacteria, Plectonema boryanum* strain, *Lactobacillus* sp., *Plectonema boryanum, Klebsiella pneumonia, Bacillus licheniformis, Klebsiella aerogenes* are used to synthesis metal NPs, while fungal species include *Verticillium sp., Aspergillus flavus, Aspergillus furnigatus, Phanerochaete chrysoparium, Fusarium oxysporum, Alternaria alternate*, *Penicillium fellutanum, Cladosporium cladosporioides* (Shang et al. 2019, Adetunji and Ugbenyen 2019). However, microorganisms mediated green synthesis of NPs is a challenging process which is stimulated by enzymatic activity of microorganisms (Meena et al. 2021). Synthesis of metal NPs Ag, Zn, Co, Cu, Au can be facilitated by use of plant species like *Brassica juncea, Medicago sativa* and *Helianthus annus*, *Sesbania* plants (Panpatte et al. 2016, Ghormade et al. 2011, Kitching et al. 2015, Iravani 2011). Plants synthesize different types of metabolites and other compounds that facilitate reduction of metal ions in synergistic manner to their metallic counterparts and stabilization of metal nanoparticles (Iravani 2011). Several studies have been made regarding synthesis of metal nanoparticles from plants and its parts (Bonatto and Silva 2014, Kumar and Yadav 2009, Song and Kim 2009, Vinod et al. 2011, Silva et al. 2015, Gholami-Shabani et al. 2017). Nanotechnology has diverse applications in pharmaceuticals, medicine, the environment, food processing and agriculture. The biogenic reduction of elements such as Ag, Au, Cu, Cd, Al, Se, Zn, Ce, Ti and Fe with plant extracts and microorganisms is most widely utilized for production of nanoparticles (NPs) and gained wide attention. In agricultural system it mainly focused on plant growth and crop protection. Hussain et al. 2018 reported that green synthesis of Ag NPs promotes fruit quality and productivity in Kinnow against *Xanthomonas axonopodis* pv. *citri.* Silver nanoparticles augmented *Calothrix elenkinii* (AgNPs-Ce) enhances biocontrol efficiency against *Alternaria blight* in tomato plants (Mahawar et al. 2020). AgNPs synthesized using *Stenotrophomonas* sp. BHU-S7 (MTCC 5978) to ameliorate soil borne and foliar pathogens (Mishra et al. 2017) (Table 2.2).

The constraints of conventional approaches of pest management instigate the increased use of nano-formulation of pesticides and it is widely used to increase crop production worldwide. The nano-formulations like nanogels, nano-emulsions, nano-encapsulation, nano-suspension, polymer coated nano-formulation pesticides & fungicide formulations, herbicide formulations etc. stimulate the controlled release of active encapsulated materials inside plants which thereby avoids the chances of toxicity of ingredient inside plant system. Engineered nanostructures also possess insecticidal properties (Haq and Ijaz 2019). Similarly, the engineered nanostructure also plays potential part in disease and weed management. NPs like nano Cu, ZnO, SiO_2, TiO_2 are promising tool to strengthen plant defense against biotic stress. NPs can also be used to prepare formulations of weedicides & insecticides. Nanoherbicides effectively enter root system, get translocated and interfere with metabolism of weeds and parasitic plants (Abigail and Chidambaram 2017).

Table 2.2 Functions of NPs in plant system

Nanoparticles	Plant species	Biotic stresses	Physiological responses	References
Se &Cu NPs	Tomato	*Alternaria solani*	Reduces disease severity, increased enzymatic and non-enzymatic compounds.	Quiterio-Gutierrez et al. (2019)
AgNPs	Cocoa	*A. flavus, F. solani*	Fungicidal properties	Villamizar-Gallardo et al. (2016)
AgNPs	Cucumber, pumpkin	*Powdery mildew*	Antifungal effects	Lamsal et al. (2011b)
Nanosilver	Gerbera	*Stem- end bacteria*	Inhibits bacterial infection	Liu et al. (2009)
Silica NPs	Arabidopsis	*Pseudomonas syringae*	Induces local resistance and SAR, pathogen tolerance	El-Shetchy et al. (2020)
CuNPs + potassium silicate	Tomato	*Clavibacter michiganensis*	Production of enzymatic and non-enzymatic defense system, confers pathogen tolerance	Cumplido-Najera et al. (2019)
Nanosilica	Maize	*Fusarium oxysporum & Aspergillus niger*	Increased disease resistance	Suriyaprabha et al. 2014
ZnO NPs	Tomato	*Tomato mosaic virus (ToMV)*	Induces resistance against ToMV by enhancing growth parameters and antioxidant defense	Sofy et al., (2021)
Chitosan NPs	Fingermillet	*Pyricularia grisea (Cke.) Sacc.*	Induces resistance against blast diseases using accumulation of ROS, peroxidase	Sathiyabama & Manikandam (2016)
MnO NPs	Cowpea	*Meloidgyne incognitia*	Improves plant physiology and growth, nematicidal effect.	Ahmed et al. (2020)
AgNPs	*Arabidopsis*	*Alternaria brassicola*	Combat blackspot diseases, enhance plant immunity	Kumari et al. (2016)
AgNPs	Tomato	*Alternaria solani*	Physiological and biochemical parameters, increased disease resistance	Kumari et al. (2017a, b)
AgNPs	Tomato	*ToMV & PVY*	Decreased disease incidence	El-Dougdoug et al., (2018)
ZnO, CuO, AgNPs	*Prunus domestica*	*B.cinerea*	Reduced grey mould symptoms	Malandrakis et al., (2019)

(continued)

Table 2.2 (continued)

Nanoparticles	Plant species	Biotic stresses	Physiological responses	References
Al_2O_3 NPs	Tomato	*Fusarium sp.*	Reduced root rot symptom	Shenashen et al. (2017)
AgNPs	*Vigna unguiculata*	*X. Axonopodis pv. Malvacearum & X. campestris pv. Campestris*	Antibacterial effect	Vanti et al. (2019)
CuO NPs	Tomato	*P. infestans*	Control late blight disease	Giannousi et al., (2013)
MgO NPs	Tomato	*Ralstonia solanacearum*	Control wilt disease	Imada et al., (2016)
Chitosan & Ag NPs	Faba bean	*A.alternata*	Disease resistance	Ahmad, (2017)
TiO_2, ZnO, Al_2O_3, Ag	–	*Caenorhabditis elegans*	Nematode toxicity	Roh et al. 2009, Wang et al. (2009)
Nanoparticles		Insect-pest species	References	
Polyethelene glycol-coated NPs		*Tribolium castaneum*	Yang et al. (2009)	
AgNPs		*Spodoptera litura*	Jafir et al. (2021)	
SiO_2 NPs		*Spodoptera littoralis*	Helaly et al. (2016)	
Ag & Zn NPs		*Aphis nerii*	Rouhani et al. (2012)	
SiO_2 NPs		*Plutella xylostella*	Shoaib et al. (2018)	
Amorphous silica		*S.oryzae, Tribolium castaneum*	Goswami et al. (2010)	
Chitosan NPs coated with fungal metabolite		*Aphis gossypii*	Chandra et al. (2013)	
Chitosan		*Callosobruchus maculatus*	Sahab et al. (2015)	
Nanoparticles		Weeds	References	
SiO_2 NPs		*Phelipanche aegyptiaca*	Shabbaj et al. (2021)	
AgNPs		*Lemna minor*	Gubbins et al. (2011)	
CuO NPs		*Lolium perenne & Lolium rigidum*	Atha et al. (2012)	

2.4.2 Mechanism of Action of Nanoparticles under Biotic Stress

NPs are extremely tiny particles which play vital role in growth and metabolic function of plant system. As discussed, NPs passes through series of cell organelles once it is uptake by plant roots or other parts. It generally includes apoplastic & symplastic pathways of vascular tissues. There are several mechanism underlying regarding the mode of action of NPs with biotic stresses are needed to be studied thoroughly.

Silver NPs breach inside bacterial & fungal cell, thus causing damage to nucleic acid. Kumari et al., 2017 studied that Ag NPs ameliorates the negative effect of *A. solani* by direct inhibiting, reduced spore count, prevention of infection establishment. Ag NPs showed direct attachment to *F. oxysporum* & is lethal to fungal spores affecting their permeability and respiration (Abkhoo and Panjehkeh 2017). AgNPs cause plasmolysis fungal hyphae (Min et al. 2009). Ahmed et al. 2020 reported that AgNPs synthesized using bacillus cereus SZT1 mitigate the leaf blight pathogen by decreasing cellular ROS & increasing antioxidant enzymes. Chitosan NPs enhance tolerance to *Fusarium andiyazi* in tomato by up-regulation of chitinases & β-1,3-glucanase genes & antioxidant enzymes (Chun et al. 2018). Carbon nanomaterials also showed resistance to fungal pathogen *F. graminearum* by communting cell surface of pathogen, inhibiting water uptake & inducing plasmolysis (Wang et al. 2014). Se & Cu NPs mitigate the negative effect of *Alternaria solani* by enhancing defense mechanism (Quiterio-Gutierrez et al. 2019). Studies conducted by Divya et al. 2020 revealed that chitosan NPs ameliorate *Rhizoctonia solani* by enhancing defense mechanism & total phenol content. Cu- chitosan NPs enhance the growth parameters against *Curvularia* leaf spot in maize by increasing antioxidant & defense mechanism (Choudhary et al. 2017). NPs induce resistance mechanism & stimulate disease suppression (Lamsal et al. 2011). NPs like AgO & ZnO increased the expression of genes of glutathione S-transferase (Shgst1) & superoxide dismutase 2 (ShSOD2) in response to tolerance against *Sclerotinia homoeocarpa* (Li et al. 2017a, b). NPs also stimulate nematode tolerance to plants by disrupting cellular & metabolic machinery, ATP synthesis, membrane permeability & ROS production (Ahamed et al. 2010; Lim et al. 2012). ZnO NPs adversely affect cellular machinery & hyphal formation of *B.cinerea* & *Penicillum expansum* in plants (He et al. 2011). Nano-silica can be used as insecticides to kill insect-pests by absorbing the particles into cuticular lipids by physio sorption. It is observed that nanosilica increases resistance to plants against *S. littoralis* (Barik et al. 2008; Hajong et al. 2019). Although various studies regarding mechanism of NPs against plant defense mechanism have been made, but the pathways are still remain less understood. NPs when translocated within plants stimulate ROS production & interfere with oxidative metabolism of plant system. The interaction of plant with NPs is highly complex phenomenon. NPs stimulate oxidative burst which in turn activate the antioxidant defense enzyme synthesis in plant system & is associated with antimicrobial activity (Ogunyemi et al. 2020). ZnO also stimulate antimicrobial activity by production of ROS (Patra and Goswami 2012). Corral-Diaz et al. 2014 studied that CeO_2 NPs stimulate production of antioxidant enzymes. ROS has signaling role in plant system inspite of being hazardous nature for plant metabolic processes. nMgO NPs causes oxidative stress in cells of *Ralstonia solanacearum* (Cai et al. 2018a) & antifungal mechanism against *Phytophthora nicotianae* & *Thielaviopsis basicola* in nMgO – exposed fungal cells (Chen et al. 2020). The tripartite interaction of NPs with plant & pathogen also stimulate the production of certain proteins & metabolic to increase disease resistant in *A. thaliana* (Kumari et al. 2020). Nanosensors are also a potential tool to detect plant pathogens & monitor metabolic production (Abbas et al. 2016) (Fig. 2.2).

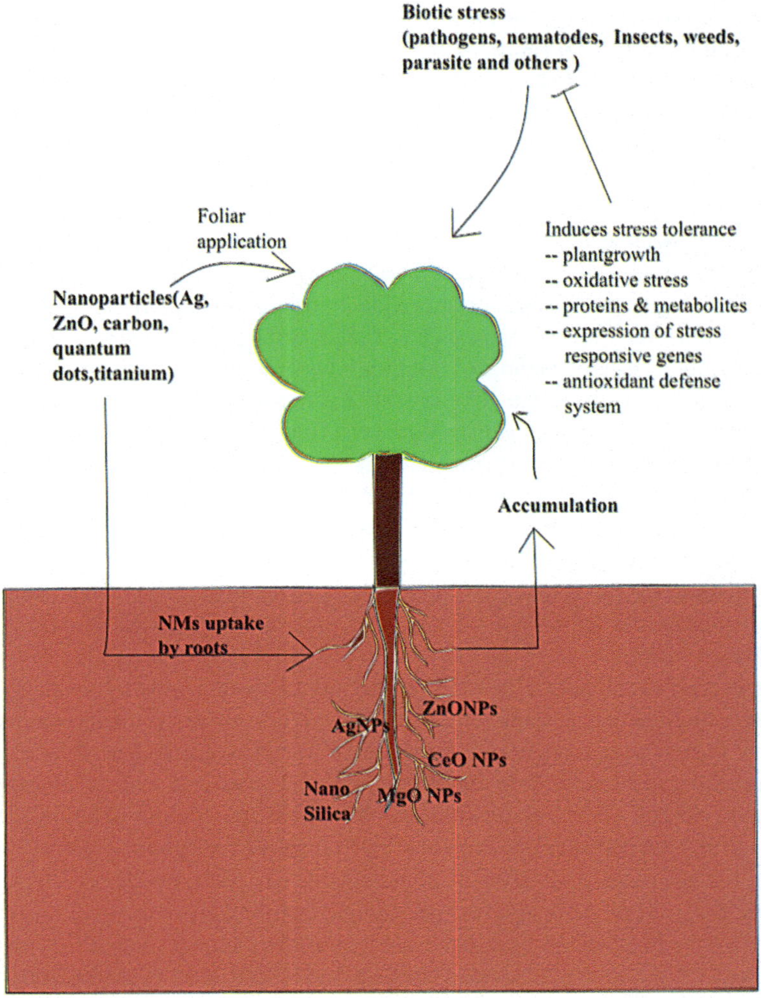

Fig. 2.2 Effect of nano-particles on plants

2.5 Toxicity of Nanoparticles

Nano-materials are potential tool which plays a crucial role in plant growth and metabolic functions. However, its influence on plant system is highly dependent on its size, shape, dosage of application, concentration and others. Although their high concentration may causes toxic effects in plant cells. Toxicity of NPs can interfere with transpiration & photosynthesis, cell wall blockage, closes stomata, damage to nucleic acid and may leads to ROS production (Rasool et al. 2019). Lopez-Moreno et al. (2010) studied that CeO2 nanoparticles cause DNA damage in soybean. Similarly, Zn and ZnO NPs showed toxicity symptoms in radish, rape, and ryegrass

(Lin and Xing 2007). High concentration of TiO_2 NPs induces toxic responses in tomato, cucumber and spinach Gohari et al. (2020). Mukherjee et al. (2014).

studied that ZnO NPs at higher concentrations decreased chlorophyll content in green peas (*Pisum sativum*) and disrupts translocation in cowpea (Wang et al. 2013). AgNPs at high concentration causes chromosomal aberrations in *Vicia faba* (Patlolla et al. 2012).Nanoparticles possess both direct and indirect effects on plant metabolism and growth. The concept of nano-toxicology is a fundamental area of research for agriculturalists and researchers which are needed to be addressed.

2.6 Conclusion

Nanotechnology has the potential of reducing greenhouse gases emissions considerably and thereby mitigating Climate change. The speed at which global warming is increasing may endanger our planet in near future. There has been a dramatic increase in greenhouse gases, notably carbon dioxide in recent times from cars and industries. Engineering science addresses the worldwide warming issue by minimizing/eliminating fossil fuels, decreasing energy consumption, and increasing the potency of existing technologies. On the opposite hand, engineering science consistently involves the utilization of nano-materials (nano-membranes, nanocatalysts, aerogels, nanoparticles, etc.), and their manufacture needs a major quantity of energy input. Additionally, the science of applying nanotechnologies to resolve the matter of worldwide climate change should be closely determined and thus the necessity for an in depth analytic thinking. Needless to say, firm commitments from major industries and governments alike to reduce greenhouse gases are also essential for our future.

References

Abbas SS, Haneef M, Lohani M, Tabassum H, Khan AF (2016) Nano-materials used as a plants growth enhancer: an update. Int J Pharm Sci Rev Res 5:17–23

Abdelmalek GAM, Salaheldin TA (2016) Silver nanoparticles as a potent fungicide for citrus phytopathogenic fungi. J Nanomed Res 3:1–8

Abigail EA, Chidambaram R (2017) Nanotechnology in herbicide resistance. In Nanostructured Mat Fabrication Appl In Tech. https://doi.org/10.5772/intechopen.68355

Abkhoo J, Panjehkeh N (2017) Evaluation of antifungal activity of silver nanoparticles on Fusarium oxysporum. Int J Inf Secur 4:41126

Acquaah G (2007) Principles of plant genetics and breeding. Blackwell, Oxford

Adrees M, Khan ZS, Ali S, Hafeez M, Khalid S, ur Rehman MZ, Hussain A, Hussain K, Chatha SAS, Rizwan M (2019) Simultaneous mitigation of cadmium and drought stress in wheat by soil application of iron nanoparticles. Chemosphere 238

Adetunji C, Ugbenyen A (2019) Mechanism of action of nanopesticide derived from microorganism for the alleviation of abiotic and biotic stress affecting crop productivity. https://doi.org/10.1007/978-981-32-9374-8_7

Aghdam MTB, Mohammadi H, Ghorbanpour M, SpringerLink (Online service) (2015) Effects of nanoparticulate anatase titanium dioxide on physiological and biochemical performance of *Linum usitatissimum* (Linaceae) under well-watered and drought stress conditions. Rev Bras Bot 39:139–146

Ahamed M, Posgai R, Gorey TJ, Nielsen M, Hussain SM, Rowe JJ (2010) Silver nanoparticles induced heat shock protein 70, oxidative stress and apoptosis in drosophila melanogaster. Toxicol Appl Pharmacol 242(3):263–269

Ahmed AI (2017) Chitosan and silver nanoparticles as control agents of some Faba bean spot diseases. J Plant Pathol Microbiol 8(9). https://doi.org/10.4172/2157-7471.1000421

Ahmed T, Shahid M, Noman M, Niazi MBK, Mahmood F, Manzoor I, Zhang Y, Li B, Yang Y, Yan C, Chen J (2020) Silver nanoparticles synthesized by using Bacillus cereus SZT1 ameliorated the damage of bacterial leaf blight pathogen in Rice. Pathogens 9:160. https://doi.org/10.3390/pathogens9030160

Ahmed T, Ya HH, Khan R, Lubis AMHS, Mahadzir S (2020) Pseudo-ductility, morphology and fractography resulting from the synergistic effect of $CaCO_{(3)}$ and bentonite in HDPE polymer nano composite. Materials 13(3333):3333

Ali S, Mehmood A, Khan N (2021) Uptake, translocation, and consequences of nanomaterials on plant growth and stress adaptation. J Nanomater. https://doi.org/10.1155/2021/6677616

Almutairi ZM (2016) Effect of nano-silicon application on the expression of salt tolerance genes in germinating tomato (Solanum lycopersicum L.) seedlings under salt stress. Plant Omics J 9(1):106–114

Amira SS, Souad A El feky, Essam D (2015) Alleviation of salt stress on Moringa peregrina using foliar application of nanofertilizers. J Hortic Forest 7(2):36–47

Ashkavand P, Tabari M, Zarafshar M, Tomášková I, Struve D (2015) Effect of SiO_2 nanoparticles on drought resistance in hawthorn seedlings. Leśne Prace Badawcze/Forest Research Papers 76(4):350–359

Askary M, Talebi SM, Amini F, Bangan ADB (2017) Effects of iron nanoparticles on Mentha piperita L. under salinity stress. Biologija 63(1):65–75

Atha DH, Wang H, Petersen EJ, Cleveland D, Holbrook RD, Jaruga P, Nelson BC (2012) Copper oxide nanoparticle mediated DNA damage in terrestrial plant models. Environ Sci Technol 46(3):1819–1827

Baig MA, Qamar S, Ali AA, Ahmad J, Qureshi MI (2020) Heavy metal toxicity and tolerance in crop plants. In: Naeem M, Ansari A, Gill S (eds) Contaminants in agriculture. Springer, Cham. https://doi.org/10.1007/978-3-030-41552-5_9

Banerjee J, Kole C (2016) Plant nanotechnology: an overview on concepts, strategies, and tools. https://doi.org/10.1007/978-3-319-42154-4_1

Barik TK, Sahu B, Swain V (2008) Nanosilica-from medicine to pest control. Parasitol Res 103:253–258

Berger S, Sinha Alok K, Thomas R (2007) Plant physiology meets phytopathology: plant primary metabolism and plant–pathogen interactions. J Exp Bot 58(15–16):4019–4026. https://doi.org/10.1093/jxb/erm298

Bonatto CC, Silva LP (2014) Higher temperatures speed up the growth and control the size and optoelectrical properties of silver nanoparticles greenly synthesized by cashew nutshells. Ind Crop Prod 58:46–54. https://doi.org/10.1016/j.indcrop.2014.04.007

Borai IH, Hendawey MH, Salah Eldin TA, Hassan NS, Mahdi AA (2017) Magnetite (Fe_3O_4) nanoparticles mitigated heat stress hazards in wheat. J Biol Chem Environ Sci 12(2):557–574. http://biochenv.blogspot.com.eg

Boyer JS (1982) Plant productivity and environment. Science 218(4571):443–448

Cai L, Liu M, Liu Z, Yang H, Sun X, Chen J, Ding W(2018a) MgONPs can boost plant growth: Evidence from increased seedling growth, morpho-physiological activities, and Mg uptake in tobacco (*Nicotiana tabacum* L.). Molecules, 23: 3375

Carmen IU, Chithra P, Huang Q, Takhistov P, Liu S, Kokini JL (2003) Nanotechnology: a new frontier in food science. Food Technol 57:24–29

Chandra JH, Raj LA, Namasivayam SKR, Bharani RA (2013) Improved pesticidal activity of fungal metabolite from Nomureae rileyi with chitosan nanoparticles. In: Paper presented at international conference. In advanced nanomaterials and emerging engineering technologies (ICANMEET), pp 387–390

Chen W-Y, Suzuki T, Lackner M (2017) Handbook of climate change mitigation and adaptation. Springer International Publishing Switzerland. https://doi.org/10.1007/978-3-319-14409-2

Chen J, Wu L, Lu M, Lu S, Li Z, Ding W (2020) Comparative study on the fungicidal activity of metallic MgO nanoparticles and macroscale MgO against Soilborne fungal Phytopathogens. Front Microbiol 11. https://doi.org/10.3389/fmicb.2020.00365

Choudhary RC, Kumaraswamy RV, Kumari S et al (2017) Cu-chitosan nanoparticle boost defense responses and plant growth in maize (Zea mays L.). Sci Rep 7:9754. https://doi.org/10.1038/s41598-017-08571-0

Chun S, Murugesan C (2018) Chitosan and chitosan nanoparticles induced expression of pathogenesis-related proteins genes enhances biotic stress tolerance in tomato. Int J Biol Macromol 125. https://doi.org/10.1016/j.ijbiomac.2018.12.167

Corral-Diaz B, Peralta-Videa JR, Alvarez-Parrilla E, Rodrigo-Garcia J, Morales MI, Osuna-Avila P, Niu G, Hernandez-Viezcas JA, Gardea-Torresdey JL (2014) Cerium oxide nanoparticles alter the antioxidant capacity but do not impact tuber ionome in Raphanus sativus (L). Plant Physiol Biochem 84:277–285

Cui H, Zhang P, Gu W, Jiang J (2009) Application of anatasa TiO2 sol derived from peroxotitannic acid in crop diseases control and growth regulation. NSTI Nanotech 2:286–289

Cumplido-Nájera CF, González-Morales S, Ortega-Ortíz H, Cadenas-Pliego G, Benavides-Mendoza A, Juárez-Maldonado A (2019) The application of copper nanoparticles and potassium silicate stimulate the tolerance to Clavibacter michiganensis in tomato plants. Sci Hortic 245:82–89

Da Costa MVJ, Sharma PK (2016) Effect of copper oxide nanoparticles on growth, morphology, photosynthesis, and antioxidant response in Oryza sativa. Photosynthetica 54(1):110–119

Danish M, Altaf M, Robab MI, Shahid M, Manoharadas S, Hussain SA, Shaikh H (2021) Green synthesized silver nanoparticles mitigate biotic stress induced by Meloidogyne incognita in Trachyspermum ammi (L.) by improving growth, biochemical, and antioxidant enzyme activities. ACS Omega 6(17):11389–11403. https://doi.org/10.1021/acsomega.1c00375

Das A, Das B (2019) Nanotechnology a potential tool to mitigate abiotic stress in crop plants. Abiotic Biot Stress Plants:1–13. https://doi.org/10.5772/intechopen.83562

de la Rosa G, García-Castañeda C, Vázquez-Núñez E, Alonso-Castro ÁJ, Basurto-Islas G, Mendoza Á, Cruz-Jiménez G, Molina C (2017) Physiological and biochemical response of plants to engineered NMs: implications on future design. Plant Physiol Biochem 110:226–235

De Sousa A, Saleh A, Habeeb T, Hassan Y, Zrieq R, Wadaan M, Hozzein W, Selim S, Matos M, Abdelgawad H (2019) Silicon dioxide nanoparticles ameliorate the phytotoxic hazards of aluminum in maize grown on acidic soil. Sci Total Environ 693:133636. https://doi.org/10.1016/j.scitotenv.2019.133636

Dimkpa BO, Singh U, Bindraban PS, Elmer WH, Gardea-Torresdey JL, White JC (2019) Zinc oxide nanoparticles alleviate drought-induced alterations in sorghum performance, nutrient acquisition, and grain fortification. Sci Total Environ 688:926–934

Divya K, Thampi M, Vijayan S, Varghese S, Jisha MS (2020) Induction of defence response in Oryza sativa L. against Rhizoctonia solani (Kuhn) by chitosan nanoparticles. Microb Pathog 149:104525

Djanaguiraman M, Belliraj N, Stefan H, Bossmann, Prasad PVV (2018) High-temperature stress alleviation by selenium nanoparticle treatment in grain sorghum. ACS Omega 3(3):2479–2491. https://doi.org/10.1021/acsomega.7b01934

Dubchak S, Ogar A, de La Rosa G, García-Castañeda C, Vázquez-Núñez E, Alonso-Castro ÁJ, Basurto-Islas G, Mendoza Á, Cruz-Jiménez G, Molina C (2010) Physiological and biochemical response of plants to engineered NMs: implications on future design. Plant Physiol Biochem 110:226–235

El-Dougdoug NK, Bondok AM, El-Dougdoug KA (2018) Evaluation of silver nanoparticles as antiviral agent against ToMV and PVY. Tomato Plants 8(1):100–111

Elsakhawy T, Omara AED, Alshaal T, El-Ramady H (2018) Nanomaterials and plant abiotic stress in agroecosystems. Env Biodiv Soil Security 2:73–94

Elsheerya NI, Sunoja VSJ, Wena Y, Zhua JJ, Muralidharana G, Cao KF (2020) Foliar application of nanoparticles mitigates the chilling effect on photosynthesis and photoprotection in sugarcane. Plant Physiol Biochem 149:50–60. https://doi.org/10.1016/j.plaphy.2020.01.035

Gao FQ, Hong FS, Liu C, Zheng L, Su MY, Wu X et al (2006a) Mechanism of nanoanatase TiO2 on promoting photosynthetic carbon reaction of spinach: inducing complex of rubisco- rubisco activase. Biol Trace Elem Res 11:239–254

Gao X, Zou CH, Wang L, Zhang F (2006b) Silicon decreases transpiration rate and conductance from stomata of maize plants. J Plant Nutr 29:1637–1647

Gholami-Shabani M, Gholami-Shabani Z, Shams-Ghahfarokhi M, Jamzivar F, Razzaghi-Abyaneh M. (2017) Green Nanotechnology: Biomimetic Synthesis of Metal Nanoparticles Using Plants and Their Application in Agriculture and Forestrytitle. https://doi.org/10.1007/978-981-10-4573-8_8

Ghormade V, Deshpande MV, Paknikar KM (2011) Perspectives for nano-biotechnology enabled protection and nutrition of plants. Biotechnol Adv 29:792–803

Giannousi K, Avramidis I, Dendrinou-Samara C (2013) Synthesis, characterization and evaluation of copper based nanoparticles as agrochemicals against Phytophthora infestans. RSC Adv 3(44):21743–21752

Giraldo JP, Landry MP, Faltermeier SM, McNicholas TP, Iverson NM, Boghossian AA et al (2014) Plant nanobionics approach to augment photosynthesis and biochemical sensing. Nat Mater 13

Gohari G, Mohammadi A, Akbari A, Panahirad S, Dadpour MR, Fotopoulos V, Kimura S (2020) Titanium dioxide nanoparticles (TiO2 NPs) promote growth and ameliorate salinity stress effects on essential oil profile and biochemical attributes of Dracocephalum moldavica. Sci Rep 10(1):912

Goswami A, Roy I, Sengupta S, Debnath N (2010) Novel applications of solid and liquid formulations of nanoparticles against insect pests and pathogens. Thin Solid Films 519(3):1252–1257

Goyal D et al (2020) Effect of heavy metals on plant growth: an overview. In: Naeem M, Ansari A, Gill S (eds) Contaminants in agriculture. Springer, Cham. https://doi.org/10.1007/978-3-030-41552-5_4

Grover M, Bodhankar S, Maheswari M, Srinivasarao C (2016) Actinomycetes as Mitigators of Climate Change and Abiotic Stress. In: G. Subramaniam et al. (eds.), Plant Growth Promoting Actinobacteria, pp: 203–212, Springer Science + Business Media Singapore. https://doi.org/10.1007/978-981-10-0707-1_13

Gubbins EJ, Batty LC, Lead JR (2011) Phytotoxicity of silver nanoparticles to Lemna minor L. Environ Pollut 159(6):1551–1559

Haghighi M, Pessarakli M (2013) Influence of silicon and nano-silicon on salinity tolerance of cherry tomatoes (Solanum lycopersicum L.) at early growth stage. Sci Hortic 161:111–117

Haghighi M, Afifipour Z, Mozafarian M (2012) The effect of N-Si on tomato seed germination under salinity levels. J Biol Environ Sci 6(16):87–90

Haghighi M, Abolghasemi R, Teixeira da Silva JA (2014) Low and high temperature stress affect the growth characteristics of tomato in hydroponic culture with se and nano-se amendment. Sci Hortic 178:231–240

Hajong M, Devi NO, Debbarma M, Majumder D (2019) Nanotechnology: an emerging tool for management of biotic stresses in plants. Springer Nature. In: Prasad R (ed) Plant nanobionics. Nanotechnology in the Life Sciences. https://doi.org/10.1007/978-3-030-16379-2_11

Hasanpour H, Maali-Amiri R, Zeinali H (2015) Effect of TiO_2 nanoparticles on metabolic limitations to photosynthesis under cold in chickpea. Russ J Plant Physiol 62:779–787

Hatami M, Kariman K, Ghorbanpour M (2016) Engineered nanomaterial-mediated changes in the metabolism of terrestrial plants. Sci Total Environ 571:275–291

Haq IU, Ijaz S (2019) Use of metallic nanoparticles and nanoformulations as nanofungicides for sustainable disease management in plants. In: Prasad R, Kumar V, Kumar M, Choudhary D (eds) Nanobiotechnology in Bioformulations. Springer, Cham, pp 289–316.

He L, Liu Y, Mustapha A, Lin M (2011) Antifungal activity of zinc oxide nanoparticles against Botrytis cinerea and Penicillium expansum. Microbiol Res 166(3):207–215

Hedayati M, Sharma P, Katyal D, Fagerlund F (2016) Transport and retention of carbon-based engineered and natural nanoparticles through saturated porous media. J Nanopart Res 18:57. https://doi.org/10.1007/s11051-016-3365-6

Helaly MN, El-Metwally MA, El-Hoseiny H, Omar SA, El-Sheery NI (2014) Effect of nanoparticles on biological contamination of 'in vitro' cultures and organogenic regeneration of banana. Aust J Crop Sci 8(4):612–624

Helaly AA, Bendary HME, Abdel-Wahab AS, El-Sheikh MAK, Elnagar S (2016) The silica nanoparticles treatment of squash foliage and survival and development of Spodoptera littoralis (Bosid.) larvae. J Entomol Zool Stud 4(1):175–180
Huang GT, Ma SL, Bai LP et al (2012) Signal transduction during cold, salt, and drought stresses in plants. Mol Biol Rep 39(2):969–987
Hussain Mubashir, Raja Naveed, Mashwani Zia-ur-Rehman, Iqbal Muhammad, Ejaz Muhammad, Aslam Sumaira (2018) Green Synthesis and evaluation of silver nanoparticles for antimicrobial and biochemical profiling in kinnow (Citrus Reticulata L.) to enhance fruit quality and productivity under biotic stress. IET Nanobiotechnol 13. https://doi.org/10.1049/iet-nbt.2018.5049
Imada K, Sakai S, Kajihara H, Tanaka S, Ito S (2016) Magnesium oxide nanoparticles induce systemic resistance in tomato against bacterial wilt disease. Plant Pathol 65(4):551–560
Iravani S (2011) Green synthesis of metal nanoparticles using plants. Green Chem 13:2638–2650
Iqbal M, Raja NI, Wattoo FH, Hussain M, Ejaz M, Saira H (2018) Assessment of AgNPs exposure on physiological and biochemical changes and antioxidative defence system in wheat (*Triticum aestivum* L.) under heat stress. IET NanobioTechnol16:230–236
Iqbal M, Raja NI, Mashwani Z-U-R, Hussain M, Ejaz M, Yasmeen F (2019) Effect of silver nanoparticles on growth of wheat under heat stress. Iranian J Sci & Technol Transaction A: Sci 43(2):387–395
Jaberzadeh A, Tohidi Moghadam HR, Moaveni P, Zahedi H (2013) Influence of bulk and nanoparticles titanium foliar application on some agronomic traits, seed gluten and starch contents of wheat subjected to water deficit stress. Notulae Botanicae Horti Agrobotanici Cluj-Napoca 41(1):201–207
Jafir M, Ahmad JN, Arif MJ, Ali S, Ahmad SJN (2021) Characterization of Ocimum basilicum synthesized silver nanoparticles and its relative toxicity to some insecticides against tobacco cutworm, Spodoptera litura Feb. (Lepidoptera; Noctuidae). Ecotoxicol Environ Saf 218:112278
Jalali M, Ghanati F, Modarres-Sanavi AM, Khoshgoftarmanesh AH (2017) Physiological effects of repeated foliar application of magnetite nanoparticles on maize plants. J Agron Crop Sci 203(6):593–602
Jalil S, Ansari MI (2019) Nanoparticles and abiotic stress tolerance in plants: synthesis, action and signaling mechanisms (Elsevier). https://doi.org/10.1016/B978-0-12-816451-8.00034-4
Jiang HS, Li M, Chang FY, Li W, Yin LY (2012) Physiological analysis of silver nanoparticles and AgNO3 toxicity to Spirodela polyrhiza. Environ Toxicol Chem 31:1880–1886
Jiang M, Dai S, Wang B, Xie Z, Li J, Wang L, Li S, Tan Y, Tian B, Shu Q, Huang J (2021) Gold nanoparticles synthesized using melatonin suppress cadmium uptake and alleviate its toxicity in rice. Environ Sci Nano. https://doi.org/10.1039/D0EN01172J
Kalteh M, Alipour ZT, Ashraf S, Aliabadi MM, Nosratabadi AF (2014) Effect of silica nanoparticles on basil (Ocimum basilicum) under salinity stress. J Chem Health Risk 4(3):49–55
Kerry RG, Gouda S, Das G, Chethala NV, Patra JK (2017) Agricultural nanotechnologies: Current applications and future prospects. https://doi.org/10.1007/978-981-10-6847-8_1
Khan MA, Akhtar MS (2015) Agricultural adaptation and climate change policy for crop production in Africa. In: Hakeem KR (ed) Crop production and global environmental issues. Springer International Publishing, Cham. https://doi.org/10.1007/978-3-319-23162-4_18
Khan MN, Mobin M, Abbas ZK, AlMutairi KA, Siddiqui ZH (2017) Role of nanomaterials in plants under challenging environments. Plant Physiol Biochem 110:194–209. https://doi.org/10.1016/j.plaphy.2016.05.038
Khan, A U; Khan, M.; Khan, M M. (2019). Antifungal and antibacterial assay by silver nanoparticles synthesized from aqueous leaf extract of Trigonella foenum-graecum. BioNanoScience
Khan M, Khan AU, Hasan MA, Yadav KK, Pinto MMC, Malik N, Yadav VK, Khan AH, Islam S, Sharma GK (2021) Agro-nanotechnology as an emerging field: a novel sustainable approach for improving plant growth by reducing biotic stress. Appl Sci 11:2282. https://doi.org/10.3390/app11052282
Khodakovskaya MV, de Silva K, Nedosekin DA, Dervishi E, Biris AS, Shashkov EV, Ekaterina IG, Zharov VP (2011) Complex genetic, photo thermal, and photo acoustic analysis of nanoparticle-plant interactions. Proc Natl Acad Sci U S A 108:1028–1033

Kiapour H, Moaveni P, Habibi D, Sani B (2015) Evaluation of the application of gibbrellic acid and titanium dioxide nanoparticles under drought stress on some traits of basil (*Ocimum basilicum L.*). Int J Agron Agric Res 6:138–150
Kim SW, Jung JH, Lamsal K, Kim YS, Min JS, Lee YS (2012) Antifungal effects of silver nanoparticles (AgNPs) against various plant pathogenic fungi. Mycobiology 4:53–58
Kim J-H, Oh Y, Yoon H, Hwang I, Chang Y-S (2015) Iron nanoparticle-induced activation of plasma membrane H+ – ATPase promotes stomatal opening in Arabidopsis thaliana. Environ Sci Technol 49(2):1113–1119
Kitching M, Ramani M, Marsili E (2015) Fungal biosynthesis of gold nanoparticles: mechanism and scale up. J Microbial Biotechnol 8:904–917
Koelmel J, Leland T, Wang H, Amarasiriwardena D, Xing B (2013) Investigation of gold nanoparticles uptake and their tissue level distribution in rice plants by laser ablation inductively coupled-mass spectrometry. Environ Pollut 174:222–228
Konate A, He X, Zhang Z, Ma Y, Zhang P, Alugongo G, Rui Y (2017) Magnetic (Fe3O4) nanoparticles reduce heavy metals uptake and mitigate their toxicity in wheat seedling. Sustainability 9(5):790. https://doi.org/10.3390/su9050790
Krishnaraj C, Ramachandran R, Mohan K, Kalaichelvan PT (2012) Optimization for rapid synthesis of silver nanoparticles and its effect on phytopathogenic fungi. Spectrochim. Acta Part A: Mol Biomol Spectrosc 93:95–99
Kumar V, Yadav SK (2009) Plant mediated synthesis of silver and gold nanoparticles and their applications. J Chem Technol Biotechnol 84(2):151–157. https://doi.org/10.1002/jctb.2023
Kumari M, Agarwal L, Pandey S, Mishra A, Nautiyal C (2016) Insight into the plant stress adaptation strategies during tripartite interaction of plant-pathogen and nanoparticles using comparative proteomics approach. International conference on plant synthetic biology and bioengineering. Miami, USA
Kumari M, Pandey S, Bhattacharya A, Mishra A, Nautiyal CS (2017a) Protective role of biosynthesized silver nanoparticles against early blight disease in Solanum lycopersicum. Plant Physiol & Biochem:121. https://doi.org/10.1016/j.plaphy.2017.11.004
Kumari M, Pandey S, Bhattacharya A, Aradhana M, Nautiyal CS (2017b) Protective role of biosynthesized silver nanoparticles against early blight disease in Solanum lycopersicum. Plant Physiol Biochem:121
Kumari M, Pandey S, Mishra SK, Giri VP, Agarwal L, Dwivedi S, Pandey AK, Nautiyal CS, Mishra A (2020) Omics-based mechanistic insight in to the role of bioengineered nanoparticles for biotic stress amelioration by modulating plant metabolic pathways. Front Bioeng Biotechnol 8:242. https://doi.org/10.3389/fbioe.2020.00242
Kurepa J, Paunesku T, Vogt S, Arora H, Rabatic BM, Lu J, Wanzer MB, Woloschak GE, Smalle JA (2010) Uptake and distribution of ultra small anatase TiO2 alizarin red S nanoconjugates in Arabidopsis thaliana. Nano Lett 10:2296–2302
Lamsal K, Kim SW, Jung JH, Kim YS, Kim KS, Lee YS (2011) Application of silver nanoparticles for the control of Colletotrichum species in vitro and pepper anthracnose disease in field. Mycobiology 39:194–199
Lamsal K, Kim SW, Jung JH, Kim YS, Kim KS, Lee YS (2011b) Inhibition effects of silver nanoparticles against powdery mildews on cucumber and pumpkin. Mycobiology 39:26–32
Latef AHA, Alhmad MFA, Abdelfattah KE (2017) The possible roles of priming with ZnO nanoparticles in mitigation of salinity stress in lupine (Lupinus termis) plants. J Plant Growth Regul 36(1):60–70
Laware SL, Raskar S (2014) Effect of titanium dioxide nanoparticles on hydrolytic and antioxidant enzymes during seed germination in onion. Int J Curr Microbiol App Sci 3(7):749–760
Li W, Zhu X, Chen H, He Y, Xu J (2013) Enhancement of extraction amount and dispersibility of soil nanoparticles by natural organic matter in soils. In: J. Xu et al. (eds.), Functions of Natural Organic Matter in Changing Environment, Zhejiang University Press and Springer Science+Business Media Dordrecht. https://doi.org/10.1007/978-94-007-5634-2_139
Li J, Sang H, Guo H, Popko JT, He L, White JC, Dhankher OP, Jung G, Xing B (2017a) Antifungal mechanisms of ZnO and ag nanoparticles to Sclerotinia homoeocarpa. Nanotechnology 28(15):155101. https://doi.org/10.1088/1361-6528/aa61f3

Li L, Zhao C, Zhang Y, Yao J, Yang W, Hu Q, Wang C, Cao C (2017b) Effect of stable antimicrobial nano-silver packaging on inhibiting mildew and in storage of rice. Food Chem 215:477–482
Lian J, Zhao L, Wu J, Xiong H, Bao Y, Zeb A, Tang J, Liu W (2019) Foliar spray of TiO_2 nanoparticles prevails over root application in reducing Cd accumulation and mitigating Cd-induced phytotoxicity in maize (Zea mays L.). Chemosphere 239(124794)
Lian J, Zhao L, Wu J, Xiong H, Bao Y, Zeb A, Tang J, Liu W (2020) Foliar spray of TiO_2 nanoparticles prevails over root application in reducing Cd accumulation and mitigating Cd-induced phytotoxicity in maize (Zea mays L.). Chemosphere 239:124794. https://doi.org/10.1016/j.chemosphere.2019.124794
Lim D, Roh JY, Eom HJ, Choi JY, Hyun J, Choi J (2012) Oxidative stress related PMK 1 P38 MAPK activation as a mechanism for toxicity of silver nanoparticles to reproduction in the nematode Caenorhabditis elegans. Environ Toxicol Chem 31(3):585–592
Lin D, Xing B (2007) Phyto-toxicity of nanoparticles: inhibition of seed germination and root growth. Environ Pollut 150(2):243–250
Linh, Tran M, Chi MN, Thi HP, Le Quynh L, Khac BN, Le Thi TH, Hoai CN, Tuong VN (2020) Metal-based nanoparticles enhance drought tolerance in soybean. J Nanomater., Article ID 4056563, 13 pages. https://doi.org/10.1155/2020/4056563
Liu J, He S, Zhang Z, Cao J, Lv P, He S, Cheng G, Joyce DC (2009) Nano-silver pulse treatments inhibit stem-end bacteria on cut gerbera cv. Ruikou flowers. Postharvest Biol Technol 54:59–62
Lopez-Moreno ML, de la Rosa G, Hernández-Viezcas J, Castillo-Michel H, Botez CE, Peralta-Videa JR, Gardea-Torresdey JL (2010) Evidence of the differential biotransformation and Geno-toxicity of ZnO and CeO2 NPs on soybean (Glycine max) plants. Environ Sci Technol 44:7315–7320
Mahawar H, Prasanna R, Gogoi R, Singh S, Chawla G, Kumar A (2020) Synergistic effects of silver nanoparticles augmented Calothrix elenkinii for enhanced biocontrol efficacy against Alternaria blight challenged tomato plants. 3 Biotech 10:102. https://doi.org/10.1007/s13205-020-2074-0
Malandrakis AA, Kavroulakis N, Chrysikopoulos CV (2019) Use of copper, silver and zinc nanoparticles against foliar and soil-borne plant pathogens. Sci Total Environ 670:292–299
Maleki M, Mansour G, Khalil K (2017) Physiological and antioxidative responses of medicinal plants exposed to heavy metals stress. Plant Gene. https://doi.org/10.1016/j.plgene.2017.04.006
Meena M, Zehra A, Swapnil P, Harish MA, Yadav G, Sonigra P (2021) Endophytic nanotechnology: an approach to study scope and potential applications. Front Chem 9:613343. https://doi.org/10.3389/fchem.2021.613343
Min JS, Kim KS, Kim SW, Jung JH, Lamsal K, Kim SB, Jung M, Lee YS (2009) Effects of colloidal silver nanoparticles on sclerotium-forming phytopathogenic fungi. J Plant Pathol 25:376–380
Mishra S, Singh B, Naqvi A, Singh H (2017) Potential of biosynthesized silver nanoparticles using Stenotrophomonas sp. BHU-S7 (MTCC 5978) for management of soil-borne and foliar phytopathogens. Sci Rep 7:45154. https://doi.org/10.1038/srep45154
Mitra D (2021) Emerging plant diseases: research status and challenges. In: Singh KP, Jahagirdar S, Sarma BK (eds) Emerging trends in plant pathology. Springer, Singapore. https://doi.org/10.1007/978-981-15-6275-4_1
Mohammadi R, Maali-Amiri R, Abbasi A (2013) Effect of TiO_2 nanoparticles on chickpea response to cold stress. Biol Trace Elem Res 152:403–410
Mohammadi R, MaaliAmiri R, Mantri N (2014) Effect of TiO_2 nanoparticles on oxidative damage and antioxidant defence systems in chickpea seedlings during cold stress. Russ J Plant Physiol 61:768–775
Monica RC, Cremonini R (2009) Nanoparticles and higher plants. Caryologia Int J Cytol Cytosyst Cytogenet 62(2):161–165. https://doi.org/10.1080/00087114.2004.10589681
Mustafa G, Sakata K, Komatsu S (2015) Proteomic analysis of flooded soybean root exposed to aluminum oxide nanoparticles. J Proteome 128:280–297
Mukherjee A, Peralta-Videa JR, Bandyopadhyay S, Rico CM, Zhao L, Gardea-Torresdey JL (2014) Physiological effects of nanoparticulate ZnO in green peas (Pisum sativum L.) cultivated in soil. Metallomics

Mukherjee et al (2016) Differential Toxicity of Bare and Hybrid ZnO Nanoparticles in Green Pea (*Pisum sativum* L.): A Life Cycle. https://doi.org/10.3389/fpls.2015.01242

Neves-Borges AC, Guimarães-Dias F, Cruz F et al (2012) Expression pattern of drought stress marker genes in soybean roots under two water deficit systems. Genet Mol Biol 35(suppl 1):212–221

Ogunyemi SO, Zhang M, Abdallah Y, Ahmed T, Qiu W, Ali MA, Li B (2020) The bio-synthesis of three metal oxide nanoparticles (ZnO, MnO2, and MgO) and their antibacterial activity against the bacterial leaf blight pathogen. Front Microbiol 11. https://doi.org/10.3389/fmicb.2020.588326

Panpatte DG, Jhala YK, Shelat HN, Vyas RV (2016) Nanoparticles: the next generation technology for sustainable agriculture. In: Microbial inoculants in sustainable agricultural productivity. Springer, New Delhi, India, pp 289–300

Patlolla AK, Berry A, May L, Tchounwou PB (2012) Genotoxicity of Ag NPs in Vicia faba: a pilot study on the environmental monitoring of NPs. Int J Environ Res Public Health 9:1649–1662

Patra P, Goswami A (2012) Zinc nitrate derived nano ZnO: fungicide for disease management of horticultural crops. Int J Innov Hort 1:79–84

Perez-de-Luque A (2017) Interaction of nanomaterials with plants: what do we need for real applications in agriculture? Front Environ Sci 5:12

Pérez-Labrada F, López-Vargas ER, Ortega-Ortiz H, Cadenas-Pliego G, Benavides-Mendoza A, JuárezMaldonado A (2019) Responses of tomato plants under saline stress to foliar application of copper nanoparticles. Plan Theory 8(6)

Prażak R, Święciło A, Krzepiłko A, Michałek S, Arczewska M (2020) Impact of ag nanoparticles on seed germination and seedling growth of green beans in normal and chill temperatures. Agriculture 10:312. https://doi.org/10.3390/agriculture10080312

Qados A, Moftah A (2015) Influence of silicon and nano-silicon on germination, growth and yield of faba bean (*Vicia faba* L.) under salt stress conditions. J Exp Agric Int 5(6):509–524

Qi M, Liu Y, Li T (2013) Nano-TiO2 improve the photosynthesis of tomato leaves under mild heat stress. Biol Trace Elem Res 156:323–328

Quiterio-Gutiérrez T, Ortega-Ortiz H, Cadenas-Pliego G, Hernández-Fuentes AD, Sandoval-Rangel A, Benavides-Mendoza A, Cabrera-de la Fuente M, Juárez-Maldonado A (2019) The application of selenium and copper nanoparticles modifies the biochemical responses of tomato plants under stress by Alternaria solani. Int J Mol Sci 20:1950. https://doi.org/10.3390/ijms20081950

Rahmatizadeh R, Arvin SMJ, Jamei R, Mozaffari H, Nejhad FR (2019) Response of tomato plants to interaction effects of magnetic (Fe_3O_4) nanoparticles and cadmium stress. J Plant Interact 14(1):474–481. https://doi.org/10.1080/17429145.2019.1626922

Rasool A, Hafiz SW, Inayatullah T, Reiaz Ul R (2019) Application of nanoparticles in crop production and protection. Springer nature Switzerland AG.235. In: Prasad R (ed) Plant nanobionics. Nanotechnology in the Life Sciences. https://doi.org/10.1007/978-3-030-16379-2_9

Reddy PVL, Hernandez-Viezcas JA, Peralta-Videa JR, Gardea-Torresdey JL (2016) Lessons learned: are engineered nanomaterials toxic to terrestrial plants? Sci Total Environ 568:470–479

Rezvani N, Sorooshzadeh A (2012) Effect of nano-silver on growth of saffron in flooding stress. Int J Biol Biomol Agric Food Biotechnol Eng 6:1–16

Rizwan M, Ali S, Ali B, Adrees M, Arshad M, Hussain A, Zia ur Rehman M, Waris AA (2019) Zinc and iron oxide nanoparticles improved the plant growth and reduced the oxidative stress and cadmium concentration in wheat. Chemosphere 214:269–277

Roh JY, Sim SJ, Yi J, Park K, Chung KH, Ryu DY, Choi J (2009) Ecotoxicity of silver nanoparticles on the soil nematode Caenorhabditis elegans using functional ecotoxicogenomics. Environ Sci Technol 43(10):3933–3940

Rouhani M, Samih MA, Kalantari S (2012) Insecticide effect of silver and zinc nanoparticles against Aphis nerii Boyer De Fonscolombe (Hemiptera: Aphididae). Chilean J Agric Res 72(4):590–594

Sabaghnia N, Janmohammadi M (2014) Effect of nano-silicon particles application on salinity tolerance in early growth of some lentil genotypes. Biologia 69(2)

Sahab AF, Waly AI, Sabbour MM, Nawar LS (2015) Synthesis, antifungal and insecticidal potential of chitosan (CS)-g-poly (acrylic acid)(PAA) nanoparticles against some seed borne fungi and insects of soybean. Int J Chem Technol Res 8:589–598

Sathiyabama M, Manikandan A (2016) Chitosan nanoparticle induced defense responses in finger-millet plants against blast disease caused by Pyricularia grisea (Cke.)Sacc. Carbohydr Polym 154:241–246

Satti SH, Raja NI, Javed B, Akram A, Mashwani Z-u-R, Ahmad MS et al (2021) Titanium dioxide nanoparticles elicited agro-morphological and physicochemical modifications in wheat plants to control Bipolaris sorokiniana. PLoS One 16(2):e0246880. https://doi.org/10.1371/journal.pone.0246880

Saxena R, Tomar RS, Kumar M (2016) Exploring nano biotechnology to mitigate abiotic stress in crop plants. J Pharm Sci Res 8(9):974–980

Scrinis G, Lyons K (2007) The emerging nano-corporate paradigm: nanotechnology and the transformation of nature, food and Agri-food systems. Int J Sociol Agric & Food 15:22–44

Sedghi M, Mitra H, Sahar T (2013) Effect of nano zinc oxide on the germination of soybean seeds under drought stress. Annals of West University of Timisoara: Series of Biology 16(2):73–78

Shabbaj II, Madany M, Tammar A, Balkhyour MA, AbdElgawad H (2021) Silicon dioxide nanoparticles orchestrate carbon and nitrogen metabolism in pea seedlings to cope with broomrape infection. https://doi.org/10.1039/D0EN01278E

Shang Y, Hasan MK, Ahammed GJ, Li M, Yin H, Zhou J (2019) Applications of nanotechnology in plant growth and crop protection: a review. Molecules 24:2558. https://doi.org/10.3390/molecules24142558

Sharifan H, Moore J, Ma X (2020) Zinc oxide (ZnO) nanoparticles elevated iron and copper contents and mitigated the bioavailability of lead and cadmium in different leafy greens. Ecotoxicol Environ Saf 191:110177. https://doi.org/10.1016/j.ecoenv.2020.110177

Sharma P, Bhatt D, Zaidi MGH, Saradhi PP, Khanna PK, Arora S (2012) Silver nanoparticle mediated enhancement in growth and antioxidant status of brassica juncea. Appl Biochem Biotechnol 167:2225–2233

Shenashen M, Derbalah A, Hamza A, Mohamed A, El Safty S (2017) Antifungal activity of fabricated mesoporous alumina nanoparticles against root rot disease of tomato caused by fusarium oxysporium. Pest Manag Sci 73:1121–1126

Shoaib A, Asem E, Muhammad W, Lulu L, Xinlai C, Qianqian Z, Zu-hua S (2018) Entomotoxic effect of silicon dioxide nanoparticles on Plutella xylostella (L.) (Lepidoptera: Plutellidae) under laboratory conditions. Toxicol & Environ Chem. https://doi.org/10.1080/02772248.2017.1387786

Siddiqui MH, Al-Whaibi MH (2014) Role of nano-SiO_2 in germination of tomato (*Lycopersicum esculentum* seeds Mill.). Saudi J Biol Sci 21(1):13–17

Luciano Paulino Silva, Ivy Garcez Reis, Cínthia Caetano Bonatto (2015). Green Synthesis of Metal Nanoparticles by Plants: Current Trends and Challenges. Springer International Publishing Switzerland V.A. Basiuk, E.V. Basiuk (eds.), Green Processes for Nanotechnology. https://doi.org/10.1007/978-3-319-15461-9_9

Singh S, Husen A (2019) Role of nanomaterials in the mitigation of abiotic stress in plants. Nanomaterials & Plant Potential:441–471. https://doi.org/10.1007/978-3-030-05569-1_18

Sofy AR, Sofy MR, Hmed AA, Dawoud RA, Alnaggar AE, Soliman AM, El-Dougdoug NK (2021) Ameliorating the adverse effects of tomato mosaic tobamovirus infecting tomato plants in Egypt by boosting immunity in tomato plants using zinc oxide nanoparticles. Molecules 26:1337. https://doi.org/10.3390/molecules26051337

Song JY, Kim BS (2009) Rapid biological synthesis of silver nanoparticles using plant leaf extracts. Bioprocess Biosyst Eng 32:79–84. https://doi.org/10.1007/s00449-008-0224-6

Song Y, Jiang M, Zhang H, Li R (2021) Zinc oxide nanoparticles alleviate chilling stress in Rice (Oryza Sativa L.) by regulating Antioxidative system and chilling response transcription factors. Molecules 26:2196. https://doi.org/10.3390/molecules26082196

Stolf-Moreira R, Lemos EGM, Carareto-Alves L et al (2011) Transcriptional profiles of roots of different soybean genotypes subjected to drought stress. Plant Mol Biol Report 29(1):19–34

Suriyaprabha R, Karunakaran G, Kavitha K, Yuvakkumar R, Rajendran V, Narayanasamy (2014) Application of silica nanoparticles in maize to enhance fungal resistance. Nanobiotechnol 8(3):133–137

Taran N, Storozhenko V, Svietlova N, Batsmanova L, Shvartau V, Kovalenko M (2017) Effect of zinc and copper nanoparticles on drought resistance of wheat seedlings. Nanoscale Res Lett 12(1):60
Thapa G, Dey M, Sahoo L, Panda SK (2011) An insight into the drought stress induced alterations in plants. Biol Plant 55(4):603–613
Torabian S, Zahedi M, Khoshgoftar AH (2016) Effects of foliar spray of two kinds of zinc oxide on the growth and ion concentration of sunflower cultivars under salt stress. J Plant Nutr 39(2):172–180
Tripathi DK, Singh S, Singh VP, Prasad SM, Dubey NK, Chauhan DK (2017) Silicon nanoparticles more effectively alleviated UV-B stress than silicon in wheat (Triticum aestivum) seedlings. Plant Physiol Biochem 110:70–81
Vanti GL, Nargund VB, Basavesha KN, Vanarchi R, Kurjogi M, Mulla SI, Tubaki S, Patil RR (2019) Synthesis of Gossypium hirsutum-derived silver nanoparticles and their antibacterial efficacy against plant pathogens. Appl Organomet Chem 33:e4630
Venkatachalam P, Jayaraj M, Manikandan R, Geetha N, Rene ER, Sharma NC, Sahi SV (2017) Zinc oxide nanoparticles (ZnONPs) alleviate heavy metal-induced toxicity in Leucaena leucocephala seedlings: a physiochemical analysis. Plant Physiol Biochem 110:59–69
Villamizar-Gallardo R, Cruz JFO, Ortíz-Rodriguez OO (2016) Fungicidal effect of silver nanoparticles on toxigenic fungi in cocoa. Pesquisa Agropecuaria Brasileira 51(12):1939
Vinod VT, Saravanan P, Sreedhar B et al (2011) A facile synthesis and characterization of ag, au and Pt nanoparticles using a natural hydrocolloid gum kondagogu (Cochlospermum gossypium). Colloids Surf B Biointerfaces 83(2):291–298. https://doi.org/10.1016/j.colsurfb.2010.11.035
Wang H, Wick RL, Xing B (2009) Toxicity of nanoparticulate and bulk ZnO, Al2O3 and TiO2 to the nematode Caenorhabditis elegans. Environ Pollut 157(4):1171–1177
Wang X, Liu X, Chen J, Han H, Yuan Z (2014) Evaluation and mechanism of antifungal effects of carbon nanomaterials in controlling plant fungal pathogen. Carbon 68:798–806
Wang P, Menzies NW, Lombi E, McKenna BA, Johannessen B, Glover CJ et. al (2013) Fate of ZnO nanoparticles in soils and cowpea (*Vigna unguiculata*) Environ. Sci. Technol. 47:13822–13830. https://doi.org/10.1021/Es403466p
Wani SH, Kumar V, Shriram V, Sah SK (2016) Phytohormones and their metabolic engineering for abiotic stress tolerance in crop plants. The Crop Journal 4(3):162–176
Xu J, Yang J, Duan X, Jiang Y, Zhang P (2014) Increased expression of native cytosolic cu/Zn superoxide dismutase and ascorbate peroxidase improves tolerance to oxidative and chilling stresses in cassava (Manihot esculenta Crantz). BMC Plant Biol 14:208
Yadav SK (2010) Heavy metals toxicity in plants: an overview on the role of glutathione and phytochelatins in heavy metal stress tolerance of plants. S Afr J Bot 76:167–179
Yang FL, Li XG, Zhu F, Lei CL (2009) Structural characterization of nanoparticles loaded with garlic essential oil and their insecticidal activity against Tribolium castaneum (Herbst) (Coleoptera: Tenebrionidae). J Agric Food Chem 57(21):10156–10162
Yousefi Z, Mohmadpour R-A, Zarei E, Pour MB (2014) Lignin degradation from synthetic wastewater of pulp and paperndustries by using of UV/Fe-doped TiO_2 photocatalytic process. J Mazandaran Univ Med Sci 23(2):96–106
Ze Y, Liu C, Wang L, Hong M, Hong F (2011) The regulation of TiO_2 nanoparticles on the expression of light-harvesting complex II and photosynthesis of chloroplasts of Arabidopsis thaliana. Biol Trace Elem Res 143:1131–1141
Zhao L, Peng B, Hernandez-Viezcas JA, Rico C, Sun Y, Peralta-Videa JR, Tang X, Niu G, Jin L, Ramirez AV, Zhang JY, Gardea-Torresdey JL (2012) Stress response and tolerance of Zea mays to CeO2 nanoparticles: cross talk among H2O2, heat shock protein and lipid peroxidation. ACS Nano 6:9615–9622
Zhu ZJ, Wang H, Yan B et al (2012) Effect of surface charge on the uptake and distribution of gold nanoparticles in four plant species. Environ Sci Technol 46(22):12391–12398
Zuverza-Mena N, Martínez-Fernández D, Du W, Hernandez-Viezcas JA, BonillaBird N, López-Moreno ML, Komárek M, Peralta-Videa JR, Gardea-Torresdey JL (2017) Exposure of engineered nanomaterials to plants: insights into the physiological and biochemical responses-a review. Plant Physiol Biochem 110:236–264. https://doi.org/10.1016/j.plaphy.2016.05.037

Chapter 3
Nanoparticles As a New Promising Tool to Increase Plant Immunity Against Abiotic Stress

Radwan Khalil, Nesma ElSayed, and Hanan A. Hashem

Abstract According to FAO reports, the scientific community faces a real challenge in increasing food crop production by 60% to 100% for an additional 2.3 billion people worldwide by 2050. Abiotic stress such as, salinity, water deficit, extreme temperatures, and heavy metal toxicity is a major constraint influencing plant growth criteria and productivity. Therefore, there is a global eagerness for new technologies that improve plant tolerance against different stresses. Recently, nano-biotechnology has gained traction as a potential solution to such constraints associated with environmental challenges, as well as to ensure a sustainable and secure future for agriculture globally. Plants' responses to abiotic stress are complex, involving changes in growth criteria as well as transcriptomic and metabolomic changes that have an adverse effect on the plant development and crop yield. The application of nanotechnology has been proved to play a key role in many physiological and metabolic activities in plants, such as, photosynthesis, proline content, hormonal pathway, carbohydrates, the antioxidant system, and the nitrogen metabolism; as well as regulating a wide variety of stress-responsive genes, to mention but few. Likewise, recent studies proved that nanoparticles (NPs) can be successfully used in plant genetic engineering as a genome editor. All of these factors make NPs a perfect candidate to be used to increase plant immunity against abiotic stress.

In this chapter, we presented a detailed overview of the types and properties of NPs, and their role as an efficient technology to increase the adaptation potential of plants cultivated in stressful conditions. Moreover, the safety assessment of the toxicity associated with NPs as well as the possible alterations in biomolecules they interact with, in living systems and environment, are also highlighted.

R. Khalil (✉) · N. ElSayed
Botany Department, Faculty of Science, Benha University, Benha, Egypt
e-mail: Radwan.aboelabbas@fsc.bu.edu.eg

H. A. Hashem
Botany Department, Faculty of Science, Ain Shams University, Cairo, Egypt
e-mail: hananhashem@Sci.asu.edu.eg

M. Faizan et al. (eds.), *Sustainable Agriculture Reviews 53*, Sustainable Agriculture Reviews 53, https://doi.org/10.1007/978-3-030-86876-5_3

Keywords Antioxidant enzymes · Biostimulants · Crop production · Nanoparticles · Phytohormones · Stress tolerance · Sustainable agriculture

3.1 Introduction

The global population is enhancing by the day and is expected to reach 9.1 billion by 2050, but food production is struggling to keep up. Raising output is difficult because the area under cultivation is expected to remain stagnant or even reduce due to rising demand for land for non-agricultural purposes. Abiotic stresses are regarded as a significant constraint to crop productivity. Abiotic stresses are responsible for approximately 70% of crop yield reductions, either directly or indirectly (Acquaah 2009).

Abiotic stress causes a chain reaction of anatomical, physiological, biochemical, and molecular changes that all have a negative impact on plant development and productivity. The most common abiotic stresses that endanger global food safety are water shortage, salt, and high temperatures. Growing plants that are resistant to stress could be a worthwhile strategy for addressing the issue of dwindling global food supply. Traditional breeding methods and inter-specific or inter-generic hybridization have had limited success in improving crop plant stress tolerance. Traditional breeding methods are limited due to the degree of complication in the characteristics of stress tolerance, as well as a lack of genetic diversity among yield components under environmental stress conditions and the absence of effective selection criteria. As a result, it is critical to seek out alternative strategies for growing stress-tolerant crops. For the genetic improvement of crop plants, all conventional breeding methods, such as selection, hybridization, polyploidy, and mutation, was used. Despite the fact that there has been additional progress in the history of agricultural crop improvement, their productivity has now hit a plateau, and food shortages and hunger remain in many developing countries. In addition to conventional and advanced breeding tools already available, the discovery of novel strategies and their exploitation is crucial for this purpose. Today's global demand is to improve food production using limited resources and the least amount of fertilizer and pesticides while still controlling environmental pollution, leading to emerging agricultural technology that can reshape current agriculture. In the era of the plant agriculture and plant biotechnology nanotechnology is the most promising recent breakthrough (Scrinis and Lyons 2007). Nanotechnology incorporates new nanomaterial properties that enable agricultural research in crop improvement programs to mitigate stresses (Moraru et al. 2003). Lynn W. Jelinski, biophysicist at Cornell University in the United States, first introduced the term "nanobiotechnology" (Saxena et al. 2016). Materials, structures, and processes that operate on a scale of 100 nanometers (nm) or less are considered nanotechnology..

The term 'nano' is commonly used to describe a size scale ranging from 1 to 100 nm.Nanomaterials are made up of very small components, and the properties of macro-level materials are affected by these components.. In addition to many other advantages, as in comparison to the same mass of substance formed into a larger shape, nanomaterials (NMs) have a comparatively greater surface area. Nano particles can increase the chemical reactivity of materials and influence their force or electrical characteristics. Because the particles have a high surface-to-volume ratio, their reactivity and potential biochemical activity are increased (Dubchak et al. 2010; Janmohammadi et al. 2016). Agriculture's use of nanotechnology has provided new opportunities for the production of nanosized agrochemicals in recent years, which have the potential to improve performance, improve stability, prolong the effective period and minimize environmental loads at the same time (Chhipa 2017; Pourzahedi et al. 2018). Nanotechnology, a constantly emerging and rapidly expanding science, can help to improve some of these stress factors by using different processes, such as the antioxidant defense system, and delivering less toxic and more efficient fertilizers (Zuverza-Mena et al. 2016).

3.2 Synthesis, Types and Properties of Nanoparticles

Recently, nanotechnology applied in many areas led to synthesis of nanaotechnology from metals or their oxides like Ag, Au, Pd, SiO2, ZnO, and TiO_2, etc. Various techniques are used for NPs synthesis including biological, physical and chemical methods (Singh et al. 2016a). Plant extracts are used for some mineral NPs biosynthesis due to their production of sugar, proteins, enzymes, phenols, flavonoids, alcohol, terpenoids and cofactors. Siddiqui et al. 2014 reported that metallic NPs synthesized from metal salts act as reducing and stabilizing agent. Furthermore, as assistant to detect promising and environmentally friendly nanoparticle synthesis solutions that can deal with well-regulated size and shape while preventing atmospheric pollution (Sharma et al. 2009).

Chemical and physical features of nanomaterials used in nanotechnology differ from those of ordinary chemicals (i.e., fullerene, carbon nanotube, nano silver, photocatalyst, silica, nano carbon). NMs enclose dendrimers, fullerenes, metals as Au, Ag, etc.) and metal oxide like titanium oxide, zinc oxide, etc.), quantum dots, nanotubes padded with carbon; whether single, double or multi walled). Nature contains nano-scale materials; both man-made processes (as diesel combustion or volcanic activity) and natural can produce materials with nanometer-size. Two types of nanomaterial processing processes are present: "top-down" processes (as milling) from which nanoscale materials produced from macro-scale counterparts and "bottom-up" processes (like self-assembly) which form nanoscale atomic and molecular materials. The nanoscale of NMs differs as it can be one dimension (for example, Superficial films), two dimensions (for example, fibers or strands), as well as three dimensions (for example, surface films) (e.g. particles). Different forms for these particles can be detected as monocular, merged, aggregated, or mass forms with tubular, annular and unequal shapes. The five types of engineered NPs are metal-based NPs (MB NPs), carbon-based NPs (CB NPs), magnetic NPs, dendrimers, and composite NPs. Fullerene (C70), fullerol [C60(OH)20], single-walled carbon nanotubes (SWCNTs), multi-walled carbon nanotubes (MWCNTs), and single-walled carbon nanohorns (SWCNHs) are examples of carbon-based NPs, whereas metal-based NPs include nanomaterials based on gold (Au), silver (Ag), copper (Cu), and iron (Fe).

Application of NPs for various metal oxides as TiO_2, FeO, $Al2O_3$, ZnO and CeO_2 are extensively studied in medicine as well as agriculture. Lu et al. (2007) reported that, magnetic NPs can be manipulated via application magnetic field where these particles are generally composed of cobalt, nickel, Fe as well as their complexes. Furthermore, it was observed that, Ferrites (iron oxide Fe_2O_3) particles become super-paramagnetic between different magnetic NPs as their size decreases below 128 nm. The quantum dot is a unique subset of NPs. These are typically particles with a diameter of lower than 10 nm, though sometimes the particle size can be as big as 50 nm. They may have the extreme distinctive properties for all NPs in terms of scale. Quantum dots are semiconducting NPs with measurements so small that the particle size affects the intrinsic band distance of the semiconductor. The photoluminescent properties of quantum dots are demonstrated by the majority of them. When a photon strikes a semiconductor, it excites an electron from the valence band to the conduction band; it results in a valence band hole. This produced photon could have a variety of outcomes: trapping of electrons and holes in a crystal disorder, electron/hole recombination with the emission of light, reaction to form a radical with the solvent, and the capping agent's reaction leading to the induction of a radical. Trindade et al. (2001) declared that, many of these activities will also occur in molecules that do not display confinement at the quantum level; however, as the molecules size lowers in a quantum-constrained setting, the least amount of energy needed to form the excited state will rise. This minimum energy is the particle's band gap.

3.3 Nanoparticle Uptake, Mobilization, and Accumulation in Plants

Recently, the adoption of carbon- and metal-based nanomaterial's (MB NMs) by plants is studied. The most studied carbon-based nanomaterials (CB NMs) are fulllerol (C60 (OH) 20), fullerene C70, and carbon nanotubes (CNTs), while Au, TiO_2, Cu, Ag, FeO, CeO_2, and ZnO NPs are the most studied metal-based nanomaterials (MB NMs). Plant species influence NP uptake, translocation, and accumulation, as well as the NPs' size, shape, chemical structure, functionalization, and stability. Dietz and Herth, (2011) found that, NPs typically enter the plant root system via the lateral root and move to the xylem through cortex then to pericycle. The interactions of NP with the plant depend primarily on chemical processes that produce the transport activity of the ion through cell membrane, oxidative damage and peroxidation of lipids, as well as reactive oxygen species (ROS). Watanabe et al. (2008)

discovered that, NPs react with sulfhydryl; carboxyl groups once entering the plant cells and finally change the activity of the protein. Membrane transporter or root exudate complexes can be formed when NPs combine, which are then transported to the plants. Xylem is the major absorption and transfer pathway of nanomaterials to the shoot and plant leaves, where they move from one root to another and from roots to leaves, stem, and grains (Miralles et al. 2012). Sharif et al. (2013) showed that, the nanomaterials have the capability to penetrate the cell cytoplasm as well as the leaf cuticle. On the other hand, NMs can bind to organelles found in the cytoplasm and damage metabolic reactions at that location (Zhang and Monteiro-Riviere 2009). Ag NPs with a molecule dimension of 20 nm can also be moved into cells via plasmodesmata in one of the previously mentioned pathways (Unrine et al. 2012). Plant techniques of nanotechnology involve a delicate preventive estimation of nanoparticle-plant reactions, as well as an understanding of the reactions of their absorption, translocation, and accumulation, as well as an appreciation of the potential harmful influences on growth of plant and development. It is difficult to predict the interactions with environmental components for plant uptake of NPs, and this because they rely on numerous factors concerning with the application method, as well as the actual nanoparticle (size, net charge, chemical structure and surface properties). González-Melendi et al. (2008) reported that, Because of their facility of detection and monitoring provided by microscopy techniques and the widespread use in industry, the majority of preliminary plant research have focused on metal and its oxide nanoparticles uptake. Although, great amount of knowledge known in metazoans, only a few integrated comparative studies have been performed to estimate the contribution of the physic-chemical characters (for example., charge, mass and coatings and so on.) of NPs in plant-nanoparticle interactivity (Song et al. 2013; Moon et al. 2016; Vidyalakshmi et al. 2017). Apoplastic transfer takes place via the cell wall and extracellular spaces beyond the plasma membrane; Water and solutes are transported between the cytoplasms of neighbouring cells via plasmodesmata and sieve layer pores in symplastic movements. It was reported that, Apoplastic transport elevate the upward movement of the aerial component via facilitating the radial movement of NMs that can pass NPs to the vascular tissues and the central root cylinder (González-Melendi et al. 2008; Sun et al. 2014; Zhao et al. 2017). By this appropriate method of NP translocation, many uses applied including systematic NP distribution. The Casparian strip which defined as a longitudinally aligned stratum formed from lignin-analogous structures prevents root endodermis from completing this radial movement (Sun et al. 2014; Lv et al. 2015). For avoiding this natural hindrance, water and another solvent turn from apoplasty to the symplastic route. Similar abilities to bypass the block at Casparian strip have been reported for various types of NPs, as reported by Schwab et al. (2016).

This is particularly widespread in anatomical parts as the Casparian strip has not yet fully developed, like root tips as well as root lateral junctions (Lv et al. 2019). The symplastic transport of NPs demands that the NPs enter the cells at certain stage. Carpita et al. 1979 showed that, the intracellular transmission of NPs much more difficult in plant cells than in animal cells, and this due to the existence of a hard plant cell wall that produces a physical barrier to cell entrance. The multilayer cell wall composed mainly of scaffold proteins as well as cellulose/hemicellulose microfibrils, providing a permeable environment that serves as a strict selective filter with average diameter of 10 nm and some exceptions of up to 20 nm. This is a pivotal point, from which the design and use of bioengineering in plants are applied (Cunningham et al. 2018). However, various kinds of NPs with average diameters ranging from 3 to 50 nm as well as carbon nanotubes have been shown to readily pass via the cell wall in several plant species (Liu et al. 2009; Chang et al. 2013; Etxeberria et al. 2019). Although other cell entrance mechanisms, as those relied on membrane translocation, pore formation, or carrier proteins, have been described in cells, endocytosis may happen preferentially (Valletta et al. 2014; Palocci et al. 2017). Endocytosis may be used preferentially for subsequent cell internalization (Nel et al. 2009; Lin et al. 2010; Wang et al. 2012). More research in plant cells and models of invertebrates is required (Marchesano et al. 2013). MWCNTs, for example, have been demonstrated to connect protoplasts of Catharanthus roseus via an endosome attempting to avoid uptake mode (Serag et al. 2011). Plasmodesmata are membrane-lined cytoplasmic bridges that have a variable diameter (20–50 nm) that maintain membrane and cytoplasmic consistency throughout plant tissues, allow NPs to move from cell to cell once inside the cytoplasm. Plasmodesta have been found to transport NPs of varying sizes in Arabidopsis, rice, and poplar plant species (Lin et al. 2009a, b; Geisler-Lee et al. 2013; Zhai et al. 2014). The symplastic and apoplastic pathways allow tiny particles to enter the xylem and phloem vessels and travel to various tissues and organs throughout the plant. Surprisingly, organs such as flowers, fruits, and seeds appear to accumulate NMs and have a high capacity to import phloem (sink activity) fluids. NP accumulation in specialized organs, in addition to plant toxicity, poses a significant problem in terms of human and animal safety (Pérez-de-Luque 2017). In terms of application, studies in maize, spinach, and cabbage have shown that metal-NPs can sneak seeds and move into seedlings with no considerable effects on seed viability, germination rate, or shoot growth. These findings imply that functional NPs can be used for seed priming and plant development stimulation even in harsh environmental conditions (Zheng et al. 2005; Răcuciu and Creangă 2009; Pokhrel and Dubey 2013).

3.4 Nanoparticles' Effects on Plants

3.4.1 Effect of Nanoparticles on Growth and Bio-Productivity

Several metal or metal oxide-based nanoparticles are being studied for their role in plant growth and development, defence against biotic and abiotic stresses, and modulation of various plant processes. However, in terms of technological development, we still have a long way to go before we can achieve sustainable agriculture (Saxena et al. 2016).Currently, NPs have the potential to be efficient plant growth and development promoters, herbicides, nano-pesticides, and nano fertilisers because they can effectively deliver the required amount of content to the target cellular organelles in plants Nanotechnology has a wide range of agricultural applications, and the potential applications of NPs, particularly their mechanism and function in plant growth and development, are unknown (Siddiqui et al. 2015a).

Nanoparticle fertilisers are widely used in agriculture to increase productivity and meet rising food demand. Fertilizers are necessary for crop development, growth, and production. As a result, there is always a need to develop new applications based on nanotechnology and nanomaterials. This not only increases crop yield and production, but also reduces fertiliser nutrient loss and improves fertiliser availability to plants. Nonfertilizer or nano-encapsulated nutrients may be a useful tool in the development of sustainable agriculture because they can significantly regulate plant growth and production through effective nutrient release and availability (Saxena et al. 2016). Researchers are increasingly interested in using ex vivo NP synthesis for a variety of agricultural purposes, particularly to improve plant physiological processes such as metabolic activities, also growth and development by leveraging NPs' unique properties (Giraldo et al. 2014). Furthermore, in a field experiment, It was discovered that using an extra low dose of nanocrystalline powders (no more than 300 mg of each metal per hectare) improved the plant growth and production, including chlorophyll pigments, total number of nodules per root, pod weight, number of pods per plant, weight of 1000 seeds, and crop production. Another research on soybeans showed that applying nano-SiO_2 and nano-TiO_2 mixtures improved germination and growth parameters (Lu et al. 2002). Nanotechnology is a rapidly emerging technology that is advancing in a number of fields. However, nanotechnology applications and nanoparticle use in sustainable agriculture and crop enhancement are still in their infancy.

3.4.2 *Effect of Nanoparticles on Photosynthesis and Plant Water Relations*

At low concentrations, Plant growth stages such as seed germination, seedling growth, root initiation, photosynthesis, and flowering are all improved by NPs (Banerjee and Kole 2016). Photosynthetic quantum performance and chlorophyll content are also improved by NPs (Sharma et al. 2012a, b; Hatami and Ghorbanpour 2013). NPs like Nano-TiO_2 enhanced germination, light absorbance, photosynthetic function, and Rubisco activation (Gao et al. 2006). Nano-TiO2 was also discovered to promote spinach growth by improving photosynthetic activities in plant as well as nitrogen metabolism (Yang et al. 2006). TiO2 NPs treated seeds provided 73% above dry weight, photosynthetic rates increased three times, as well as chlorophyll a content increased 45% more than control seeds at 30-days germination cycle (Mingfang et al. 2013). Engineered nanomaterials (ENs) are applied to the surface of leaves and penetrate through stomata openings or trichome bases, where they are translocated to different tissues. he accumulation of ENs on photosynthetic surfaces, on the other hand, causes foliar heating, which changes gas exchange due to stomata obstruction, which changes plant physiological and cellular functions (Aslani et al. 2014). Changes in photosynthetic activity disrupt ATP and NADPH synthesis, as well as other biochemical reactions and physiological processes like cellular development. As a consequence, cellular growth changes caused by NPs may indicate a change in Photosystem II (PSII) activity, which is measured by variable chlorophyll fluorescence emission (Seaton and Walker 1990; Stirbet 2012). Overexpression of the PSII subunit S, a component of the main light-harvesting complex that also performs water oxidation, can result in an increase in photosynthesis water use efficiency. This is a significant limiting component in many agricultural settings that drives higher yields in Nicotiana tabacum (Glowacka et al. 2018). Researchers are currently working to determine the connections between these different Photonic structures and photosynthesis are two topics that have received a lot of attention recently. Jacobs et al. (2016) recently demonstrated that iridescent modified chloroplasts (iridoplasts) found in Begonia epidermis increase photosynthesis light harvesting. By simulating these systems, it may be possible to create nanomaterials that mimic these natural photonic structures and modulate how light is obtained by plants. Recent research has begun to demonstrate that NPs can be engineered to interact with photosynthesis and photoprotection via proven transgenic and naturally occurring pathways, resulting in their augmentation and much-needed crop yield increases (Giraldo et al. 2014). NPs have been demonstrated to be a new, efficient, and promising method for enhancing photosynthesis in plants that has yet to be fully exploited. The effect of NPs on photosynthesis is still being studied, and further researches are needed to cover this topic in detail.

3.4.3 *Effect of Nanoparticles on Plant Antioxidant Machinery*

NPs may interact chemically or mechanically with biological systems, plants for example and these particular interactions are largely due to their low size, large surface area, and intrinsic catalytic reactivity. Just a few studies identify the effect of NPs on antioxidants and molecular levels. The treatment of *Brassica juncea* with silver NPs (Sharma et al. 2012a, b) increased antioxidant enzyme activity (ascorbate peroxidase, guaiacol peroxidase and catalase), leading to a reduction in the level of ROS. The activation of the Spirodela polyrhiza antioxidant system was triggered by the application of 6 nm Ag NPs at a concentration of 5 mg/l, as evidenced by the induced activity of superoxide dismutase (SOD), peroxidase (POX), and catalase (CAT) (Jiang et al. 2012). Furthermore, the concentrations of ROS, glutathione, and malondialdehyde have skyrocketed. The use of gold nanoparticles (GNPs) on Brassica juncea seedlings resulted in an increase in antioxidant enzyme activity as well as higher concentrations of H2O2 and proline (Gunjan and Zaidi 2014). The content of H_2O_2 and proline has been found to increase with increasing concentrations of GNPs.. In particular, the activities of glutathione reductase (GR), guaiacol peroxidase (GPX), and ascorbate peroxidase (APXincrease by increasing the concentration of GNPs up to 400 ppm, while the maximum GR activity was detected in response to 200 ppm. Exposure to CeO_2 NPs was associated with elevated antioxidant enzymes activity (APX, CAT, and GPX) in in kidney beans leaves, roots, and stems (Mazumdar and Ahmed 2011). They discovered that after prolonged exposure to 500 mg nano CeO2/l, root antioxidant enzyme activities were significantly reduced, while root soluble protein was increased.

Furthermore, nano CeO2 exposure increased GPX activity in leaves, preserving cellular homeostasis. Gene expression analyses of the model plant Arabidopsis using RT-PCR have provided new insights into the molecular mechanisms of plant responses to Ag NPs. The transcriptional response of Arabidopsis plants to Ag NPs was studied using microarrays of whole-genome cDNA expression (Kaveh et al. 2013), which revealed that 286 genes were upregulated, including metal and oxidative stress-related genes (e.g., vacuolar cation/proton exchanger, SOD, cytochrome P450-dependent oxidase, and POX).

3.4.4 Effect of Nanoparticles on Phytohormones

Plant hormones defined as organic materials that serve a purposeformed by metabolic processes that can monitor plant's physiological responses to different stimuli during plant development (Santner et al. 2009). Plant growth and production are aided by phytohormones (Wong and Osmond 1991). Plant hormone content and activity are regarded as essential indicators of plant toxicity. According to Le Van et al. (2015) CeO_2 NPs had no effect on plant hormones such as indole-3-acetic acid (IAA), abscisic acid (ABA), or gibberellic acid (GA) in the leaves of Bt-transgenic and non-transgenic cotton. Non-transgenic cotton leaves were exposed to CeO_2 NPs at a concentration of 500 mg/L., when compared to the control, the content of trans zeatin-riboside (t-ZR) in the leaves was reduced by 25%. The IAA and ABA content of transgenic and non-transgenic rice roots increased in response to Fe_2O_3, according to Gui et al. (2015). Hao et al. (2016) plant hormone content was significantly influenced by NPs in rice seedlings. They reported a decrease in phytohormone concentrations in response to nanotubes of carbon application. Bleecker and Kende (2000) reported that, Ag ions inhibit the formation of ethylene, the interaction between IAA and ethylene is significantly weakened. Regulation of 81 genes, including those involved in plant defense and hormonal stimuli (for example, auxin-regulated gene, ethylene signaling pathway, and Systemic Acquired Resistance (SAR) against pathogens)., As revealed by a proteomic analysis of rice that identified silver NPs sensitive proteins, Ca^{2+} regulation and signalling, protein degradation, cell wall synthesis, cell division, and apoptosis were primarily associated with the effects of silver NPs (Mirzajani et al. 2014).

3.5 Nanoparticles Increase Plant Immunity to Abiotic Stress

Abiotic stress causes a cascade of morphological, physiological, biochemical, and molecular changes in plants. All these things are bad for plant development and production. The much more common abiotic stressors impacting universal food security are drought, salt, and extreme heat. The creation of stress-tolerant plants might be a viable method for addressing global food production decline. In addition to current traditional and sophisticated breeding methods, researchers aim to enhance agricultural yields in the face of environmental stress by harnessing the promise of nanotechnology. Nanotechnology has the potential to revolutionize agriculture, making it essential for sustainable food and crop production (Servin et al. 2015). Several studies have evaluated NPs-mediated responses in plants subjected to various stresses, such as Ahmed et al. (2011), who examined the outcome of silicon (Si) application on sorghum grown under drought stress; Almutairi (2016) who explored the effect of nano SiO_2 on tomato plants subjected to salt-stress; and Rezvani et al. (2012) who studied the impact of nano-silver (Ag) on saffron grown under flooding stress. The researchers have shown that NPs suppressed the negative impact of stresses on plants.

TiO_2 NPs are believed to have pro-oxidant and antioxidant activities. Latef et al. (2017) Broad beans can withstand salty soil and grow faster in it after being treated with TiO2 NPs, according to research. Additionally, they discovered that promoting growth outcomes were linked to higher chlorophyll b, proline, and soluble carbohydrates and enhanced antioxidant enzyme activity. According to Lei et al. TiO2 NPs can decrease abiotic stress in "*Spinacia oleracea*" by decreasing the injurious effect of different ROS and boosting the quenching effect of antioxidant enzymes (2008). The effect of NPs on ROS and antioxidant metabolism varies depending on the form and the concentration of the NP.

By augmentation, the activity of stress-related genes, carbon nanotubes with multi-walled dramatically impacted the development and germination of tomatoes (Khodakovskaya et al. 2009).

3.5.1 The Effect of Nanoparticles on Salt-Stressed Plants

A range of biological functions relevant to plant development and production are harmed by salt stress. Specific mutual effects of stress-salinity on plants include decreased soil osmotic potential, nutritional imbalances, increased ionic toxicity, or a combination of these factors (Ashraf 1994).

Barley (Karami and Sepehri 2018), maize (Fathi et al. 2017), Arabidopsis (Wu et al. 2018), berseem clover (Abd El-Naby et al. 2018), wheat (Abou-Zeid and Ismail 2018), cotton (Hussein and Abou-Baker 2018), and tomato (Almutairi 2016) grown under salt stress have all benefited from the use of NPs. Elamawi et al. (2016) discovered that ZnO NPs boost rice development and yield characteristics with stress-salinity. In developing rice seeds, nano-priming stimulates and activates alpha-amylase by gene overexpression of aquaporin. It should be investigated for enhancing rice germinating seeds and seedling vigor under salt (Mahakham et al. 2017).

Under salinity stress, silicon NPs and silicon fertilizer were found to have promising properties on physiological and morphological traits of basil "*Ocimum basilicum* L." vegetative characteristics. The results indicated a substantial increase in growth and development indices, chlorophyll value, and proline level when basil was supplemented with silicon NPs and silicon fertilizer under salt stress (Kalteh et al. 2014). Under salt stress, application of Nano-SiO_2 particles increased chlorophyll content, leaf fresh and dry weights, proline concentration, and antioxidant enzyme activity (Haghighi et al. 2012).

When nano-SiO_2 is sprayed to maize plants under salt stress, other investigations have demonstrated a rise in fresh soot freshness and weight (Gao et al. 2006). One method employed by silica NPs to alleviate salt stress in plants is decreased Na+ ion uptake by plant tissues (Raven 1982). CeO_2 (cerium oxide) NPs improved plant growth and photosynthesis rate in "*Glycine max* L." under salt stress by controlling photosynthesis, water useability, and Rubisco carboxylase (Cao et al. 2017) and modifying the stress-salinity responses by inhibiting salt intake in *Brassica napus* L. while maintaining nutritional value (Rossi et al. 2016).

3.5.2 The Effect of Nanoparticles on Drought-Stressed Plants

In the plant, the vitality is highly dependent on water, which is primarily essential for nutrient transportation; thus, a lack of water causes drought stress, which results in depleted plant vitality. Nanotechnology promises to make a significant contribution to reducing drought stress.

Plant resistance to dryness stress in Hawthorns "*Crataegus* sp." is improved by using various levels of silica NPs. At various levels of water stress, ranging from mild to extreme stress, hawthorn seedlings' physiological and biochemical responses to different concentrations of silica NPs differ. SNP pre-treatment improved photosynthesis parameters, relative water content (RWC), chlorophyll, carotenoid, carbohydrate, and proline contents (Ashkavand et al. 2015). Maybe silicon NPs play a role in maintaining important physiological and biochemical characteristics in drought-stressed hawthorn seedlings, but the exact mechanism is unknown. When compared to unstressed plants of faba bean (Kalteh et al. 2014), tomato (Siddiqui et al. 2014), and alfalfa (Siddiqui et al. 2014), nano-Si significantly increased the activities of catalase (CAT) and peroxidase (POD) in plant leaves (Zheng et al. 2005). Furthermore, silica NPs influence xylem humidity, water immobilization, and turgor pressure, which increases the RWC of leaf and water use efficiency in plants.

The addition of Si to two sorghum cultivars with various dryness sensitivity enhanced dryness resistance by reducing the shoot-root ratio (S/R), indicating enhanced root development and photosynthetic efficiency conservation (Hattori et al. 2005). Applying Si NPs to drought-stressed sorghum additionally improved the root dry weights, leaf dry weight, specific leaf weight, chlorophyll content (SPAD), shoot, and leaf area index, according to Ahmed et al. (2011).

Drought stress was found to be mitigated by nano-titanium oxide (TiO_2 NPs) in wheat by increasing yield, starch content, and development (Jaberzadeh et al. 2013), flax by increasing chlorophyll and carotenoids content while decreasing H_2O_2 and malondialdehyde (MDA) content (Aghdam et al. 2016), and basil by reducing the destructive impact of drought on plants (Kiapour et al. 2015).

Another research found that iron (Fe) NPs had a significant impact on plant characteristics like the number of bolls per branch, number of seeds per boll, seed index, and seed weight at a probability level of one %. Drought stress was alleviated by foliar application of iron NPs to Goldasht spring safflower cultivars' yield components and oil percentage. Compared to drought stress without iron nanoparticles (Fe NPs) implementation, Fe NPs application improved yield components at both flowering and granulation stages, though it was better at flowering than seed forming (Davar et al. 2014).

3.5.3 The Effect of Nanoparticles on Heat-Stressed Plants

Heat stress has been described as temperature spikes in this manner, a dramatic rise for a prolonged time that results in progressive deterioration of plant development and growth (Wahid 2007). Heat stress boosts the development of oxidative byproducts "reactive oxygen species", causing oxidative damage due to membrane lipid degradation and membrane ion leaking. Protein degradation occurred (Savicka and Skute 2010) and a decrease in photosynthesis rate and chlorophyll content (Prasad and Djanaguiraman 2011). Low concentrations of Se NPs boosted hydrating efficiency, chlorophyll level, and plant development, minimizing the impacts of heat stress (Haghighi et al. 2014). Plants produced heat shock proteins (HSPs) and molecular chaperones in response to heat stress (Schulze et al. 2005). In stressful conditions, HSPs enable other proteins maintain their stability. (Wahid 2007) and are engaged in heat stress immunity. MWCNTs have been shown to have a function in boosting the production of heat shock proteins like HSP90 (Khodakovskaya et al. 2011). In addition, exposure to CeO_2 NPs in maize results in increased H_2O_2 production and HSP70 upregulation (Zhao et al. 2012). Furthermore, TiO_2 NPs reduced the influence of the high temperature by regulating stomatal opening (Qi et al. 2013). When compared to controls, AgNPs safeguard wheat (*T. aestivum* L.) plants from heat stress by increasing root and shoot length, fresh and dry weight, leaf size, leaf number, leaf fresh and dry weights (Iqbal et al. 2019). AgNPs increase shoot development by blocking ethylene signaling, which is a shoot growth inhibitor, as previously described by Syu et al. (2014).

3.5.4 The Effect of Nanoparticles on Plants Exposed to Heavy Metal Stress

Heavy metals are metals with a density greater than 5 g/m; 17 of the 53 heavy metals are crucial for living organisms, including iron (Fe), zinc (Zn), manganese (Mn), copper (Cu), and molybdenum (MO) (Miransari 2016). The residual heavy metals, such as cadmium (Cd), chromium (Cr), mercury (Hg), and lead (Pd), are hazardous to the environment and can interfere with plant biochemical pathways (Kumar et al. 2015).

Plants suffer from heavy metal stress because of nutrient deficiency, which is caused by the interruption of essential supplement absorption along with the inhibition of enzyme activities (Capuana 2011). Synthesized NPs have been shown to be very efficient in reducing heavy metal-induced phytotoxicity (Tripathi et al. 2015). Because of these features, NPs are very tiny and have a high surface area; they may easily infiltrate plant cells and have a significant affinity for heavy metals. According to research, exposure to TiO_2 NPs lowers cadmium toxicity while improving photosynthetic rate and plant development (Singh and Lee 2016). Furthermore, the treatment of *Brassica juncea* with hydroxyapatite NPs decreases cadmium toxicity (Li and Huang 2014). The addition of Si NPs to pea growth medium has been found to reduce chromium toxicity (Tripathi et al. 2015).

Polyhydroxyfullerene (PHF) is a carbon nano-particle with a good water solubility, that reduces Cd bioavailability by binding Cd to the extracellular medium, decreasing Cd-induced oxidative damage. Cr phytotoxicity in pea seedlings can be reduced by using Si NPs, which decrease oxidation damage by decreasing Cr deposition and boosting antioxidant defense machinery (Tripathi et al. 2015). Wang et al. (2016) proved that application of organic and inorganic nano-Si decreased BCF (bio-concentration factor) and TF (translocation factor) of heavy metals (Zn, Cu, Pb, and Cd) in rice cultivars grown in artificially contaminated soil. Heavy metal TF decreased from roots to shoots and then to grains. The co-precipitation of heavy metals with Si in the apoplast may be the primary mechanism for reducing heavy metal toxicity in rice plants and accumulation in grains. Mustafa and Komatsu (2016) used a proteomics research approach to identify and understand the cell pathways impacted by heavy metals. They found that NPs protect plant cells from metal-induced oxidative damage by modulating ROS, antioxidants, and energy metabolism.

The latest knowledge of stress responses associated with the use of NPs in abiotically stressed plants is summarized in Fig. 3.1.

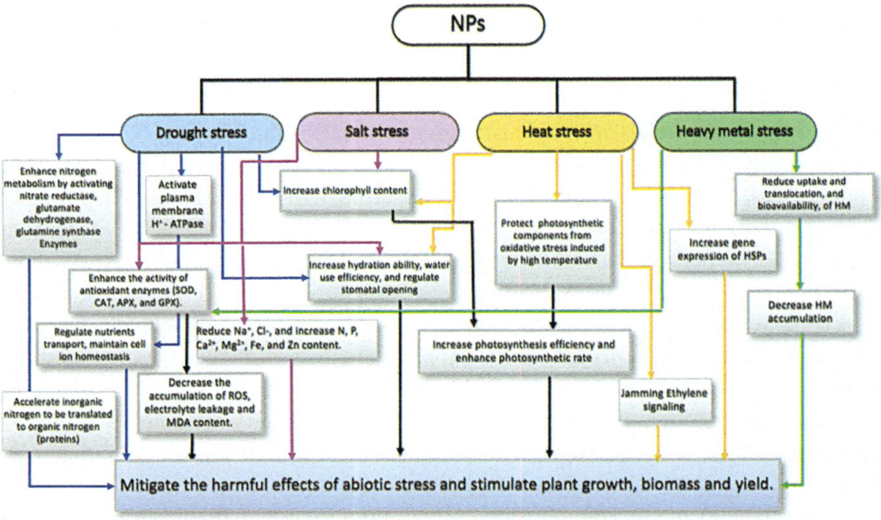

Fig. 3.1 A summary of the possible modes of action of NPs in plants to alleviate environmental stress

3.6 Nanoparticles as Genome Editors

The challenge of introducing external macromolecules into plant cells across cell walls has delayed the development of agricultural biotechnology genome-editing applications. The cell wall acts as a barrier to exogenous biomolecule delivery into plant cells. Different techniques based on *Agrobacterium* transformation or biolistic gun for DNA delivery in plant cells are used worldwide to get through this barrier and obtain plant genetic transformation. These approaches have limitations due to a limited host range and extensive plant damage. Plant cell cultures were used for the majority of the early studies on nanomaterial-based plant genetic engineering. For instance, Si carbide-aided delivery was found as a potential technique for delivering DNA in a variety of calli (cotton, soybean, rice, and tobacco) (Lau et al. 2017, Asad and Arsh 2012, Serik et al. 1996, Armstrong and Green 1985, respectively).

NMs can overcome the cell wall barrier in mature plants, reducing the disadvantages related to delivery systems of DNA targeting. Magnetic NPs (MNPs) were used to achieve a better steady genetic delivery in cotton plants. Magnetic force was used to infiltrate a glucuronidase (GUS) reporter gene-MNP complex into cotton pollen grains without compromising pollen growth. Its transgenic was created by pollinating them with external genetic material, and magneto-fected pollen was effectively inserted into its genetic material, translated to proteins, and steadily passes to the next generation (Zhao et al. 2017). Other work employed carbon nanotube scaffolds to deliver linear DNA, plasmid DNA, and siRNA to *Nicotiana benthamiana*, *Gossypium hirsutum* leaves, and protoplasts, leading to elevated transient Green Fluorescent Protein (GFP) expression (Demirer et al. 2018). According to

Valenstein et al. (2013), mesoporous silica NPs (MSNs) are the first and most effective way to alter genomes. MSNs, which have loxP sites built into their chromosomal DNA, have been utilized to transport Cre recombinase to immature *Zea mays* embryos. The loxP was properly recombined after biolistic application of modified MSNs in plant tissues, resulting in effective genome editing.

"CRISPR/Cas9 (Clustered Regularly Interspaced Palindromic Repeats/CRISPR-associated Protein 9" is a cutting-edge tool for site-directed genome editing that is relatively simple and efficient. This system is made up of two core constituents: "Cas9 RNA-guided nuclease" and "CRISPR RNA (crRNA)", which directs the Cas9 enzyme to a specific sequence in the genome. The activity of Cas9 nuclease causes double-stranded interruptions in the goal genetic sequences (Jinek et al. 2012). CRISPR–Cas9 genome engineering has the potential to boost crop yields and strength. One of the most promising avenues for making agriculture more sustainable is nanocarriers in gene editing (Landry and Mitter 2019). However, research into nanoparticle delivery across the plant cell wall is limited. Many unanswered questions remain, including how shape, size, aspect ratios, tensile strengths, and other physicochemical parameters affect NPs ability to internalize into plant cells (Zhang et al. 2019). Furthermore, to determine the effects on plants, a deep study of the effects of engineered nanomaterials on plant physiology must be performed. In addition, consumer adoption of this technology should be considered in light of the rejection and distrust of many consumers against genetically modified crops, which has resulted in the more cautious launch of new genomic editing technologies, such as nanotechnology, in the food sector.

3.7 Are Nanoparticles Safe?

NPs have many beneficial effects on plants; as plant growth modulators, they may help you gain weight and absorb nutrients, such as when nano-fertilizers are added to the soil. However, it has been noted that NPs have both beneficial and undesirable outcomes on a variety of plant species due to their configuration, varying size, changing physical and chemical characteristics, and the amounts of various NPs utilized (Ma et al. 2010; Tripathi et al. 2017).

3.7.1 Nanoparticle's Toxicity

The toxicity of NPs is primarily due to their small size, large surface area to volume ratio, and reactive facets. Because NPs are extremely tiny, they react with ecosystem elements after entering the environment through photolysis and oxidative processes, releasing harmful materials hazardous to plants (Lowry et al. 2010). At specific doses, many NPs cause toxicity, lowering crop production by changing their biochemical, genetic content, and morpho-physiological (Tripathi et al. 2016). Noxious metals like Hg, Cd, nickel (Ni), Cu, and Zn react with sulfhydryl, carboxyl, or imidazole groups on DNA or cellular structure in metal NPs, changing their functions (Asgari Lajayer et al. 2017). Interruption of cell functions in metal-exposed plants often results in redox imbalances and oxidative stress (Ghorbanpour et al. 2016). The investigators have concentrated on the potential genotoxicity outcome of nanoparticles and their ability to pass in the nuclear membrane (Karlsson 2007). However, certain materials, for example, nitric oxide (NO), can minimize the toxic effect of NPs in "*O. sativa* L." seedlings (Chen et al. 2015). The use of nitroprusside on "*Brassica nigra*" exposed to nano-Ag and Ag nitroprusside stress accelerated seed germination.

3.7.2 What Makes Some Nanoparticles More Toxic Than Others?

In terms of toxicology, the proportion of surface area to a particle size of a nanoparticle is an essential physical property; as the particle size decreases, the surface area becomes more prominent. Consequently, From the exterior, many particles can be detected from the inside (Begum et al. 2011). Several experiments have shown that the surface properties of NPs are more toxic than the surface properties of non-nanoparticles of the same substance (Kashem and Kawai 2007): The usage of several types of NPs has been used in experiments as demonstrated previously, such as cobalt, Ni, and TiO_2. TiO_2 NPs with a size of less than 30 nm were found to cause inflammation 43-fold higher than NPs $\geq$ 200 nm (Feizi et al. 2012; Qiu et al. 2013). Currie and Perry (2007) have shown that nano-particle sizes are significantly further harmful than micro-particle sizes. The surface area of NPs was discovered to be an essential parameter of their noxiousness. In general, the current phytotoxic outline of NPs is somewhat speculative at first. However, the impact of nano-particle properties is poorly understood, and further research on harmful effects, especially on essential food crops, is needed (Groppa et al. 2008).

The toxic effect of NPs has also been linked to exposure. Tolaymat et al. (2017) discovered that treating peanuts of non-zero valency Fe surges root development up to a level of 4.18 ppm nevertheless decreases root growth beyond this point. Size, reactivity, chemical structure, surface coating, and other variables govern the pace at which NPs affect plants, which is why NPs have both good and adverse effects on living organisms (Khodakovskaya et al. 2012).

As a result, multiple studies revealed that direct contact to a specific form of nano-particle induced a significant phytotoxic impact, reinforcing the need for environmental transparency when disposing of waste-containing NPs. More research is vital on the effects of NPs on plant food and the milieu.

3.7.3 Nanoparticles' Toxicity in Plants

3.7.3.1 Nanoparticles Can Be Stress Elicitors As Well As Stress Mitigators

In Sect.3.5, we discussed the function of NPs in stressor factor mitigation in plants. However, many other studies have highlighted the effect of NPs as stress-inducing agents (Table 3.1) In this regard, the toxicity mechanism of Ag nano-particles was studied in the initial phases of "*Glycine max* L." plants under flooding stress, revealing which proteins related to routes of signaling, as well as cell metabolism and stress response, were affected. Moreover, Ag nano-particle treatment degraded the glyoxalase enzyme involved in the detoxify mechanism (Mustafa et al. 2015a). In other research, Mustafa et al. (2015b) found that treatment with Ag, ZnO, and Al_2O_3 NPs depleted protein synthesis in "*G. max* L." plants under flooding stress. The metabolic processes of lipid metabolism and glycolysis were also affected. These experiments show how NPs interact with plants; however, more study is needed to understand these discoveries properly. As a result, NPs may have both positive and negative impacts on plants. Future studies on the interplay of NPs with plants will help to provide a piece of better knowledge, implying that specific NPs may be administered to plants for desired results, perhaps leading to higher agricultural yields.

Root elongation in plants has been found to be influenced by non-functionalized single-walled carbon nanotubes (SWCNTs) and functionalized multi-walled carbon nanotubes (fSWCNTs) in "tomato, *Lycopersicon esculentum*", "lettuce, *Lactuca sativa*", "onion, *Allium cepa*", "carrot, *Daucus carota*", "cucumber, *Cucumis sativus*", and "cabbage, *Brassica oleracea*" (Cañas et al. 2008). The dried mass of pumpkin "*Cucurbita pepo*" reduced by 60%, accompanied by 1000 ppm multi-walled carbon nanotubes (MWCNTs) (Stampoulis et al. 2009). MWCNTs and C70 NPs can delay the blooming of "*Oryza sativa*" by roughly a month, according to Lin et al. (2009a, b). MWCNTs and C60 fullerenes, according to De La Torre-Roche et al. (2013), reduce the biomass of soybean and maize by about 45%.

Plants react with nanoparticles in various ways, not only directly but also physically, and electrochemically; the result of these impacts is the production of reactive oxygen species, which are significant contributors to DNA damage. ROS has been revealed in studies to be an effective messenger in plants under both abiotic and biotic stresses (Takahashi et al. 2011).

Plants have an intrinsic protection mechanism against toxic effects caused by stressors such as nano-particles. Plants have several enzymatic and nonenzymatic stress response mechanisms (Asgari Lajayer et al. 2017). These protection schemes are frequently incapable of alleviating and detoxifying the stressors generated by stress stimuli, resulting in plant cell death (Hossain et al. 2016). As a result, in plants, the examining of ROS generation is a powerful tool in evaluating the toxic effects of toxic metals and NPs (Sharma et al. 2012a, b; Rico et al. 2013). To assess oxidative damage in plants, electrolyte discharge, propidium iodide fluorescence

Table 3.1 Nano-particles induced stress-like responses in plants

Types of nano-particles used in the study	Plants	Nano-particles-induced response in plants	References
Aluminum oxide	*Glycine max* L.	Low seed germination rate, decreased raw and dries weight and shoot and root height, DNA damage, decreased transpiration rate, enhance lipid peroxidation, up-and down-regulation of many stress-related genes, and apoptosis.	Hossain et al. (2016)
Carbon nanotube (multi-walled)	*Solanum Lycopersicum* L.	Stress-related gene expression is up-regulated, and seedling development is negatively affected.	Singh et al. (2016b) and Koul et al. (2018)
Copper oxide	*Oryza sativa* L.	Stimulated enzymatic antioxidant values.	Shaw and Hossain (2013)
	Lolium perenne L., *Lolium rigidum* Gaudin and *Raphanus sativus* L.	Reduced growth and DNA damage.	
	Fagopyrum esculentum Moench	Cu is released, causing metal stress in plants.	Atha et al. (2012)
		Root growth and biomass inhibition, root morphology changes, and DNA polymorphism.	Lee et al. (2013)
Fullerenes (nC_{60})	*Scenedesmus obliquus*	Chlorophyll a & b accumulation was inhibited; magnesium deficiency was caused by inhibition of Mg2 + -ATPase activity.	Tao et al. (2015)
Gold	Phytoplanktonic alga (*Scenedesmus subspicatus*)	Progressive intra-cellular and cell wall disruption.	Renault et al. (2008)
Graphene oxide	*Malus domestica*	Inhibition of lateral roots, increased CAT, POD, and SOD activities, and increased transcription of auxin efflux carrier and auxin influx genes.	Li et al. (2018)
	Triticum aestivum L.	Mineral elements decreased, sugar content increased, proteins decreased, and globulin, prolamin, amylose, and amylopectin decreased.	Gao et al. (2020)
Multi-walled carbon nanotubes (MWCNTs)	*Oryza sativa* L.	ROS generation eventually affects cell proliferation and, consequently, plant cell death, chromatin condensation, and shrinkage.	Tan et al. (2009)

(continued)

Table 3.1 (continued)

Types of nano-particles used in the study	Plants	Nano-particles-induced response in plants	References
Nickel oxide	*Solanum Lycopersicum* L.	The activity of SOD, CAT increased as well as glutathione content and lipid peroxidation.	Faisal et al. (2013)
	Coriandrum sativum L.	Concentration-dependent decreases in RWC, photosynthetic pigments, and root and shoot elongation were observed.	Miri et al. (2017)
	Brassica rapa L.	Chlorophyll, carotenoid, and sugar content were declined, while proline and anthocyanins were up-regulated.	Chung et al. (2019)
Silver	*Allium cepa* L.	Chromotoxic outcome on mitotic cell division.	Kumari et al. (2009)
	Oryza sativa L.	Significant decrease in root and shoot weights, leaf area, and chlorophyll and carotenoids amounts.	Nair and Chung (2014)
	Arabidopsis thaliana L.	Elevation of oxidative stress-related proteins.	Mirzajani et al. (2014)
		Increased ROS generation and oxidative stress.	Kaveh et al. (2013)
Titanium dioxide	*Arabidopsis thaliana* L.	Genes that are triggered in abiotic and biotic stress conditions are up-and down-regulated.	Landa et al. (2012)
Zinc oxide	Aquatic plant *Spirodela punctuta*	Formation of ROS/RNS and H_2O_2	Thwala et al. (2013)
	Arabidopsis thaliana L.	Reduced germination rate, as well as up and down-regulation of genes that are triggered in abiotic and biotic stresses.	Wang et al. (2004) and Landa et al. (2012)
	Cucumis sativus L.	Degradation of the epidermis, roots, cortical cells, and vacuole's structure.	Zhao et al. (2013)

assays, and lipid peroxidation can all be employed. For example, treating "*O. sativa*" with CeO_2 NPs (ranging from 62.5 to 500 ppm) boosted oxidative byproducts accumulation and caused oxidation-induced damage (Rico et al. 2013).

Other research has revealed that metal-based or carbon NPs can cause damage to plant carbohydrates, DNA, lipids, and proteins, increasing oxidative byproduct generation. Carbon nanotubes are a kind of carbon-based NP that forms oxidation-induced damage and increases oxidative byproduct production *in vivo* and *in vitro* studies (Liu et al. 2010, 2013).

3.7.3.2 Nanoparticles and Genotoxicity in Plants

Plant NPs are being examined as a good delivery strategy since they can reach plants as targeted-aid genes in specific cellular components (Siddiqui et al. 2015b). NPs may pass their compartments via various routes, such as via carriers (Sahebi et al. 2015). Investigators discovered Si carriers (Lsi1, Lsi2, and Lsi6 genes) in the root of "*O. sativa*" (Ma and Yamaji 2015). The NPs attach to macromolecules like polysaccharides, genetic material, and protein when they penetrate plant cells. TiO_2 nanoparticle in "*C. pepo*" has been shown to cause genetic damage and alter genetic up-regulation (Moreno-Olivas et al. 2014).

Ag and TiO_2 nanoparticles were found to have phytotoxic effects on the chromosomes of "*A. cepa*" and "*Z. mays*" (Kumari et al. 2009; Castiglione et al. 2011; respectively). Random amplified polymorphic DNA techniques (RAPD) revealed the damaging effect of engineered CeO_2 NPs (2000 and 4000 ppm) to the DNA structure of "*G. max*" (López-Moreno et al. 2010). Similarly, the harmful effect of TiO_2 nano-particles on the genetic content of "*A. cepa*" at 4 mM was observed, as was bridge forming at the anaphase and telophase stages when these NPs are present at the same concentration (Ghosh et al. 2010). Furthermore, DNA damage was observed in "*Nicotiana tabacum*" and "*Z. mays*" in response to TiO_2 NPs at 2 and 10 mM, respectively (Ghosh et al. 2010; Castiglione et al. 2011).

There is currently very little research on the molecular basis of NPs-induced phytotoxicity in vascular plants. On the other hand, Transcriptomics technologies have disclosed details about NPs-intermediated toxic effect in advanced plants by serving as a relation among gene regulation (Landa et al. 2015).

3.7.4 Risks of Nanoparticles on Humans, Soil, and Environment

Because of the random release of NPs into natural habitats caused by industrial effluents, a massive accumulation occurs (Brunner et al. 2006; Owen and Handy 2007). The majority of manufactured NPs are heavy metals that contaminate water and soil, causing environmental deterioration. Plant interplays with additional NPs in soil and water can result in nano-particle uptake and accumulation in the plant, eventually leading to physical and chemical injury to various plant parts (Dietz and Herth 2011). Though the cell wall prevents NPs from entering the plant's body, the pore size of the cell walls is fixed, enabling the movement of NPs lesser than the pore size of the walls (Navarro et al. 2008). As engineered nanomaterials react with carbon-based acids in roots, the pH decreases, causing nutrient quantity and engineered material characteristics to change (Marschner 1995). NPs have been widely used as antimicrobial agents against pathogenic bacteria, but this indiscriminate use harms communities of soil microbes that play essential roles in the ecosystem, such as stimulating plant development, element recycling, and pollutant degradation

(Remédios et al. 2012). Adding NPs to the soil will pollute the groundwater when they combine with organic matter (soluble organic matter). Nano-fertilizers and pesticides used on farms can end up in lakes (due to erosion) and create a dangerous ecosystem, such as eutrophication (Moghadam et al. 2019). Therefore, soil and, consequently, plants may represent the most critical exposure avenue for evaluating the environmental risk and raising questions about NPs entrance into food chains and human and animal access to contaminated water and farmlands.

Metallic NPs have been shown to trigger oxidation damage in a living organism by producing reactive oxygen species. Oxidation byproducts formed by NPs dislocate electron and ion influx and efflux, impair membrane permeability, and decrease glutathione content within human lung epithelial cells (Limbach et al. 2007). ROS upsurge cell membrane penetrability by oxidizing fatty acid double bonds. TiO_2 NPs have been confirmed to possess photocatalytic characteristics (Khus et al. 2006) and produce ROS when exposed to ultraviolet radiation (Zhao et al. 2007), generating genetic injury. Photoactive silver NPs were demonstrated to break the DNA strands when subjected to UV light (Badireddy et al. 2007). CeO_2 NPs have now been found to oxidize membrane-bound respiratory electron transport chain complexes and to be hazardous in living organisms (Thill et al. 2006). Ag^+ released from NPs interplay with the thiol groups of enzymes and suppressed respiratory enzymes (Kim et al. 2007).

Generally, the impact of nanoparticles in living organisms can be caused by (1) chemical effects associated with metal ions during dissolution; (2) mechanical changes induced by hard spheres; (3) the catalytic function of NPs; (4) attaching with macromolecules via noncovalent/covalent mechanisms; and (5) ROS formation in cells. (Ma et al. 2010).

3.8 Conclusions and Future Perspectives

Previous research has shown that plants respond to NPs by increasing growth, yield, and photosynthesis efficiency and displaying an abundance of ROS-associated proteins, stress signaling, plant hormone pathways, and detoxification. More research is required to properly understand the factors that influence how plants react to specific NPs. Furthermore, metabolomics and transcriptomics analyses may hold great promise to elucidate the complete picture of plant response to NPs. All this information would provide us with a wealth of information to clarify the response mechanism of plants to nanoparticle-induced stress and the protective role of NPs in improving plant tolerance to environmental challenges. The interplay among nanoparticles and plants is highly sophisticated, and it is affected by several factors, including NP form, specific surface area, crystalline chemistry, concentration, plant species, growth stage, and application methods. Before the widespread use of NPs, an information gap must be filled, such as the threshold of plant tolerance to nanoparticles and the threshold of antioxidant enzyme system activation in the plant. Phytotoxicity of nanoparticles should also be considered a red flag, and the risk of nanoparticles being transferred into edible plant parts should be explored further. Also, there is an urgent need to define the best standard phytotoxicity test that should be used in evaluating NPs toxicity because it is necessary to precisely determine the efficient and sufficient NPs concentrations to achieve the desired positive effect.

Acknowledgments The current work is dedicated to Prof. Raifa Hassanein, former head of the Botany department, Faculty of Science, Ain Shams University, who died on February 4, 2021; the authors will be eternally grateful for her great guidance, compassion, and immense scientific support.

References

Abd El-Naby ZA, Hafez WA, Hashem HA (2018) Remediation of salt-affected soil by natural and chemical amendments to improve berseem clover yield and nutritive quality. Afr J Range Forage Sci 2018:1–12

Abou-Zeid HM, Ismail GS (2018) The role of priming with biosynthesized silver nanoparticles in the response of *Triticum aestivum* L. to salt stress. Egypt J Bot 58:73–85

Acquaah G (2009) Principles of plant genetics and breeding. Wiley

Aghdam MTB, Mohammadi H, Ghorbanpour M (2016) Effects of nanoparticulate anatase titanium dioxide on physiological and biochemical performance of *Linum usitatissimum* (Linaceae) under well-watered and drought stress conditions. Rev Bras Bot 39:139–146. https://doi.org/10.1007/s40415-015-0227-x

Ahmed M, Hassen F, Qadeer U, Aslam MA (2011) Silicon application and drought tolerance mechanism of sorghum. Afr J Agric Res 6(3):594–607

Almutairi ZM (2016) Effect of nano- silicon application on the expression of salt tolerance genes in germinating tomato (*Solanum lycopersicum* L.) seedlings under salt stress. Plan Omics J 9(1):106–114

Armstrong CL, Green CE (1985) Establishment and maintenance of friable, embryogenic maize callus and the involvement of L-proline. Planta 164:207–214. https://doi.org/10.1007/BF00396083

Asad S, Arsh M (2012) Silicon carbide whisker-mediated plant transformation. In: Gerhardt R (ed) Properties and applications of silicon carbide. BoD-Books on Deman, Rijeka, pp 1–16. https://doi.org/10.5772/15721

Asgari Lajayer B, Ghorbanpour M, Nikabadi S (2017) Heavy metals in contaminated environment: destiny of secondary metabolite biosynthesis, oxidative status and phytoextraction in medicinal plants. Ecotoxicol Environ Saf 145:377–390

Ashkavand P, Tabari M, Zarafshar M, Tomášková I, Struve D (2015) Effect of SiO_2 nanoparticles on drought resistance in hawthorn seedlings. Leśne Prace Badawcze/For Res Paper Grudzień 76(4):350–359

Ashraf M (1994) Organic substances responsible for salt tolerance in *Eruca sativa*. Biol Plant 36:255–259

Aslani F, Bagheri S, Muhd Julkapli N, Juraimi AS, Hashemi FSG, Baghdadi A (2014) Effects of engineered nanomaterials on plants growth: an overview. Sci World J

Atha DH, Wang H, Petersen EJ, Cleveland D, Holbrook RD, Jaruga P, Dizdaroglu M, Xing B, Nelson BC (2012) Copper oxide nanoparticle mediated DNA damage in terrestrial plant models. Environ Sci Technol 46:1819–1827

Badireddy AR, Hotze EM, Chellam S, Alvarez P, Wiesner MR (2007) Inactivation of bacteriophages via photosensitization of fullerol nanoparticles. Environ Sci Technol 41:6627–6632

Banerjee J, Kole C (2016) Plant nanotechnology: an overview on concepts, strategies, and tools. In: Kole C et al (eds) Plant nanotechnology. Springer, Cham, pp 1–14. https://doi.org/10.1007/978-3-319-42154-4_1

Begum P, Ikhtiari R, Fugetsu B (2011) Graphene phytotoxicity in the seedling stage of cabbage, tomato, red spinach, and lettuce. Carbon 49(12):3907–3919

Bleecker AB, Kende H (2000) Ethylene: a gaseous signal molecule in plants. Annu Rev Cell Dev Biol 16:1–18

Brunner TJ, Wick P, Manser P, Spohn P, Grass RN, Limbach LK, Bruinink A, Stark WJ (2006) *In vitro* cytotoxicity of oxide nanoparticles: comparison to asbestos, silica, and the effect of particle solubility. Environ Sci Technol 40:4374–4381

Cañas M, Long S, Nations R, Vadan L, Dai M, Luo R, Ambikapathi EH, Lee DO (2008) Effects of function- alized and nonfunctionalized single-walled carbon nanotubes on root elongation of select crop species. Environ Toxicol Chem 27:1922–1931

Cao Z, Stowers C, Rossi L, Zhang W, Lombardini L, Ma X (2017) Physiological effects of cerium oxide nanoparticles on the photosynthesis and water use efficiency of soybean (Glycine max (L.) Merr.). Environ Sci Nano 4(5):1086e1094

Capuana M (2011) Heavy metals and woody plants biotechnologies for phytoremediation. J Biogeo Sci For 4:715

Castiglione MR, Giorgetti L, Geri C, Cremonini R (2011) The effects of nano-TiO_2 on seed germination, develop- ment and mitosis of root tip cells of *Vicia narbonensis* L. and *Zea mays* L. J Nanopart Res 13:2443–2449

Chang FP, Kuang LY, Huang CA, Jane WN, Hung Y, Hsing YIC et al (2013) A simple plant gene delivery system using mesoporous silica nanoparticles as carriers. J Mater Chem B 1:5279–5287. https://doi.org/10.1039/c3tb20529k

Chen J, Liu X, Wang C, Yin SS, Li XL, Hu WJ, Simon M, Shen ZJ, Xiao Q, Chu CC, Peng XX (2015) Nitric oxide ameliorates zinc oxide nanoparticle s-induced phytotoxicity in rice seedlings. J Hazard Mater 297:173–182

Chhipa H (2017) Nanofertilizers and nanopesticides for agriculture. Environ Chem Lett 15:15–22

Chung IM, Venkidasamy B, Thiruvengadam M (2019) Nickel oxide nanoparticles cause substantial physiological, phytochemical, and molecular-level changes in Chinese cabbage seedlings. Plant Physiol Biochem 139:92–101

Cunningham FJ, Goh NS, Demirer GS, Matos JL, Landry MP (2018) Nanoparticle-mediated delivery towards advancing plant genetic engineering. Trends Biotechnol 36:882–897. https://doi.org/10.1016/j.tibtech.2018.03.009

Currie HA, Perry CC (2007) Silica in plants: biological, biochemical, and chemical studies. Ann Bot 100(7):1383–1389

Davar F, Zareii AR, Amir H (2014) Evaluation the effect of water stress and foliar application of Fe nanoparticles on yield, yield components and oil percentage of safflower (*Carthamus tinctorious* L.). Int J Adv Biol Biomed Res 2(4):1150–1159

De La Torre-Roche R, Hawthorne J, Deng Y, Xing B, Cai W, Newman LA, Wang Q, Ma X, Hamdi H, White JC (2013) Multiwalled carbon nanotubes and C60 fullerenes differentially impact the accumulation of weathered pesticides in four agricultural plants. Environ Sci Technol 47:12539–12547

Demirer GS, Zhang H, Matos J, Goh N, Cunningham FJ, Sung Y et al (2018) High aspect ratio nanomaterials enable delivery of functional genetic material without DNA integration in mature plants. bioRxiv 10:1–32. https://doi.org/10.1101/179549

Dietz KJ, Herth S (2011) Plant nanotoxicology. Trends Plant Sci 16:582–589

Dubchak S, Ogar A, Mietelski JW, Turnau K (2010) Influence of silver and titanium nanoparticles on arbuscular mycorrhiza colonization and accumulation of radiocaesium in Helianthus annuus. Span J Agric Res 1:103–108

Elamawi RM, Bassiouni SM, Elkhoby WM, Zayed BA (2016) Effect of zinc oxide nanoparticles on brown spot disease and rice productivity under saline soil. J Plant Prot Path Mansoura Univ 7:171–181

Etxeberria E, Gonzalez P, Bhattacharya P, Sharma P, Ke PC (2019) Determining the size exclusion for nanoparticles in citrus leaves. Hort Sci 51:732–737. https://doi.org/10.21273/HORTSCI.51.6.732

Faisal M, Saquib Q, Alatar AA, Al-Khedhairy AA, Hegazy AK, Musarrat J (2013) Phytotoxic hazards of NiO- nanoparticles in tomato: a study on mechanism of cell death. J Hazard Mater 250:318–332

Fathi A, Zahedi M, Torabian S (2017) Effect of interaction between salinity and nanoparticles (Fe_2O_3 and ZnO) on physiological parameters of *Zea mays* L. J Plant Nutr 40:2745–2755

Feizi H, Rezvani MP, Shahtahmassebi N, Fotovat A (2012) Impact of bulk and nanosized titanium dioxide (TiO_2) on wheat seed germination and seedling growth. Biol Trace Elem Res 146(1):101–106

Gao F, Hong F, Liu C, Zheng L, Su M, Wu X et al (2006) Mechanism of nano-anatase TiO_2 on promoting photosynthetic carbon reaction of spinach. Biol Trace Elem Res 111:239–253

Gao M, Xu Y, Chang X, Dong Y, Song Z (2020) Effects of foliar application of graphene oxide on cadmium uptake by lettuce. J Hazard Mater 398:122859

Ghorbanpour M, Asgari Lajayer H, Hadian J (2016) Influence of copper and zinc on growth, metal accumulation and chemical composition of essential oils in sweet basil (*Ocimum basilicum* L.). J Med Plant 3:132–144

Ghosh M, Bandyopadhyay M, Mukherjee A (2010) Genotoxicity of titanium dioxide (TiO_2) nanoparticles at two trophic levels: plant and human lymphocytes. Chemosphere 81:1253–1262

Giraldo JP, Landry MP, Faltermeier SM, McNicholas TP, Iverson NM, Boghossian AA et al (2014) Plant nanobionics approach to augment photosynthesis and biochemical sensing. Nat Mater 13(4):400–408. https://doi.org/10.1038/nmat3890

Głowacka K et al (2018) Photosystem II subunit S overexpression increases the efficiency of water use in a field-grown crop. Nat Commun 9:868. https://doi.org/10.1038/s41467-018-03231-x

Groppa MD, Rosales EP, Iannone MF, Benavides MP (2008) Nitric oxide, polyamines and cd-induced phytotoxicity in wheat roots. Phytochemistry 69(14):2609–2615

González-Melendi P, Fernández-Pacheco R, Coronado MJ, Corredor E, Testillano PS, Risueño MC et al (2008) Nanoparticles as smart treatment-delivery systems in plants: assessment of different techniques of microscopy for their visualization in plant tissues. Ann Bot 101:187–195. https://doi.org/10.1093/aob/mcm283

Gui X, Deng Y, Rui Y, Gao B, Luo W, Chen S, Van Nhan L, Li X, Liu S, Han Y et al (2015) Response difference of transgenic and conventional rice (Oryza sativa) to nanoparticles (γFe2O3). Environ Sci Pollut Res 22:17716–17723

Gunjan B, Zaidi MGH (2014) Impact of gold nanoparticles on physiological and biochemical characteristics of Brassica juncea. J Plant Biochem Physiol

Haghighi M, Afifipour Z, Mozafarian M (2012) The effect of N-Si on tomato seed germination under salinity levels. J Biol Environ Sci 6(16):87–90

Haghighi M, Abolghasemi R, Teixeira da Silva JA (2014) Low and high temperature stress affect the growth characteristics of tomato in hydroponic culture with Se and nano-Se amendment. Sci Hortic 178:231240

Hao Y, Yu F, Lv R, Ma C, Zhang Z, Rui Y, Liu L, Cao W, Xing B, Choi J (2016) Carbon nanotubes filled with different ferromagnetic alloys affect the growth and development of rice seedlings by changing the C:N ratio and plant hormones concentrations. PLoS One 11:e0157264

Hatami M, Ghorbanpour M (2013) Effect of nano-silver on physiological performance of pelargonium plants exposed to dark storage. J Horti Res 21(1):15–20

Hattori T, Inanaga S, Araki H, An P, Morita S, Luxová M et al (2005) Application of silicon enhanced drought tolerance in *Sorghum bicolour*. Physiol Plant 123:459–466

Hossain Z, Mustafa G, Sakata K, Komatsu S (2016) Insights into the proteomic response of soybean towards $Al_2 O_3$, ZnO, and Ag nanoparticles stress. J Hazard Mater 304:291–305

Hussein MM, Abou-Baker NH (2018) The contribution of nano-zinc to alleviate salinity stress on cotton plants. R Soc Open Sci 5:171809

Iqbal M, Raja NI, Mashwani ZUR, Hussain M, Ejaz M, Yasmeen F (2019) Effect of silver nanoparticles on growth of wheat under heat stress. Iran J Sci Technol Trans A Sci 43:387–395. https://doi.org/10.1007/s40995-017-0417-4

Jaberzadeh A, Moaveni P, Moghadam HRT, Zahedi H (2013) Influence of bulk and nanoparticles titanium foliar application on some agronomic traits, seed gluten and starch contents of wheat subjected to water deficit stress. Notulae Botanicae Horti Agrobotanici Cluj- Napoca 41:201–207

Jacobs M, Lopez-Garcia M, Phrathep O-P, Lawson T, Oulton R, Whitney HM (2016) Photonic multilayer structure of begonia chloroplasts enhances photosynthetic efficiency. Nat Plants 2:16162. https://doi.org/10.1038/nplants.2016.162

Janmohammadi M, Amanzadeh T, Sabaghnia N, Dashti S (2016) Impact of foliar application of nano micronutrient fertilizers and titanium dioxide nanoparticles on the growth and yield components of barley under supplemental irrigation. Acta Agri Slovenica 107:265–276

Jiang HS, Li M, Chang FY, Li W, Yin LY (2012) Physiological analysis of silver nanoparticles and AgNO3 toxicity to Spirodela polyrhiza. Environ Toxicol Chem 31(8):1880–1886

Jinek M, Chylinski K, Fonfara I et al (2012) A programmable dual-RNA-guided DNA endonuclease in adaptive bacterial immunity. Science 337:816–821

Kalteh M, Alipour ZT, Ashraf S, Aliabadi MM, Nosratabadi AF (2014) Effect of silica nanoparticles on basil (*Ocimum basilicum*) under salinity stress. J Chem Heal Risks 4:49–55

Karami A, Sepehri A (2018) Nano titanium dioxide and nitric oxide alleviate salt induced changes in seedling growth, physiological and photosynthesis attributes of barley. Zemdirbyste-Agric 105:123–132

Karlsson HL (2007) The comet assay in nanotoxicology research. Anal Bioanal Chem 398:651–666

Kashem MA, Kawai S (2007) Alleviation of cadmium phytotoxicity by magnesium in Japanese mustard spinach. Soil Sci Plant Nutr 53(3):246–251

Kaveh R, Li YS, Ranjbar S, Tehrani R, Brueck CL, Van Aken B (2013) Changes in Arabidopsis thaliana gene expression in response to silver nanoparticles and silver ions. Environ Sci Technol 47(18):10637–10644

Khodakovskaya M, Dervishi E, Mahmood M, Xu Y, Li Z, Watanabe F, Biris AS (2009) Carbon nanotubes are able to penetrate plant seed coat and dramatically affect seed germination and plant growth. ACS Nano 3(10):3221–3227

Khodakovskaya MV, de Silva K, Nedosekin DA, Dervishi E, Biris AS, Shashkov EV et al (2011) Complex genetic, photo- thermal, and photoacoustic analysis of nanoparticle-plant interactions. Proc Natl Acad Sci U S A 108:10281033

Khodakovskaya MV, de Silva K, Biris AS, Dervishi E, Villagarcia H (2012) Carbon nanotubes induce growth enhancement of tobacco cells. ACS Nano 6:2128–2135

Khus M, Gernjak W, Ibanez PF, Rodriguez SM, Galvez JB, Icli S (2006) A comparative study of supported TiO2 as photocatalyst in water decontamination at solar pilot plant scale. J Sol Energy Eng 128:331–337

Kiapour H, Moaveni P, Habibi D, Sani B (2015) Evaluation of the application of gibbrellic acid and titanium dioxide nanoparticles under drought stress on some traits of basil (*Ocimum basilicum* L.). Int J Agro Agri Res 6(4):138–150

Kim JS, Kuk E, Yu KN, Kim JH, Park SJ, Lee HJ, Kim SH, Park YK, Park YH, Hwang CY, Kim YK, Lee YS, Jeong DH, Cho MH (2007) Antimicrobial effects of silver nanoparticles. Nanomed Nanotechnol Biol Med 3:95–101

Koul A, Kumar A, Singh VK, Tripathi DK, Mallubhotla S (2018) Exploring plant-mediated copper, iron, titanium, and cerium oxide nanoparticles and their impacts. In: Nanomaterials in plants, algae, and microorganisms. Academic, San Diego, pp 175–194

Kumar R, Mishra RK, Mishra V, Qidwai A, Pandey A, Shukla SK, Pandey M, Pathak A, Dikshit A (2015) Detoxification and tolerance of heavy metals in plants. In: Ahmad P (ed) Plant metal interaction: Emerging remediation techniques. Elsevier, Amsterdam, pp 335–359. ISBN 978-0-12-803158-2

Kumari M, Mukherjee A, Chandrasekaran N (2009) Genotoxicity of silver nanoparticles in Allium cepa. Sci Total Environ 407:5243–5246

Landa P, Vankova R, Andrlova J, Hodek J, Marsik P, Storchova H, White JC, Vanek T (2012) Nanoparticle- specific changes in Arabidopsis thaliana gene expression after exposure to ZnO, TiO_2, and fullerene soot. J Hazard Mater 241–242:55–62

Landa P, Prerostova S, Petrova S, Knirsch V, Vankova R, Vanek T (2015) The transcriptomic response of *Arabidopsis thaliana* to zinc oxide: a comparison of the impact of nanoparticle, bulk, and ionic zinc. Environ Sci Technol 49:14537–14545

Landry MP, Mitter N (2019) How nanocarriers delivering cargos in plants can change the GMO landscape. Nat Nanotechnol 14:512–514

Latef A, Hamed AA, Srivastava AK, El-sadek MSA, Kordrostami M, Tran LSP (2017) Titanium dioxide nanoparticles improve growth and enhance tolerance of broad bean plants under saline soil conditions. Land Degrad Dev. https://doi.org/10.1002/ldr.2780

Lau HY, Wu H, Wee EJH, Trau M, Wang Y, Botella JR (2017) Specific and sensitive isothermal electrochemical biosensor for plant pathogen DNA detection with colloidal gold nanoparticles as probes. Sci Rep 7:38896. https://doi.org/10.1038/srep38896

Le Van N, Ma C, Rui Y, Liu S, Li X, Xing B, Liu L (2015) Phytotoxic mechanism of nanoparticles: destruction of chloroplasts and vascular bundles and alteration of nutrient absorption. Sci Rep 5:116–118

Lee S, Chung H, Kim S (2013) The genotoxic effect of ZnO and CuO nanoparticles on early growth of buckwheat, *Fagopyrum esculentum*. Water Air Soil Pollut 224:1668–1678

Lei Z, Su MY, Wu X, Liu C, Qu CX, Chen L, Huang H, Liu XQ, Hong FS (2008) Antioxidant stress is promoted by nano-anatase in spinach chloroplasts under UV-Beta radiation. Biol Trace Elem Res 121:69e79

Li Z, Huang J (2014) Effects of nanoparticle hydroxyapatite on growth and antioxidant system in pakchoi (Brassica chinensis L.) from cadmium-contaminated soil. J Nanomater 2014:17

Li F, Sun C, Li X, Yu X, Luo C, Shen Y, Qu S (2018) The effect of graphene oxide on adventitious root formation and growth in apple. Plant Physiol Biochem 129:122–129

Limbach LK, Wick P, Manser P, Grass RN, Bruinink A, Stark WJ (2007) Exposure of engineered nanoparticles to human lung epithelial cells: influence of chemical composition and catalytic activity on oxidative stress. Environ Sci Technol 41(11):4158–4163

Lin S, Bhattacharya P, Rajapakse NC (2009a) Effects of quantum dots adsorption on algal photosynthesis. J Phys Chem C 113:10962–10966

Lin S, Reppert J, Hu Q, Hudson JS, Reid ML, Ratnikova TA et al (2009b) Uptake, translocation, and transmission of carbon nanomaterials in rice plants. Small 5(10):1128–1132

Lin J, Zhang H, Chen Z, Zheng Y (2010) Penetration of lipid membranes by gold nanoparticles: insights into cellular uptake, cytotoxicity, and their relationship. ACS Nano 4:5421–5429. https://doi.org/10.1021/nn1010792

Liu Q, Zhao Y, Wan Y (2010) Study of the inhibitory effect of water-soluble fullerenes on plant growth at the cellular level. ACS Nano 4:5743–5748

Liu Q, Zhang X, Zhao Y (2013) Fullerene-induced increase of glycosyl residue on living plant cell wall. Environ Sci Technol 47:7490–7498

Liu Q, Chen B, Wang Q, Shi X, Xiao Z, Lin J et al (2009) Carbon nanotubes as molecular transporters for walled plant cells. Nano Lett 9:1007–1010. https://doi.org/10.1021/nl803083u

López-Moreno ML, de la Rosa G, Hernández-Viezcas JÁ, Castillo-Michel H, Botez CE, Peralta-Videa JR, Gardea-Torresdey JL (2010) Evidence of the differential biotransformation and genotoxicity of ZnO and CeO_2 nanoparticles on soybean (*Glycine max*) plants. Environ Sci Technol 44:7315–7320

Lowry GV, Hotze EM, Bernhardt ES, Dionysiou DD, Pedersen JA, Wiesner MR, Xing B (2010) Environmental occurrences, behavior, fate, and ecological effects of nanomaterials: an introduction to the special series. J Environ Qual 39:1867–1874

Lu CM, Zhang CY, Wen JQ, Wu GR, Tao MX (2002) Research on the effect of nanometer materials on germination and growth enhancement of Glycine max and its mechanism. Soybean Sci 21(3):68–172

Lu A-H, Hui A, Salabas EL, Schüth F (2007) Magnetic nanoparticles: synthesis, protection, functionalization, and application. Angew Chem Int Ed Engl 46(8):1222–1244. https://doi.org/10.1002/anie.200602866

Lv J, Christie P, Zhang S (2019) Uptake, translocation, and transformation of metal-based nanoparticles in plants: recent advances and methodological challenges. Environ Sci Nano 6:41–59. https://doi.org/10.1039/C8EN00645H

Lv J, Zhang S, Luo L, Zhang J, Yang K, Christied P (2015) Accumulation, speciation and uptake pathway of ZnO nanoparticles in maize. Environ Sci Nano 2:68–77. https://doi.org/10.1039/C.4E.N.00064A

Ma JF, Yamaji NA (2015) Cooperative system of silicon transport in plants. Trends Plant Sci 20:435–442

Ma X, Geisler-Lee J, Deng Y, Kolmakov A (2010) Interactions between engineered nanoparticles (ENPs) and plants: phytotoxicity, uptake and accumulation. Sci Total Environ 408:3053–3061

Mahakham W, Sarmah AK, Maensiri S, Theerakulpisut P (2017) Nanopriming technology for enhancing germination and starch metabolism of aged rice seeds using photosynthesized silver nanoparticles. Sci Rep 7:8263

Marchesano V, Hernandez Y, Salvenmoser W, Ambrosone A, Tino A, Hobmayer B et al (2013) Imaging inward and outward trafficking of gold nanoparticles in whole animals. ACS Nano 7:2431–2442. https://doi.org/10.1021/nn305747e

Marschner HM (1995) Nutrition of higher plants, 2nd edn. Academic, San Diego

Mazumdar H, Ahmed GU (2011) Phytotoxicity effect of silver nanoparticles on Oryza sativa. Int J ChemTech Res 3(3):1494–1500

Mingfang Q, Yufeng L, Tianlai L (2013) Nano-TiO2 improve the photosynthesis of tomato leaves under mild heat stress, biological trace element research. Biol Trace Elem Res 156(1):323–328

Miralles P, Church TL, Harris AT (2012) Toxicity, uptake, and translocation of engineered nanomaterials in vascular plants. Environ Sci Technol 46(17):9224–9239

Miransari M (2016) Soybean production and heavy metal stress. In: Abiotic and biotic stresses in soybean production. Elsevier, Amsterdam, pp 198–216. ISBN 978-0-12-801536-0

Miri A, Shakib E, Ebrahimi O, Sharifi RJ (2017) Impacts of nickel nanoparticles on grow characteristics, photosynthetic pigment content and antioxidant activity of *Coriandrum Sativum* L. Orient J Chem 33:1297–1303

Mirzajani F, Askari H, Hamzelou S, Schober Y, Römpp A, Ghassempour A, Spengler B (2014) Proteomics study of silver nanoparticles toxicity on *Oryza sativa* L. Ecotoxicol Environ Saf 108:335–339

Moghadam NK, Hatami M, Rezaei S, Bayat M, Lajayer BA (2019) Induction of plant defense machinery against nanoparticles exposure. In: Ghorbanpour M, Wani SH (eds) Advances of Phytonanotechnology from synthesis to applications. Elsevier Inc. Publisher. https://doi.org/10.1016/B978-0-12-815322-2.00010-9

Moon JW, Phelps TJ, Fitzgerald CL, Lind RF, Elkins JG, Jang GG et al (2016) Manufacturing demonstration of microbially mediated zinc sulfide nanoparticles in pilot-plant scale reactors. Appl Microbiol Biotechnol 100:7921–7931. https://doi.org/10.1007/s00253-016-7556-y

Moraru CI, Panchapakesan CP, Huang Q, Takhistov P, Liu S, Kokini JL (2003) Nanotechnology: a new frontier in food science understanding the special properties of materials of nanometer size will allow food scientists to design new, healthier, tastier, and safer foods. Nanotechnology 57(12):24–29

Moreno-Olivas F, Gant VU Jr, Johnson KL, Peralta-Videa JR, Gardea-Torresdey JL (2014) Random amplified polymorphic DNA reveals that TiO_2 nanoparticles are genotoxic to Cucurbitapepo. J Zhejiang Univ Sci:618–623

Mustafa G, Komatsu S (2016) Toxicity of heavy metals and metal-containing nanoparticles on plants. Biochim Biophys Acta Proteins Proteom 1864(8):932e944

Mustafa G, Sakata K, Komatsu S (2015a) Proteomic analysis of flooded soybean root exposed to aluminum oxide nanoparticles. J Proteome 128:280–297
Mustafa G, Sakata K, Hossain Z, Komatsu S (2015b) Proteomic study on the effects of silver nanoparticles on soybean under flooding stress. J Proteome 122:100–118
Nair PMG, Chung IM (2014) Physiological and molecular level effects of silver nanoparticles exposure in rice (*Oryza sativa* L.) seedlings. Chemosphere 112:105–113
Navarro E, Baun A, Behra R, Hartmann NB, Filser J, Miao AJ, Quigg A, Santschi PH, Sigg L (2008) Environmental behavior and ecotoxicity of engineered nanoparticles to algae, plants, and fungi. Ecotoxicology 17:372–386
Nel AE, Mädler L, Velegol D, Xia T, Hoek EMV, Somasundaran P et al (2009) Understanding biophysicochemical interactions at the nano-bio interface. Nat Mater 8:543–557. https://doi.org/10.1038/nmat2442
Owen R, Handy R (2007) Formulating the problems for environmental risk assessment of nanomaterials. Environ Sci Technol 41:5582–5588
Palocci C, Valletta A, Chronopoulou L, Donati L, Bramosanti M, Brasili E et al (2017) Endocytic pathways involved in PLGA nanoparticle uptake by grapevine cells and role of cell wall and membrane in size selection. Plant Cell Rep 36:1917–1928. https://doi.org/10.1007/s00299-017-2206-0
Pérez-de-Luque A (2017) Interaction of nanomaterials with plants: what do we need for real applications in agriculture? Front Environ Sci 5:12. https://doi.org/10.3389/fenvs.2017.00012
Pokhrel LR, Dubey B (2013) Evaluation of developmental responses of two crop plants exposed to silver and zinc oxide nanoparticles. Sci Total Environ 452–453:321–332. https://doi.org/10.1016/j.scitotenv.2013.02.059
Pourzahedi L, Pandorf M, Ravikumar D, Zimmerman JB, Seager TP, Theis TL, Westerhoff P, Gilbertson LM, Lowry GV (2018) Life cycle considerations of nanoenabled agrochemicals: are today's tools up to the task? Environ Sci Nano 5:1057–1069
Prasad PV, Djanaguiraman M (2011) High night temperature decreases leaf photosynthesis and pollen function in grain sorghum. Funct Plant Biol 12(38):993–1003
Qi M, Liu Y, Li T (2013) Nano-TiO_2 improves the photosynthesis of tomato leaves under mild heat stress. Biol Trace Elem Res 156:323328
Qiu Z, Yang Q, Liu W (2013) Photocatalytic degradation of phytotoxic substances in waste nutrient solution by various immobilized levels of nano-TiO_2. Water Air Soil Pollut 224(3):1–10
Răcuciu, M., and Creangă, D. E. (2009) Cytogenetical changes induced by b-cyclodextrin coated nanoparticles in plant seeds. Rom Reports Phys 54:125–131
Raven JA (1982) Transport and function of silicon in plants. Biol Rev 58:179–207
Remédios C, Rosário F, Bastos V (2012) Environmental nanoparticles interactions with plants: morphological, physiological, and genotoxic aspects. J Bot 212. https://doi.org/10.1155/2012/751686
Renault S, Baudrimont M, Mesmer-Dudons N, Gonzalez P, Mornet S, Brisson A (2008) Impacts of gold nanoparticle exposure on two freshwater species: a phytoplanktonic alga (*Scenedesmussubspicatus*) and a benthic bivalve (*Corbiculafluminea*). Gold Bull 41(2):116–126
Rezvani N, Sorooshzadeh A, Farhadi N (2012) Effect of Nano-silver on growth of saffron in flooding stress world academy of science, engineering and technology. Int J Biol Biomol Agri Food Biotechnol Eng 6(1):10–16
Rico CM, Hong J, Morales MI (2013) Effect of cerium oxide nanoparticles on rice: a study involving the antioxidant defense system and in vivo fluorescence imaging. Environ Sci Technol 47:5635–5642
Rossi L, Zhang W, Lombardini L, Ma X (2016) The impact of cerium oxide nanoparticles on the salt stress responses of Brassica napus L. Environ Pollut 219:28e36
Sahebi M, Hanafi MM, Siti Nor Akmar A, Rafii MY, Azizi P, Tengoua FF, Nurul Mayzaitul Azwa J, Shabanimofrad M (2015) Importance of silicon and mechanisms of biosilica formation in plants. Biomed Res Int 396010:16. https://doi.org/10.1155/2015/396010

Santner A, Calderon-Villalobos LIA, Estelle M (2009) Plant hormones are versatile chemical regulators of plant growth. Nat Chem Biol 5:301–307
Savicka M, Skute N (2010) Effects of high temperature on malon- dialdehyde content, superoxide production and growth changes in wheat seedlings (*Triticum aestivum* L.). Ekologija 56:2633
Saxena R, Tomar RS, Kumar M (2016) Exploring nanobiotechnology to mitigate abiotic stress in crop plants. J Pharm Sci Res 8(9):974
Schulze E-D, Beck E, Muller-Hohenstein K (2005) Plant ecology. Springer, Berlin
Schwab F, Zhai G, Kern M, Turner A, Schnoor JL, Wiesner MR (2016) Barriers, pathways and processes for uptake, translocation and accumulation of nanomaterials in plants – critical review. Nanotoxicology 10:257–278. https://doi.org/10.3109/17435390.2015.1048326
Scrinis G, Lyons K (2007) The emerging nano-corporate paradigm: nanotechnology and the transformation of nature, food and agri-food systems. Int J Sociol Agri Food 15(2):22–44
Seaton GG, Walker DA (1990) Chlorophyll fluorescence as a measure of photosynthetic carbon assimilation. Proc Royal Soc London B Biol Sci 242(1303):29–35
Serag MF, Kaji N, Gaillard C, Okamoto Y, Terasaka K, Jabasini M et al (2011) Trafficking and subcellular localization of multiwalled carbon nanotubes in plant cells. ACS Nano 5:493–499. https://doi.org/10.1021/nn102344t
Serik O, Ainur I, Murat K, Tetsuo M, Masaki I (1996) Silicon carbide fiber-mediated DNA delivery into cells of wheat (*Triticum aestivum* L.) mature embryos. Plant Cell Rep 16:133–136. https://doi.org/10.1007/BF01890853
Servin A, Elmer W, Mukherjee A, De la Torre-Roche R, Hamdi H, White JC et al (2015) A review of the use of engineered nano- materials to suppress plant disease and enhance crop yield. J Nanopart Res 17:92
Sharif F, Westerhoff P, Herckes P (2013) Sorption of trace organics and engineered nanomaterials onto wetland plant material. Environ Sci: Processes Impacts 15(1):267–274
Sharma VK, Yngard RA, Lin Y (2009) Silver nanoparticles: Green synthesis and their antimicrobial activities. Adv Colloid Interf Sci 145:83–96
Sharma P, Jha AB, Dubey RS, Pessarakli M (2012a) Reactive oxygen species, oxidative damage, and antioxidative defense mechanism in plants under stressful conditions. J Bot 1–27
Sharma P, Bhatt D, Zaidi MGH, Saradhi PP, Khanna PK, Arora S (2012b) Silver nanoparticle-mediated enhancement in growth and antioxidant status of Brassica juncea. Appl Biochem Biotechnol 167(8):2225–2233
Shaw AK, Hossain Z (2013) Impact of nano-CuO stress on rice (*Oryza sativa* L.) seedlings. Chemosphere 93:906–915
Siddiqui MH, Al-Whaibi MH, Faisal M, Al Sahli AA (2014) Nano-silicon dioxide mitigates the adverse effects of salt stress on Cucurbita pepo L. Environ Toxicol Chem 33(11):2429–2437
Siddiqui MH, Al-Whaibi MH, Firoz M, Al-Khaishany MY (2015a) Role of nanoparticles in plants. In: Book nanotechnology and plant science, pp 19–35
Siddiqui MH, Al-Whaibi MH, Mohammad F (2015b) Nanotechnology and plant sciences: Nanoparticles and their impact on plants. Springer
Singh J, Lee BK (2016) Influence of nano-TiO_2 particles on the bioaccumulation of Cd in soybean plants (Glycine max): a possible mechanism for the removal of Cd from the contaminated soil. J Environ Manag 170:8896
Singh A, Singh S, Prasad SM (2016a) Scope of nanotechnology in crop science: profit or loss. RRJBS 5(1)
Singh S, Tripathi DK, Dubey NK, Chauhan DK (2016b) Effects of nanomaterials on seed germination and seedling growth: striking the slight balance between the concepts and controversies. Mater Focus 5(3):195–201
Song U, Jun H, Waldman B, Roh J, Kim Y, Yi J, Lee EJ (2013) Functional analyses of nanoparticle toxicity: a comparative study of the effects of TiO2 and Ag on tomatoes (Lycopersicon esculentum). Ecotoxicol Environ Saf 93:60–67
Stampoulis D, Sinha SK, White JC (2009) Assay-dependent phytotoxicity of nanoparticles to plants. Environ Sci Technol 43:9473–9479

Stirbet A (2012) Chlorophyll a fluorescence induction: a personal perspective of the thermal phase, the J–I–P rise. Photosynth Res 113(1):15–61

Sun D, Hussain HI, Yi Z, Siegele R, Cresswell T, Kong L et al (2014) Uptake and cellular distribution, in four plant species, of fluorescently labeled mesoporous silica nanoparticles. Plant Cell Rep 33:1389–1402. https://doi.org/10.1007/s00299-014-1624-5

Syu YY, Hung JH, Chen JC, Chuang HW (2014) Impacts of size and shape of silver nanoparticles on Arabidopsis plant growth and gene expression. Plant Physiol Biochem 83:57–64

Takahashi T, Mizoguchi R, Yoshida K, Ichimura K, Shinozaki A (2011) Calmodulin dependent activation of MAP kinase for ROS homeostasis in *Arabidopsis*. Mol Cell 4:649–660

Tan X, Lin C, Fugetsu B (2009) Studies on toxicity of multi-walled carbon nanotubes on suspension rice cells. Carbon 47:3479–3487. The Royal Society and the Royal Academy of Engineering

Tao X, Yu Y, Fortner JD, He Y, Chen Y, Hughes JB (2015) Effects of aqueous stable fullerene nanocrystal (nC_{60}) on Scenedesmus obliquus: evaluation of the sub-lethal photosynthetic responses and inhibition mechanism. Chemosphere 122:162e167

Thill A, Zeyons O, Spalla O, Chauvat F, Rose J, Auffan M, Flank AM (2006) Cytotoxicity of CeO2 nanoparticles for Escherichia coli. Physico-chemical insight of the cytotoxicity mechanism. Environ Scitechnol 40(19):6151–6156

Thwala M, Musee N, Sikhwivhilu L, Wepener V (2013) The oxidative toxicity of ag and ZnO nanoparticles towards the aquatic plant Spirodela punctuta and the role of testing media parameters. Environ Sci Process Impacts 15:1830–1843

Tolaymat T, Genaidy A, Abdelraheem W, Dionysiou D, Andersen C (2017) The effects of metallic engineered nanoparticles upon plant systems: an analytic examination of scientific evidence. Sci Total Environ 579:93–106

Trindade T, O'Brien P, Pickett N (2001) Nanocrystalline semicon-ductors: synthesis, properties, and perspectives. Chem Mater 13:3843–3858

Tripathi DK, Singh VP, Prasad SM, Chauhan DK, Dubey NK (2015) Silicon nanoparticles (SiNp) alleviate chromium (VI) phytotoxicity in Pisum sativum (L.) seedlings. Plant Physiol Biochem 96:189198

Tripathi DK, Singh VP, Singh S, Prasad SM, Chauhan DK, Dubey NK (2016) Effect of silicon and silicon nanoparticle (SiNp) on seedlings of maize cultivar and hybrid differing in arsenate tolerance. Front Environ Sci 4:46. https://doi.org/10.3389/fenvs.2016.00046

Tripathi DK, Ahmad P, Sharma S, Chauhan DK, Dubey NK (eds) (2017) Nanomaterials in plants, algae, and microorganisms: concepts and controversies, vol 1. Academic Press, London

Unrine JM, Colman BP, Bone AJ, Gondikas AP, Matson CW (2012) Biotic and abiotic interactions in aquatic microcosms determine fate and toxicity of Ag nanoparticles. Part 1. Aggregation and dissolution. Environ Sci Technol 46(13):6915–6924

Valenstein JS, Lin VS-Y, Lyznik LA, Martin-Ortigosa S, Wang K, Peterson DJ et al (2013) Mesoporous silica nanoparticle-mediated intracellular cre protein delivery for maize genome editing via loxP site excision. Plant Physiol 164:537–547. https://doi.org/10.1104/pp.113.233650

Valletta A, Chronopoulou L, Palocci C, Baldan B, Donati L, Pasqua G (2014) Poly(lactic-co-glycolic) acid nanoparticles uptake by Vitis vinifera and grapevine-pathogenic fungi. J Nanopart Res 16:1917–1928. https://doi.org/10.1007/s11051-014-2744-0

Vidyalakshmi N, Thomas R, Aswani R, Gayatri GP, Radhakrishnan EK, Remakanthan A (2017) Comparative analysis of the effect of silver nanoparticle and silver nitrate on morphological and anatomical parameters of banana under in vitro conditions. Inorg Nano Metal Chem 47:1530–1536. https://doi.org/10.1080/24701556.2017.1357605

Wahid A (2007) Physiological implications of metabolites biosynthe- sis in net assimilation and heat stress tolerance of sugarcane (*Saccharum officinarum*) sprouts. J Plant Res 120:219228

Wang X, Summers CJ, Wang ZL (2004) Large-scale hexagonal-patterned growth of aligned ZnO nanorods for nano-optoelectronics and nano sensor arrays. Nano Lett 4(3):423–426

Wang T, Bai J, Jiang X, Nienhaus GU (2012) Cellular uptake of nanoparticles by membrane penetration: a study combining confocal microscopy with FTIR spectroelectrochemistry. ACS Nano 6:1251–1259. https://doi.org/10.1021/nn203892h

Wang S, Wang F, Gao S, Wang X (2016) Heavy metal accumulation in different rice cultivars as influenced by foliar application of nano-silicon. Water Air Soil Pollut 227:228

Watanabe T, Misawa S, Hiradate S, Osaki M (2008) Root mucilage enhances aluminum accumulation in Melastoma malabathricum, an aluminum accumulator. Plant Signal Behav 3(8):603–605

Wong SC, Osmond CB (1991) Elevated atmosphere partial pressure of CO2 and plant growth. III. Interactions between Triticum aestivum (C3) and Echinochloa frumentacea (C4) during growth in mixed culture under different CO2, N nutrition and irradiance treatments, with emphasis on below- ground responses estimated using the δ13C value of root biomass. Aus J Plant Physiol 18:137–152

Wu H, Shabala L, Shabala S, Giraldo JP (2018) Hydroxyl radical scavenging by cerium oxide nanoparticles improves Arabidopsis salinity tolerance by enhancing leaf mesophyll potassium retention. Environ Sci Nano. https://doi.org/10.1039/c8en00323h

Yang F, Hong F, You W, Liu C, Gao F, Wu C et al (2006) Influence of nano-anatase TiO2 on the nitrogen metabolism of growing spinach. Biol Trace Elem Res 110(2):179–190

Zhai G, Walters KS, Peate DW, Alvarez PJJ, Schnoor JL (2014) Transport of gold nanoparticles through plasmodesmata and precipitation of gold ions in woody poplar. Environ Sci Technol Lett 1:146–151. https://doi.org/10.1021/ez400202b

Zhang LW, Monteiro-Riviere NA (2009) Mechanisms of quantum dot nanoparticle cellular uptake. Toxicol Sci 110(1):138–155

Zhang H et al (2019) DNA nanostructures coordinate gene silencing in mature plants. Proc Natl Acad Sci U S A 116:7543–7548

Zhao XU, Liz W, Chen Y, Ahi LY, Zhu YF (2007) Solid-phase photocatalytic degradation of polyethylene plastic under UV and solar light irradiation. J Mol Catal A Chem 268:101–106

Zhao L, Peng B, Hernandez-Viezcas JA, Rico C, Sun Y, Peralta- Videa, JR et al (2012) Stress response and tolerance of *Zea mays* to CeO_2 nanoparticles: cross talk among H_2O_2, heat shock protein and lipid peroxidation. ACS Nano 6:9615–9622

Zhao L, Sun Y, Hernandez-Viezcas JA, Servin AD, Hong J, Niu G (2013) Influence of CeO_2 and ZnO nanoparticles on cucumber physiological markers and bioaccumulation of Ce and Zn: a life cycle study. J Agric Food Chem 61:11945–11951

Zhao X, Meng Z, Wang Y, Chen W, Sun C, Cui B et al (2017) Pollen magnetofection for genetic modification with magnetic nanoparticles as gene carriers. Nat Plants 3:956–964. https://doi.org/10.1038/s41477-017-0063-z

Zheng L, Hong F, Lu S, Liu C (2005) Effect of Nano-TiO_2 on strength of naturally aged seeds and growth of spinach. Biol Trace Elem Res 104(1):83–91

Zuverza-Mena N, Martínez-Fernández D, Du W, Hernandez-Viezcas JA, Bonilla-Bird N, López-Moreno ML, Komárek M, Peralta-Videa JR, Gardea-Torresdey JL (2016) Exposure of engineered nanomaterials to plants: Insights into the physiological and biochemical responses-A review. Plant Physiol Biochem 110:236–264. https://doi.org/10.1016/j.plaphy.2016.05.037

Chapter 4
Exploring Nanotechnology to Reduce Stress: Mechanism of Nanomaterial-Mediated Alleviation

Sidra tul Muntha, Mohammad Faizan, Saba Mehreen, and Shareen

Abstract Global food production does not meet the demand of ever increasing population which is expected to reach around 9 billion by the first half of twenty-first century. Plants being sessile organisms are under detrimental effects of environment as well as plant diseases. Abiotic stresses including drought, heat, flooding, salinity and frost are major factors, adversely affect plant growth that may minimize the productivity to 70%. These stresses lead to morphological, physiological, biochemical and molecular changes in plants resulting crop loss. Due to unavailability of adequate arable land and severe environmental issues, there is an immediate need of novel avenues of research to meet global food supply. Progress in plant sciences and genetics has revealed new technologies to develop stress tolerant plants and investigate the better ways to grow plants under detrimental conditions. Nanotechnology comprises nanoparticles (NP) gained high impulse to mitigate the limitations related to abiotic stresses resulting high plant growth. Nanoparticles are metal or metal oxide molecules synthesize by physiochemical or biological approaches, with small size of 1–100 nm dimensions. Nanoparticles due to their exclusive properties of small size, high reaction potential, increased surface area, tensile pore width and divergent morphology opens new doors in agriculture research. Hence, the current chapter will focus on the role of nanoparticles against plant environmental challenges and how can nanoparticles be used in growth improvement of plants under stressed conditions.

S. t. Muntha (✉)
Laboratory of Germplasm Innovation and Molecular Breeding, College of Agriculture and Biotechnology, Zhejiang University, Hangzhou, China

M. Faizan
Collaborative Innovation Center of Sustainable Forestry in Southern China, College of Forest Science, Nanjing University, Nanjing, China

S. Mehreen
Department of Chemistry, Quaid-i-Azam University, Islamabad, Pakistan

Shareen
Department of Agricultural Microbiology, Faculty of Agriculture, Aligarh Munslim University, Aligarh, India

M. Faizan et al. (eds.), *Sustainable Agriculture Reviews 53*, Sustainable Agriculture Reviews 53, https://doi.org/10.1007/978-3-030-86876-5_4

Keywords Abiotic stress · Nanotechnology · Nanoparticles

4.1 Introduction

4.1.1 Cascade of Signaling Behind Plant-NPs Interaction and Stress Tolerance

Plants have a complex system for defense against environmental stresses, yet the mechanism of perception of stimulus to the transduction and activation of responsive units against stimulant before stress prompted destruction is decisive to cellular machinery. As per available information about plant-NPs interaction against abiotic stresses, it is obvious that synthesis of ROS is vital phenomenon in plant cell against environmental stimuli (Khan et al. 2017). Although the mechanism of action of plants by the application of NPs is not well understood, however the omic approaches help to conceive the signaling pathways in plant cells.

Signaling molecules predominantly activate a defense mechanism which triggers molecular network to encounter a particular challenge. Calcium ions (Ca^{2+}) being secondary messenger molecules act as vital components in signal transduction. Upon receiving a stress signal Ca^{2+} start transport from stores to cytosol via Ca2+ gated channels and ultimately increase the cytoplasmic level of secondary messenger, which is identified my calcium binding proteins (CaBPs) that start a cascade of reaction downstream to alter the gene expression of plant against that particular stress (Khan et al. 2014a, b; Tuteja and Mahajan 2007). It has been evident that nitric oxide (NO) promotes the cytosolic Ca2+ in cell upon the onset of environmental stress and pathogen attack (Khan et al. 2012a, b; Lamotte et al. 2006) and inevitably Ca2+ promote the NO synthesis (Del Rio et al. 2004; Corpas et al. 2004).

Proteomic studies on *Oryza sativa* roots have declared the role of Ag NPs-responsive proteins in oxidative stress pathway, second messenger (Ca2+) regulation, signal transduction and expression and post transcription changes, cellular growth and programmed cell death (Mirzajani et al. 2014). The study is a prove of previous work by Goyer (1995) who anticipated the Ag-NPs role in cellular metabolism by binding Ca2+ gated channels and Ca/Na pumps (Goyer 1995). It has been further observed that C60 nano-crystals prompted the functional regulation of Ca2+/calmodulin- dependent protein kinase II (CaMKII) (Miao et al. 2014). Further studies on Arabidopsis thaliana investigated the overexpressed Ca2+ dependent protein (CML45) and CaMPK23 caused by the activity of cadmium sulfide QDs (Marmiroli et al. 2015). These overexpressed proteins play a pivotal role in stress resistance in various plants (Delk et al. 2005; Xu et al. 2011; Boudsocq and Sheen 2013) however Ca2+ in the proteins can be switched with NPs (Cheung 1980).

NPs have been reported to involve in increased nitrate reductase activity that is key enzyme in nitric oxide (NO) biosynthesis to regulate plant immune system (Carpenter et al. 2012; Shahrokh et al. 2014; Chandra et al. 2015). On the other hand

NO interacts NP-induced toxicity mechanism and antioxidant genes to increase their transcript level and decrease ROS and lipid degradation (Chen et al. 2015). NPs are involved in upregulated expression of genes related to stress responsive and cell growth and development (Almutairi 2016; Khodakovskaya et al. 2011a, b, 2012).

The studies on NPs-plant interaction have exhibited the increased generation of ROS that play not only signaling molecules in plant but also increase phytotoxicity (Oukarroum et al. 2012; Qi et al. 2013; Ma et al. 2010; Van Hoecke et al. 2008; Simon et al. 2013; Wei and Wang 2013).

The dual action of NPs needs a comprehensive study to evaluate how NPs protect plants against ROS while acting as inducers of oxidative stress at the same time. Genomics and proteomics approaches aid to existing knowledge by investigating the role of NPs in plants under environmental stress. A proteomic study using Ag NPs and AgNO3 in *Eruca sativa* demonstrated the altered level of proteins involved in redox regulation and metabolism of sulfur as a result of physiochemical nature of NPs (Vannini et al. 2013). A differential expression of genes realted to abiotic stress has been observed in Arabidopsis thaliana by the application of Ag NPs and Ag+ covered with polyvinylpyrrolidone (PVP) (Kaveh et al. 2013). The study concluded that the Ag NPs induced stress is partially due to Ag+ toxicity and partially it is the consequence of nanoparticle-specific effects.

Expression of miRNA in response to NPs paves another way to understand the mechanism of action of NPs against environment in plants. Increased expression of miRNA in Tobacco plant on application of optimum concentration of TiO2 and Al2O3 was demonstrated against metal stress however increased level of NPs negatively affected the growth and development of plant (Frazier et al. 2014; Burklew et al. 2012). AHA2 (gene involved in stomatal opening) in Arabidopsis is upregulated by the treatment of zero valent NP, which cause tolerance to drought stress (Kim et al. 2015). Moreover studies conducted on the application of TiO2 and MWCNTs on Arabidopsis showed the downregulated gene expression involved in development and phosphate deficiency (García-Sanchez et al. 2015).

4.2 Nanoparticles and Abiotic Stress Resistance

4.2.1 Salinity Stress

Salinity, due to the deposition of anions (chloride and sulfate) or cations (primarily sodium but occasionally of calcium and magnesium) in arid to coastal areas soil, is one of the major abiotic stresses limiting food production. Over 20% of agricultural land is affected by salinity and the limit is ever increasing. Salinity is detrimental as it causes reduction in growth and development by influencing physiological, biochemical and molecular pathways in plants. It not only shifts the osmotic stress but also ionic imbalance in plants due to high accumulation of salts. Due to Osmotic stress plants' nutrients and water uptake is decreased while ionic stress engenders decreased proportion of K+/Na+ (Khan et al. 2012) and over production of ROS that affects molecular mechanisms in plants leading to electrolytes splits and damages metabolic pathways such as protein and lipid metabolism, and photosynthesis in cytol (Sharma et al. 2012; Ismail et al. 2014; Khan et al. 2010).

Current advancement in nanotechnology highlighted that the NP of Silica (SiO2) the second most abundant natural element and titanium dioxide (TiO2) contributed much to enhance vegetative growth and overall crop production under salinity. The tolerance is attributed to silicon NP that might better absorbed by maize roots than its micro or bulk counterparts Suriyaprabha et al. (2012) and form a thin layer in cell wall to enhance resistance against stress to maintain yield (Latef et al. 2018; Derosa et al. 2010).

The stressed tomato and squash plants exhibited better seed germination, the anti-oxidative enzyme activities, photosynthetic rate and leaf water absorption rate when treated with Si NP (Haghighi and Pourkhaloee 2013).

Na+ ion toxicity led a reduced yield in maize; however SiO2NPs alleviated the plant response under salinity stress by reducing Na+ ion concentration in cell wall through lower absorption of the ions (Gao et al. 2006a, b). Similar studies were conducted in tomato plants that lead to better plant growth (Savvasd et al. 2009). A remarkable elevation in germination and seedling growth was observed in *Lens culinaris* under salinity stress by application of Si NP (Sabaghnia and Janmohammadi 2014). A promising effect of Si nano fertilizer was highlighted by Kalteh et al. (2014) in *Ocimum basilicum*, where increased chlorophyll content, proline level and other physiological traits were recorded under salinity stress. Under high concentration of salt Squash (Cucurbita pepo) showed a lethal reduction in roots and shoots growth, vigor length and yield of plant (Siddiqui et al. 2014). Use of SiO2 NP ameliorated the traits by decreasing the electrolyte leakage and level of hydrogen peroxide (H2O2), malondialdehyde (MDA) and chlorophyll degradation.

Fe2 NPs proved to be auspicious addition in Nano biotechnology. The application revealed that it positively affected foliar fresh and dry weights and mineral contents of peppermint (Mentha piperita) however it didn't show any effect on sodium content. Maximum activities of anti- oxidant enzymes were recorded under salinity stress but masked by the application of Fe_2O_3 NP (Askary et al. 2017). Torabian et al. (2016) have described high level of chlorophyll content, photosynthesis rate, CO2 concentration, osmotic regulation and reduced Na content in *Helianthus annuus* by the use of nanoZnO under salt stress.

Contemporary studies on the significance of chitosan NPs (maize and tomato), multi-walled carbon nanotubes (broccoli) and silver NPs (wheat) have demonstrated their mitigating effect on salinity (Bruna et al. 2016; Hernandez-Hernandez et al. 2018; Martinez-Ballesta et al. 2016; Abou-Zeid and Ismail 2018; Mohamed et al. 2017).

4.2.2 Drought Stress

Water is the vital component for life on the planet and its deficiency leads to severe conditions (Drought stress) in living organisms including plants. Drought is most commonly occurring environmental stress which affects almost 45% of global agriculture area (Dos Reis et al. 2016). Water scarcity in plants leads to decrease in water potential and turgor of cell, which increase the level of molecules in the cytol and extracellular surfaces. Later, decreased cell size elicits the retarded growth and reproduction failure in plants. Ultimately cell starts to accumulate abscisic acid (ABA) and proline (Osmotic regulator), which leads to excessive production of ROS, glutathione and ascorbate (radical scavengers) which exasperates the severity (Hussain et al. 2019; Ahmad et al. 2017). Drought not only affects cell water potential but it also influences the stomatal closure, gaseous and ionic exchange, photosynthesis and transpiration rate (Schulze et al. 2019).

During the past decade tremendous efforts have been made to counter the water induced stress in plants using NP. The nanoparticles of TiO (Rutile) exhibited intense effect by exogenous application in spinach plant. The morphological, biochemical and physiological changes occurred in plant resulted in high rubisco activase activity, chlorophyll synthesis and promoted photosynthesis which leads to increase in dry weight of plant (Gao et al. 2008). Foliar application of TiO2 NP might cause an increase photosynthesis rate which augmented the overall seed yield in cow pea (Vigna unguiculata L.) (Owolade et al. 2008). As the effect of TiO2 NP varies among the species and with different applied environments, 0.02% of foliar spray of TiO2 NP enhanced the vigor of wheat plants by improving yield traits such as plant height, ear number and weight, 1000-kernal weight and seeds/plant, harvest index, and starch and gluten content of plant under water scarcity (Jaberzadeh et al. 2013). Dragonhead (Dracocephalum moldavica) plants treated with TiO2 NP (10 ppm) exhibited more proline level with less H2O2 and MAD content as compared to control plant under water deficit state (Mohammadi et al. 2014a, b). It was established that drought-prompted mutilations in plants like membrane damage and oxidative stress can be mitigated by optimal concentrations of TiO2 NPs. Silica NPs protects cell wall during water deficit conditions by reducing cell wall permeability of leaves resulting low lipid peroxidation (Zhu et al. 2004). SiO2 has proved to increase proline content with addition of escalated CAT and POD activities in plants under stress vs. controlled plants of tomato (Siddiqui and Al-Whaibi 2014), faba bean (Qados and Moftah 2015; Qados 2015) and alfalfa (Cakmak et al. 1996). Reative water content (RWC), water use proficiency and turgor pressure in leaf cells are the physio-chemical processes effected by Silica NPs that directly influence xylum transport plant. At different level of water deficiencies the response of Silica NPs varies in Hawthorns (Crataegus sp.). Enhanced tolerance was observed in plant against drought at different concentrations of NPs by positively effecting physio-biochemical processes (chlorophyll, carotenoid, carbohydrateand proline contents, and increased photosynthesis rate, MDA, (RWC) and membrane electrolyte leakage (ELI)) within the cell (Ashkavand et al. 2015).Silicon NP posed a positive effect on two sorghum cultivars with relatively different drought vulnerability irrespective of level of stress by maintaining photosynthesis rate and improved root growth (Hattori et al. 2005).

ZnO and CuO NPs act as fertilizers as these are source of Zn and Cu to plants. At different doses the NPs react on different parts of roots as Zn NPs causes increased lateral roots whereas Cu NPs induce proliferated and elongated root hair close to root tip in *Triticum aestivum* seedlings under drought (Yang et al. 2017). The short root length may reduce access to water. CuO possibly change the water supply thus shrink the cell wall in Arabidopsis and mustard thus increase lignification. The altered water transport may also be the reason of Cu-pectin association in cell wall (Nair and Chung 2017). The continuous drought stress elevates proline and anthrocynin in cell. The high level of ROS during the stress which leads to increased ABA may cause differential gene expression for drought tolerance (Dimkpa et al. 2012). Silver nanoparticles (AgNPs) are one of the most abundant NPs in use to mitigate abiotic as well as biotic stresses. AgNPs provided an inhibitory role against microorganisms (Beyene et al. 2017). In water deficit lentil plants, application AgNPs resulted in high germination rate and high growth and production parameters (Hojjat and Ganjali 2016).

Sodium nitroprusside (SN) along with Multi walled carbon nanotubes (MWCNTs) provide tolerance in *Hordeum vulgare* against water and salt stress by not only improving water absorption capacity of seed as well as seedling water concentration (Karami and Sepehri 2017). Increased antioxidants and high germination rate was recorded in *Hordeum vulgare*, *Glycine max* and *Zea mays* using MWCNTs (Lahiani et al. 2013; Liu et al. 2016). High root and shoot growth in *Triticum aestivum* suggested the drought tolerance though MWCNTs (Srivastava and Rao 2014).cerium oxide (CeO2) provided another source of NP to enhance crop production under water defict condition in Glycine max (Cao et al. 2018). In addition the *in vitro* use of iron (FeO2) NPs alongside salicylic acid manifested to be an effective tool against water deficiency in strawberry at pre transplantation to soil (Mozafari et al. 2018).

A comprehensive knowledge of metabolic and molecular mechanisms of plant through NPs to ameliorate abiotic stress will pave a way to develop stress resistant crops (Singh et al. 2017). Syntheses of dehydrins in susceptible plants, by application of NPs cause mitigation of drought stress (Lopez et al. 2003). Production of compatible molecules like proline, betaine, etc., is initiated by dehydrins which in turn maintains cell integrity and water deficiency (Paleg et al. 1984). Once it is known that at which stage of metabolic pathways NPs counter abiotic stresses, the massive increase in their use will be possible.

4.2.3 *Temperature Stress*

Temperature is vital factor which determines plant growth, development and crop yield. It is characterized by the ideal point beyond that plant growth is effected badly, though the optimum temperature varies between species and genera. Temperature stress includes high temperature stress (above the threshold temperature) and low temperature or chilling stress (very low than ideal) for a certain time span to cause a permanent damage to plant (Wahid 2007).

4.2.3.1 Heat Stress

Thermal stress implies the increase in temperature beyond the optimum level for a longer time span that causes permanent loss to development and vegetative growth (Wahid 2007). The stress negatively affects growth and yield of crop globally. High temperature elevates the synthesis of ROS and cause oxidative imbalance which leads to breakdown of organic molecules (proteins), degradation of lipids and escape of ions in cell membrane (Karuppanapandian et al. 2011; Moller et al. 2007; Savicka and Skute 2010;) that may affect chlorophyll content ultimately photosynthesis (Prasad et al. 2011).

Selenium (Se) nanoparticles provide an alleviated response of high temperature when used in low concentration by modulation of hydration potential and chlorophyll content (Haghighi et al. 2014). High level of Se is associated with oxidative stress while the low concentration is considered to be responsible for antioxidative response (Hasanuzzaman et al. 2014; Hartikainen et al. 2000). Heat shock proteins (molecular chaperones) are produced by plant during high temperature stress which along with other proteins cause stress tolerance by retaining their stability under challenging condition (Wahid et al. 2007). It has been reported that MWCNTs are involved in up regulation of gene expression associated with stress tolerance including HSP 90 (Khodakovskaya et al. 2011a, b). Furthermore a study in susceptible corn plant confirmed the role of heat shock proteins by application of cerium oxide (CeO2) NPs that lead to the higher synthesis of H2O2 and high expression of HSP70 (Zhao et al. 2012). TiO2 NPs also play role in heat stress by enhanced photosynthesis by regulation of stomata opening in plant leaves (Qi et al. 2013).

4.2.3.2 Cold Stress

When plants are exposed to the temperature very lower than their optimum temperature, the cell and tissues are damaged due to physiological and morphological changes, the phenomenon is known as Cold stress (Hasanuzzaman et al. 2013). Electrolytes imbalance and degradation in cell membrane are the adversities related to cold stress which eventually leads to decreased germination, reduced growth and crop production (Welti et al. 2002; Suzuki et al. 2008). Nevertheless, sensitivity to the stress may differs inter species and inter genera with tolerant plants showing least membrane damage than the susceptible (Maali Amiri et al. 2010; Heidarvand et al. 2011). Despite plants vulnerability to stress, NP like TiO2NPs possesses the potential to mitigate the chilling effect by reducing membrane degeneration and maintain electrolyte imbalance (Mohammadi et al. 2013). However its accumulation ratio is more in sensitive (thinner membrane layer and wide stomata) to tolerant genotype (Giacomo et al. 2010). Photosynthesis is essential process of plant that is affected by the chilling stress. Plants subjected to cold result in photosystem damage by decreasing in chlorophyll content, transpiration rate, deterioration in photosystem enzymes (Liu et al. 2012; Yordanova and Popova 2007).

Ameliorations of NPs on photosystem have been inferred by elevated synthesis of Rubisco (photosystem enzymes) (Gao et al. 2006a, b), chlorophyll capacity to absorb light, (Ze et al. 2011), rate of electron movement and and suppressed ROS synthesis in chloroplast (Giraldo et al. 2014). TiO2 NPs manifested the increased expression level in genes associated with Rubisco and chlorophyll binding proteins (Hasanpour et al. 2015), improved activities of antioxidant enzyme such as CAT, APX and SOD (Mohammadi et al. 2014a, b), finally increase resistance against chilling stress.

When plants are subjected to cold stress the transcript level of antioxidant genes like MeAPX2 and MeCu/ ZnSOD is up regulated, dehydroascorbate reductase, monodehydroascorbate reductase and glutathione reductase activities are elevated. As a result ROS scavenging which leads to repressed oxidative stress factors (lipid peroxidation, pigment degradation and H2O2 production) ultimately develop tolerance (Xu et al. 2014). While the application of NPs on Chiling stress have been reported with enhanced growth and physiochemical processes in plants under cold (Azimi et al. 2014; Hawrylak-Nowak et al. 2010; Kohan-Baghkheirati and Geisler-Lee 2015; Haghighi et al. 2014).

4.2.4 Heavy Metals Stress

Metal with high molecular weight and toxic at very low concentrations are termed as heavy metals. Metal stress has become one of the alarming environmental issues plants are facing worldwide that cause toxicity and serious crop loss (Chibuike and Obiora 2014; Rahimi et al. 2012). Heavy metal causes reduction in plant growth by disruption of important plant activities like reduced up take of vital elements, repressed enzyme activities which results into deprivation of important element (Capuana 2011). Plant Growth medium augmented with metals accelerates ROS synthesis, resulting in to oxidative stress in plant cell by disruption of cell structure including organic molecules and plasma membrane (Sharma et al. 2012; Rascio and Navari-Izzo 2011).

To combat heavy metal stress plant evolve a special defense mechanism by producing metal chelate, polyphosphates and organic acids which restrain the uptake of metal ions regulation of anti-oxidative pathways (CAT, POD and SOD)and ultimately ROS scavenging. Although plant defense mechanism is pivotal to counter heavy metals, artificially induced NPs play key role in reducing heavy metal phyto-toxicity (Gunjan et al. 2014; Tripathi et al. 2015).

ZnO NPs along with other micronutrients (Zn, Cu, Mn) are reported to play crucial response to reduce efflux of cadmium (Cd) in plants (Baybordi 2005; Venkatachalam et al. 2017). River tamarind (*Leucaena leucocephala*) possesses Cd and Pb phytotoxicity, which can be ameliorated by the foliar spray of ZnO NPs. The NPs are responsible for elevated soluble proteins, chlorophyll and carotenoid content in leaves, and decrease in oxidative damaged to lipids membrane (Venkatachalam et al. 2017). The induced level of antioxidative enzymes (CAT, SOD, POD) in leaves of *L. leucocephala* and lipid perodixadation was confirmed at seedling stage. Similar effects were recorded by Si NPs application to reduce Cr toxicity through activation of anti-oxidative pathways in pea plant (Tripathi et al. 2015). Enormous studies have been conducted using TiO2 against abiotic stresses in plants. Besides mitigating effects on environmental stress, TiO2 NPs proved to limit Cadmium (Cd) phytotoxicity by augmented growth and increased energy driven pathways (Singh and Lee 2016). In addition TiO2 multiple NPs evinced positive role against heavy metal challenge in plants (Table 4.1). Li and Huang (2014) exhibited that nano- hydroxyapatite ($Ca_5(PO_4)_3$) may regulate toxicity of Cd in *Brassica chinensis.*

4.3 Conclusion and Future Perspectives

Nanoparticles minimize the damage caused by environmental stresses by activating the defense system of plants. The activated defense system is the result of high ROS activities, which exhibited the toxic effects. Taking advantage of their size NPs become permeable and modulate ion channels to promote growth and germination of plants. The large surface area helps in absorption and delivery of molecules. The exact mechanism of action of MPs is still not very well studied however the omic studies revealed that NPs mimic secondary messengers and related proteins are activated. A cascade of reactions starts that leads to altered gene expression responsible for abiotic stress tolerance and plant growth. In conclusion it is merely important to further study the exact role of NPs at molecular level to confirm whether these molecules are involved in stress tolerance or stress induction.

Table 4.1 physiochemical effects of nanoparticles in plants against various Abiotic stresses

Stress Type	NPs.	Plant Name	Family	Physiological Effects on Host	References
Salinity	SiO_2	*Zea mays*	Poaceae	Formation of thing layer in cell wall, enhance resistance	Derosa et al. (2010) and Latef et al. (2018)
		Lycopersicon esculentum	Solanaceae	Better seed germination, the anti-oxidative enzyme activities, photosynthetic rate and water absorption capacity	Haghighi and Pourkhaloee (2013)
		Lycopersicon esculentum	Solanaceae	Reduced Na + ion concentration in cell wall, lower absorption of the ions	Savvasd et al. (2009)
		Cucurbita pepo	Cucurbitaceae	Better seed germination, the anti-oxidative enzyme activities, water absorption capacity and photosynthetic rate	Haghighi and Pourkhaloee (2013)
		Zea mays	Poaceae	Reducing Na + ion concentration in cell wall, lower absorption of the ions	Gao et al. (2006a, 2006b)
		Lens culinaris	Fabaceae	Increased germination rate and seedling growth	Sabaghnia and Janmohammadi (2014)
		Ocimum basilicum	Lamiaceae	Increased chlorophyll content, proline level and physiological traits	Kalteh et al. (2014)
		Cucurbita pepo	Cucurbitaceae	Decreased the electrolyte leakage,level of H_2O_2, MDA and chlorophyll degradation	Siddiqui et al. (2014)
	Fe_2O_3 NP	*Mentha piperita*	Lamiaceae	Increased fresh and dry weights and mineral contents, masked antioxidative pathway	Askary et al. (2017)
	Fe_2O_3 + ZnO	*Moringa peregrina*	Moringaceae	Increased leaf pigments, proline, carbohydrates, biomolecules and antioxidants	Soliman et al. (2015)
	ZnO NP	*Helianthus annuus*	Asteraceae	Increased chlorophyll content, photosynthesis rate, CO_2 concentration, osmotic regulation and decreased Na content	Torabian et al. (2016)
	Chitosan (S-nitroso-MSA-CS)	*Zea mays*	Poaceae	Ameliorating deleterious effects of salinity in photosystem II activity, and increased chlorophyll content and growth even at lower doses	Oliveira et al. (2016)

Drought	TiO_2 (rutile)	*Spinacia oleracea*	Amaranthaceae	Increased rubisco activase activity, chlorophyll synthesis, photosynthesis, increased dry weight	Gao et al. (2008)
		Vigna unguiculata L.	Fabaceae	Increased photosynthesis,	Owolade et al. (2008)
				Augmented overall seed yield	
		Triticum aestivum	Poaceae	Increase in plant height, ear number and weight, 1000-kernal weight and seeds/plant, harvest index, and starch and gluten content	Jaberzadeh et al. (2013)
		Dracocephalum moldavica	Lamiaceae	More proline level with less H_2O_2 and MAD	Mohammadi et al. (2014a, b)
				Mitigation of membrane damage and oxidative stress	
	SiO_2	*Vicia faba*	Fabaceae	Increase proline content,	Qados and Moftah (2015) and Qados (2015)
				Escalated catalase (CAT) and peroxidase (POD) activities	
				Alter physio-chemical processes	
		Solanum lycopersicum	Solanaceae	Increase proline content,	Siddiqui and Al-Whaibi (2014)
				Escalated catalase (CAT) and peroxidase (POD) activities	
				Alter physio-chemical processes	
		Medicago sativa	Fabaceae	Increase proline content,	Cakmak et al. (1996)
				Escalated catalase (CAT) and peroxidase (POD) activities	
				Alter physio-chemical processes	
		Crataegus monogyna	Rosaceae	Improved chlorophyll, carotenoid, carbohydrate and proline contents, and increased photosynthesis rate, MDA, (RWC) and membrane electrolyte leakage (ELI))	Ashkavand et al. (2015)
		Sorghum bicolor	Poaceae	Maintaining photosynthesis rate and improved root growth	Hattori et al. (2005)

(continued)

Table 4.1 (continued)

Stress Type	NPs.	Plant Name	Family	Physiological Effects on Host	References
	CuO	*Triticum aestivum*	Poaceae	Multiplication and elongation of root hair close to root tip	Yang et al. (2017)
		Arabidopsis thaliana	Brassicaceae	Change the water supply thus shrink the cell wall, increase lignification, cu-pectin association in cell wall	Nair and Chung (2017)
		Brassica juncea	Brassicaceae	Change the water supply thus shrink the cell wall, increase lignification, cu-pectin association in cell wall	Nair and Chung (2017)
	ZnO	*Triticum aestivum*	Poaceae	Increased lateral roots formation	Yang et al. (2017)
	AgNPs	*Lens culinaris*	Fabaceae	High germination rate and high growth and production parameters	Hojjat and Ganjali (2016)
	MWCNTs	*Hordeum vulgare*	Poaceae	Improved water absorption capacity, seedling water content	Karami and Sepehri (2017)
		Glycine max	Fabaceae	Increased antioxidants and high germination rate	Lahiani et al. (2013)
		Zea mays	Poaceae	Increased antioxidants and high germination rate	Liu et al. (2016)
		Triticum aestivum	Poaceae	High root and shoot growth	Srivastava and Rao (2014)
	CeO2	*Glycine max*	Fabaceae	Enhanced crop production	Cao et al. (2018)
	FeO2+ salicylic acid	*Fragaria ananassa*	Rosaceae	Improved growth parameters and increased leaf pigments level, RWC, MSI, iron and potassium level	Mozafari et al. (2018)

Heat	Ag NP	*Triticum aestivum*	Poaceae	Protected plants against thermal stress and improved plant growth	Husen et al. (2017)
	CeO2	*Zea mays*	Poaceae	Enhanced degeneration of H2O2 and upregulation of HSP70	Zhao et al. (2012)
	MWCNTs	*Lycopersicon esculentum*	Solanaceae	Up regulated the transcript level of various stress-related genes including HSP90	Khodakovskaya et al. (2011a, b)
	Se	*Lycopersicon esculentum*	Solanaceae	Increased chlorophyll content, hydration of plants, and growth	Haghighi et al. (2014) and Djanaguiraman et al. (2018)
	TiO2	*Lycopersicon esculentum*	Solanaceae	Promoted photosynthesis by regulating energy depletion, induced stomatal opening resulted into cooling of leaves	Qi et al. (2013)
Cold	TiO2	*Cicer arietinum*	Fabaceae	Increased antioxidative enzymes activities, decreased H2O2 level and electrolyte leakage	Mohammadi et al. (2013)
			Fabaceae	Upregulation of chlorophyll binding protein and rubisco genes, decreased in H2O2 level, enhanced activity of PEP carboxylase	Hasanpour et al. (2015)
	SiO2	*Agropyron elongatum*	Poaceae	Breakage of seed dormancy, enhanced germination and seedling weight	Azimi et al. (2014)
	Ag	*Arabidopsis thaliana*	Brassicaceae	Activated and enriched antioxidant genes	Kohan-Baghkheirati and Geisler-Lee (2015)
	Se	*Cucumis sativus L.*	Cucurbitaceae	Increased proline content in leaves, reduced lipid peroxidation in roots,	Hawrylak-Nowak et al. (2010)

(continued)

Table 4.1 (continued)

Stress Type	NPs.	Plant Name	Family	Physiological Effects on Host	References
Heavy metals Cd	ZnO	*Leucaena leucocephala*	Fabaceae	Elevated soluble proteins, chlorophyll and carotenoid content in leaves, and decrease in oxidative damaged to lipids membrane, induced level of antioxidative enzymes (CAT, SOD, POD) in leaves, lipid perodixadation	Venkatachalam et al. (2017)
	TiO2	*Leucaena leucocephala*	Fabaceae	Augmented growth and increased energy driven pathways, increased leaf pigments, relative water content,	Singh and Lee (2016)
	$Ca_5(PO_4)_3$	*Brassica juncea*	Brassicaceae	Nano scale protection against cd stress	Li and Huang (2014)
	Hydroxyapatite	*Brassica chinensis L.*	Brassicaceae	High biomass, level of chlorophyll and ascorbic acid, increased antioxidant activities (SOD, CAT, POD) decreased level of MDA	Li and Huang (2014)
Cr	Si NPs	*Pisum sativum*	Fabaceae	Activation of anti-oxidative pathways	Tripathi et al. (2015)
Cr (VI)	Na2SiO3	*Pisum sativum*	Fabaceae	Reduced assimilation of Cr(VI) and oxidative stress, enhanced antioxidative defense systems, and enriched accumulation of nutrient elements, improved growth	Tripathi et al. (2015)
Pb and Cu	ZnS quantum dots	*Chlorella kesslerii, Chlamydomonas reinhardtii*	Graphidaceae, Chlamydomonadaceae	Decreased intracellular Cu and Pb in walled strains, wall-less strains contained elevated Cu and Pb	Worms et al. (2012)

References

Abou-Zeid HM, Ismail GSM (2018) The role of priming with biosynthesized silver nanoparticles in the response of Triticum aestivum L. to salt stress. Egypt J Bot 58:73–85

Ahmad J, Bashir H, Bagheri R, Baig A, Al-Huqail A, Ibrahim MM, Qureshi MI (2017) Drought and salinity induced changes in ecophysiology and proteomic profile of Parthenium hysterophorus. PLoS One 12(9):0185118

Almutairi ZM (2016) Effect of nano-silicon application on the expression of salt tolerance genes in germinating tomato (Solanum lycopersicum L.) seedlings under salt stress. Plant Omics J 9:106–114

Ashkavand P, Tabari M, Zarafshar M, Tomášková I, Struve D (2015) Effect of SiO2 nanoparticles on drought resistance in hawthorn seedlings. Leśne Prace Badawcze/Forest Research Papers Grudzień 76(4):350–359

Askary M, Talebi SM, Amini F, Bangan ADB (2017) Effects of iron nanoparticles on Mentha piperita under salinity stress. Biologija 63:65–75

Azimi R, Borzelabad MJ, Feizi H, Azimi A (2014) Interaction of SiO2 nanoparticles with seed prechilling on germination and early seedling growth of tall wheatgrass (Agropyron elongatum L.). Pol J Chem Tech 16:25–29

Baybordi A (2005) Effect of zinc, iron, manganese and copper on wheat quality under salt stress conditions. J Water Soil 140:150–170

Beyene HD, Werkneh AA, Bezabh HK, Ambaye TG (2017) Synthesis paradigm and applications of silver nanoparticles (AgNPs), a review. Sustain Mater Technol 13:18–23

Boudsocq M, Sheen J (2013) CDPKs in immune and stress signaling. Trends Plant Sci 18:30–40

Bruna HCO, Gomes CR, Milena T, Pelegrino A, Seabra B (2016) Nitric oxide-releasing chitosan nanoparticles alleviate the effects of salt stress in maize plants. Nitric Oxide 61:10–19

Burklew CE, Ashlock J, Winfrey WB, Zhang B (2012) Effects of aluminum oxide nanoparticles on the growth, development, and microRNA expression of tobacco (Nicotiana tabacum). PLoS One 7:34783

Cakmak I, Yilmaz A, Torun B, Erenoglu B, Broun HJ (1996) Zinc deficiency as a critical nutritional problem in wheat production in Central Anatolia. Plant Soil 180:165–172

Cao Z, Rossi L, Stowers C, Zhang W, Lombardini L, Ma X (2018) The impact of cerium oxide nanoparticles on the physiology of soybean (Glycine max (L.) Merr.) under different soil moisture conditions. Environ Sci Pollut Res 25:930–939

Capuana M (2011) Heavy metals and woody plants biotechnologies for phytoremediation. J Biogeo Sci For 4:7–15

Carpenter AW, Worley BV, Slomberg DL, Schoenfisch MH (2012) Dual action antimicrobials: nitric oxide release from quaternary ammonium-functionalized silica nanoparticles. Biomacromolecules 13(10):3334–3342. https://doi.org/10.1021/bm301108x

Chandra S, Chakraborty N, Dasgupta A, Sarkar J, Panda K, Acharya K (2015) Chitosan nanoparticles: a positive modulator of innate immune responses in plants. Sci Rep 5:15195

Chen J, Liu X, Wang C, Yin SS, Li XL, Hu WJ, Simona M, Shen ZJ, Xiao Q, Chu CC, Peng XX, Zheng HL (2015) Nitric oxide ameliorates zinc oxide nanoparticles-induced phytotoxicity in rice seedlings. J Hazard Mater 297:173–182

Cheung WY (1980) Calmodulin plays a pivotal role in cellular regulation. Science 207:19–27

Chibuike GU, Obiora SC (2014) Heavy metal polluted soils: effect on plants and bioremediation methods. Appl Environ Soil Sci:1–13

Corpas FJ, Barroso JB, Carreras A, Quiros M, Leon AM, Romero-Puertas MC et al (2004) Cellular and subcellular localization of endogenous nitric oxide in young and senescent pea plants. Plant Physiol 136:2722–2733

Del Rio LA, Corpas FJ, Barroso JB (2004) Nitric oxide and nitric oxide synthase activity in plants. Phytochemistry 65:783–792

Delk NA, Johnson KA, Chowdhury NI, Braam J (2005) CML24, regulated in expression by diverse stimuli, encodes a potential Ca2þ sensor that functions in response to abscisic acid, daylength, and ion stress. Plant Physiol 139:240–253

Derosa MR, Monreal C, Schmitzer M, Walsh R, Sultan Y (2010) Nanotechnology in fertilizers. Nat Nanotechnol 1:193–225

Dimkpa CO, McLean JE, Latta DE, Manang6n E, Britt DW, Johnson WP, Anderson AJ (2012) CuO and ZnO nanoparticles: phytotoxicity, metal speciation, and induction of oxidative stress in sand-grown wheat. J Nanopart Res 14(9):1125

Djanaguiraman M, Boyle DL, Welti R, Jagadish SVK, Prasad PVV (2018) Decreased photosynthetic rate under high temperature in wheat is due to lipid desaturation, oxidation, acylation, and damage of organelles. BMC plant biology 18(1):1–17

Dos Reis SP, Marques DN, Lima AM, de Souza CRB (2016) Plant molecular adaptations and strategies under drought stress. In: Drought stress tolerance in plants. Springer, Cham, pp 91–122

Frazier TP, Burklew CE, Zhang B (2014) Titanium dioxide nanoparticles affect the growth and microRNA expression of tobacco (Nicotiana tabacum). Funct Integr Genom 14:75–83

Gao FQ, Hong FS, Liu C, Zheng L, Su MY, Wu X, Yang F, Wu C, Yang P (2006a) Mechanism of nanoanatase TiO2 on promoting photosynthetic carbon reaction of spinach: inducing complex of Rubisco-Rubisco activase. Biol Trace Elem Res 11:239–254

Gao X, Zou CH, Wang L, Zhang F (2006b) Silicon decreases transpiration rate and conductance from stomata of maize plants. J Plant Nutr 29:1637–1647

Gao F, Liu C, Qu C, Zheng L, Yang F, Su M, Hong F (2008) Was improvement of spinach growth by nano-TiO2 treatment related to the changes of rubisco activase? BioMetals 21(2):211–217

García-Sanchez S, Bernales I, Cristobal S (2015) Early response to nanoparticles in the Arabidopsis transcriptome compromises plant defence and root-hair development through salicylic acid signaling. BMC Genomics 16:341

Giacomo B, Forino LMC, Tagliasacchi AM, Bernardi R, Durante M (2010) Ozone damage and tolerance in leaves of two poplar genotypes. Caryologia 63:422–434

Giraldo JP, Landry MP, Faltermeier SM, McNicholas TP, Iverson NM, Boghossian AA, Reuel NF, Hilmer AJ, Sen F, Brew JA, Strano MS (2014) Plant nanobionics approach to augment photosynthesis and biochemical sensing. Nat Mater 13. https://doi.org/10.1038/NMAT3890

Goyer RA (1995) Nutrition and metal toxicity. Am J Clin Nutr 61:646S–650S

Gunjan B, Zaidi MGH, Sandeep A (2014) Impact of gold nanoparticles on physiological and biochemical characteristics of Brassica juncea. J Plant Biochem Physiol 2:133

Haghighi M, Pourkhaloee A (2013) Nanoparticles in agricultural soils: their risks and benefits for seed germination. Minerva Biotecnol 25:123–132

Haghighi M, Abolghasemi R, Teixeira da Silva JA (2014) Low and high temperature stress affect the growth characteristics of tomato in hydroponic culture with Se and nano-Se amendment. Sci Hortic 178:231–240

Hartikainen H, Xue T, Piironen V (2000) Selenium as an antioxidant and prooxidant in rye grass. Plant Soil 225:193–200

Hasanpour H, Maali-Amiri R, Zeinali H (2015) Effect of TiO2 nanoparticles on metabolic limitations to photosynthesis under cold in chickpea. Russ J Plant Physiol 62:779–787

Hasanuzzaman M, Nahar K, Fujita M (2013) Extreme temperature responses, oxidative stress and antioxidant defense in plants. In: Vahdati K, Leslie C (eds) Abiotic stress – plant responses and applications in agriculture. InTech Open Access Publisher

Hasanuzzaman M, Nahar K, Fujita M (2014) Silicon and selenium: two vital trace elements that confer abiotic stress tolerance to plants. Emerging technologies and management of crop stress tolerance. Elsevier, The Netherlands, pp 377–422

Hattori T, Inanaga S, Araki H, An P, Morita S, Luxová M et al (2005) Application of silicon enhanced drought tolerance in Sorghum bicolour. Physiol Plant 123:459–466

Hawrylak-Nowak B, Matraszek R, Szymanska M (2010) Selenium modifies the effect of short-term chilling stress on cucumber plants. Biol Trace Elem Res 138:307–315

Heidarvand L, Maali-Amiri R, Naghavi MR, Farayedi Y, Sadeghzadeh B, Alizadeh KH (2011) Physiological and morphological characteristics of chickpea accessions under low temperature stress. Russ J Plant Physiol 58:157–163

Hernandez-Hernandez H, Gonzalez-Morales S, Benavides-Mendoza A, Ortega-Ortiz H, Cadenas-Pliego G, Juarez-Maldonado A (2018) Effects of chitosan–PVA and cu nanoparticles on the growth and antioxidant capacity of tomato under saline stress. Molecules 23:178

Hojjat SS, Ganjali A (2016) The effect of silver nanoparticle on lentil seed germination under drought stress. Int J Farm Allied Sci 5(3):208–212

Husen A, Iqbal M, Aref IM (2017) Plant growth and foliar characteristics of faba bean (Vicia faba L.) as affected by indole-acetic acid under water-sufficient and water-deficient conditions. Journal of Environmental Biology, 38(2):179

Hussain S, Hussain S, Qadir T, Khaliq A, Ashraf U, Parveen A, Rafiq M (2019) Drought stress in plants: an overview on implications, tolerance mechanisms and agronomic mitigation strategies. Plant Sci Today 6(4):389–402

Ismail A, Takeda S, Nick P (2014) Life and death under salt stress: same players, different timing? J Exp Bot 65:2963–2979

Jaberzadeh A, Payam M, Hamid R, Tohidi M, Hossein Z (2013) Influence of bulk and nanoparticles titanium foliar application on some agronomic traits, seed gluten and starch contents of wheat subjected to water deficit stress. Not Bot Horti Agrobot Cluj-Na 41:201–207

Kalteh M, Alipour ZT, Ashraf S, Aliabadi MM, Nosratabadi AF (2014) Effect of silica nanoparticles on basil (ocimum basilicum) under salinity stress. J Chem Health Risk 4:49–55

Karami A, Sepehri A (2017) Multiwalled carbon nanotubes and nitric oxide modulate the germination and early seedling growth of barley under drought and salinity. Agric Conspec Sci 82:331–339

Karuppanapandian T, Wang HW, Prabakaran N, Jeyalakshmi K, Kwon M, Manoharan K, Kim W (2011) 2,4-dichlorophenoxyacetic acid-induced leafsenescence in mung bean (Vigna radiate (L.) Wilczek) and senescence inhibition by co-treatment with silver nanoparticles. Plant Physiol Biochem 49:168–217

Kaveh R, Li YS, Ranjbar S, Tehrani R, Brueck CL, Aken BV (2013) Changes in Arabidopsis thaliana gene expression in response to silver nanoparticles and silver ions. Environ Sci Technol 47:10637–10,644

Khan MN, Siddiqui MH, Mohammad F, Naeem M, Khan MMA (2010) Calcium chloride and gibberellic acid protect Linseed (Linum usitatissimum L.) from NaCl stress by inducing antioxidative defence system and osmoprotectant accumulation. Acta Physiol Plant 32:121–132

Khan MN, Siddiqui MH, Mohammad F, Naeem M (2012) Interactive role of nitric oxide and calcium chloride in the tolerance of plants to salt stress. Nitric Oxide 27:210–218

Khan MIR, Syeez S, Nazar R, Anjum NA (2012a) An insight into the role of salicylic acid and jasmonic acid in salt stress tolerance. In: Khan NA, Nazar R, Iqbal N, Anjum NA (eds) Phytohormones and abiotic stress tolerance in plants. Springer, pp 277–300

Khan MIR, Asgher M, Khan NA (2014a) Alleviation of salt induced photosynthesis and growth inhibition by salicylic acid involves glycinebetaine and ethylene in mungbean (Vigna radiate L.). Plant Physiol Biochem 80:67–74

Khan MN, Mohammad F, Mobin M, Saqib MA (2014b) Tolerance of plants to abiotic stress: a role of nitric oxide and calcium. In: Khan MN, Mobin M, Mohammad F, Corpas FJ (eds) Nitric oxide in plants: Metabolism and role. Springer

Khan MN, Mobin M, Abbas ZK, AlMutairi KA, Siddiqui ZH (2017) Role of nanomaterials in plants under challenging environments. Plant Physiol Biochem 110:194–209

Khodakovskaya MV, de Silva K, Nedosekin DA, Dervishi E, Biris AS, Shashkov EV, Ekaterina IG, Zharov VP (2011a) Complex genetic, photo thermal, and photo acoustic analysis of nanoparticle-plant interactions. Proc Natl Acad Sci U S A 108:1028–1033

Khodakovskaya MV, de Silva K, Nedosekin DA, Dervishi E, Biris AS, Shashkov EV, Galanzha EI, Zharov VP (2011b) Complex genetic, photothermal, and photoacoustic analysis of nanoparticle-plant interactions. Proc Natl Acad Sci U S A 108:1028–1033

Khodakovskaya MV, de Silva K, Biris AS, Dervishi E, Villagarcia H (2012) Carbon nanotubes induce growth enhancement of tobacco cells. ACS Nano 6:2128–2135

Kim JH, Oh Y, Yoon H, Hwang I, Chang YS (2015) Iron nanoparticle-induced activation of plasma membrane Hþ-ATPase promotes stomatal opening in Arabidopsis thaliana. Environ Sci Technol 49:1113–1119

Kohan-Baghkheirati E, Geisler-Lee J (2015) Gene expression, protein function and pathways of Arabidopsis thaliana responding to silver nanoparticles in comparison to silver ions, cold, salt, drought, and heat. Nanomaterials 5:436–467

Lahiani MH, Dervishi E, Chen J, Nima Z, Gaume A, Biris AS, Khodakovskaya MV (2013) Impact of carbon nanotube exposure to seeds of valuable crops. ACS Appl Mater Interfaces 5(16):7965–7973

Lamotte O, Courtois C, Dobrowolska G, Besson A, Pugin A, Wendehenne D (2006) Mechanism of nitric-oxide-induced increase of free cytosolic Ca2þ concentration in Nicotiana plumbaginifolia cells. Free Radic Biol Med 40:1369–1376

Latef AAHA, Srivastava AK, El-Sadek MSA, Kordrostami M, Tran LP (2018) Titanium dioxide nanoparticles improve growth and enhance tolerance of broad bean plants under saline soil conditions. Land Degrad Dev 29:1065–1073

Li Z, Huang J (2014) Effects of nanoparticle hydroxyapatite on growth and antioxidant system in pakchoi (Brassica chinensis L.) from cadmium-contaminated soil. J Nanomater:1–7

Liu YF, Qi MF, Li TL (2012) Photosynthesis, photoinhibition, and antioxidant system in tomato leaves stressed by low night temperature and their subsequent recovery. Plant Sci 196:8–17

Liu FY, Xiong FX, Fan YK, Li J, Wang HZ, Xing GM, He R (2016) Facile and scalable fabrication engineering of fullerenol nanoparticles by improved alkaline-oxidation approach and its antioxidant potential in maize. J Nanopart Res 18(11):338

Lopez CJ, Banowetz GM, Peterson CJ, Kronstad WE (2003) Dehydrin expression and drought tolerance in seven wheat cultivars. Crop Sci 43:577–582

Ma X, Geiser-Lee J, Deng Y, Kolmakov A (2010) Interactions between engineered nanoparticles (ENPs) and plants: phytotoxicity, uptake and accumulation. Sci Total Environ 408:3053–3061

Maali Amiri R, Yur'eva NO, Shimshilashvili KR, Goldenkova-Pavlova IV, Pchelkin VP, Kuznitsova EI et al (2010) Expression of acyl-lipid D 12-desaturase gene in prokaryotic and eukaryotic cells and its effect on cold stress tolerance of potato. J Integr Plant Biol 52:289–297

Marmiroli M, Imperiale D, Pagano L, Villani M, Zappettini A, Marmiroli N (2015) The Proteomic response of Arabidopsis thaliana to cadmium sulphide quantum dots, and its correlation with the transcriptomic response. Front Plant Sci 6:1104

Martinez-Ballesta MC, Zapata L, Chalbi N, Carvajal M (2016) Multiwalled carbon nanotubes enter broccoli cells enhancing growth and water uptake of plants exposed to salinity. J Nanobiotech 14:42

Miao Y, Xu J, Shen Y, Chen L, Bian Y, Hu Y et al (2014) Nanoparticle as signalling protein mimic: robust structural and functional modulation of CaMKII upon specific binding to fullerene C60 nanocrystals. ACS Nano 8:6131–6144

Mirzajani F, Askari H, Hamzelou S, Schober Y, Rhompp A, Ghassempour A et al (2014) Proteomics study of silver nanoparticles toxicity on Oryza sativa L. Ecotoxicol Environ Saf 108:335–339

Mohamed AKSH, Qayyum MF, Abdel-Hadi AM, Rehman RA, Ali S, Rizwan M (2017) Interactive effect of salinity and silver nanoparticles on photosynthetic and biochemical parameters of wheat. Arch Agron Soil Sci 63(12):1736–1747

Mohammadi R, Maali-Amiri R, Abbasi A (2013) Effect of TiO2 nanoparticles on chickpea response to cold stress. Biol Trace Elem Res 152:403–410

Mohammadi H, Esmailpour M, Gheranpaye A (2014a) Effects of TiO2 nanoparticles and water-deficit stress on morpho-physiological characteristics of dragonhead (Dracocephalum moldavica L.) plants. Acta Agri Slovenica 107(2):385–396

Mohammadi R, MaaliAmiri R, Mantri N (2014b) Effect of TiO2 nanoparticles on oxidative damage and antioxidant defense systems in chickpea seedlings during cold stress. Russ J Plant Physiol 61:768–775

Moller IM, Jensen PE, Hansson A (2007) Oxidative modifications to cellular components in plants. Ann Rev Plant Biol 58:459–481

Mozafari AA, Havas F, Ghaderi N (2018) Application of iron nanoparticles and salicylic acid in in vitro culture of strawberries (Fragaria × ananassa Duch.) to cope with drought stress. Plant Cell Tissue Organ Cult 132:511–523

Nair PMG, Chung IM (2017) Regulation of morphological, molecular and nutrient status in Arabidopsis thaliana seedlings in response to ZnO nanoparticles and Zn ion exposure. Sci Total Environ 575:187–198

Oliveira HC, Gomes BC, Pelegrino MT, Seabra AB (2016) Nitric oxide-releasing chitosan nanoparticles alleviate the effects of salt stress in maize plants. Nitric Oxide 61:10–19

Oukarroum A, Bras S, Perreault F, Popovic R (2012) Inhibitory effects of silver nanoparticles in two green algae, Chlorella vulgaris and Dunaliella tertiolecta. Ecotoxicol Environ Saf 78:80–85

Owolade OF, Ogunleti DO, Adenekan MO (2008) Titanium dioxide affected diseases, development and yield of edible cowpea. Elec J Environ Agricult Food Chem 7(5):2942–2947

Paleg LG, Stewart GR, Bradbeer JW (1984) Proline and glycine betaine influence protein solvation. Plant Physiol 75:974–978

Prasad PVV, Pisipati SR, Mom I, Ristic Z (2011) Independent and combined effects of high temperature and drought stress during grain filling on plant yield and chloroplast EF-Tu expression in spring wheat. J Agron Crop Sci 197:430–441

Qados AMSA (2015) Mechanism of nanosilicon-mediated alleviation of salinity stress in faba bean (Vicia faba L.) plants. Am J Exp Agric 7:78–95

Qados AMSA, Moftah AE (2015) Influence of silicon and nano-silicon on germination, growth and yield of faba bean (Vicia faba L.) under salt stress conditions. Am J Exp Agric 5:509–524

Qi M, Liu Y, Li T (2013) Nano-TiO2 improve the photosynthesis of tomato leaves under mild heat stress. Biol Trace Elem Res 156:323–328

Rahimi R, Mohammakhani A, Roohi V, Armand N (2012) Effects of salt stress and silicon nutrition on cholorophyll content, yield and yield components in fennel (Foeniculum vulgare Mill.). Int J Agric Crop Sci 4:1591–1595

Rascio N, Navari-Izzo F (2011) Heavy metal hyperaccumulating plants: how and why do they do it? and what makes them so interesting? Plant Sci 180:169–181

Sabaghnia N, Janmohammadi M (2014) Effect of nano-silicon particles application on salinity tolerance in early growth of some lentil genotypes. Ann UMCS Biol 69:39–55

Savicka M, Skute N (2010) Effects of high temperature on malondialdehyde content, superoxide production and growth changes in wheat seedlings (Triticum aestivum L.). Ekologija 56:26–33

Savvasd G, Giotes D, Chatzieustratiou E, Bakea M, Patakioutad G (2009) Silicon supply in soilless cultivation of Zucchini alleviates stressinduced by salinity and powdery mildew infection. Environ Exp Bot 65:11–17

Schulze ED, Beck E, Buchmann N, Clemens S, Müller-Hohenstein K, Scherer-Lorenzen M (2019) Water deficiency (Drought). In: Plant ecology. Springer, Berlin, pp 165–202

Shahrokh S, Hosseinkhani B, Emtiazi G (2014) The impact of silver nanoparticles on bacterial aerobic nitrate reduction process. J Bioprocess Biotech 4:152

Sharma P, Jha AB, Dubey RS, Pessarakli M (2012) Reactive oxygen species, oxidative damage, and antioxidant defense mechanisms in plants under stressful conditions. J Bot:1–26

Siddiqui MH, Al-Whaibi MH (2014) Role of nano-SiO2 in germination of tomato (Lycopersicum esculentum seeds Mill.). Saudi J Biolog Sci 21:13–17

Siddiqui MH, Al-Whaibi MH, Faisal M, Al Sahli AA (2014) Nano-silicon dioxide mitigates the adverse effects of salt stress on Cucurbita pepo L. Environ Toxicol Chem 33:2429–2437

Simon DF, Domingos RF, Hauser C, Hutchins CM, Zerges W, Wilkinson KJ (2013) Transcriptome sequencing (RNA-seq) analysis of the effects of metal nanoparticle exposure on the transcriptome of Chlamydomonas reinhardtii. Appl Environ Microbiol 79:4774–4785

Singh J, Lee BK (2016) Influence of nano-TiO2 particles on the bioaccumulation of Cd in soybean plants (Glycine max): a possible mechanism for the removal of Cd from the contaminated soil. J Environ Manag 170:88–96

Singh S, Vishwakarma K, Singh S, Sharma S, Dubey NK, Singh VK, Liu S, Tripathi DK, Chauhan DK (2017) Understanding the plant and nanoparticle interface at transcriptomic and proteomic level: a concentric overview. Plant Gene 11:265–272

Soliman AS, El-feky SA, Darwish E (2015) Alleviation of salt stress on Moringa peregrina using foliar application of nanofertilizers. J Hortic For 7:36–47

Srivastava A, Rao DP (2014) Enhancement of seed germination and plant growth of wheat, maize, peanut and garlic using multiwalled carbon nanotubes. Eur Chem Bull 3(5):502–504

Suriyaprabha R, Karunakaran G, Yuvakkumar R, Prabu P, Rajendran V, Kannan N (2012) Growth and physiological responses of maize (Zea mays L.) to porous silica nanoparticles in soil. J Nanopart Res 14:1–14

Suzuki K, Nagasuga K, Okada M (2008) The chilling injury induced by high root temperature in the leaves of rice seedlings. Plant Cell Physiol 49:433–442

Torabian S, Zahedi M, Khoshgoftar AH (2016) Effects of foliar spray of two kinds of zinc oxide on the growth and ion concentration of sunflower cultivars under salt stress. J Plant Nutr 39:172–180

Tripathi DK, Singh VP, Prasad SM, Chauhan DK, Dubey NK (2015) Silicon nanoparticles (SiNp) alleviate chromium (VI) phytotoxicity in Pisum sativum (L.) seedlings. Plant Physiol Biochemist 96:189–198

Tuteja N, Mahajan S (2007) Calcium signalling network in plants. Plant Sig Behav 2:79–85

Van Hoecke K, De Schamphelaere KA, Van der Meeren P, Lucas S, Janssen CR (2008) Ecotoxicity of silica nanoparticles to the green alga Pseudokirchneriella subcapitata: importance of surface area. Environ Toxicol Chem 9:1948–1957

Vannini C, Domingo G, Onelli E, Prinsi B, Marsoni M, Espen L, Bracale M (2013) Morphological and proteomic responses of Eruca sativa exposed to silver nanoparticles or silver nitrate. PLoS One 8:68752

Venkatachalam P, Jayaraj M, Manikandan R, Geetha N, Rene ER, Sharma NC, Sahi SV (2017) Zinc oxide nanoparticles (ZnONPs) alleviate heavy metal-induced toxicity in Leucaena leucocephala seedlings: a physiochemical analysis. Plant Physiol Biochem 110:59–69

Wahid A (2007) Physiological implications of metabolites biosynthesis in net assimilation and heat stress tolerance of sugarcane (Saccharum officinarum) sprouts. J Plant Res 120:219–228

Wahid A, Gelani S, Ashraf M, Foolad M (2007) Heat tolerance in plants: an overview. Environ Exp Bot 61(3):199–223

Wei H, Wang E (2013) Nanomaterials with enzyme-like characteristics (nanozymes): next-generation artificial enzymes. Chem Soc Rev 42:6060–6093

Welti R, Li W, Li M, Sang Y, Biesiada H, Zhou HE, Rajashekar C, Williams TD, Wang X (2002) Profiling membrane lipids in plant stress responses. Role of phospholipase Da in freezing induced lipid changes in Arabidopsis. J Biol Chem 277:31994–32,002

Worms IAM, Boltzman J, Garcia M, Slaveykova VI (2012) Cell-wall-dependent effect of carboxyl-CdSe/ZnS quantum dots on lead and copper availability to green microalgae. Environmental pollution 167:27–33

Xu GY, Rocha P, Wang ML, Xu ML, Cui YC, Li LY, Zhu YX, Xia X (2011) A novel rice calmodulin-like gene, OsMSR2, enhances drought and salt tolerance and increases ABA sensitivity in Arabidopsis. Planta 234:47–59

Xu J, Yang J, Duan X, Jiang Y, Zhang P (2014) Increased expression of native cytosolic Cu/Zn superoxide dismutase and ascorbate peroxidase improves tolerance to oxidative and chilling stresses in cassava (Manihot esculenta Crantz). BMC Plant Biol 14:208

Yang KY, Doxey S, McLean JE, Britt D, Watson A, Al Qassy D, Jacobson AR, Anderson A (2017) Remodeling of root morphology by CuO and ZnO nanoparticles: effects on drought tolerance for plants colonized by a beneficial pseudomonad. Botany 96(3):175–186

Yordanova R, Popova L (2007) Effect of exogenous treatment with salicylic acid on photosynthetic activity and antioxidant capacity of chilled wheat plants. Gen Appl Plant Physiol 33:155–170

Ze Y, Liu C, Wang L, Hong M, Hong F (2011) The regulation of TiO2 nanoparticles on the expression of light-harvesting complex II and photosynthesis of chloroplasts of Arabidopsis thaliana. Biol Trace Elem Res 143:1131–1141

Zhao L, Peng B, Hernandez-Viezcas JA, Rico C, Sun Y, Peralta-Videa JR, Tang X, Niu G, Jin L, Ramirez AV, Zhang JY, Gardea-Torresdey JL (2012) Stress response and tolerance of Zea mays to CeO2 nanoparticles: cross talk among H2O2, heat shock protein and lipid peroxidation. ACS Nano 6:9615–9622

Zhu J, Wei G, Li J, Qian Q, Yu J (2004) Silicon alleviates salt stress and increases antioxidant enzymes activity in leaves of salt-stressed cucumber (Cucumis sativus L.). Plant Sci 167:527–533

Chapter 5
Alleviation Mechanism of Drought Stress in Plants Using Metal Nanoparticles – A Perspective Analysis

Iqra Naseer, Sumera Javad, Ajit Singh, Saba Maqsood, Sumera Iqbal, and Khajista Jabeen

Abstract Drought stress is the dilemma and most of the human population will be exposed to its worse effects in coming years. Global production of food crops has been reduced due to drought stress. This is an alarming situation for economies to feed their population. Researchers are working on different parameters to address this problem for ever-increasing food demand of world population. One of them is the use of metallic nanoparticles e.g., zinc, iron, titanium etc. (size less than 100 nm) which are center of interest at time for solving the issue of drought stress for plants. These nanoparticles have higher surface volume as compared to their size which make them more active and more penetrative. These particles are not only used as readily absorbable nutritive elements for plants but also reduce the oxidative stress, produced due to drought by increasing the photosynthetic efficiency of drought affected plants. NPs have also been known to induce gene expression to tolerate drought stress. These nanoparticles can be supplied to plants in form of seed coating, soil drench or foliar spray according to their mode of action. In present chapter role of metal nanoparticles will be discussed for alleviation of drought stress from food crops.

Keywords Drought · Fertilizer · Iron nanoparticles · Nanoparticles · Water stress · Zinc nanoparticles

---

I. Naseer · S. Javad (✉) · S. Maqsood · S. Iqbal · K. Jabeen
Department of Botany, Lahore College for Women University, Lahore, Pakistan

A. Singh
Malaysia School of Biosciences, University of Nottingham, Semenyih, Malaysia
e-mail: ajit.singh@nottingham.edu.my

M. Faizan et al. (eds.), *Sustainable Agriculture Reviews 53*, Sustainable Agriculture Reviews 53, https://doi.org/10.1007/978-3-030-86876-5_5

5.1 Introduction

In the field of agronomy, high quality agricultural applications and related technology, not only increase the product, but also facilitate the associated individuals and farmers economically, by lessening the inflation rate and encouraging food security. This results in financial stability and food security of a region. While, wrong choices of agricultural techniques like fertilizers and cheap pesticides which can reduce crop yield and production on large scale in very short time with pollution effects (Aissa et al. 2018; Barnabás et al. 2008). Certain climatic problems such as increased temperatures, unusual long winters/summers, salinity, water logging, pest attack and water stress also affect the agriculture. These climatic changes or disasters can't be avoided but can be addressed with proper use of knowledge and technology.

Water stress or drought condition is one of the major climatic catastrophe affecting crops globally. Drought can be defined as a water deficit phenomenon in which loss of water exceeds from aerial parts of the plant in comparison to the water absorbed by the roots. The reasons of this abiotic stress could be that

1. water is not retained in the soil particles,
2. lower water level in soil and
3. lesser precipitation (Salehi-Lisar and Bakhshayeshan-Agdam 2016; Lambers et al. 2008).

Hence, plants have to come up with this water deficiency to preserve their hydric status by changing their morphology and physiology (Chaves et al. 2009).

About one third of the land on globe is semi-arid and arid and most of the other land is affected with drought. Drought is also known to be one of the main reasons of low grain quality and lesser crop production. Water deficiency is detrimental to growth of plants and may lead to many social and economic problems. Different aspects of plant growth and related parameters such as photosynthetic and respiratory machinery, cell membranes, lipid and protein production have been badly affected by the drought. It also affects DNA, carbohydrates and production of reactive oxygen species (Caverzan et al. 2016).

All over the world, problem of water scarcity has worsened due to continuous change in climate, rapidly increasing global population and water pollution. Research and techniques are required in forms of smart agriculture to cope with such disasters. Advancement in science and technology has improved human life with the enhanced potential of exploring natural resources (Trenberth et al. 2014). Nanotechnology is one of the extension of science which has the potential to help mankind to cope with disasters like water stress. In this chapter, hazards related to drought stress on plants and their possible solution with nanotechnology is discussed.

5.2 Drought as Limiting Factor for Crop Production

In the past, drought was the cause of great famines. Water scarcity and a rapid increase in the world population are two main serious problems which have endangered the world food security on one hand and on the other hand, has provoked added to the limiting effects of the drought (Somerville and Briscoe 2001). Drought severity is quite an unpredictable response of an environment sometimes in some areas which have never faced the problem before. Because it is dependent upon rainfall, water storage capacity of soils and rate of evaporation (Wery et al. 1994).

Early stages of plant growth and development such as seed germination, coleoptiles length and health are reduced due to drought stress. It limits and alters the many characteristics of the plants. One of the main factors of plant progression is seed germination which is sensitive to drought stress. Significant changes at the germination stage of many plants including maize (Queiroz et al. 2019), wheat (Qayyum et al. 2011) and sorghum (Patanè et al. 2013) were observed. Plants which usually bear exposure to drought stress have shown decreased plant height, wilting in leaves and delayed buds and flower formation (Fig. 5.1). Stem length is also reduced due to the limited availability of the nutrients and soil water to the plants in water stress conditions (Razmjoo et al. 2008).

Drought conditions make the plants to change their root structure for water absorption from the deep soil layers (Asadi et al. 2012). Roots are part of many important activities of plants such as nutrient and water absorption and interaction with microorganisms as symbiosis in rhizosphere. In addition, water availability is mainly recognized by the roots, as a result, root growth and characters (length, spread, number, lateral roots) are regulated (Salazar et al. 2015). Thus, water scarcity increases root length in plants. Strong root system of the plant give rise to better plant growth especially in earlier stages of plant development (Smith and Smet 2012). Increased root growth in drought, if possible, causes soil water retention and enhanced nutrient accumulation which in turn increases plant biomass production (Zulfiqar et al. 2020). Drought significantly decreases plant biomass but plant root to shoot ratio improves to cope up stress conditions which can be further modulated by using different techniques.

The leaf is photosynthetic part of the plant and another important part which is badly affected by the drought condition usually. It has been reported in literature that number of leaves are decreased under drought stress as reported in *Andrographi spaniculate* (Kapoor et al. 2020). Photosynthesis requires good maintained leaf area and development which is the key factor for plant growth. During drought conditions, rate of photosynthesis is usually reduced as a result of lesser water absorption in roots (Srivastava et al. 2014). Leaf area is also reduced after facing a long term drought situation because of decrease in cell division and cell turgidity. Reduction in leaf area as a result of drought also helps to reduce the rate of transpiration from leaves, thus overall reducing the negative effects of drought (Bangar et al. 2019).

There are a lot of treatments to help plants in competing stresses like drought. These include breeding techniques, genetic engineering, chemical additions and a

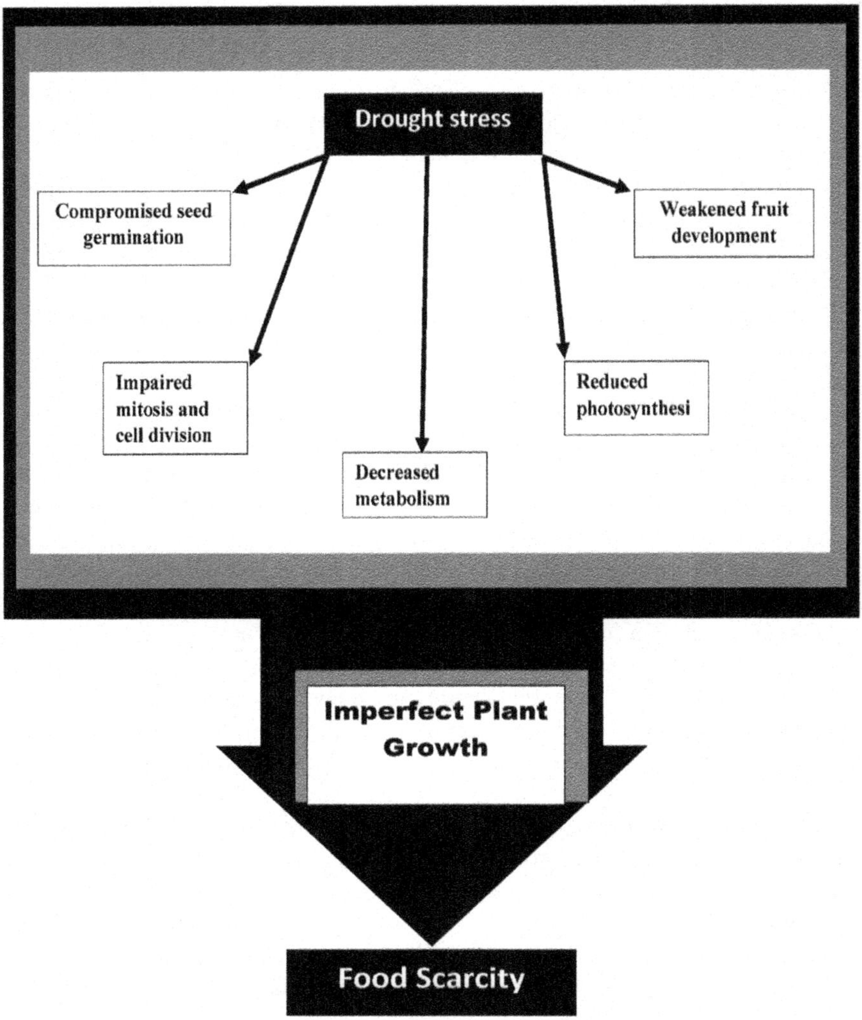

Fig. 5.1 Drought stress, a channel towards food scarcity

lot more with their own advantages and disadvantages. Nanotechnology is an easy to opt modern technology, showing its deep benefits. For most of the times, nanoparticle treatment help to enhance the existing mechanism of plants to fight against stresses. So here an overview has been presented about role of nanoparticles in enhancing and boosting the defense mechanism of plants.

5.3 Nanotechnology as Drought Resistant Technique

Water, being an important element for survival of plants and having an essential role in transportation of nutrients, can bring drought stress if deficient, which can further result in weakened strength of plants. A significant effort is promised by nanotechnology in the mitigation of drought stresses. Various novel studies have assessed mediation of nanoparticle in variety of stresses (Barrena et al. 2009). These particles are engineered compositions that have a diameter or length in nanometer measurement and at least one side measurement is less than 100 nm (Khan et al. 2019). These nanoparticles can be synthesized by inorganic, organic or biological ingredients. With regard to their distinctive physical and chemical characteristics, NPs are widely utilized in agriculture and industry, which has paved the way for enhanced number of processes and products. There are various uses of NPs in different areas i.e., for industrial (Aziz et al. 2015), human health (Aziz et al. 2016), food science processes (Saxena 2009), and agricultural (Mishra et al. 2014). Yet, nanoparticles can interact with biological systems depending upon their surface characteristics, mechanical, physical and chemical properties (Klaine et al. 2008). This versatile area of nanotechnology has retained its usage in almost all the current disciplines of science.

For instance, in the area of agriculture, NPs are being utilized as nano-fertilizers, nano-pesticides, nano-sensors, etc. (Iqbal et al. 2020). NPs application as nano-fertilizers can increase various growth parameters of plant like enhanced germination and growth of seedling. These have also positive effects on physiological activities of plants as well like nitrogen metabolism and photosynthesis and including actions of APX, POX and CAT inside the leaf tissue. Moreover, nanoparticles can cause positive alterations in gene expression of plants which is an evidence that NPs can be used in improvement of crops. There are a number of researches showing significant impact of NPs conc. Size and shape on sugar, chlorophyll and protein content.

Nanoparticles (NPs) exhibit large surface area, enhanced reactivity, and changing morphology of particles. Many metal NPs or their oxides have been detected to prevent biotic and abiotic stresses. Still, there is much research needed to create the technology for the achievement of viable agriculture (Saxena et al. 2016).

NPs also seem to increase the tolerance to water stress through the enhancement of hydraulic conductance of root and uptake of water in plants. NPs also assist other proteins which fight drought by oxidation-reduction, signaling of stress, ROS detoxification, and hormonal pathways. Nanoparticles possess very high mobility, which allows quick nutrient transport to all parts of plant. Particularly, concrete investigation and exploration of facts and mechanisms is required to find the roles of nanoparticles in stress conditions of plants (Das and Das 2019).

5.3.1 Mechanism Involved

It is already a known fact that supplementation of minerals like zinc, copper, magnesium, cobalt, iron and nickel, can significantly enhance crop yields in trauma environments of stress (Ashraf et al. 2012; Jaleel et al. 2008). Nanoparticles with their larger surface area and smaller size present a better treatment for drought stress to plants as compared to their macro-size salts, being conventionally used. Nanomaterials when used as nanofertilizers or nanopesticides, they present a great solution for efficient use of resources. If they are used in a targeted manner, they can give an outstanding output in terms of plant yield and growth. They can also prove their effective role for plants in coping with issues related to soil plant interaction in drought resistance (Ragab and Saad-Allah 2020; Prasad et al. 2014).

5.3.2 Role of Nanoparticles

Some of the essential functions of nanoparticles to reduce the effects of stress are given below (Dimkpa et al. 2019; Khan et al. 2017; Taran et al. 2017)

1. Boosting of antioxidant structures and mechanisms in plant cells and tissues
2. Enhancing seed germination
3. Toxic metal Co-precipitation with Silicon
4. Ions of toxic metal are Immobilized in growth media
5. Enhanced Nitrogen Acquisition
6. Stabilizing photosynthetic components
7. Plasma membrane ATPase activation
8. Stomatal activity maintenance
9. Activation of stress related genes
10. Change in uptake processes for water and minerals
11. Compartmentation of metal ions within plants

5.4 Plant Adaptations to Drought Stress

Plants have tendency to adapt their own pathways at physiological, biochemical and molecular stages as a defense mechanism to cope with climatic changes throughout their life cycle (Fig. 5.2). Plants alter gene expression at molecular level to overcome these environmental stresses. Drought affects growth and yield parameters of the plant very badly. Breeding is done to get plant varieties with drought tolerance at physiological and morphological level. Thus, primary purpose of breeding crops is to enhance the survival rate, growth and development of plants in water stress situations. Water Use Efficiency (WUE) is considered as significant selection quality to measure the performance and quality of the crop. As a matter of fact, a range

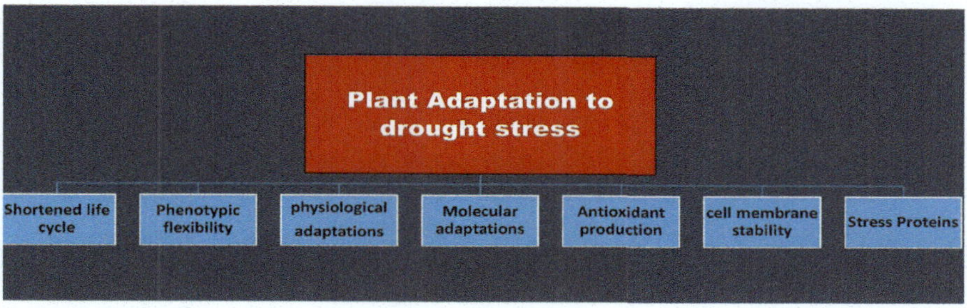

Fig. 5.2 Various adaptations in plants to cope drought stress

of methods have been developed by the plants to lessen the usage of their resources and adjust their growth to unfavorable climate change (Verma et al. 2020; Nishiyama et al. 2013).

Plants endure many environmental stresses with the variety of responses at morphological biochemical and physiological stages to survive (Huber and Bauerle 2016). At physiological level, Plants show two significant responses as a result of water deficiency and heat stress:

1. Avoidance mechanisms

   Avoidance mechanism mainly provides escape to the drought and heat stress. These are basically modifications (morphology and physiology) of certain plant parts which includes enhanced root systems, lesser number of stomatal conductance, reduced leaf area, increased leaf thickness and curling to reduce the transpiration rate (Goufo et al. 2017). Moreover, biosynthesis of the cuticle wax on the aerial parts of the plant is another adaptive response as well (Lee and Suh 2013). No doubt, this is going to decrease in production.

2. Tolerance mechanisms

   Tolerance mechanisms are associated with osmotic adjustments. They preserve the hydrostatic pressure of the tissues. This is done by making changes at by cellular and biochemical level (Blum 2017). Many sensors which are part of response signaling helps plants to recognize stress under water scarcity. These signaling pathway make survival possible without morphological mechanism.

5.4.1 Role of Phenotypic Flexibility to Cope Drought and Related Role of Nanoparticles

As drought adversely affect the plant growth, so plant develop morphological adaptations such as changes in roots and shoots because these are mostly affected plant parts.

(a) Leaf area along with number of leaves per plant are reduced during water deficit conditions to minimize the water loss as a result yield becomes low. It is a known fact that plants having small leaves such as xeric plants can endure drought very well even though their biomass and growth rate is low (Ball et al. 1994). A phenomenon of leaf pubescence protects the plants from bearing extreme temperatures. Also, hairs on leaves reduces the transpiration rate and temperature but differences can be present under inter and intra species (Sandquist and Ehleringer 2003). Leaf hair reduces water loss from the leaf surface by causing hindrance in the movements of vapors from the surface into the environment during heat and temperature stress. Drought stress stimulates formation of trichrome on both side of wheat leaves but no noteworthy effect was observed on boundary layer resistance. Mature stems under water stress showed decreased water content and water potential i.e., 4% - 0.25 MPa respectively.

During the water stress condition, mature stem of *Hylocereus undatus* actively provided younger stem with phloem supply of water resources and other assimilates to maintain growth. Furthermore, sucrose containing nectar was preserved during water deficiency maintained but stem elongation was stopped by the girdling phloem of younger stems. Insignificant axial hydraulic conductivity was observed. Also, for xylem transportation the water potential gradient was moving in alternate direction i.e., older stem to the younger ones (Nerd and Neumann 2004).

Metal nanoparticles have shown progressive effects related to shoot and leaf growth and development in plants. Nano zinc oxide can modulate or alter the effects of drought from plants. Researchers have reported that effect of nanoparticles is usually size, morphology, size and type dependent. Yusefi-Tanha et al. (2020) used three different types of zinc nanoparticles with a different combination of size and shape. It included 38 nm, 59 nm, 500 nm with spherical, flower like and rod like shapes respectively. They reported that 38 nm spherical shape ZnO nanoparticles were most efficient to increase the growth of soya bean, but up to a maximum concentration of 400 mg. During the NPs treatment during water stress on plants, mineral uptake by plants is increased mainly which can increase chlorophyll content of plant cells. This can lead to increased leaf and shoot growth (Yusefi-Tanha et al. 2020). Nanoparticles are also increased the microorganism in the rhizome, thus positively enhancing mineral exchange between soil and plant roots. This directly or indirectly increases the plant shoot and leaf growth (Cota-Ruiz et al. 2020). Dimkpa et al. (2020) also reported that in drought affected wheat plants, chlorophyll content is decreased where nanoparticle treatment to wheat seedlings can positively enhance chlorophyll content and shoot growth (Dimkpa et al. 2020). Titanium oxide nanoparticles cause the loosening of the cell wall and cause the increase of the shoot growth in drought stress. These nanoparticles also cause an increase the cell division and cell development in plants like the effects of gibberellins and cytokinins (Hojjat 2020; Mohammadi et al. 2016; Sauret-Gueto et al. 2012).

(b) Root growth and other characters such as its size, thickness, and production are important during water stress because they are basis of water absorbance from the soil. In plants roots are significant organs to adapt under water stress. In extended periods of vegetative stress where leaf area and growth are maintained, modifications in root architecture maintain the plant water status. In rice, broad and deep root system has significantly showed tolerance to the water stress (Kavar et al. 2008). During crop growing season, root features such as structure and its spread rather than its quantity matters in water absorbance. Root development and its function enhanced the drought resistance in cotton, tea and onion. In legumes yield has been increased under water stress conditions due to the wide spread root system (Subbarao et al. 1995).

Table 5.1 shows the influence of different nanoparticles on plants related to their root system health.

It can be sum up that plant might avoid water stress by two main procedures i.e., with upholding the high water potential and making an effort to enhance water uptake or mixture of both procedures. They maintain their water potential by reduced water loss and increase water absorption by morphological and physiological changes. Moreover, plants can lessen their growth phase and surface area with the formation of small leaves and leaf shedding.

Table 5.1 Reports of positive influence of different nanoparticles on root growth of plants

Sr. #	Crop/Plant	Type of nanoparticles	References
1	*Anthemis silanica*	Silicon oxide nanoparticles	Ahmadi et al. (2020)
2	*Brassica oleraceae*	Zinc oxide nanoparticles	Awan et al. (2021)
3	*Capsicum annuum*	Zinc oxide nanoparticles	Afrayeem and Chaurasia (2017)
4	*Daucus carota*	Zinc oxide combined with iron oxide nanoparticles	Elizabath et al. (2017)
5	*Lemna minor*	Aluminium nanoparticles	Juhel et al. (2011)
6	*Lycopersicum esculentum*	Iron nanoparticles	Brasili et al. (2020)
7	*Mentha Piperita*	Titanium oxide nanoparticles	Samadi et al. (2014)
8	*Oryza sativa*	Iron oxide and titanium oxide nanoparticles, carbon nanotubes	Hao et al. (2016)
		Silver nanoparticles	Ejaz et al. (2018)
		Zinc nanoparticles	Guha et al. (2018)
9	*Glycine max*	AgNPs + ZnONPS	Afsheen et al. (2020)
10	*Spinacia oleraceae*	Silver oxide nanoparticles + sewage sludge	Singh and Kumar (2020)
11	*Triticum aestivum*	Silver nanoparticles	Iqbal et al. (2019)
12	*Vigna radiata*	Titanium oxide nanoparticles	Mathew et al. (2021)
13	*Zea mays*	MgO	Shinde et al. (2020)
		ZnO	Itroutwar et al. (2020)

5.4.2 Physiological Mechanisms

Some key mechanisms are of significant importance for drought resistance which includes: osmo-protection, osmotic adjustment, anti-oxidation and scavenging defense system. Genetic changes for physiological mechanisms under drought are vague but some other complex methods have been suggested. Nanoparticles of various types, origin, shapes and sizes have been reported to directly or indirectly affecting these phenomenon, thus helping plants to cope water stress.

A few of these methods are explained below (Fig. 5.3).

A) ***Role of osmolyte production***

The general approach for stress tolerance is excessive synthesis of various harmonious solutes, organic in nature (Serraj and Sinclair 2002). These osmlytes are even nontoxic to plants cell at higher cytosolic concentrations. They are known possess higher solubility with low molecular weight. Usually these organic solutes save plants from stresses by different mechanisms which includes:

- Causes changes in osmotic pressure
- Work against reactive oxygen species thus minimizing the loss
- Help in membrane stability along with the structure Maintenance of proteins, membranes and enzymes

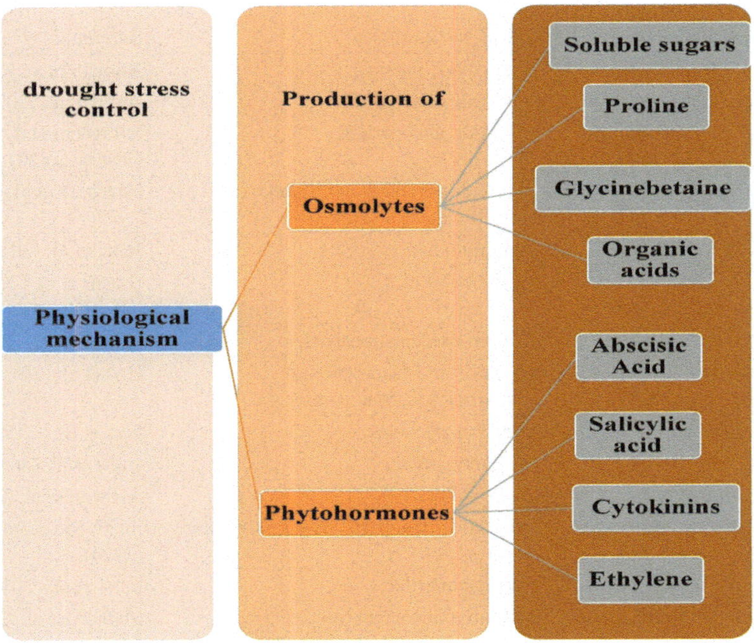

Fig. 5.3 Physiological mechanisms in plants to cope water stress

When water relations are maintained under osmotic stress it is known as osmotic adjustment. It contains many molecules or ions which are osmotically active such as, sugar alcohols, organic acids, proline, soluble sugar, glycine betaine, chloride ions, calcium, potassium and many more. During drought, solute accumulation leads to reduced osmotic potential as a result water is attracted to the cell and turgor pressure is maintained. Osmotic modifications help plant to carry out its functions like photosynthesis, incorporate partitioning and grain filling at normal rate (Subbarao et al. 2000).

Proline is one of the significant cyto-solute which is freely accumulated in bacteria, algae, animals and higher plants under lower water potential. When water potential is low, increased bio synthesis and slow oxidation in mitochondria leads to synthesis of proline. Proline has many roles which include establishment of macromolecules, storage of carbon and nitrogen after water deficiency and a sink for enhanced reductant (Zhu 2002). In peas, proline content was improved during water stress (Alexieva et al. 2001). When there is drought situation, many herbaceous plants are more threatened than other stronger plants like trees and shrubs. There are research reports about such plants as well that during drought plants like petunia (*Petunia hybrida*) drought resistant cultivars varieties fought against water stress by accumulating free proline which was osmo-protectants and stimulated drought resistance (Yamada et al. 2005).

There are number of reports showing enhanced proline content in plants treated with nanoparticles. Latef et al. (2018) reported higher content of proline in broad bean plants when treated with titanium oxide. Faizan et al. (2018) documented improved plant growth in tomato plants under the effects of zinc oxide nanoparticles. Zahedi et al. (2018) worked on strawberry plants under the effects of selenium nanoparticles. They also reported the enhanced conc. of osmolytes in presence of nanoparticles.

One of the widely studied solute and ammonium compound in bacteria, plants and animals is **Glycinebetaine**, N, N, N-trimethyl glycine (Fig. 5.4a). It has been reported that it has significant importance in improving plant resistance to many stresses and drought as well (Quan et al. 2004). Plants which don't form glycinebetaine, when were engineered with glycinebetaine producing genes, they showed higher tolerance to various types of stresses. In drought tolerant cotton varieties glycinebetaine found to be accumulated as compared to the non-tolerant ones (Naidu et al. 1998). Moreover, it not only works as an osmo-protectant or helpful in enzyme or membrane integrity but glycine betaine is also effective in ameliorating various stresses in plant cells by taking part in signal transduction pathways (Subbarao et al. 2000). There are also reports of an increased levels of glycine betaine in nanoparticles treated plants, thus helping and assisting plant mechanisms in tolerating the stress situation (Najafi et al. 2020).

Citrulline is an amino acid (Fig. 5.4b) and its named is derived from *Citrullus* (Latin name of watermelon). Various proteins have citrulline though these are not encoded by the genes and during formation not build into proteins (Kawasaki et al. 2000). It is proposed that citrulline buildup improves the fighting ability of wild watermelon (*Citrullus lanatus*) to face extreme water stress in spite of having

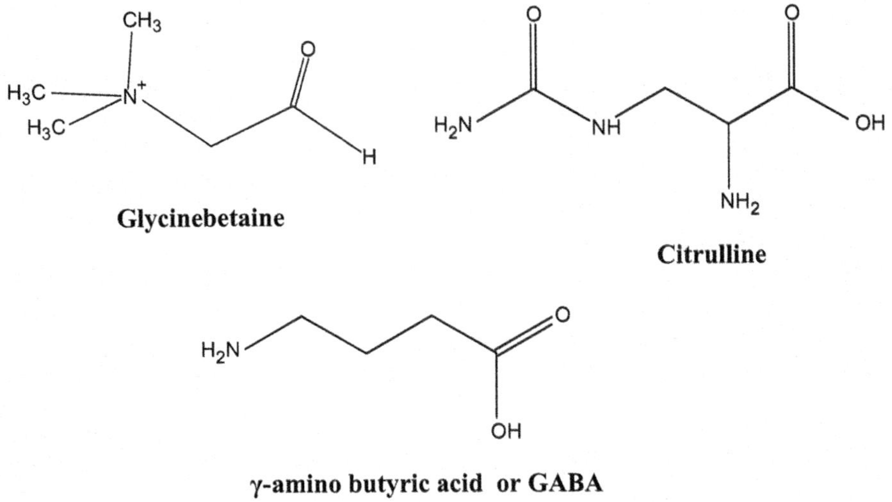

Fig. 5.4 **(a-c)**: Phytochemicals being produced in plants during water stress

normal C3-type photosynthesis (Akashi et al. 2001). Citrulline replaced other osmolytes in wild watermelon as their leaves contained higher amount of citrulline (Kawasaki et al. 2000). It is presumed that citrulline makes only 5% of the mesophyll cells and present in cytosol only. In addition, it is a novel solute known as an efficient hydroxyl ion scavenger observed till now. Also, citrulline protects the essential enzymes and DNA from injuries caused by oxidative species or ROS (Bektasoglu et al. 2006). There is not a valid and significant research report to the date, reporting higher level of citrulline in plants on exposure to nanoparticles. But this aspect may also be worked with.

γ-amino butyric acid is a non-protein amino acid (Fig. 5.4c) which acts as zwitter ion . It is found in free form. Also, it has flexible cylic-molecular structure which is like proline. A quick buildup of γ-aminobutyric acid was observed in tissues during exposure to stress. It is highly soluble in water at physiological pH. In higher plants, it might work as signaling molecule during stress condition (Serraj et al. 1998). In water stress conditions, γ-aminobutyric acid act as osmoregulator, ROS detoxifier. It plays its role in proline metabolism by enhancing its accumulation by its conversion from putrescine to proline, as a mediator of a signal transduction. It also causes detoxification of reactive oxygen radicals (Kinnersley and Turano 2000). γ-aminobutyric acid synthesis takes place under water stress by initiation of signal transduction mechanism in which elevated levels of cytosolic Ca^{2+} stimulates $Ca^{2+/}$calmodulin dependent glutamate decarboxylase action. γ-aminobutyric acid is involved in plant growth, pH regulation, nitrogen storage, defense, glutamate utilization and a compatible osmolyte (Wahid 2007). Study of relationship of nanoparticles and resultant γ-amino butyric acid content of plants can help in treating water stress areas and crop selection there along with type of nanoparticles.

As a matter of fact, main point for plants is to preserve water in cells and tissues to combat water stress for plants. In this way they can sustain anti-oxidant shield and can maintain cell membranes. There are number of compounds which play an important role in drought tolerance mechanism of plants like free amino acids, polyamines, sugars, γ-aminobutyric acid and plant growth regulators etc. They can perform their action by controlling various basic metabolic pathways of plants like stomatal regulation to control water movement, protection of DNA, proteins, attack against ROS and maintaining integrity of cell membranes by maintain the water balance.

B) ***Role of Phytohormones***

(i) ***Abscisic acid (ABA)***

Abscisic acid (ABA) is known as stress hormone because it has involvement in both biotic and abiotic stress response (Kiba et al. 2011). It acts as a negotiator in plant responses to various stresses and drought stress as well (Keskin 2012). In water deficiency, ABA regulates the guard cell for maintenance of water status and stimulation of genes for proteins and enzymes linked with drought tolerance. There are many variants present which are ABA deficient such as maize, tomato and *Arabidopsis* (Zhu 2001). In continuous drought, abscisic acid lacking varieties of *Arabidopsis* wilt and die. ABA preserve root and shoot growth under drought conditions. It also stops over production of ethylene.

Vanvoka et al. reported in 2017, that stress hormones like abscisic acid are upregulated during nanoparticle treatment. Higher exposure of plants to nanoparticles (ZnO) could produce very higher amounts of this growth inhibitor. But it is also plant species dependent. If doses and time of application of nanoparticles, this characteristic or negative impact of nanoparticles on plants may be utilized positively during stress management. This can help plants to cope better with drought stress as enhanced quantities of ABA can help plant in stomatal regulation and better water management. But extensive research projects are needed to explore effects of these individual nanoparticles and their combinations in cash crops. One of such research was carried out by Zahedi et al. (2021) who reported increased drought resistance in pomegranate trees when treated with foliar spray of selenium nanoparticles. One of the main parameters studied in the trees was the amount of ABA. Foliar spray of 10 nm size selenium nanoparticles had a positive effect and a positive direct correlation between growth of trees and ABA content of the plants.

(ii) ***Salicylic acid (SA)***

Although, salicylic acid has various physiological roles but it has also significant importance in reducing many stresses like heat, chilling and water stress (Hussain et al. 2008). It's a phenolic compound and plays an important role in signal transduction mechanism of the plant to fight with stresses (Vlot et al. 2009; Malamy et al. 1992). Application of SA to tomatoes under drought, enhanced certain parameters which included photosynthesis, chlorophyll contents, relative water content, carbonic anhydrase activity, nitrate reductase activity and over all leaf water potential (Hayat et al. 2007).

A different aspect for application of nanoparticles through salicylic mediated treatment of drought in plants has been investigated. Salicylic loaded nanoparticles have been synthesized and caused a slow release of SA when applied to plants which help them to better control their immune response against stress and continue normal growth (Kumaraswamy et al. 2019). SA acid in combination with nanoparticles has also shown positive impact on stress tolerance of plants by enhancing their growth indicators (Abdoli et al. 2020). Particularly a combination of iron oxide nanoparticles and SA is a useful combination as used in drought affected strawberries by Mozafari et al. (2018). Strawberries plantlets were grown with different level of drought stress and SA improved growth parameters in all applied drought levels.

(iii) ***Jasmonic acid (JA)***

Jasmonic acid regulates the development of the plant. It protects the plant from pathogen invasion and various climatic stresses including drought (Cheong and Do 2003). Ascorbate metabolism is an antioxidant response which is induced by the JA. Under drought conditions jasmonic acid starts to form in roots and leaves of maize and pear respectively (Aimar et al. 2011). Fujikawa et al. (2021) reported the anti-stress action mechanism of magnesium oxide nanoparticles through jasmonic acid signaling pathway. This can be further tunneled with drought resistant action of jasmonic acid.

(iv) ***Cytokinins (CKs)***

Stress stimulated production of cytokinins reduces leaf aging process and hence increases drought tolerance in crop plants (Peleg and Blumwald 2011). Water stress limits the synthesis of cytokinins and usually stressed plants contain low levels of cytokinins in xylem exudates (Pospisilova et al. 2000).

In plants who have a reduced production of cytokinins for example Arabidiopsis and transgenic tobacco, root growth was enhanced which helped the plants to fight against drought stress. There is a direct relationship between root increase cytokinin degradation. And then root structure and formation is directly related to water stress resistance. This has been proven true for a number of crops and plants including Arabidiopsis and tobacco plant. Studies in transgenic plants have proven these facts. Therefore, to develop drought stress tolerance in plants, attempts are being done to develop such plant breeds which have extensive root systems. Because plants which have extensive root system can better absorb water and nutrients, thus can fight the stress more easily. Their survival in drought is probably more expected (Tuberosa 2012; Werner et al. 2010).

Silicon oxide nanoparticles have been reported to show progressive effects on cytokinins signaling genes, thus promoting growth in treated plants (Azhar et al. 2021). Cytokinins in leaves and other tissues of capsicum have been reported to be increased by application of silver nanoparticles in a dose reliant mode (Vinkovic et al. 2017; Zuverza-Mena et al. 2016). Particularly, production of zeatin type cytokinins is enhanced when nanoparticles are applied to them (Vankova et al. 2017;

Vinkovic et al. 2017). These researches put the base for use of nanoparticles in struggling against drought stress through cytokinins pathway.

(v) ***Ethylene***

Stomatà are closed when ethylene produce hydrogen peroxide in guard cells. Ethylene also expresses SodERF3 gene which enhanced osmotic potential in tobacco plants to tolerate drought (Trujillo et al. 2008).

C) ***Role of Antioxidant defense***

Plant defense system is mostly triggered as a result of drought induced reactive oxygen species (ROS). To overcome damages caused by the ROS, a powerful, rapid and effective antioxidant system is essential which also provides drought resistance. It involves enzymes and non-enzymatic factors which help in detoxification of the plant by repairing injuries caused by the reactive oxygen species. An oxidative stress is induced during water deficiency by the singlet oxygen, hydrogen peroxide (H_2O_2), hydroxyl radicals (OH^-) and free oxygen radicals O^{2-} (Impa et al. 2012). Strength and integrity of cell membrane and other organs are maintained by enhancing antioxidant machinery. This reduces the electrolyte leakage, lipid peroxidation and ROS scavenging in plants. ROS activity is controlled by the production of antioxidant enzymes which activates the cell redox reactions as a result, cell injury and cell death is avoided. Antioxidant enzymes which include superoxide dismutase (SOD), catalase (CAT), ascorbate peroxidase (APX) and peroxidase (POD) were found to very active under drought (Cao et al. 2017). Often, drought resistant plants show increased antioxidant activity as compared to non-resistant varieties (Table 5.2).

APX is an essential enzyme which is commonly found in chloroplast and cytoplasm. In chloroplasts, it scavenges the hydrogen peroxide. Change in its activity is widespread in fibrous roots than in cytoplasm (Caverzan et al. 2016). In poeny plant, change in functions of antioxidant enzymes (CAT, APX, SOD, POD) protected it from drought stress (Wang et al. 2019). During water scarcity *Coleus plectranthus* showed enhanced activity of alpha tocopherol, ascorbic acid, glutathione (non-enzymatic antioxidants) and APX, SOD, CAT (antioxidant enzymes) (Prathyusha and Chaitanya 2019). In *Vicia faba*, antioxidants enzymes were formed during

Table 5.2 Antioxidant enzymes and their role in plants during stress

Sr. #	Enzyme	Function
1	Superoxide dismutase (SOD)	In leaf tissues, this is responsible to control production of hydrogen per oxide and singlet oxygen and their diffusion in leaf tissues to compete with stress (Moussa and Abdel-Aziz 2008)
2	Peroxidase (POX),) and catalase (CAT	To control the making of singlet reactive oxygen and hydrogen peroxide in plants during stress
3	Ascorbate peroxidase (APX)	Its substrate is ascorbate, for conversion of hydrogen peroxide to neutral molecules of water and its main participation in the process is as ROS scavenger (Hossain et al. 2012).

drought (Abid et al. 2017). In drought resistant variety of faba bean, antioxidant enzymes activity was increased as compared to non-resistant variety (Siddiqui et al. 2015).

Drought conditions, whether mild or acute, enhance the production of non-enzymatic and enzymatic antioxidants compounds in plants. In *Adonis amurensis* and *Adonis pseudo-amurensis* actions of APX, CAT, POD and SOD enhanced which caused a prominent reduction in the oxidative damage during water stress conditions, by lowering the reactive oxygen species inside plants (Gao et al. 2020). In a study of *Vigna mungo* POD and SOD activities were improved which increased the resistance to drought to limit its adverse effects during water deficiency (Gurumurthy et al. 2019). Further, to neutralize the effects of H_2O_2 antioxidants (SOD, CAT, POX) were enhanced in *Glycyrrhiza glabra* L during drought stress (Hosseini et al. 2018).

Nanoparticles play a function in up regulation of the activities of enzymes that are antioxidant like, CAT, SOD and POD. Nanoparticles of TiO_2 increase the activity of SOD and also, the range of enhancement was considered as a role played by concentration of NPs (Laware and Raskar 2014). Silicon Nanoparticles (SNPs) can efficiently alleviate the different abiotic stresses i.e., salinity, heavy metal toxicities, drought, freezing stresses, and chilling.

5.4.3 Role of Cell Membrane Stability

Various abiotic stresses have biological membranes as their first target. Generally, it is suggested that, under water stress, the preservation of stability and integrity of membranes is a foremost factor of plants' drought tolerance (Bajji et al. 2002). According to Premachandra et al. (1991), the stability of cell membrane, is considered as a physiological key which is a good indicator of measuring drought tolerance in individual plants. Moreover, it is also suggested that this is related to plant genetic makeup, since in drought stressed rice the quantitative trait loci have been drawn at various stages of growth (Tripathy et al. 2000).

Dhanda et al. (2004) elaborated that for the screening of germplasm in drought tolerance, the highly important characteristics found to be the stability of membrane of the leaf segment. Hence, the stability of cell membrane deteriorated swiftly in Kentucky bluegrass with simultaneous exposure of heat and drought stress (Wang and Huang 2004). An analysis was done by Gnanasiri et al. (1991) on maize, according to which, drought tolerance is improved by potassium nutrition, mainly because of enhanced stability of cell membrane. Moreover, tolerance to drought is estimated as improvement in the stability of cell membrane. In the seedlings of holm oak (*Quercusilex*), hardening enhanced the tolerance to drought mainly by decreasing stomatal regulation and osmotic potential, improved the growth capacity of new root and increased the stability of cell membrane. Among the seedlings which were given treatment, the most responsive were those subjected to hardening at moderate level. Variation in the stability of cell membrane, capacity of root growth and stomatal regulation had negative correlation with osmotic change (Villar-Salvador et al. 2004). Factors which are considered as a base for disruption of membrane are not well known; nevertheless, if the volume of cell decreases, it produces crowding and enhances the thickness of the components of cytoplasm. It leads to the increase in possibility of molecular interactions which can induce membrane fusion and protein denaturation. For systems of protein and model membrane, a wide range of certain compounds have been recognized that can reduce the above mentioned deleterious effects of molecular interactions. These chemical compounds include, fructans, glutamate, carnitine, proline, polyols, mannitol, glycinebetaine, sorbitol, carnitine, sucrose, oligosaccharides and trehalose etc. An additional chance for the leakage of ions from the cell could be because of inhibition of enzymes which are bound to membranes under the effects of heat. These enzymes are accountable for retaining cell's chemical gradients. However, in a study by Gigon et al. (2004) leaf membranes of *Arabidopsis* were found to be very repellent to the deficiency of water. They show their strong ability to retain the contents of polar lipid. In this way they can retain their composition and stability even when drought conditions are really acute.

Titanium oxide and zinc oxide nanoparticles treated plants have shown improved membrane stability indices in various stresses (Satti et al. 2021; Semida et al. 2021). These areas can be further exploited in terms of drought stress in particular doses of nanoparticles.

5.4.4 Molecular Mechanisms

Deficiency in cellular water of plant can arise due to the reduction in soil water content. In conditions like this, variations occur in expression of genes which include both upregulation and down regulation of genes. A variety of genes are transcribed as a result of drought, and gene products thus formed are playing an important role towards creating tolerance to drought (Kavar et al. 2008). Stress conditions can openly trigger gene expression, or it may be the result of secondary stresses or responses of injury. Furthermore, it should be made clear that resistance to drought or water stress is not a simple phenomenon at gene level, it is performed by a number of genes as a joint venture in a complicated way (Cattivelli et al. 2002).

A. ***Aquaporins***

Aquaporins are able to regulate and facilitate the water exchange passively across the membranes. Moreover, they are a part of family of intrinsic type of proteins present in cell membranes (Tyerman et al. 2002). Plasma membranes and vacuolar membranes of plant cells contain an abundant number of these proteins in them. An analysis of their structure revealed the general system of water transport mediated by proteins. While aquaporins were discovered in plants, it resulted in a shift in prototype of the understanding of water relations in plant. The relationship of drought resistance in plant with aquaporins is still not clearly definable. Nonetheless, it is considered that the hydraulic conductivity of membranes can be regulated by the demand. Aquaporins can also potentiate an increase often to 20-fold in the permeability of water (Maurel and Chrispeels 2001).

For many years, the research has been done on plant water relations and aquaporins. Mercury is considered to be a likely inhibitor of aquaporins. It became evident from numerous reports on deterioration of conductivity of root hydraulic induced by mercury, which proved the role of aquaporins in overall water uptake through root and also, in root cells that are highly compartmented, they perform a function in osmoregulation of cell (Javot et al. 2003). Reversal of genetics provide a refined method of exploration of role played by aquaporin in plant water interactions.

B. ***Stress proteins***

To cope with the widespread stressful conditions like deficiency of water stress proteins are produced, which is a pervasive response. Most of these proteins are water soluble and hence, contribute to the phenomena of stress tolerance by hydration of structures of cell (Wahid 2007). In tolerance to drought, production of various transcription components and stress proteins is completely involved. The genes that bind dehydration-alert component are included in the signaling pathway of abiotic stress. There was a possibility to control the dehydration-alert component binding genes' expression, which could further help in producing stress tolerance in transgenic plants.

When anew dehydration-alert component-binding gene is introduced as a transcriptional factor, it essentially improves the tolerance to drought ability of rice (Yamaguchi-Shinozaki and Shinozaki 2004) and groundnut (Mathur et al. 2004). Various *Capsella bursa pastoris* like genes are expressed as a result of drought stress in plant species which include rice (Dubouzet et al. 2003), tomato and rye (Jaglo et al. 2001), cotton (Huang and Liu 2006), wheat (Shen et al. 2003), soybean (Chen et al. 2007) and Brassica (Zhao et al. 2006). When the dehydration alert component-binding gene1A were introduced in transgenic tall fescue (*Festuca arundinacea*), there appeared an increase in drought resistance by the growth of an enhanced level of proline. Also, it shed some light on the capability of *Capsella bursa-pastoris* genes for stimulating tolerance to water stress (Zhao et al. 2007).

With help of genetic and molecular studies various genes playing regulatory functions have been recognized for osmotic stress as well as drought response. There are two key pathways involved in water stress response:

- ABA dependent
- ABA independent pathway

5.5 Metal nanoparticles and Drought Resistance

Metal and metal oxide nanoparticles ((TiO_2, FeO, ZnO, Al_2O_3, CuO and SeO_3), non metals (Phosphorous) and metalloids (Si and SiO_2) have been utilized to improve the toxic effects of water stress in many agricultural crops.

5.5.1 *Titanium Dioxide (TiO_2) Nanoparticles*

Although, it is not an important element for plants, however titanium has many significant physiological roles even in small amount (Tiwari et al. 2017). In water stress, the yield, starch and gluten contents were improved by applying 0.02% of TiO_2 NPs as a foliar spray (Jaberzadeh et al. 2013). Also, in flax *(Linum usitatissimum)* lower concentrations of titanium nanoparticles up to 10 mg/L reduced the water stress stimulated damage (Aghdam et al. 2016). The relative water content, CAT and anthocyanin production of *Ocimum basilicum* was also enhanced by applying 0.03% of these NPs during water stress (Kiapour et al. 2015). In cotton plants, by applying 50 mg/L of nano particles of titanium oxide at pre-flowering stage improved yield and biochemical parameters which included soluble proteins, antioxidant enzymes activity, total phenols, soluble sugars and plant pigments under water stress (Shallan et al. 2016). The toxic effects of drought were reduced in *Eruca sativa* plants with the application of 20 mg/L of TiO_2-NPs which as result increased formation of cysteine and H_2S to enhance antioxidant activity, buildup of osmolytes and relative water content with immediate decrease in hydrogen per oxide and lipid peroxidation (Khan and Alzuaibr 2018). Similarly, in moldavian dragonhead plant the lipid peroxidation, and ROS activity was reduced while proline content was increased with the 10 mg/L application of nanoparticles of titanium under drought (Mohammadi et al. 2016). Kamalizadeh et al. (2019) also demonstrated TiO_2-NPs effects on Moldavian dragonhead plant. They observed enhanced oil content, phenolic compounds but no effect on dry biomass of the plant. Drought stress, application of TiO_2-NPs (30–50 mg/L) in this plant can be used to enhance the phenolic contents.

5.5.2 *Iron Oxide (FeO) Nanoparticles*

Iron (Fe) is an important component in plant growth and development as it is the part of chlorophyll, development of chloroplast, nucleic acid metabolism, respiration and redox reactions. It is also essential for many enzymes because it is involved in co factors (Mimmo et al. 2014). Iron in the form of nano zero-valent iron (nZVI), nano-goethite (α-FeOOH), nano-hematite (α-Fe_2O_3), nano-maghemite (γ-Fe_2O_3), nano-magnetite (Fe_2O_4), and nano-iron pyrite (FeS_2) is of significant value because they have magnetic characteristics. Researchers have fascinated by their positive effects on plants (Zuverza-Mena et al. 2017; Srivastava et al. 2014). In cow pea and safflower foliar spray of 1.5 mg/L of iron nanoparticles enhanced seeds per pod, seed nitrogen content, yield and oil content simultaneously under water stress (Zareii et al. 2014). The growth of sun flower, planted during drought in contaminated mine soil, was increased by applying 1% of maghemite, γ-Fe_2O_3 nanoparticles in soil (Martínez-Fernandez et al. 2015). In the same way, the yield and development and antioxidant activity of *Brassica napus* and strawberry was enhanced even at low concentrations such as 3.4 and 0.8 mg/L of Fe_2O_3 and Fe_3O_4 respectively (Palmqvist et al. 2017; Mozafari et al. 2018), correspondingly in drought. Soil application of Fe-NPs (100 mg/kg) in cadmium contaminated soil enhanced iron uptake, improved photosynthesis, grain production and reduced the ROS and cadmium concentrations in wheat plants grown in water stress (Adrees et al. 2020).

5.5.3 Zinc Oxide (ZnO) Nanoparticles

For plant cells, zinc is a vital element because it is involved in the biosynthesis of tryptophan (precursor of indole acetic acid). IAA is a phytohormone and take part in various biochemical and physiological activities including cell division. It is essential as well to reduce the unfavorable effects of the various stresses (Cakmak 2008; Hafeez et al. 2013). Zinc NPs strengthens the photosynthesis, enhances the water use efficacy, improves the stomatal movement and hence ameliorate the adverse effects of drought. Activities of various enzymes like Phosphoglucoisomerase, UDP-glucose pyrophosphorylase and invertase are improved by the zinc nanoparticles application. These enzymes help in sucrose and starch biosynthesis (Sun et al. 2021).

Zinc oxide nanoparticles have different effects on various plants because they are dependent upon the plant species. Their role is determined by their size and concentration. Like, 10 mg/L of ZnO NPs enhanced plant growth in canola *(Brassica napus)* but 1000 mg/L showed toxicity (Rahmani et al. 2016). Similarly, foliar spray of 10 mg/L of zinc NPs to coffee plants improved the quality of fruit, photosynthesis and dry weight (Rossi et al. 2019). Concerning water stress, 1000 mg/L of zinc NPs enhanced the germination, seedling growth, crop productivity and water usage efficacy in soybean, rice and sunflower. Nano particles of zinc known to increase the zinc uptake, quick usage of seed reservoirs and hence show positive effects (Rameshraddy et al. 2017). Lately, Dimkpa et al. (2019) reported that in sorghum plants, harmful effects of drought were reduced by amending the soil with ZnO-NPs. It was observed that 5 mg/kg of zinc nanoparticles enhanced the grain production, transportation of sodium, potassium and zinc nutrient in grain while reducing the delay of flag leaf and grain head emergence.

5.5.4 Silicon Oxide (SiO_2) Nanoparticles

Silicon as an element has positive effects on plant growth and development by reducing the adverse effects of abiotic stresses from the past two decades. It alters the expression of genes and modifies the production of reactive oxygen species (Kim et al. 2017). When hawthorn (*Crataegus aronia*) seeds were primed with SiO_2-NPs (10–30 nm), biochemical parameters such as chlorophyll, sugar, proline contents as well as biomass, photosynthesis and stomatal conductance were improved at 100 mg/L under water stress (Ashkavand et al. 2015). In rye grass, 1 mM of silicon oxide nanoparticles enhanced mineral nutrients and additional quality parameters during extreme drought stress (Mahdavi et al. 2016). Similarly, in drought grown tomato seeds, 1 or 2 mM SiO_2 NPs enhanced germination rate of the plants (Haghighi et al. 2013). Alsaeedi et al. (2019) amended soil with silica nanoparticles (200 mg/kg of soil). They grew cucumber in this soil with drought stress. They reported that adverse effects of drought were reduced in cucumber due to high silicon and potassium. In fact high levels of potassium and silicon helped plants to regulate the transpiration and to maintain ion balance.

5.5.5 Selenium Oxide (SeO_3) Nanoparticles

Selenium acts as a cofactor of enzymes important in antioxidant metabolism (Rayman 2008); nonetheless selenium is important for plants due its antioxidant activity. It is reported that Se NPs enhances antioxidant capacity of plants thereby improving the stress tolerance ability of plants as stimulants (Hussein et al. 2019). Se NPs as compared to their bulk size are less toxic and have remarkable biological characteristics. High concentrations of bulk selenium have negative effects on plants because they induce oxidative stress (Gupta and Gupta 2017). For instance, in *Nicotiana tabacum* plant growth and photosynthesis was reduced by applying 10 mg L^{-1} of selenate but in the same plants positive effects of 100 mg/L of Se NPs were observed (Zsiros et al. 2019). Hence, it can be applied in nano form to the crops and better results can be expected in comparison to its bulk form. During water stress, foliar application of Se NPs was applied to wheat plants which resulted in enhanced root and shoot length, leaf area, leaf length, plant height and shoot fresh and dry weights (Ikram et al. 2020).

5.5.6 Aluminium Oxide (Al_2O_3) Nanoparticles

Aluminum oxide nanoparticles are being synthesized in large amounts at commercial and industrial scale. These are also being utilized in military and commercial fields such as in catalyst support and microelectronics (Jakubiak et al. 2014). In a study by Lee et al. (2010) aluminum nanoparticles showed no toxicity in Arabidopsis plant when exposed to different treatments i.e. 400, 2000, 4000 mg/L. Positive effects on root growth of lettuce, radish, ryegrass have also been studied and rape plant (Lin and Xing 2007). Aluminum nanoparticles (50–2500 mg/L) significantly increased antioxidants (APX, CAT, SOD and POD), saponins, phenolics, DPPH activity and plant biomass (Chahardoli et al. 2020) which can be really helpful in reducing the effects of drought.

5.5.7 Copper Oxide (CuO) Nanoparticles

Copper (Cu) being a micronutrient is required by plants in very small amounts. It has various functions in plant growth and development including seed formation, chlorophyll production (Viera et al. 2019). Cu deficiency is quite common in various important crops (Karamanos et al. 2004). Various enzymes that stimulates the biochemical processes of the plant involves copper (Singh et al. 2017; Drazkiewicz et al. 2004). It is the part of photosynthesis and helps in plant metabolism of carbohydrates and proteins (Ambrosini et al. 2018; Din et al. 2017; Singh et al. 2017). In a study by Nguyen et al. (2020), effect of nano-CuO priming was observed on maize under drought. It contributed to the enhancement of plant growth and biomass, maintained leaf water status, chlorophyll and carotenoid content. It also increased the antioxidant activity of enzymes which included; APX, SOD. Similarly, Taran et al. (2017) showed that negative effects of drought action can be slowed down or decreased by using Copper and zinc nanoparticles, when applied on plants of steppe ecotype. These nanoparticles positively affected the metabolism of these plants by increasing the activity of anti-oxidative enzymes. Increased concentration or release of these enzymes in cell sap, decrease the production of TBARS (thiobarbituric acid reactive substances). They also helped the plants in stability of photosynthetic pigments including chlorophylls and carotenoids and also increased relative water content of leaves.

5.6 Methods of applications of nanoparticles for Drought Resistance

5.6.1 Nanoparticles Pretreatment of Seeds or Seed Priming

Seed priming is a very common technique often used to activate the plant growth particularly seed germination. It is a technique of pretreatment of seeds with different chemicals in order to enhance the uniformity in germination, germination percentage and germination rate etc. It is usually done before sowing the seeds. This treatment can cause the initiation of early stages of germination without protruding out of radicle. It can act in a number of ways to control the emergence of seeds depending upon the chemical used for treatment. It may involve simply soaking of seeds in plain water, and where they can absorb enough water and can get enough hydration which can enable them to start earlier and healthier germination by weakening the testa, mobilizing enzymes and sugars etc. (Nascimento et al. 2004; Rehman et al. 2011).

Priming is an approach that involves treating seeds with different organic or inorganic chemicals and or with high or low temperatures (Kamithi et al. 2016). Priming includes the dipping of seeds in various solutions which causes imbibition of seeds. It is done in controlled conditions. Seeds are then dried again to the initial moisture level of seeds so that they are not having pre-sowing emerged radicles. This then involves triggering of number of metabolic processes which help in germination of seed species, particularly seeds of vegetables, small seeded grasses and ornamental species (Tavili et al. 2011). Seed priming is considered to be an easy, highly effective, low cost and low risk technique. Primed seeds are more useful because of numerous advantages such as uniformity, early and faster appearance (Musa et al. 1999), germination in broad range of temperature, crop establishment, efficient use of water, enhancing roots to grow deeper, allowing germination in dormant seeds by increasing metabolic events, to initiate growth of organs for reproduction (Soleimanzadeh 2013), early flowering and maturity (Singh et al. 2015), better competition with weeds, combat against abiotic stresses (Elouaer and Hannachi 2013) and soil-borne destructive diseases (*R. solani, Fusarium spp., Sclerotium rolfsii* etc.) (Rafi et al. 2015).

Ashkavand et al. (2018) worked on the seedlings of *Prunus mahaleb* (Mahaleb) under drought stress. They used priming of seeds with silicon dioxide nanoparticles. Their results showed that seeds which were pretreated with nanoparticles were less affected by drought stress particularly in reference to their transpiration rate, photosynthesis and stomatal conductance. They used high concentrations of silicon dioxide nanoparticles for priming. Seed priming with silicon dioxide nanoparticles helped them to maintain the same trophic or nutritional level in drought stress as those of well-watered plants. Priming treatment with nanoparticles also helped the seedlings to maintain root length and dry root mass for maximum efficiency. Therefore, they suggested that seed priming with silicon dioxide nanoparticles can be a solution for growing plants in drought hit areas of the world where they can help in agronomic practices.

Ashkavand et al. (2015) worked on hawthorn seedlings for their biochemical and physiological responses in three different experimental soil conditions including zero, moderate and acute stress. They performed seed priming with silicon dioxide nanoparticle solution. They studied plant responses like carbohydrate, carotenoid, chlorophyll content, malondialdehyde (MDA), membrane electrolyte leakage (ELI), proline content and relative water content (RWC). Results confirmed the positive effects of treating seeds with nanoparticles after growing in moderate and severe water stress conditions. It can be concluded from above studies that seed priming can be a solution for drought related issues in plants.

5.6.2 Nanoparticles as Foliar Spray

Foliar spray is another treatment which is used to feed plants but in the form of liquid nutrients sprayed to plant aerial parts or leaves. Plants can absorb these nutrients through their leaves using cuticular or stomatal pathway, apoplastic or symplastic pathway. It is particularly suitable for supplying micronutrients to plants as those are requires in lesser quantities to the plants. When there are water stress conditions and top most soil is dry, that means, soil is not going to help the plant to absorb mineral through roots, then even macronutrients can be provided through sprays. Importance of foliar spray has gained much importance in agronomy in recent years, particularly for horticulture. It needs awareness campaigns for the local farmers to increase the acceptability of this technique for all types of crops as it is already well researched and documented for achieving required results. It needs application now (Patil and Chetan 2018).

Foliar application could be considered one of the most common methods, which used to deliver the needed nutrients to plants in adequate concentrations and improve plant nutritional status as well as increase the crop yield and its quality (Smoleń 2012). Foliar spray has a number of practical applications as it can be used for treating damages in plants caused by various stresses like drought, heat and frost etc. For this purpose different nutrients can be sprayed to plants like amino acids, disaccharides, growth regulators, growth stimulators, peptide chains, pesticides, sugars as well as nanomaterials (Shalaby and El-Ramady 2014; Smoleń 2012). A combination of fertilizer and Foliar spray has a number of practical advantages for better growth of crops, optimum yield and protection from pests. Therefore, according to environmental factors (if there is any stress) and requirements of plants, doses of foliar spray can be decided to get optimum crop yield (Dordas 2009).

Nanomaterials when used as a source of foliar spray can be proved very effective as they have very small sizes as compared to pores on the surface of leaves. They can easily penetrate into the plant cells when applied as nanospray. Djanaguiraman et al. (2018) used cerium oxide nanoparticles (nanoceria) to treat the oxidative stress on plants caused by the reactive oxygen species produced in plants as a result of drought stress. They selected sorghum crop for their studies. They were successful to protect photosynthetic machinery and grain yield of sorghum by using foliar spray of nanoceria. Nanoceria foliar spray at concentration of just 10 mg/L, reduced the hydrogen peroxide levels, membrane lipid peroxidation and hydrogen peroxide levels of sorghum leaves by 36, 41 and 37% respectively under drought stress. Positive effects of nanospray were not limited to oxidative stress treatment but it also increased seed yield (31%), photosynthesis rate (38%) and pollen germination (31%) as well in drought affected plants as compared to normal watered plants. Ikram et al. (2020) synthesized selenium nanoparticles (SeNPs) using plants and applied them to wheat plants grown in normal irrigation and in drought stress in the form of foliar spray. They assessed the efficiency of foliar applications to increase the growth of wheat crop affected with drought. For this purpose various concentrations of SeNPs were prepared and applied including 10, 20, 30, and 40 mg/L. Two wheat varieties were used, one was drought-tolerant (V1) and other was drought-susceptible (V2). They observed significant increase in leaf area, leaf length, leaf number, plant height, root dry weight, root fresh weight, root length, shoot dry weight, shoot fresh weight and shoot length with foliar spray of SeNPs at 30 mg/L concentration used.

5.6.3 Soil application of Nanoparticles

When nutrients are applied in liquid form near the roots of plant in soil or injected there, that process in called soil injection or soil drenching method. It is, therefore, required that chemical to be used should be water soluble as much as possible. For optimum results, chemical may be applied near to roots and to moist soils. Soil drenching can be done just by pouring the water solution of chemical on soil near the roots of plant. Any chemical can be applied by this method to the plants like pesticides, fertilizers, insecticides, fungicides and nanoparticles as well. But in drought conditions it is not much valued, because for most of the chemicals, wet soils are required for soil drenching. Soil injectors are also used to inject the chemicals 2–4″ deep in soil according to the requirement if chemical being applied and crop under consideration (Fishel 2018). Application of nanoparticles to the plants using soil drench method should be last option as nanoparticles are leached down easily and can affect bacteria in rhizosphere of plants and can also affect absorption of other nutrients as well.

Dimkpa et al. (2019) evaluated drought effects on performance and nutrient acquisition and distribution in sorghum and ZnO nanoparticles (ZnO-NPs) might alleviate such effects. This study represents the first evidence of mitigation of drought stress in full-term plants solely by exposure to ZnO-NPs in soil. The ability of ZnO-NPs to accelerate plant development, promote yield, fortify edible grains with critically essential nutrients such as Zn, and improve N acquisition under drought stress has strong implications for increasing cropping systems resilience, sustaining human/animal food/ feed and nutrition security, and reducing nutrient losses and environmental pollution associated with fertilizers.

5.7 Conclusion

A lot of research has already been done on toxicological effects of nanoparticles on plants due to their enhanced exposure to NPs. Stress induction studies and drought tolerance mechanism of plants suggest that by controlling the dose, size and morphology of nanoparticles, their effect can be optimized by study various morphological and physiological parameters of plant stress response including root health, shoot growth, membrane stability index, production of stress and growth hormones, phenolics and stress enzymes etc. . It also varies according to the plant type, therefore, research is needed to optimize this relationship for commercial acceptance of nanoparticle based products for treatment of drought stress in plants.

References

Abdoli S, Ghassemi-Golezani K, Alizadeh-Salteh S (2020) Response of Ajwain (*Trachyspermum ammi* L.) to exogenous salicylic acid and iron oxide nanoparticles. Environ Sci Pollut Res 27:36939–36953

Abid G, Mohamdi M, Mingeot D, Aouida M, Aroua I, Muhovski Y, Sassi K, Souissi F, Mannai K, Jebara ME (2017) Effect of drought stress on chlorophyll fluorescence, antioxidant enzyme activities and gene expression patterns in faba bean (*Vicia faba* L.). Arch Agron Soil Sci 63:536–552

Adrees M, Khan ZS, Ali S, Hafeez M, Khalid S, Rehman MZ, Hussain A, Hussain K, Chatha SAS, Rizwan M (2020) Simultaneous mitigation of cadmium and drought stress in wheat by soil application of iron nanoparticles. Chemosphere 238:124681

Afrayeem SM, Chaurasia AK (2017) Effect of zinc oxide nanoparticles on seed germination and seed vigour in chilli (*Capsicum annuum* L.). J Pharmacog Phytochem 6(5):1564–1567

Afsheen S, Naseer H, Iqbal T, Abrar M, Bashir A, Ijaz M (2020) Synthesis and characterization of metal sulphide nanoparticles to investigate the effect of nanoparticles on germination of soybean and wheat seeds. Mat Chem Phy 252:123216

Aghdam MTB, Mohammadi H, Ghorbanpour M (2016) Effects of nanoparticulate anatase titanium dioxide on physiological and biochemical performance of *Linum usitatissimum* (Linaceae) under well-watered and drought stress conditions. Braz J Bot 39(1):139–146

Ahmadi N, Hassanpour M, Hekmati M, Ghanbarzadeh M (2020) Effect of SiO_2 nanoparticles on phytochemical and anatomical alterations in *Anthemis gilanica*. Iran J Plant Phys 10(3):3223–3231

Aimar D, Calafat M, Andrade AM, Carassay L, Abdala GI, Molas ML (2011) Drought tolerance and stress hormones: from model organisms to forage crops. Plant Environ:137–164

Aissa N, Malagoli M, Radhouane L (2018) An approach to alleviate the impact of drought stress with selenium amendment. Iran J Sci Tech Transac A: Sci 42(1):283–288

Akashi K, Miyake C, Yokota A (2001) Citrulline, a novel compatible solute in drought-tolerant wild watermelon leaves, is an efficient hydroxyl radical scavenger. FEBS Lett 508:438–442

Alexieva V, Sergiev I, Mapelli S, Karanov E (2001) The effect of drought and ultraviolet radiation on growth and stress markers in pea and wheat. Plant Cell Environ 24(12):1337–1344

Alsaeedi A, El-Ramady H, Alshaal T, El-Garawany M, Elhawat N, Al-Otaibi A (2019) Silica nanoparticles boost growth and productivity of cucumber under water deficit and salinity stresses by balancing nutrients uptake. Plant Physiol Biochem 139:1–10

Ambrosini VG et al (2018) High copper content in vineyard soils promotes modifications in photosynthetic parameters and morphological changes in the root system of 'Red Niagara' plantlets. Plant Physiol Biochem 128:89–98

Asadi S, Lebaschy MH, Khourgami A, Rad AS (2012) Effect of drought stress on the morphology of three Salvia sclarea populations. Ann Biol Res 3(9):4503–4507

Ashkavand P, Tabari M, Zarafshar M, Tomášková I, Struve D (2015) Effect of SiO_2 nanoparticles on drought resistance in hawthorn seedlings. Forest Res Pap 76:50–359

Ashkavand P, Zarafshar M, Tabari M, Mirzaie J, Nikpour A, Bordbar SK, Striker GG (2018) Application of SiO_2 nanoparticles as pretreatment alleviates the impact of drought on the physiological performance of *Prunus mahaleb* (Rosaceae). Boletín de la Sociedad Argentina de Botánica 53(2):207–219

Ashraf MY, Mahmood K, Ashraf M, Akhter J, Hussain F (2012) Optimal supply of micronutrients improves drought tolerance in legumes. Crop Prod Agr Imp:637–657

Awan S, Shahzadi K, Javad S, Tariq A, Ahmad A, Ilyas S (2021) A preliminary study of influence of zinc oxide nanoparticles on growth parameters of *Brassica oleraceae* var italic. J Saudu Soc Agr Sci 20:18–24

Azhar BJ, Noor A, Zulfiqar A, Zeenat A, Shakeel A, Chishti I, Abbas Z, Shakeel AN (2021) Effect of ZnO, SiO_2 and composite nanoparticles on *Arabidiopsis thaliana* and involvement of ethylene and cytokinins signaling pathway. Pak J Bot 53(2):437–446

Aziz N, Faraz M, Pandey R, Shakir M, Fatma T, Varma A, Barman I, Prasad R (2015) Facile algae-derived route to biogenic silver nanoparticles: synthesis, antibacterial, and photocatalytic properties. Langmuir 31:11605–11612

Aziz N, Pandey R, Barman I, Prasad R (2016) Leveraging the attributes of *Mucor hiemalis*-derived silver nanoparticles for a synergistic broad-spectrum antimicrobial platform. Front Microbiol 7:1984

Bajji M, Kinet J, Lutts S (2002) The use of the electrolyte leakage method for assessing cell membrane stability as a water stress tolerance test in durum wheat. Plant Growth Regul 36:61–70

Ball RA, Oosterhuis DM, Mauromoustakos A (1994) Growth dynamics of the cotton plant during water-deficit stress. Agron J 86(5):788–795

Bangar P, Chaudhury A, Tiwari B, Kumar S, Kumari R, Bhat KV (2019) Morphophysiological and biochemical response of mungbean [*Vigna radiata* (L.) Wilczek] varieties at different developmental stages under drought stress. Turk J Biol 43(1):58–69

Barnabás B, Jäger K, Fehér A (2008) The effect of drought and heat stress on reproductive processes in cereals. Plant Cell Env 31(1):11–38

Barrena R, Casals E, Colon J, Font X, Sanchez A, Puntes V (2009) Evaluation of the ecotoxicity of model nanoparticles. Chemosphere 75:850–857

Bektaşoĝlu B, Esin CS, Ozyürek MO, Kubilay G, Resat A (2006) Novel hydroxyl radical scavenging antioxidant activity assay for water-soluble antioxidants using a modified CUPRAC method. Biochem Bioph Res Co 345:1194–2000

Blum A (2017) Osmotic adjustment is a prime drought stress adaptive engine in support of plant production. Plant Cell Env 40(1):4–10

Brasili E, Bavasso I, Petrucceli V, Vilardi G, Valletta A, Bosco CD, Gentili A, Pasqua G, Palma LD (2020) Remediation of hexavalent chromium contaminated water through zerovalent iron nanoparticles and effects on tomato plant growth performance. Sci Rep 10:1920

Cakmak I (2008) Enrichment of cereal grains with zinc: agronomic or genetic biofortification. Plant Soil 302:1–17

Cao Y, Luo Q, Tian Y, Meng F (2017) Physiological and proteomic analyses of the drought stress response in *Amygdalus Mira* (Koehne) Yü et Lu roots. BMC Plant Biol 17:53

Cattivelli L, Baldi P, Crosetti C, Di Fonzo N, Faccioli P, Grassi M, Mastrangelo AM, Pecchioni N, Stanca AM (2002) Chromosome regions ans stress-related sequences involved in resistance to abiotic stress in triticeae. Plant Mol Biol 48:649–665

Caverzan A, Casassola A, Brammer SP (2016) Reactive oxygen species and antioxidant enzymes involved in plant tolerance to stress. In book: Abiotic and biotic stress in plants-recent advances and future perspectives. https://doi.org/10.5772/61368

Chahardoli A, Karimi N, Ma X, Qalekahani F (2020) Effects of engineered aluminium and nickel oxide nanoparticles on the growth and antioxidant defense system of *Nigella arvensis* L. Sci Rep 10:3847

Chaves MM, Flexas J, Pinheiro C (2009) Photosynthesis under drought and salt stress regulation mechanisms from whole plant to cell. Ann Bot 103:551–560

Chen M, Wang QY, Cheng XG, Xu ZS, Li LC, Ye XG (2007) GmDREB2, a soybean DRE-binding transcription factor, conferred drought and high-salt tolerance in transgenic plants. Biochem Biophys Res Commun 353:299–305

Cheong JJ, Do CY (2003) Methyl jasmonate as a vital substance in plants. Trends Genet 19(7):409–413

Cota-Ruiz K, Ye Y, Valdes C, Deng C, Wang Y, Wang Y, Jose AHV, Duarte-Gardea M, Gardea-Torresdey JL (2020) Copper nanowires as nanofertilizers of alfalfa plants: understanding nano-bio systems interactions from microbial genomics, plant molecular responses and spectroscopic studies. Sci Total Env 742:140572

Das A, Das B (2019) Nanotechnology a potential tool to mitigate abiotic stress in crop plants Abiotic and Biotic Stress in Plants. IntechOpen

Dhanda SS, Sethi GS, Behl RK (2004) Indices of drought tolerance in wheat genotypes at early stages of plant growth. J Agron Crop Sci 190:6–12

Dimkpa CO, Singh U, Bindraban PS, Elmer WH, Gardea-Torresdey JL, White JC (2019) Zinc oxide nanoparticles alleviate drought-induced alterations in sorghum performance, nutrient acquisition, and grain fortification. Sci Total Environ 688:926–934

Dimkpa CO, Andrews J, Sanabria J, Bindraban PS, Singh U, Elmer WH, Jorge LGT, White JC (2020) Interactive effects of drought, organic fertilizer, and zinc oxide nanoscale and bulk particles on wheat performance and grain nutrient accumulation. Sci Total Env 722(20):137808

Din MI, Arshad F, Hussain Z, Mukhtar M (2017) Green adeptness in the synthesis and stabilization of copper nanoparticles: catalytic, antibacterial, cytotoxicity, and antioxidant activities. Nanoscale Res Lett 12:638

Djanaguiraman M, Nair R, Giraldo JP, Prasad PVV (2018) Cerium oxide nanoparticles decrease drought-induced oxidative damage in sorghum leading to higher photosynthesis and grain yield. ACS omega 3(10):14406–14416

Dordas C (2009) Role of nutrients in controlling plant diseases in sustainable agriculture: a review. In: E. Lichtfouse et al. (Eds.), Sustainable Agriculture, Springer Science 443–460. https://doi.org/10.1007/978-90-481-2666-8-28

Drazkiewicz M, Skorzynska-Polit E, Krupa Z (2004) Copper-induced oxidative stress and antioxidant defense in *Arabidopsis thaliana*. Biometals 17:379–387

Dubouzet JG, Sakuma Y, Ito Y, Kasuga M, Dubouzet EG, Miura S, Seki M, Shinozaki K, Yamaguchi-Shinozaki K (2003) OsDREB genes in rice, *Oryza sativa* L. encode transcription activators that function in drought-, high-salt- and cold-responsive gene expression. Plant J 33:751–763

Ejaz M, Raja NI, Mashwani ZR, Ahmad MS, Hussain M, Iqbal M (2018) Effect of silver nanoparticles and silver nitrate on growth of rice under biotic stress. IET Nanobiotech 12(7):927–932

Elizabath A, Bahadur V, Misra P, Prasad VM, Thomas T (2017) Effect of different concentrations of iron xide and zinc oxide nanoparticles on growth and yield of carrot (*Daucus carota* L.). J Pharmacog Phytochem 6(4):1266–1269

Elouaer MA, Hannachi C (2013) Influence of seed priming on emergence and growth of coriander (*Coriandrum sativum* L.) seedlings grown under salt stress. Acta Agriculturae Slovenica 101(1):42–47

Faizan M, Faraz A, Yusuf M, Khan ST, Hayat S (2018) Zinc oxide nanoparticles-mediated changes in photosynthetic efficiency and antioxidant system of tomato plants. Photosynthetica 56(2):678–686

Fishel FM (2018) Pesticide injection and drenching. EDIS 2018(1)

Fujikawa I, Takehara Y, Ota M, Imada K, Sasaki K, Kajihara H, Sakai S, Jogaiah S, Ito S (2021) Magnesium oxide induces immunity against Fusarium induced wilt by triggering the jasmonic acid signaling pathway. J Biotech 325:100–108

Gao S, Wang Y, Yu S, Huang Y, Liu H, Chen W, He XE (2020) Effects of drought stress on growth, physiology and secondary metabolites of two Adonis species in Northeast China. Sci Hortic 259:108795

Gigon A, Matos A, Laffray D, Zuily-fodil Y, Pham-Thi A (2004) Effect of drought stress on lipid metabolism in the leaves of *Arabidopsis thaliana* (ecotype Columbia). Ann Bot 94:345–351

Gnanasiri SP, Saneoka H, Ogata S (1991) Cell membrane stability and leaf water relations as affected by potassium nutrition of water-stressed maize. J Exp Bot 42:739–745

Goufo P, Moutinho-Pereira JM, Jorge TF, Correia CM, Oliveira MR, Rosa EA, Trindade H (2017) Cowpea (*Vigna unguiculata* L. Walp.) metabolomics: osmoprotection as a physiological strategy for drought stress resistance and improved yield. Front Plant Sci 8:586

Guha T, Ravikumar KVG, Mukherjee A, Mukherjee A, Kundu R (2018) Nanopriming with zerovalent iron (nZVI) enhances germination percentage and growth in aromatic rice cultivar (*Oryza sativa* cv. Gobindabhog L.). Plant Physiol Biochem 127:403–413

Gupta M, Gupta S (2017) An overview of selenium uptake, metabolism, and toxicity in plants. Front Plant Sci 7:1–14

Gurumurthy S, Sarkar B, Vanaja M, Lakshmi J, Yadav S, Maheswari M (2019) Morpho-physiological and biochemical changes in black gram (*Vigna mungo* L. Hepper) genotypes under drought stress at flowering stage. Acta Physiol Plant 41:42

Hafeez B, Khanif YM, Saleem M (2013) Role of zinc in plant nutrition- a review. Am J Exp Agr 3:374–391

Haghighi M, Da Silva JAT, Mozafarian M, Afifipour Z (2013) Can Si and nano-Si alleviate the effect of drought stress induced by PEG in seed germination and seedling growth of tomato. Minerva Biotechnol 25:17–22

Hao Y, Zhang Z, Rui Y, Ren J, Hou T, Wu S, Rui M, Jiang F, Liu L (2016) Effect of different nanoparticles on seed germination and seedling growth in rice. 2nd annual International Conference on Advanced Material Engineering (AME), Atlantis press

Hayat S, Ali B, Ahmad A (2007) Salicylic acid: biosynthesis, metabolism and physiological role in plants. Springer, Dordrecht, pp 1–14

Hojjat SS (2020) Effects of TiO_2 nanoparticles on germination and growth characteristics of grass pea (*Lathyrus odoratus* L.) seed under drought stress. Naotech Russia 15(2):204–221

Hossain Z, Nouri MZ, Komatsu S (2012) Plant cell organelle proteomics in response to abiotic stress. J Proteome Res 11:37–48

Hosseini MS, Samsampour D, Ebrahimi M, Abadía J, Khanahmadi ME (2018) Effect of drought stress on growth parameters, osmolyte contents, antioxidant enzymes and glycyrrhizin synthesis in licorice (*Glycyrrhiza glabra* L.) grown in the field. Phytochemistry 156:124–134

Huang B, Liu JY (2006) Cloning and functional analysis of the novel gene GhDBP3 encoding a DRE-binding transcription factor from *Gossypium hirsutum*. Biochim Biophys Acta 1759:263–269

Huber AE, Bauerle TL (2016) Long-distance plant signaling pathways in response to multiple stressors: the gap in knowledge. J Exp Bot 67(7):2063–2079

Hussain M, Malik MA, Farooq M, Ashraf MY, Cheema MA (2008) Improving drought tolerance by exogenous application of glycinebetaine and salicylic acid in sunflower. J Agron Crop Sci 194(3):193–199

Hussein HA, Darwesh OM, Mekki BB (2019) Environmentally friendly nano-selenium to improve antioxidant system and growth of groundnut cultivars under sandy soil conditions. Biocatal Agric Biotechnol 18:101080

Ikram M, Raja NI, Javed B, Mashwani ZR, Hussain M, Hussain M, Ehsan M, Rafique N, Malik K, Sultana T, Akram A (2020) Foliar applications of bio-fabricated selenium nanoparticles to improve the growth of wheat plants under drought stress. Green Processing and Synthesis 9:706–714

Impa SM, Nadaradjan S, Jagadish SVK (2012) Drought stress induced reactive oxygen species and anti-oxidants in plants. In: Abiotic stress responses in plants. Springer, New York, pp 131–147

Iqbal M, Raja NI, Mashwani Z, Hussain M, Ejaz M, Yasmeen F (2019) Effect of silver nanoparticles on growth of wheat under heat stress. Iran J Sci Tech Trans Sci 43:387–395

Iqbal S, Waheed Z, Naseem A (2020) Nanotechnology and abiotic stress. In: Nanoagronomy, Springer Nature

Itroutwar PD, Kasivelu G, Raguraman V, Malaichamy K, Sevathapandian SK (2020) Effects of biogenic zinc oxide nanoparticles on seed germination and seedling vigor of maize (*Zea mays*). Biocat Agricult Biotech 29:101778

Jaberzadeh A, Moaveni P, Moghadam HRT, Zahedi H (2013) Influence of bulk and nanoparticles titanium foliar application on some agronomic traits, seed gluten and starch contents of wheat subjected to water deficit stress. Notulae Botanicae Horti Agrobotanici Cluj-Napoca 41(1):201–207

Jaglo KR, Kleff S, Amundsen KL, Zhang X, Haake V, Zhang JZ, Deits T, Thomashow MF (2001) Components of the Arabidopsis C-repeat/dehydration-responsive element binding factor cold response pathway are conserved in *Brassica napus* and other plant species. Plant Physiol 127:910–917

Jakubiak M, Giska I, Asztemborska M, Bystrzejewska-Piotrowska G (2014) Bioaccumulation and biosorption of inorganic nanoparticles: factors affecting the efficiency of nanoparticle mycoextraction by liquid-grown mycelia of *Pleurotus eryngii* and *Trametes versicolor*. Mycol Prog 13:525–532

Jaleel CA, Changxing Z, Jayakumar K, Iqbal M (2008) Low concentration of cobalt increases growth, biochemical constituents, mineral status and yield in *Zea mays*. J Sci Res 1(1):128–137

Javot H, Lauvergeat V, Santoni V, Martin-Laurent F, Guclu J, Vinh J, Heyes J, Franck KI, Schaffner AR, Bouchez D, Maurel C (2003) Role of a single aquaporin isoform in root water uptake. Plant Cell 15:509–522

Juhel G, Batisse E, Hugues Q, Daly D, Pelt FV, Halloran JO, Jansen MAK (2011) Alumina nanoparticles enhance growth of *Lemna minor*. Aq Toxicol 105(3–4):328–336

Kamalizadeh M, Bihamta M, Zarei A (2019) Drought stress and TiO_2 nanoparticles affect the composition of different active compounds in the Moldavian dragonhead plant. Acta Physiol Plantarum 41(2):21

Kamithi KD, Wachira F, Kibe AM (2016) Effects of different priming methods and priming durations on enzyme activities in germinating chickpea (*Cicer arietinum* L.). Amr J Nat App Sci 1:A1–A9

Kapoor D, Bhardwaj S, Landi M, Sharma A, Ramakrishnan M, Sharma A (2020) The impact of drought in plant metabolism: how to exploit tolerance mechanisms to increase crop production. Appl Sci 10(16):5692

Karamanos RE, Pomarenski Q, Goh TB, Flore NA (2004) The effect of foliar copper application on grain yield and quality of wheat. Canad J Plant Sci 84:47–56

Kavar T, Maras M, Kidrič M, Šuštar-Vozlič J, Meglič V (2008) Identification of genes involved in the response of leaves of *Phaseolus vulgaris* to drought stress. Mol Breed 21(2):159–172

Kawasaki S, Miyake C, Kouchi T, Yokota A (2000) Responses of wild watermelon to drought stress: accumulation of an ArgE homologue and citrulline in leaves during water deficit. Plant Cell Phys 41:864–873

Keskin H (2012) Physiological and biochemical characterization of drought tolerance in chickpea (Doctoral dissertation, İzmir Institute of Technology)

Khan MN, Alzuaibr F (2018) Nano-titanium dioxide-induced synthesis of hydrogen sulfide and cysteine augment drought tolerance in *Eruca sativa*. Asian J Plant Sci 17(4):213–221

Khan MN, Mobin M, Abbas ZK, AlMutairi KA, Siddiqui ZH (2017) Role of nanomaterials in plants under challenging environments. Plant Phys Biochem 110:194–209

Khan I, Saeed K, Khan I (2019) Nanoparticles: properties, applications and toxicities. Arab J Chem 12(7):908–931

Kiapour H, Moaveni P, Habibi D (2015) Evaluation of the application of gibbrellic acid and titanium dioxide nanoparticles under drought stress on some traits of basil (*Ocimum basilicum* L.). Int J Agron Agr Res 6(4):138–150

Kiba T, Kudo T, Kojima M, Sakakibara H (2011) Hormonal control of nitrogen acquisition: roles of auxin, abscisic acid, and cytokinin. J Exp Bot 62(4):1399–1409

Kim YH, Khan AL, Waqas M, Lee IJ (2017) Silicon regulates antioxidant activities of crop plants under abiotic-induced oxidative stress: a review. Front Plant Sci 8:510

Kinnersley AM, Turano FJ (2000) Gama aminobutyric acid (GABA) and plant responses to stress. Crit Rev Plant Sci 19:479–509

Klaine SJ, Alvarez PJJ, Batley GE, Fernandes TF, Handy RD, Lyon DY, Mahendra S, McLaughlin MJ, Lead JR (2008) Nanomaterials in the environment: behavior, fate, bioavailability, and effects. Environ Toxicol Chem 27:1825–1851

Kumaraswamy RV, Kumari S, Choudhary RC, Sharma SS, Pal A, Raliya R, Biswas P, Saharan V (2019) Salicylic acid functionalized chitosan nanoparticle: a sustainable bio stimulant for plant. Int J Biol Macromol 123:59–69

Lambers H, Chapin FS, Pons TL (2008) Plant physiological ecology. Springer Science & Business Media

Latef AAHA, Srivastava AK, El Sadek MSA, Kordrostami M, Tran LSP (2018) Titanium dioxide nanoparticles improve growth and enhance tolerance of broad bean plants under saline soil conditions. Land Degrad Dev 29:1065–1073

Laware S, Raskar S (2014) Effect of titanium dioxide nanoparticles on hydrolytic and antioxidant enzymes during seed germination in onion. Int J Curr Microbiol App Sci 3(7):749–760

Lee SB, Suh MC (2013) Recent advances in cuticular wax biosynthesis and its regulation in Arabidopsis. Mol Plant 6(2):246–249

Lee CW, Mahendra S, Zodrow K, Li D, Tsai YC, Braam J (2010) Developmental phytotoxicity of metal oxide nanoparticles to *Arabidopsis thaliana*. Environ Toxicol Chem 29:669–675

Lin D, Xing B (2007) Phytotoxicity of nanoparticles: inhibition of seed germination and root growth. Environ Pollut 150:243–250

Mahdavi S, Kafi M, Fallahi E, Shokrpour M, Tabrizi L (2016) Water stress, nano silica, and digoxin effects on minerals, chlorophyll index, and growth in ryegrass. Int J Plant Prod 10:251–264

Malamy J, Henning J, Klessig DF (1992) Temperature dependent induction of salicylic acid and its conjugates during the resistance response to tobacco mosaic virus infection. Plant Cell 4(3):359–366

Martínez-Fernández D, Vítková M, Bernal MP, Komárek M (2015) Effects of nano-maghemite on trace element accumulation and drought response of *Helianthus annuus* L. in a contaminated mine soil. Water Air Soil Pollut 226:1–9

Mathew SS, Sunny NE, Shanmugam V (2021) Green synthesis of anatase titanium dioxide nanoparticles using *Cuminum cyminum* seed extract; effect on mung bean (*Vigna radiate)* seed germination. Inorg Chem Comm 126:108485

Mathur PB, Devi MJ, Serraj R, Yamaguchi-Shinozaki K, Vadez V, Sharma KK (2004) Evaluation of transgenic groundnut lines under water limited conditions. Int Archis Newslett 24:33–34

Maurel C, Chrispeels MJ (2001) Aquaporins: a molecular entry into plant water relations. Plant Physiol 125:135–138

Mimmo T, Del Buono D, Terzano R, Tomasi N, Vigani G, Crecchio C et al (2014) Rhizospheric organic compounds in the soil–microorganism–plant system: their role in iron availability. Eur J Soil Sci 65(5):629–642

Mishra V, Mishra RK, Dikshit A, Pandey AC (2014) Chapter 8-Interactions of nanoparticles with Plants: an emerging prospective in the agriculture Industry. Emerging Technologies & Management of Crop Stress Tolerance. Elsevier Inc. 159–180

Mohammadi H, Esmailpour M, Gheranpaye A (2016) Effects of TiO_2 nanoparticles and water-deficit stress on morpho-physiological characteristics of dragonhead (*Dracocephalum moldavica* L.) plants. Acta Agriculturae Slovenica 107(2):385–396

Moussa HR, Abdel-Aziz SM (2008) Comparative response of drought tolerant and drought sensitive maize genotypes to water stress. Aust J Crop Sci 1:31–36

Mozafari AA, Havas F, Ghaderi N (2018) Application of iron nanoparticles and salicylic acid in in vitro culture of strawberries (*Fragaria X ananassa* Duch.) to cope with drought stress. Plant Cell Tissue Org Cult 132:511–532

Musa AM, Johansen C, Kumar J, Haris D (1999) Response of chickpea to seed priming in the high Barind tract of Bangladesh. Int Chickpea Pigeonpea Newslett 6:20–22

Naidu B, Cameron D, Konduri S (1998) Improving drought tolerance of cotton by glycinebetaine application and selection. Paper presented at the Proceedings of the 9th Australian agronomy conference, Wagga

Najafi S, Razavi SM, Khoshkam M, Asadi A (2020) Effects of green synthesis of sulfur nanoparticles from *Cinnamomum zeylanicum* barks on physiological and biochemical factors of lettuce (*Lactuca sativa*). Physiol Mol Biol Plants 26(5):1055–1066

Nascimento WM, Cantliffe DJ, Huber DJ (2004) Ethylene evolution and endo-β-mannanase activity during lettuce seed germination at high temperature. Sci Agric 61(2):156–163

Nerd A, Neumann PM (2004) Phloem water transport maintains stem growth in a drought-stressed crop cactus (*Hylocereus undatus*). J Amer Soc Horticul Sci 129(4):486–490

Nguyen DV, Nguyen HM, Le NT, Nguyen KH, Le HM, Nguyen A, Dinh NTT, Hoang SA, Ha CV (2020) Copper nanoparticles application enhances plant growth and grain yield in maize under drought stress conditions. J Plant Growth Reg. https://doi.org/10.1007/s00344-021-10301-w

Nishiyama R, Watanabe Y, Leyva-Gonzalez MA, Van Ha C, Fujita Y, Tanaka M, Herrera-Estrella L (2013) Arabidopsis AHP2, AHP3, and AHP5 histidine phosphotransfer proteins function as redundant negative regulators of drought stress response. Proceed Nat Acad Sci 110(12):4840–4845

Palmqvist NGM, Gulaim A, Seisenbaeva GA, Svedlindh P, Kessler VG (2017) Maghemite nanoparticles acts as nanozymes, improving growth and abiotic stress tolerance in *Brassica napus*. Nanoscale Res Lett 12:631

Patanè C, Saita A, Sortino O (2013) Comparative effects of salt and water stress on seed germination and early embryo growth in two cultivars of sweet sorghum. J Agron Crop Sci 199(1):30–37

Patil B, Chetan HT (2018) Foliar fertilization of nutrients. Marumegh 3(1):49–53

Peleg Z, Blumwald E (2011) Hormone balance and abiotic stress tolerance in crop plants. Curr Opin Plant Biol 14(3):290–295

Pospisilova J, Synkova H, Rulcova J (2000) Cytokinins and water stress. Biol Plantarum 43(3):321–328

Prasad R, Kumar V, Prasad KS (2014) Nanotechnology in sustainable agriculture: present concerns and future aspects. Afr J Biotechnol 13(6):705–713

Prathyusha IVSN, Chaitanya KV (2019) Effect of water stress on the physiological and biochemical responses of two different Coleus (Plectranthus) species. Biol Fut 70:312–322

Premachandra GS, Saneoka H, Kanaya M, Ogata S (1991) Cell membrane stability and leaf surface wax content as affected by increasing water deficits in maize. J Exp Bot 42:167–171

Qayyum A, Razzaq A, Ahmad M, Jenks MA (2011) Water stress causes differential effects on germination indices, total soluble sugar and proline content in wheat (*Triticum aestivum* L.) genotypes. Afr J Biotech 10(64):14038–14045

Quan R, Shang M, Zhang H, Zhao Y, Zhang J (2004) Improved chilling tolerance by transformation with betA gene for the enhancement of glycinebetaine synthesis in maize. Plant Sci 166(1):141–149

Queiroz MS, Oliveira CE, Steiner F, Zuffo AM, Zoz T, Vendruscolo EP, Menis FT (2019) Drought stresses on seed germination and early growth of maize and sorghum. J Agr Sci 11(2):310–318

Rafi H, Dawar S, Zaki MJ (2015) Seed priming with extracts of *Acacia nilotica* (L.) Willd. Ex Delile and *Sapindus mukorossi* (L.) plant parts in the control of root rot fungi and growth of plants. Pak J Bot 47:1129–1113

Ragab GA, Saad-Allah KM (2020) Green synthesis of sulfur nanoparticles using *Ocimum basilicum* leaves and its prospective effect on manganese-stressed *Helianthus annuus* (L.) seedlings. Ecotoxicol Environ Saf 191:110242

Rahmani F, Peymani A, Daneshvand E, Biparva P (2016) Impact of zinc oxide and copper oxide nano-particles on physiological and molecular processes in *Brassica napus* L. Ind J Plant Physiol 21:122–128

Rameshraddy G, Pavithra J, Reddy BHR, Salimath M, Geetha KN, Shankar AG (2017) Zinc oxide nano particles increases Zn uptake, translocation in rice with positive effect on growth, yield and moisture stress tolerance. Ind J Plant Physiol 22:287–294

Rayman MP (2008) Food-chain selenium and human health: emphasis on intake. Br J Nutr 100:254–268

Razmjoo K, Heydarizadeh P, Sabzalian MR (2008) Effect of salinity and drought stresses on growth parameters and essential oil content of *Matricaria chamomile*. Int J Agric Biol 10(4):451–454

Rehman HU, Basra SMA, Farooq M (2011) Field appraisal of seed priming to improve the growth, yield, and quality of direct seeded rice. Turk J Agri Forestry 35:357–365

Rossi L, Fedenia LN, Sharifan H, Ma X, Lombardini L (2019) Effects of foliar application of zinc sulfate and zinc nanoparticles in coffee (*Coffea Arabica* L.) plants. Plant Physiol Biochem 135:160–166

Salazar C, Hernández C, Pino MT (2015) Plant water stress: associations between ethylene and abscisic acid response. Chilean J Agr Res 75:71–79
Salehi-Lisar SY, Bakhshayeshan-Agdam H (2016) Drought stress in plants: causes, consequences, and tolerance. Drought Stress Tolerance Plants 1:1–16
Samadi N, Yahyaabadi S, Rezayatmand Z (2014) Effect of TiO_2 and TiO_2 nanoparticles on germination, root and shoot length and photosynthetic pigments of *Mentha piperita*. Int J Plant Soil Sci 3(4):408–418
Sandquist DR, Ehleringer JR (2003) Population-and family-level variation of brittlebush (*Encelia farinosa*, Asteraceae) pubescence: its relation to drought and implications for selection in variable environments. Amer J Bot 90(10):1481–1486
Satti SH, Raja NI, Javed B, Akram A, Mashwani Z, Ahmad MS, Ikram M (2021) Titanium dioxide nanoparticles elicited agro morphological and physicochemical modifications in wheat plants to control *Biopolaris sorokiana*. Plose One. https://doi.org/10.1371/journal.pone.0246880
Sauret-Gueto S, Calder G, Harberd NP (2012) Transient gibberellin application promotes *Arabidiopsis thaliana* hypocotyl cell elongation without maintaining transverse orientation of microtubules on the outer tangential wall of epidermal cells. Plant J 69(4):628–639
Saxena N (2009) Emerging trends of nanoparticles application in food technology: safety paradigms. Nanotoxicol 3:10–18
Saxena R, Tomar RS, Kumar M (2016) Exploring nanobiotechnology to mitigate abiotic stress in crop plants. J Pharm Sci Res 8(9):974
Semida WM, Abdelkhalik A, Mohamed G, El-Mageed TAA, El-Mageed SAA, Rady MM, Ali EF (2021) Foliar application of zinc oxide nanoparticles promotes drought stress tolerance in egg plant. Plan Theory 10:421
Serraj R, Sinclair T (2002) Osmolyte accumulation: can it really help increase crop yield under drought conditions? Plant Cell Env 25(2):333–341
Serraj R, Barry JS, Sinclair TR (1998) Accumulation of γ-aminobutyric acid in nodulated soybean in response to drought stress. Physiol Plant 102:79–86
Shalaby TA, El-Ramady HR (2014) Effect of foliar application of some bio-stimulants on growth, yield and its components and storability of garlic (*Allium sativum* L.). Aus J Crop Sci 8(2):271–275
Shallan MA, Hassan HM, Namich AA, Ibrahim AA (2016) Biochemical and physiological effects of TiO_2 and SiO_2 nanoparticles on cotton plant under drought stress. Res J Pharm Biol Chem Sci 7(4):1540–1551
Shen YG, Zhang WK, He SJ, Zhang JS, Liu Q, Chen SY (2003) An EREBP/AP2-type protein in *Triticum aestivum* was a DRE-binding transcription factor induced by cold, dehydration and ABA stress. Theor Appl Genet 106:923–930
Shinde S, Paralikar P, Ingle AP, Rai M (2020) Promotion of seed germination and seedling growth of *Zea mays* by magnesium hydroxide nanoparticles synthesized by filtrate from *Asprgillus Niger*. Arab J Chem 13:3172–3182
Siddiqui MH, Al-Khaishany MY, Al-Qutami MA, Al-Whaibi MH, Grover A, Ali HM, Al-Wahibi MS, Bukhari NA (2015) Response of different genotypes of faba bean plant to drought stress. Int J Mol Sci 16:10214–10227
Singh D, Kumar A (2020) Binary mixture of nanoparticles in sewage sludge: impact on spinach growth. Chemosph 254:126794
Singh H, Jassal RK, Kang JS, Sandhu SS, Kang H, Grewal K (2015) Seed priming techniques in field crops- a review. Agr Rev 36(4):251–264
Singh A, Singh NB, Hussain I, Singh H (2017) Effect of biologically synthesized copper oxide nanoparticles on metabolism and antioxidant activity to the crop plants *Solanum lycopersicum* and *Brassica oleracea* var. botrytis. J Biotech 262:11–27
Smith S, De Smet I (2012) Root system architecture: insights from Arabidopsis and cereal crops. Philos Trans R Soc Lond Ser B Biol Sci 367(1595):1441–1452

Smoleń S (2012) Foliar nutrition: current state of knowledge and opportunities. In: Srivastava AK (Ed.), Advances in citrus nutrition, Springer Science pp 41–58. https://doi.org/10.1007/978-94-007-4171-3-4

Soleimanzadeh H (2013) Effect of seed priming on germination and yield of corn. Int J Agr Crop Sci 5(4):366–369

Somerville C, Briscoe J (2001) Genetic engineering and water. Sci 292(5525):2217–2217

Srivastava G, Das CK, Das A, Singh SK, Roy M, Kim H, Sethy N, Kumar A, Sharma RK, Singh SK, Philipij D, Das M (2014) Seed treatment with iron pyrite (FeS_2) nanoparticles increases the production of spinach. RSC Adv 4:58495–58504

Subbarao G, Johansen C, Slinkard A, Nageswara R, Saxena N, Chauhan Y, Lawn R (1995) Strategies for improving drought resistance in grain legumes. Crit Rev Plant Sci 14(6):469–523

Subbarao GV, Nam NH, Chauhan YS, Johansen C (2000) Osmotic adjustment, water relations and carbohydrate remobilization in pigeonpea under water deficits. J Plant Physiol 157(6):651–659

Sun L, Song F, Zhu X, Liu S, Liu F, Wang Y, Li X (2021) Nano-ZnO alleviates drought stress via modulating the plant water use and carbohydrate metabolism in maize. Arch Agron Soil Sci 67(2) https://doi.org/10.101080/03650340.2020.1723003

Taran N, Storozhenko V, Svietlova N, Batsmanova L, Shvartau V, Kovalenko M (2017) Effect of zinc and copper nanoparticles on drought resistance of wheat seedlings. Nanoscale Res Lett 12(1):60

Tavili A, Zare S, Moosavi SA, Enayati A (2011) Effects of seed priming on germination characteristics of Bromus species under salt and draught conditions. Amr-Euras J Agr Environ Sci 10(2):163–168

Tiwari M, Sharma NC, Fleischmann P, Burbage J, Venkatachalam P, Sahi SV (2017) Nanotitania exposure causes alterations in physiological, nutritional and stress responses in tomato (*Solanum lycopersicum*). Front Plant Sci 8:633

Trenberth KE, Dai A, Van Der Schrier G, Jones PD, Barichivich J, Briffa KR, Sheffield J (2014) Global warming and changes in drought. Nat Climate Change 4(1):17–22

Tripathy JN, Zhang J, Robin S, Nguyen TT, Nguyen HT (2000) QTLs for cell-membrane stability mapped in rice (*Oryza sativa* L.) under drought stress. Theor Appl Genet 100:1197–1202

Trujillo LE, Sotolongo M, Menendez C, Ochogavia ME, Coll Y, Hernandez I, Hernandez L (2008) SodERF3, a novel sugarcane ethylene responsive factor (ERF), enhances salt and drought tolerance when over expressed in tobacco plants. Plant Cell Physiol 49(4):512–525

Tuberosa R (2012) Phenotyping for drought tolerance of crops in the genomics era. Front Physiol 3:347

Tyerman SD, Niemietz CM, Brameley H (2002) Plant aquaporins: multifunctional water and solute channels with expanding roles. Plant Cell Environ 25:173–194

Vanvoka R, Landa P, Podlipna R, Dobrev PI, Prerostova S, Langhansova L, Gaudinova A, Motkova K, Knirsch V, Vanek T (2017) ZnO nanoparticles effects on hormonal pools in *Arabidopsis thaliana*. Sci Total Env 593-594:535–542

Verma KK, Singh P, Song XP, Malviya MK, Singh RK, Chen GL, Solomon S, Li YR (2020) Mitigating climate change for sugarcane improvement: role of silicone in alleviating abiotic stresses. Sugar Tech 22(5):741–749

Viera I, Perez-Galvez A, Roca M (2019) Green natural colorants. Molecules 24

Villar-Salvador P, Planelles R, Oliet J, Peñuelas-Rubira JL, Jacobs DF, González M (2004) Drought tolerance and transplanting performance of holm oak (*Quercus ilex*) seedlings after drought hardening in the nursery. Tree Physiol 24:1147–1155

Vinkovic T, Novák O, Strnad M, Goessler W, Jurasin DD, Paradikovic N, Vrcek IV (2017) Cytokinin response in pepper plants (*Capsicum annuum* L.) exposed to silver nanoparticles. Environ Res 156:10–18

Vlot AC, Dempsey DMA, Klessig DF (2009) Salicylic acid, a multifaceted hormone to combat disease. Annu Rev Phytopathol 47:177–206

Wahid A (2007) Physiological implications of metabolites biosynthesis in net assimilation and heat stress tolerance of sugarcane (*Saccharum officinarum*) sprouts. J Plant Res 120:219–228

Wang Z, Huang B (2004) Physiological recovery of Kentucky bluegrass from simultaneous drought and heat stress. Crop Sci 44:1729–1736

Wang Q, Zhao R, Chen Q, da Silva JAT, Chen L, Yu X (2019) Physiological and biochemical responses of two herbaceous peony cultivars to drought stress. Hort Sci 54:492–498

Werner T, Nehnevajova E, Kollmer I, Novak O, Strnad M, Kramer U, Schmulling T (2010) Root-specific reduction of cytokinin causes enhanced root growth, drought tolerance, and leaf mineral enrichment in Arabidopsis and tobacco. Plant Cell 22(12):3905–3920

Wery J, Silim S, Knights E, Malhotra R, Cousin R (1994) Screening techniques and sources of tolerance to extremes of moisture and air temperature in cool season food legumes Expanding the Production and Use of Cool Season Food Legumes, Springer Publisher: 439–456

Yamada M, Morishita H, Urano K, Shiozaki N, Yamaguchi-Shinozaki K, Shinozaki K, Yoshiba Y (2005) Effects of free proline accumulation in petunias under drought stress. J Exp Bot 56(417):1975–1981

Yamaguchi-Shinozaki K, Shinozaki K (2004) Improving drought and cold stress tolerance in transgenic rice. Proceed World Rice Res Conf Tsukuba, Japan, 5–7 November 2004

Yusefi-Tanha E, Fallah S, Ali R, Pokhrel LR (2020) Zinc oxide nanoparticles (ZnONPs) as novel nanofertilizer: influence on seed yield and antioxidant defense system in soil grown soy bean (*Glycine max* cv.Kowsar). Sci Total Env 738:140240

Zahedi SM, Abdelrahman M, Hossieni MS, Hoveizeh NF, Tran LSP (2018) Alleviation of the effect of salinity on growth and yield of strawberry by foliar spray of selenium nanoparticles. Env Poll 253:246–258

Zahedi SM, Hosseini MJ, Meybodi NDH, Peijnenburg W (2021) Mitigation of the effect of drought on growth and yield of pomegranates by foliar spraying of different sizes of selenium nanoparticles. J Sci Food Agr. https://doi.org/10.1002/jsfa.11167

Zareii FD, Roozbahani A, Hosnamidi A (2014) Evaluation the effect of water stress and foliar application of Fe nanoparticles on yield, yield components and oil percentage of safflower (*Carthamus tinctorious* L.). Int J Adv Biol Biomed Res 2:1150–1159

Zhao TJ, Sun S, Liu Y, Liu JM, Liu Q, Yan YB, Zhou HM (2006) Regulating the drought-responsive element (DRE)-mediated signaling pathway by synergic functions of trans-active and trans-inactive DRE binding factors in *Brassica napus*. J Biol Chem 281:10752–10759

Zhao J, Ren W, Zhi D, Wang L, Xia G (2007) Arabidopsis DREB1A/CBF3 bestowed transgenic tall fescue increased tolerance to drought stress. Plant Cell Rep 26:1521–1528

Zhu JK (2001) Cell signaling under salt, water and cold stresses. Curr Opin Plant Biol 4(5):401–406

Zhu JK (2002) Salt and drought stress signal transduction in plants. Ann Rev Plant Biol 53(1):247–273

Zsiros O, Nagy V, Párducz Á, Nagy G, Ünnep R, El-Ramady H, Prokisch J, Lisztes-Szabó Z, Fári M, Csajbók J (2019) Effects of selenate and red se-nanoparticles on the photosynthetic apparatus of *Nicotiana tabacum*. Photosynth Res 139:449–460

Zulfiqar F, Younis A, Riaz A, Mansoor F, Hameed M, Akram NA, Abideen Z (2020) Morpho-anatomical adaptations of two *Tagetes erecta* L. cultivars with contrasting response to drought stress. Pak J Bot 52(3):801–810

Zuverza-Mena N, Armendariz R, Peralta-Videa JR, Gardea-Torresdey JL (2016) Effects of silver nanoparticles on radish sprouts: root growth reduction and modifications in the nutritional value. Front Plant Sci:7–90

Zuverza-Mena N, Martínez-Fernández D, Du W, Hernández-Viezcas JA, Bonilla-Bird N, López-Moreno ML, Komárek M, Peralta-Videa JR, Gardea-Torresdey JL (2017) Exposure of engineered nanomaterials to plants: insights into the physiological and biochemical responses-a review. Plant Physiol Biochem 110:236–264

Chapter 6
Role of Various Nanoparticles in Countering Heavy Metal, Salt, and Drought Stress in Plants

Suliman Mohammed Alghanem, Haifa Abdulaziz S. Alhaithloul, Magdi T. Abdelhamid, and Mona H. Soliman

Abstract The various forms of nanoparticles (NPs) designed so far and their application in crop plants for sustainable crop production is receiving great attention in today's life. In plants, NPs minimize loss of nutrients, reduces various diseases and improves the growth and yields under different abiotic stresses. Application of various forms of NPs in plants affect seed germination, seedling vigor, respiration, flowering, initiation of roots, antioxidant defense responses, growth and photosynthesis and regulate responses to various abiotic stress conditions. The usage of various NPs is occupied in the plants' to protect against induced oxidative stress as they have been found to mimic the role of various antioxidative enzymes. The high doses of NPs induces phytotoxic effect in plants as they orchestrates the production of various reactive oxygen species (ROS), whereas, low doses of NPs exert beneficiary results in various plants. Therefore, keeping in view such a conflicting and ambiguous situation of NPs, the present chapter deciphers the ameliorating role of various NPs under stressful environmental conditions like salt, drought and heavy metal stresses in crop plants. A brief explanation of NPs mediated control of vital plant processes under abiotic stresses is also presented.

S. M. Alghanem
Biology Department, Faculty of Science, Tabuk University, Tabuk, Saudi Arabia
e-mail: s-alghanem@ut.edu.sa

H. A. S. Alhaithloul
Biology Department, College of Science, Jouf University, Sakaka, Kingdom of Saudi Arabia
e-mail: haifasakit@ju.edu.sa

M. T. Abdelhamid
Botany Department, National Research Centre, Cairo, Egypt

M. H. Soliman (✉)
Botany and Microbiology Department, Faculty of Science, Cairo University, Giza, Egypt

Biology Department, Faculty of Science, Taibah University, Yanbu, Kingdom of Saudi Arabia
e-mail: hmona@sci.cu.edu.eg

M. Faizan et al. (eds.), *Sustainable Agriculture Reviews 53*, Sustainable Agriculture Reviews 53, https://doi.org/10.1007/978-3-030-86876-5_6

Keywords Salt · Drought · Heavy metal stress · Nanoparticles

6.1 Introduction

Abiotic stresses such as toxic trace heavy metal/metalloid, salt, heat, cold, drought, flood or the other oxidizing factors are major factors limiting agricultural productivity worldwide thus restricting quality, quantity and the economic yield of crop plants (Li et al. 2019; Hasanuzzaman et al. 2019; Ahammed et al. 2020; Šamec et al. 2021; Alhaithloul and Soliman 2021). In the natural environment, plants experience different environmental pressures simultaneously which altogether induce physiological, morphological, biochemical and molecular alterations affecting their growth and metabolism potential to a great extent (Wani et al. 2018; Zaid et al. 2020; Sadiq et al. 2020; Alhaithloul et al. 2020a; Godoy et al. 2021). In the present decade, with the increasing global population coupled with the problem of climate change effects, there is a dire need to engineer stress-tolerance in diverse crops in the available and shrinking agricultural tracts. In an estimate of Food and Agricultural Organization (2009), global agriculture is facing a massive threat under the dynamic environmental conditions, population explosion and over-exploitation of natural resources. IPCC, (2007) has predicted a variation in global climate and proposed that the variation is supposed to cause an increase in mean temperature by of 2–4 °C by the end of this century. The agriculture sector is regarded as one of the prime areas affected by a change in climatic conditions (Karimi et al. 2018, 2021; Agovino et al. 2019; De Pinto et al. 2020; Raza et al. 2019; Zaid et al. 2021). Agriculture and climate change are inter-correlated, as a change in one factor could cause severe agricultural impacts on the particular area (Fahad and Wang 2020; Alhaithloul et al. 2020b). Nevertheless, the productivity and improvement of crop plants under diverse abiotic pressures under changing climate is one of the prime tasks faced by the nano-biotechnologists community across the globe (Khan et al. 2017; Iftikhar et al. 2019). Several efforts are currently going on to engineer tolerance in plants challenged with various environmental pressures. The induced stress resistance obtained by metabolic regulation engineering, characterization of defensive genes and transcription factors (TFs) by phytohormones' application have also yielded paramount results. The underlying abiotic stress tolerance mechanisms are complicated and multigenic, and thus through research efforts are needed to engineer stress resistance in crop plants (Wani et al. 2018; Abdel Razik et al. 2020). Nonetheless, the contribution of various kinds of NPs in plant stress physiology became the new area of interest; as they are cost effective agents as compared to the phytohormones. Nevertheless, stress resistance can also be engineered by the exogenous application of various kinds of NPs. Inspite of various extensive research efforts currently being utilized in the field, the quest for yielding abiotic stress-resistant crop plants by the application of various kinds of NPs remains very scanty.

The field of nanotechnology is emerging now day by day and has opened new dimensions of the formation, characterization and applications of various nanoparticles (NPs) in agriculture and allied sectors. The branch of nanotechnology provides sophisticated tools and brilliant technology platforms for investigating and transforming biological systems. All man-made and biological systems have nanoscale viz- such as nanocrystal, nanotubes or nanobiomotors as the first level of organization at which their basic properties and functions are characterised. In 1959, the "nanotechnology" was first put forwarded by Richard Feynman and is now regarded as one of the fastest emerging areas among research community (Reviewed by Irshad et al. 2021). Nevertheless, the term "nanotechnology" was first used by Professor Norio Taniguchi in 1974 (Bakker et al. 2008). In current times nanotechnology has increased the potentiality touse various kinds of exogenous NPs on diverse crop plants to ameliorate the abiotic pressures like salinity, water deficit and heavy metal stresses (Rajput et al. 2018; Ahmad et al. 2019; Rizwan et al. 2019; Abdel Latef et al. 2020; Tarrahi et al. 2021). The term "nanoparticle"in the field of nanotechnology is the nano-sized material having diameter not larger than 100 nm (Ahmed et al. 2021a). The NPs possessed size-dependent physico-chemical properties that differentiate them from various other sub-micron/micron-sized particles (Dasgupta et al. 2017).The high reactivity and the corresponding physicochemical dynamicity of NPs are due to the higher surface area (S)-to-volume (V) ratio, and this is considered as one of the foremost features of NPs (Mauter et al. 2018). In addition to this, the high surface energy and quantum confinement is also the distinguish characteristics of NPs than their bulk materials in the environment (Ma et al. 2010). As NPs have small size, larger surface area, high catalytic reacting potential, the features enable NPs to interact within plant tissues and correspondingly affect morpho-physiological traits of plants (Khan et al. 2017). NPs are potential signalling molecules for the delivery of various chemicals under in vitro conditions for improving growth and yield of crop plants as well as resistance against various stresses. So far as the synthesis of NPs is concerned, multiple methods like physico-chemical or biological are utilized (Singh et al. 2018). Various forms of NPs contain metals. Some of the notable metal containing NPs includes aluminum (Al), titanium (Ti), silver (Ag), copper (Cu), bismuth (Bi), zinc (Zn), gold (Au), cobalt (Co), indium (In), iron (Fe), silica (Si), molybdenum (Mo), tin (Sn) and nickel (Ni). Among these, the most commonly produced metal-oxide NPs are titanium dioxide (TiO_2), copper oxide (CuO), zinc oxide (ZnO), iron oxide (Fe_2O_3) silicon dioxide (SiO_2), aluminum oxide (Al_2O_3), cerium dioxide (CeO_2), cuprous oxide (Cu_2O), nickel oxide (NiO), magnesium oxide (MgO), zirconium dioxide (ZrO_2), indium oxide (In_2O_3) and lanthanum oxide (La_2O_3) (Rajput et al. 2018; Ameen et al. 2021). The external supply of various kinds of NPs is known to accelerate the growth, development, physio-biochemistry and mineral nutrient uptake under normal and stressful environmental conditions in crop plants (Ahmad et al. 2019, 2021; Faizan et al. 2020; Singh et al. 2021; Sharifi et al. 2021). A classical study involving *Linum usitatissimum* L. plants, the efficacy of four different types of NPs viz., ZnO, SiO_2, Fe_2O_3, and TiO_2 was evaluated and their potentiality of salt induce stress mitigation was tested on growth, physiology and biochemistry of the plant (Singh et al. 2021). The four different types of NPs at same dose (50 mg/L) were applied on stressed and non-stressed plants under sodium chloride

(NaCl) stress. The results revealed that the tested plants were affected by all the four types NPs positively and caused an improvement in growth, assimilation of carbon and nutrient, whilst imposition of salt stress accelerated the proline as well as reactive oxygen species (ROS) production. The plants grown under the salt stress and NPs application showed an increase in the antioxidant enzymatic system and other physiochemical reactions. Thus, in view of the potentiality of various kinds of NPs, we discuss their role in alleviating major abiotic stress in diverse crop plants. The existing research studies reported the ameliorating effect of various NPs under single abiotic stress. However, the role of various NPs under salt, heavy metal and drought stress receives little attention. Thus, the present chapter throws light at length on the role of various NPs in promoting plant productivity under different environment stresses.

6.2 Heavy Metal/Metalloid Stress

The contamination of heavy metal/metalloid in soil and water environment has always remained a top priority of agricultural scientists and environmentalists as their ions transfer form one food chain to other. In the present era, heavy metals (HMs), due to their toxic effects, are one of the well-known environmental pollutants. The environmental contamination by trace elements (so called heavy metals/metalloids) is increasing day by day due to rapid industrialization and urbanization across the globe. The mining and industrial use by human beings causes build up of trace element accumulation at levels above baseline concentrations in environment (Shaheen and Rinklebe 2015; Shahid et al. 2017, 2018; Shahid 2021; Arif et al. 2019; Natasha et al. 2021). The living organisms, including animals, humans and plants are directly or indirectly exposed to these toxic trace elements (Antoniadis et al. 2019; Hasanuzzaman et al. 2019; Badr et al. 2014). The effects of these ions on these organisms are driven chiefly by their bioavailability. In the current decade, the applications and production of metal-based and metal-oxide NPs have increased manifold as a result of their enhanced physicochemical properties and profound role in alleviating HM stress in diverse crop plants with respect to the bulk parent materials. In a report on rice plants, Ahmed et al. (2021c) unravelled the ameliorative role of NPs under combined drought and Cd stress toxicity by studying the nutrient acquisition and stress responsive genetic mechanisms. The iron oxide (IONPs) and hydrogel NPs (HGNPs) were applied on tested plants. *Bacillus* strain *RNT1* was used for the biofabrication of the IONPs, while HGNPs were obtained by chemical reactions. On the basis of XRD and FTIR results, it was shown that the NPs capping by different functional groups together with their crystalline state. The application of NPs was found to increase antioxidant enzymes, photosynthetic efficiency, biomass, and nutrient acquisition but caused a decrease in ROS content and rate of Cd translocation. It was revealed that the expression of the Cd transporter genes viz-*OsHMA2*, *OsHMA3* and *OsLCT1* were also curtailed by the applicationof NPs-treatment. The Cd and Pb HM pollution induced toxicity and the prime role of

ZnO-NPs application in*Gossypium hirsutum* L. seedlings were studied by unravelling the photosynthetic machinery and antioxidant enzymes mediated defense mechanisms (Priyanka et al. 2021). The combined application of ZnO-NPs along with Cd and Pb treatments significantly promoted the growth of shoot and root, biomass, levels of chlorophyll *a*, *b* and carotenoids and content of total soluble protein. A significant increment in the activities of catalase (CAT), peroxidase (POX) superoxide dismutase (SOD), and ascorbate peroxidase (APX) was noticed in plants grown under ZnO-NPs along with Cd and Pb supply. Their study concluded that the addition of Zn-ONPs protected seedlings of cotton by ameliorating the HM-induced toxic effects and increased the plant physiochemical characteristics via regulating the photosynthesis capacity inhibition as well as acting on the antioxidant defense system. Emamverdian et al. (2021) studied the effect of SiO_2-NPs (100 and 200 μM), on the germination characteristics and growth parameters of *Phyllostachys edulis* seedling under 100 μM Cd stress. The SiO_2-NPs application caused a significant improvement in germination characteristics and increased mean germination time. The results revealed that application of SiO_2-NPs at 200 μM showed more pronounced effects rather than 100 μM could ameliorate the 100 μM Cd stress during germination of seed and the seedling growth inhibition in tested plants. In wheat plants, the green synthesis, characterization, and application of FeO-NPs for joint-amelioration of salt and Cd stresses were studied (Manzoor et al. 2021). The used NPs were synthesized from *Pantoea ananatis* strain *RNT4*. The FeO-NPs were found to be a spherical with a size ranging from 19 to 40 nm as authenticated by images taken through scanning electron microscopy (SEM) and transmission electron microscopy (TEM). The application of bioengineered FeO-NPs (100 mg/kg) stimulated the plant growth, reprogrammed the morpho-physiological state and caused a significant increase in the N, P and K^+ contents, whilst simultaneously reduced the salt ions in the wheat grains. In a recent pot study, Adrees et al. (2021) tested the potentiality of foliar exposure of ZnO-NPs at four levels (0, 25, 50, 100 mg/L) in improving the growth of *Triticum aestivum* L. plants under simultaneous Cd and various moisture conditions. The two levels (70% and 35% of water holding capacity) of soil moisture regimes were started from 6 weeks of germination. The wheat growth was increased under ZnO-NPs supply and the optimum results were found with 100 mg/L ZnO-NPs under normal moisture level. It was found that the concentrations of Cd in grains decreased by 26.00%, 81.00% and 87.00% in normal moisture and 35.00%, 66.00% and 81.00% in water deficit conditions, with application of 25, 50 and 100 mg/L of ZnO-NPs. The application of ZnO-NPs increased the contents of chlorophyll, decreased the oxidative stress and boosted the activities of leaf SOD and POX. The mechanistic role of different biogenic Cu-NPs (25, 50, and 100 mg/kg soil) synthesized from *Shigella flexneri* SNT22 in reducing the Cd translocation from soil to wheat plants was studied (Noman et al. 2020) in a growth chamber under optimal environmental regimes. It was found that the treatment of 100 mg/kg Cu-NPs increased the plant length, shoot dry weight, N, P, contents, but decreased the acropetal Cd translocation. In lettuce plants, Gao et al. (2020) tested the effectiveness of exogenous application of graphene oxide (GO) treatment, which is a novel engineered nanomaterial in increasing the photosynthetic capacity and reducing the oxidative stress under Cd-stress conditions. The treatment of plants with 30 mg/L of GO were found to significantly reduced the signs of Cd-induced toxicity by increasing the net photosynthesis, stomatal conductance, transpiration rate, chlorophyll content, photosynthetic electron transport rate, maximum photochemical efficiency of photosystem II, actual quantum yield, concentrations of ribulose-1,5-bisphosphate carboxylase and oxygenase, and biomass of tested plants. Nevertheless, 30 mg/LGO also reduced the ROS accumulation and malondialdehyde (MDA) content. Hussain et al. (2021a) explored the effects of ZnO-NPs on chlorophyll contents, antioxidants and proline content in *Persicaria hydropiper* L. for Pb phytoremediation potential in a hydroponic study. The seedlings of *Persicaria hydropiper* L. were grown in Pb-challenged environment supplemented with four doses (5, 10, 15, and 20 mg/L) of ZnO-NPs. It was found that Pb significantly decreased the growth of seedlings, but ZnO-NPs application significantly alleviated Pb-induced stress by increasing the growth of tested plants, improving the contents of chlorophyll and carotenoid. ZnO-NPs decreased the oxidative stress in the tested seedlings by enhancing the synthesis of osmolytes, secondary metabolites, and modulating the activity of antioxidative defense enzymes. The analysis of Pb in plant tissues was studied by inductively coupled plasma spectroscopy (ICP) method. In Pb and Cd challenged environment, the effectiveness of multi-walled carbon nanotubes (MWCNTs) in enhancing growth traits, state of redox and total glutathione content in *Calendula officinalis* L. plants was studied (Sharifi et al. 2021).The seedlings were cultivated under 0, 50, 100, 250, 500 and 1000 mg/L MWCNT in the presence of Pb and Cd-polluted soils. It was found that the MWCNT application up to 250 mg/L alleviated Pb and Cd-induced stress by minimizing oxidative damage and increasing antioxidant defense system on one hand and promoted the phytoremediation capacity on the other hand. In rice plants, a pot study was executed to unravel the effects of the exogenous spray of different concentrations (0, 50, 75, 100 mg/L) of ZnO-NPs alone or in combination with biochar (1.0% w/w) grown on an aged Cd-contaminated soil. The combined application of ZnO-NPs and biochar was found to improve the photosynthetic potential and plant biomass, whilst significantly decreasing the concentration of Cd and in contrast to increasing concentrations of Zn in shoot and root tissues. The combined application of 100 mg/L ZnO-NPs with biochar was found to diminish the Cd concentrations in rice shoot and root by 39% and 38% respectively (Ali et al. 2019). In a classical study, Rizwan et al. (2019) applied ZnO (0, 25, 50, 75, and 100 mg/L) and Fe-NPs (0, 5, 10, 15, and 20 mg/L) through seed priming for 24 h in improving the growth and reducing the accumulation of Cd in wheat plants. The plant traits like height of plant, length of spike, dry weights of shoots, roots, spikes, and grains were increased with the application of NPs in higher doses. The NPs supply increased the photosynthetic potential, while reduced electrolyte leakage (EL), accumulation of Cd in roots, shoots, and grains and SOD and POX activities in leaves of Cd-stressed wheat. In yet another study, Hussain et al. (2019) explored the efficacy of seed priming Si-NPs (0, 300, 600, 900, 1200 mg/L) for 24 h under Cd stress conditions in wheat in terms of growth, photosynthesis, oxidative stress, yield, and accumulation of Si and Cd in wheat plants. It was observed that application of Si-NPs showed a positive effect on the wheat growth and chlorophyll contents, concentration of Si, while diminished the oxidative stress, Cd concentrations, especially in grains, and positively affected the plant antioxidant enzyme activity. Ragab and Saad-Allah (2020) tested the efficacy of seed priming with greenly synthesized sulfur-NPs for 18 h in enhancing defense mechanisms for minimizing 100 mM manganese sulphate-induced oxidative injury in *Helianthus annuus* L. seedlings. The Mn application induced oxidative stress indicators such as superoxide ion ($O_2^{\cdot-}$) and hydrogen peroxide (H_2O_2) and MDA but lowered antioxidant compounds, such as ascorbic acid, glutathione, and total flavonoids content and the activities of SOD, CAT, while induced APX, guaiacol peroxidise (GPOX), polyphenol oxidase (PPO), and glutathione reductase (GR) activities. The priming application with S-NPs significantly boosted CAT and SOD activities and enhanced the levels of antioxidant compounds and caused significant decrease in ROS levels and MDA. In order to investigate the Cd toxicity; the applicability of exogenous interventions of nano-TiO_2 on bioaccumulation of tissue Cd, enzymes associated with stress and potential dietary health risk was assessed in cowpea plants (Ogunkunle et al. 2020; Abd Ellatif et al. 2021). The plants were exposed to Cd toxicity (10 mg/kg soil) for 21 days and six regimes of foliar applications of nano-TiO_2 were given. The nano-TiO_2 promoted chlorophyll b and total chlorophyll contents under Cd stress, reduced significantly the contents of Cd in roots, shoots and grains, promoted stress enzymes activity and increased the Zn, Mn and Co levels. The impact of TiO_2-NPs (0, 100, 250 mg/L) in a hydroponic study in maize plants under Cd (0, 50 mM) was determined in order to evaluate the best mode of application of TiO_2 NPs (Lian et al. 2020). The results indicated that root application of TiO_2-NPs and 100 mg/L Cd significantly enhanced the Cd uptake and produced greater maize phytotoxicity whereas, by contrast, the foliar TiO_2-NPs markedly decreased shoot Cd contents and caused a strong effect on ameliorating the Cd-induced toxicity by boosting SOD, GST activities and up-regulated the metabolism of galactose and citrate cycle, aspartate, alanine, glutamate, serine, glycine and threonine levels. Irshad et al. (2019) in a kinetic and equilibrium study, prepared and studied the application of TiO_2-NPs for removing the Cd present in wastewater. The results based on Langmuir, Freundlich, and Dubinin-Radushkevich isotherm models successfully favored the Cd adsorption from wastewater by TiO_2-NPs supply. In *Azolla filiculoides*, Spanò et al. (2019) applied TiO_2-NPs for alleviating the Cd toxicity. It was revealed that the higher Cd contents in leaves did not cause damage to the photosynthetic machinery. The TiO_2-NPs permanence ensured an optimum antioxidant defense (proline and GPOX, and CAT activities) and consequently induced a decrease in H_2O_2 content. Singh and Lee, (2016) carried out the potentiality of nano-TiO_2 on the Cd bioaccumulation in soybean plants. Cd addition caused a significant decrement in plant growth, biomass, pigment and protein contents, increased proline and MDA content appreciably. The application of nano-TiO_2 restricted the toxicity of Cd by causing an increase in the photosynthetic potential and growth traits of the tested plants. The overall results of this study revealed that the nano-TiO_2 application could boost the uptake of Cd and ameliorate the Cd-induced stress in soybean. In a recent study, Shah et al. (2021) studied the combined efficacy of *Bacillus fortis*

IAGS 223 and ZnO-NPs (20 mg/kg) in order to minimize the Cd-induced (75 mg/kg) phytotoxicity in *Cucumis melo* plants. The application of ZnO-NPs and *B.fortis* IAGS 223 increased growth traits including photosynthetic pigments and reduced H_2O_2, MDA contents and modulated the activities of antioxidative enzymes. Raghib et al. (2020) unravelled the interactive effect of ZnO-NPsand AM (*Glomus macrocarpum*) application in mitigating the Pb-phytotoxicity in wheat plants. The combined dose of AM fungi and ZnO-NPs synergistically increased the overall growth performance and reduced the uptake of Pb in tested plants by increasing 30.66% plant height, 30.62% plant fresh weight, 54.26% plant dry weight, 45.45% chlorophyll content, 19.59% proline content, 26.65% higher activities of SOD, 15.12% CAT, and 52.09% and 58.19% decrease in Pb concentrations in root and shoot of wheat plants respectively. Under Pb stress and ZnO-NPs and AM supply, the structural changes in chloroplast of *Triticum aestivum* L. were also observed. It was found that in control plants chloroplast with regular shaped and well-arranged thylakoid systems were found, however, distorted chloroplast structure with wide intracisternal spaces, wavy appearance of the grana and stroma thylakoids were observed in plants spiked with Pb. The regained regular shaped with well-arranged thylakoid systems were observed when plants were treated with AM fungi and ZnO-NPs. In light of the above discussion, it is inferred that various kinds of NPs exert an ameliorating role in countering stress of various HMs.

6.3 Salt Stress

The problem of salt toxicity is one of the principal abiotic threats to sustainable agriculture practices that reduce yield of plants by causing impairments in various physiological, biochemical, and molecular functions to an appreciable extent. The salt ions are principally soluble in water. The presence of salt ions in the soil and water environment decreases growth and development of the plant to a considerable extent. In general, salinity is divided into primary and secondary type. Primary salinity results from the natural processes viz- rain, weathering and wind. The secondary salinity is caused by the action of various anthropogenic activities like shifting cultivation, deforestation, excessive irrigation practices and land degradation. In soil and water environment salts ions like Na^+, Cl^-, Mg^{2+}, Ca^{2+}, SO_4^{2-}, and HCO_3^- are present (Tanji 2002; Yan et al. 2015). The concentration of various salt ions is expressed usually as $mmol_c\ L^{-1}$ or mg/L. However, in scientific and analytical terms, electrical conductivity (EC) or siemens per meter (Sm^{-1}) is taken as a measurable unit. Globally large tracts of lands are affected by saline stress. It has been estimated that soil salinity harms 900 million ha of land, which is approximately 20% worldwide land area or about half of the total arable irrigated land (Shin et al. 2016; Velmurugan et al. 2020). In terms of losses due to salinity, over US $27.3 billion per annum is lost in agriculture sector (Qadir et al. 2014).The problem of soil salinization is expected to worse further globally due to inefficient irrigation practices, inappropriate usage of various agrochemicals, and the ever increasing pollution, which markedly caused a 50% loss in plant productivity globally (Zhao et al. 2020).Salt stress affects germination of seedlings, plant growth traits, and affects vital physiological and biochemical processes in plants (Misra and Gupta 2005; Khan et al. 2010; Elkelish et al. 2019; Soliman et al. 2020c; Ahanger et al. 2020; Abdel Latef et al. 2021). The presence of excess salt ions disturbs the homeostasis of K^+/Na^+ in cytoplasm, thereby the ratio of cytosolic K^+/Na^+ gets reduced (Li et al. 2020; Soliman et al. 2020b; Hamid et al. 2021; Hussain et al. 2021b). Excessive ions of Na^+ and Cl^- inside tissues of plants lead to ionic, osmotic and oxidative stress by the generation of various ROS (Khan et al. 2010; Shelke et al. 2019; Soliman et al. 2020a; Feng et al. 2020; Sami et al. 2021; Sheikhalipour et al. 2021; Alharbi et al. 2021). The ROS includes oxygen radicals, like hydroxyl (·OH), $O_2^{\cdot -}$, peroxyl (ROO·) and some non-radicals such as H_2O_2, singlet oxygen ($^1O_2$) and ozone (O_3) (Zaid and Wani 2019; Hasanuzzaman et al. 2020). The reactive nitrogen species (RNS) is a similar other collective term including radicals like nitric oxide (NO) and nitric dioxide (NO_2), and non-radicals such as nitrous acid (HNO_2) and dinitrogen tetroxide (N_2O_4). Both these ROS and RNS act as signalling agents in modulating salt stress responses in crop plants (Kohli et al. 2019; Tomar et al. 2021). Thus, engineering salinity tolerance in crop plants is vital for increasing global security of food and for sustainable development of modern agriculture. To achieve these goals, it is imperative to dissect new ways and means to ameliorate the salinity problem. In current years, the adoption of nanotechnology and the products derived from it holds ground in agriculture and has gained a considerable ground in ameliorating salt-induced effects in various crop pants. Many reports are available to show that NPs confer salinity stress tolerance in diverse species of crop plants (Avestan et al. 2019; Alabdallah and Alzahrani 2020; Ye et al. 2020; Sheikhalipour et al. 2021; Faizan et al. 2021; Mohammadiet al. 2021). It has been reviewed by Hussain et al. (2021b) that a variety of NPs like C, K, Ca, S, Ag, Cu, Fe, Zn, B, Si, and Ti under natural as well as controlled growth conditions effectively enable plants to strive well under salt stress conditions. Abou-Zeid et al. (2021) studied the influence of seed priming with ZnO-NPs;50, 100 and 500 mg/L on the salt-induced (150 mM) damages in *Triticum aestivum* L. plants. The NPs treatment resulted in a significant improvement in wheat growth traits. It was also found that salt-induced alterations of growth, photosynthetic pigments, photosynthetic efficiency, and leaf ultrastructure changes were minimized by the treatment of ZnO-NPs priming in addition to the induction of the various changes in shoot protein pattern profiles of tested plants. In pearl millet plants: a life cycle study was executed to show the efficacy of Ag-NPs induced improvement in plant growth traits and ability to reduce the salt ions accumulation (Khan et al. 2020a). The pearl millet plants were exposed to graded (0, 120, 150 mM) levels of salinity treatments. Imposition of salt stress triggered a substantial increase in the endogenous concentrations of Na^+ and Cl^- ions in different organs which in turn led to increase in H_2O_2 and MDA content, augmentation of antioxidant enzyme activities, negative effect on plant growth traits like height, root and shoot dry biomass, chlorophylls and carotenoid contents which eventually caused a negative effect in yield of plant. The Ag-NPs (0, 10, 20 and 30 mM) treatment significantly increased the morphology, physiology and yield-related traits.

The Ag-NPs application augmented the plant growth and reduced the oxidative stress and Na^+ and Cl^- uptake, maintained the Na^+/K^+ ratio. The Ag-NPs treatment was found to improve the SOD, CAT activities but decreased the POX activity, while simultaneously reduced the salt-induced oxidative stress. Unravelling the destructive impact of salt pressure in sweet basil, the role of graphene oxide NPs and glycine betaine (GB) each in three concentrations; 0, 50 and 100 mg/L were studied (Ganjavi et al. 2021).It was shown that GO-GB NPs treatment ameliorated the salinity toxicity by boosting traits related to proline, antioxidant enzymes activities, agronomy, photosynthetic pigment contents, chlorophyll fluorescence, membrane stability index, phenols, and constituents of essential oils while decreasing oxidative stress biomarkers. The best positive effects were obtained by 50 mg/L. In a completely randomized design experiment with three replications, Hezaveh et al. (2019) studied the interactive effects of salinity (0, 50 and 100 mM) and ZnO-NPs (0, 20 and 80 mg/L) on physiological and molecular parameters of *Brassica napus* L. plants. It was found that the imposition of salt stress triggered leakage of ions, lessen relative water content, stomata density and Hill reaction and reduced the expression of *ARP*, *MYC* and *SKRD2* genes but *MPK4* gene expression increased in tested plants. The application of ZnO-NPs was found to improve the Hill reaction, and reduced ion leakage and decreased the expression of *MYC*, *MPK4*, and *SKRD2* genes but increased the *ARP* gene expression. It was concluded that 20 mg/L ZnO-NPs was the optimum concentration to reduce the toxicity of salinity stress in rapeseed plants. Thus, it becomes clear that various forms of NPs in different doses increase salt stress tolerance in diverse crop plants.

6.4 Drought Stress

In the current era of changing climate, imposition of drought stress in plants is a major growth-limiting stress factor and the main reason for drought stress is the yield reduction both in quantity and quality. Drought exerts a major threat to crop plants and is seen as one of the major constraints on global agriculture as the frequency and intensity of drought stress is further expected to rise in the near future (Anjum et al. 2011; Koffler et al. 2014; Kour et al. 2020; Kirova et al. 2021). Due to tremendous anthropogenic induced-climate changes, the annual precipitation pattern is affected which leads to severe imposition of drought stress in many agricultural land tracts (Guerra et al. 2015; Gao et al. 2019). In plants drought stress imposes a negative effect in growth traits and productivity through photosynthetic capacity inhibition and closure of stomata (Saradadevi et al. 2017; Oraee and Tehranifar 2020). The conspicuous response of drought induce stress is the accumulation of ROS which can cause severe photosynthetic apparatus and chlorophyll structure damage (Sharma et al. 2017; Pakdel et al. 2020; Hasan et al. 2020).Plants in some cases are unable to absorb water the soil environment, even though the optimum moisture is present and this process is known as physiological drought (Daryanto et al. 2017; Salehi-Lisar and Bakhshayeshan-Agdam 2020). The treatment of various NPs in diverse crop plants has been reported to modulate various physio-biochemical activities. Semida et al. (2021) applied ZnO-NPs (0, 50, and 100 ppm) in *Solanum melongena* L. plants for promoting drought stress tolerance (60% of crop evapo-transpiration). It was found that imposition of drought stress significantly decreased relative water content (RWC), membrane stability index (MSI), and photosynthetic efficiency of tested plants. The foliar application of ZnO-NPs to drought-stressed plants regained the decreased values of RWC and MSI which enhanced the photosynthetic efficiency. Under field conditions, the impact of exogenous application of ZnO-NPs on physiological, yield components and quality attributes of four flax cultivars under drought stress were evaluated (Rashwan et al. 2020).The results showed that under drought stress conditions foliar spray of ZnO-NPs induced a significant increment in studied traits. In *Vicia faba* L. plants, the involvement of NO in nano-titanium dioxide (nTiO_2) induced accumulation of osmolytes and activation of antioxidant defense system towards water-deficit stress tolerance was worked out (Khan et al. 2020c). Water-deficit stress increased the concentration of ROS, intensity and contents of EL and MDA which decreased the chlorophyll content, nitrate reductase (NR) activity and growth traits of tested plants. The application of 15 mg/L nTiO_2 under water-stress significantly induced synthesis of NO and the NR activity and boosted the enzymatic and non-enzymatic defense system, enhanced the accumulation of osmolytes (proline and GB) and suppressed the generation of ROS, EL and MDA content. The nTiO_2-induced synthesis of NO in activation of defense system in faba bean was confirmed by using the scavenger of NO viz- cPTIO [2-(4-carboxyphenyl)-4,4,5,5-tetramethylimidazoline-1-oxyl-3-oxide] which re-imposed the water stress conditions. Dimkpa et al. (2019) under drought-induced stress in sorghum worked out the alleviatory role of ZnO-NPs by studying the acquisition of nutrients and fortification of grains. The plants were grown in soil amended with ZnO-NPs (1, 3, and 5 mg Zn/kg) and drought (40% of field moisture capacity) was imposed after 4 weeks of seed germination. It was found that drought significantly reduced grain yield, grain N translocation and total (root, shoot and grain) N acquisition and delayed the emergence of flag leaf and grain head. The ZnO-NPs (5 mg/kg) improved grain N after translocation and restored the total N levels, promoted the shoot uptake of P. It was concluded that the ZnO-NPs accelerated the plant development, promoted yield by increasing the critically essential nutrients and improved the N acquisition. In order to study the effect of CeO_2-NPs in decreasing the drought-induced oxidative damage in sorghum plants, Djanaguiraman et al. (2018) carried out outdoor pot-culture experiment in a randomized complete block for increasing photosynthesis and grain yield. Imposition of drought induced the oxidative stress and reduced photosynthesis and reproductive strategies. Foliar-sprayed 10 mg/L CeO_2-NPs under drought stress increased the antioxidant properties that resulted in the alleviation of drought-induced oxidative stress by causing the scavenging of ROS and reduced the leaf $O_2^{\cdot -}$, levels of $H_2O_2$and MDA content but increased leaf C assimilation rates, germination of pollens and seed yield per plant. Under the drought-induced stress conditions, in order to study the effects of various metal-based NPs like Fe, Cu, Co, and Zn, Linh et al. (2020) treated them in soybean plants. The results argued that NPs application improved the drought stress tolerance by increasing biomass reduction rate, RWC and drought tolerance index. The study got further confirmation by quantitative-PCR analysis which revealed that expression of several drought-responsive genes was up-regulated by NPs treatment. Khan et al. (2020b) in a pot study studied the impact of Si-NPs on wheat growth and physiology grown under Cd contaminated habitat under different soil moisture levels. The Si-NPs were applied in four different (0, 25, 50, and 100 mg/kg) concentrations and two water levels (70% and 35% soil water-holding capacity) were imposed to wheat plants. It was found that plants treated with Si- NPs showed an improvement in the plant growth traits and photosynthesis potential, and reduced the concentrations of metal in grains under drought stress by reducing the oxidative stress via minimizing contents of H_2O_2, EL, and MDA and increasing SOD and POX activities. In a recent work, Ali et al. (2021) studied the role of chitosan-NPs (1%) in improving drought (100 and 50% of field capacity) stress tolerance in *Catharanthus roseus* L. It was found that drought stress decreased plant growth, total chlorophyll content, RWC, stomatal conductance and but increased oxidative stress. Treating plants with NPs mitigated the drought stress effects and increased the accumulation of proline and CAT and APX activities and reduced oxidative stress. The imposition of drought stress increased accumulation of alkaloid that had an additive effect on further subjecting plants to the application of NPs. The NPs caused the high alkaloid content accumulation under drought stress conditions and was concerned with the up-regulation of expression of gene- strictosidine synthase, deacetylvindoline-4-O-acetyltransferase, peroxidase 1 and geissoschizine synthase. In wheat plants, Taran et al. (2017) applied colloidal solution of Zn and Cu-NPs for inducing drought stress resistance. It was found that both the applied NPs ameliorated the negative effects of drought stress by increasing the antioxidative enzymes activities and reducing the MDA content, stabilized the content of photosynthetic pigments and increased RWC of tested plants. In yet another study, Ahmed et al. (2021b) showed that applications of Cu and Ag-NPs in wheat induced tolerance against drought stress and resulted in increment of yield. By applying Se-NPs in pomegranates, it was found that the effect of drought stress on growth and yield was mitigated (Zahedi et al. 2021). Thus in view of the above discussion, it is inferred that various kinds of NPs applied in diverse crop plants impart drought stress tolerance by modulating functional physiological traits.

6.5 Conclusion

The present collected literature clearly indicates that various forms of NPS act as regulatory signalling molecules in plants under various abiotic pressures. It is inferred that NPs ameliorated abiotic stresses like salinity, drought, and heavy metal toxicity by regulating various principal plant metabolic processes. The modulated antioxidant defense system have been observed under abiotic circumstances, however, NPs application further boosted it. This cause entire regained and recharged

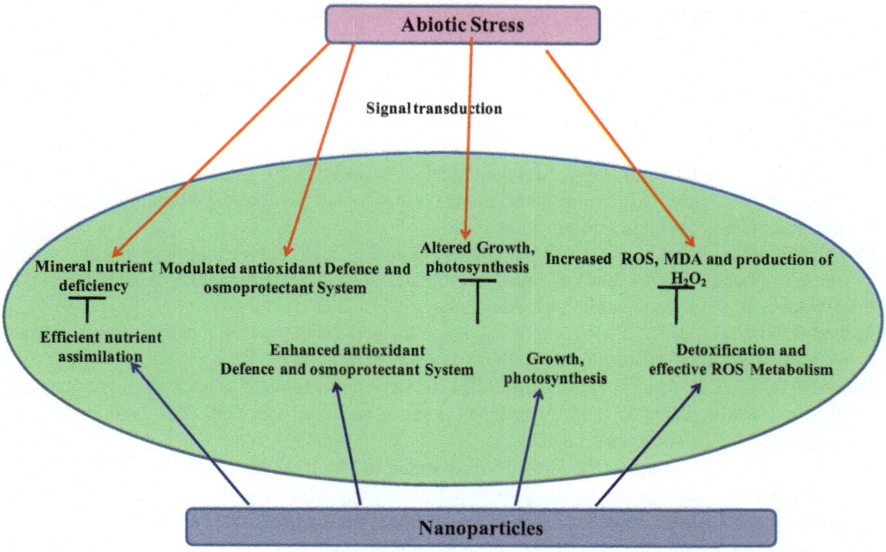

Fig. 6.1 The effect of nanoparticles induced stress tolerance in plants. For details, readers are kindly requested to go through the text in which details have been described apprehensively

antioxidant defense machinery by plants that could ameliorate various abiotic stress due to efficient detoxification various ROS. The entire sequence of events involving boosted antioxidant defense system by various kinds of NPs have diagrammatically represented in Fig. 6.1.

References

Abd Ellatif S, El-Sheekh MM, Senousy HH (2021) Role of microalgal ligninolytic enzymes in industrial dye decolorization. Int J Phytoremediation 23(1):41–52. https://doi.org/10.1080/15226514.2020.1789842

Abdel Latef AAHA, Zaid A, Alhmad MFA, Abdelfattah KE (2020) The impact of priming with Al_2O_3 nanoparticles on growth, pigments, osmolytes, and antioxidant enzymes of Egyptian Roselle (*Hibiscus sabdariffa* L.) cultivar. Agronomy 10:681

Abdel Latef AAHA, Zaid A, Alwaleed EA (2021) Influences of priming on selected physiological attributes and protein pattern responses of salinized wheat with extracts of *Hormophysa cuneiformis* and *Actinotrichia fragilis*. Agronomy 11:545

Abdel Razik ES, Alharbi BM, Pirzadah TB, Alnusairi GS, Soliman MH, Hakeem KR (2020) γ-Aminobutyric acid (GABA) mitigates drought and heat stress in sunflower (*Helianthus annuus* L.) by regulating its physiological, biochemical and molecular pathways. Physiol Plant 13216. https://doi.org/10.1111/ppl.13216

Abou-Zeid HM, Ismail GSM, Abdel-Latif SA (2021) Influence of seed priming with ZnO nanoparticles on the salt-induced damages in wheat (*Triticum aestivum* L.) plants. J Plant Nutr 44:629–643

Adrees M, Khan ZS, Hafeez M, Rizwan M, Hussain K, Asrar M, Ali S (2021) Foliar exposure of zinc oxide nanoparticles improved the growth of wheat (*Triticum aestivum* L.) and decreased cadmium concentration in grains under simultaneous cd and water deficient stress. Ecotoxicol Environ Saf 208:111627

Agovino M, Casaccia M, Ciommi M, Ferrara M, Marchesano K (2019) Agriculture, climate change and sustainability: the case of EU-28. Ecol Indic 105:525–543

Ahammed GJ, Li X, Liu A, Chen S (2020) Brassinosteroids in plant tolerance to abiotic stress. J Plant Growth Regul:1–14

Ahanger MA, Aziz U, Alsahli AA, Alyemeni MN, Ahmad P (2020) Influence of exogenous salicylic acid and nitric oxide on growth, photosynthesis, and ascorbate-glutathione cycle in salt stressed *Vigna angularis*. Biomol Ther 10:42

Ahmad B, Zaid A, Jaleel H, Khan MMA, Ghorbanpour M (2019) Nanotechnology for phytoremediation of heavy metals: mechanisms of nanomaterial-mediated alleviation of toxic metals. In: Advances in Phytonanotechnology. Academic, pp 315–327

Ahmed B, Rizvi A, Ali K (2021a) Nanoparticles in the soil–plant system: a review. Environ Chem Lett. https://doi.org/10.1007/s10311-020-01138-y

Ahmed F, Javed B, Razzaq A, Mashwani ZUR (2021b) Applications of copper and silver nanoparticles on wheat plants to induce drought tolerance and increase yield. IET Nanobiotechnol. https://doi.org/10.1049/nbt2.12002

Ahmed T, Noman M, Manzoor N, Shahid M, Abdullah M, Ali L, Li B (2021c) Nanoparticle-based amelioration of drought stress and cadmium toxicity in rice via triggering the stress responsive genetic mechanisms and nutrient acquisition. Ecotoxicol Environ Saf 209:111829

Alabdallah NM, Alzahrani HS (2020) The potential mitigation effect of ZnO nanoparticles on [*Abelmoschus esculentus* L. Moench] metabolism under salt stress conditions. Saudi J Biol Sci 27:3132–3137

Alhaithloul HAS, Soliman MH (2021) Methyl Jasmonate and Brassinosteroids: emerging plant growth regulators in plant abiotic stress tolerance and environmental changes. In: Aftab T, Hakeem KR (eds) Plant growth regulators. Springer, Cham. https://doi.org/10.1007/978-3-030-61153-8_8

Alhaithloul HAS, Abu-Elsaoud AM, Soliman MH (2020a) Abiotic Stress Tolerance in Crop Plants: Role of Phytohormones. In Abiotic Stress in Plants. IntechOpen

Alhaithloul HA, Soliman MH, Ameta KL, El-Esawi MA, Elkelish A (2020b) Changes in ecophysiology, osmolytes, and secondary metabolites of the medicinal plants of *Mentha piperita* and *Catharanthus roseus* subjected to drought and heat stress. Biomol Ther 10(1):43

Alharbi BM, Elhakem AH, Alnusairi GS, Soliman MH, Hakeem KR, Hasan MM, Abdelhamid MT (2021) Exogenous application of melatonin alleviates salt stress-induced decline in growth and photosynthesis in *Glycine max* (L.) seedlings by improving mineral uptake, antioxidant and glyoxalase system. Plant Soil Environ 67:208–220. https://doi.org/10.17221/659/2020-PSE

Ali S, Rizwan M, Noureen S, Anwar S, Ali B, Naveed M, Ahmad P (2019) Combined use of biochar and zinc oxide nanoparticle foliar spray improved the plant growth and decreased the cadmium accumulation in rice (*Oryza sativa* L.) plant. Environ Sci Pollut Res 26:11288–11299

Ali EF, El-Shehawi AM, Ibrahim OHM, Abdul-Hafeez EY, Moussa MM, Hassan FAS (2021) A vital role of chitosan nanoparticles in improvisation the drought stress tolerance in *Catharanthus roseus* (L.) through biochemical and gene expression modulation. Plant Physiol Biochem 161:166–175

Ameen F, Alsamhary K, Alabdullatif JA, Al-Nadhari S (2021) A review on metal-based nanoparticles and their toxicity to beneficial soil bacteria and fungi. Ecotoxicol Environ Saf 213:112027

Anjum S, Xie X, Wang L, Saleem M, Man C, Lei W (2011) Morphological, physiological andbiochemical responses of plants to drought stress. Afr J Agric Res 6:2026–2032

Antoniadis V, Shaheen SM, Levizou E, Shahid M, Niazi NK, Vithanage M, Ok YS, Bolan N, Rinklebe J (2019) A critical prospective analysis of the potential toxicity of trace element regulation limits in soils worldwide: are they protective concerning health risk assessment?—a review. Environ Inter 127:819–847

Arif MS, Yasmeen T, Shahzad SM, Riaz M, Rizwan M, Iqbal S, Ali S (2019) Lead toxicity induced phytotoxic effects on mung bean can be relegated by lead tolerant *Bacillus subtilis* (PbRB3). Chemosphere 234:70–80

Avestan S, Ghasemnezhad M, Esfahani M, Byrt CS (2019) Application of nano-silicon dioxide improves salt stress tolerance in strawberry plants. Agronomy 9:246

Badr AM, Shabana EF, Senousy HH, Mohammad HY (2014) Antiinflammatory and anti-cancer effects of b-carotene, extracted from Dunaliella bardawil by milking. J Food Agric Environ 12(3):24–31

Bakker RM, Yuan HK, Liu Z, Drachev VP, Kildishev AV, Shalaev VM, Boltasseva A (2008) Enhanced localized fluorescence in plasmonic nanoantennae. Appl Physics Lett 92:043101

Daryanto S, Wang L, Jacinthe PA (2017) Global synthesis of drought effects on cereal, legume, tuber and root crops production: a review. Agricul Water Manag 179:18–33

Dasgupta N, Ranjan S, Ramalingam C (2017) Applications of nanotechnology in agriculture and water quality management. Environ Chem Lett 15:591–605

De Pinto A, Cenacchi N, Kwon HY, Koo J, Dunston S (2020) Climate smart agriculture and global food-crop production. PLoS One 15(4):e0231764

Dimkpa CO, Singh U, Bindraban PS, Elmer WH, Gardea-Torresdey JL, White JC (2019) Zinc oxide nanoparticles alleviate drought-induced alterations in sorghum performance, nutrient acquisition, and grain fortification. Sci Total Environ 688:926–934

Djanaguiraman M, Nair R, Giraldo JP, Prasad PVV (2018) Cerium oxide nanoparticles decrease drought-induced oxidative damage in sorghum leading to higher photosynthesis and grain yield. ACS Omega 3:14406–14416

Elkelish AA, Soliman MH, Alhaithloul HA, El-Esawi MA (2019) Selenium protects wheat seedlings against salt stress-mediated oxidative damage by up-regulating antioxidants and osmolytes metabolism. Plant Physiol Biochem 137:144–153

Emamverdian A, Ding Y, Mokhberdoran F, Ahmad Z, Xie Y (2021) The effect of silicon nanoparticles on the seed germination and seedling growth of Moso bamboo (*Phyllostachys edulis*) under cadmium stress. Development 14:15

Fahad S, Wang J (2020) Climate change, vulnerability, and its impacts in rural Pakistan: a review. Environ Sci Pollut Res 27:1334–1338

Faizan M, Faraz A, Mir AR, Hayat S (2020) Role of zinc oxide nanoparticles in countering negative effects generated by cadmium in *Lycopersicon esculentum*. J Plant Growth Regul:1–15

Faizan M, Bhat JA, Chen C, Alyemeni MN, Wijaya L, Ahmad P, Yu F (2021) Zinc oxide nanoparticles (ZnO-NPs) induce salt tolerance by improving the antioxidant system and photosynthetic machinery in tomato. Plant Physiol Biochem 161:122–130

Feng S, Ren L, Sun H, Qiao K, Liu S, Zhou A (2020) Morphological and physiological responses of two willow species from different habitats to salt stress. Sci Rep 10:1–11

Ganjavi AS, Oraei M, Gohari G, Akbari A, Faramarzi A (2021) Glycine betaine functionalized graphene oxide as a new engineering nanoparticle lessens salt stress impacts in sweet basil (*Ocimum basilicum* L.). Plant Physiol Biochem 162:14–26

Gao J, Yang X, Zheng B, Liu Z, Zhao J, Sun S, Li K, Dong C (2019) Effects of climate change on the extension of the potential double cropping region and crop water requirements in northern China. Agric For Meteorol 268:146–155

Gao M, Chang X, Yang Y, Song Z (2020) Foliar graphene oxide treatment increases photosynthetic capacity and reduces oxidative stress in cadmium-stressed lettuce. Plant Physiol Biochem 154:287–294

Godoy F, Olivos-Hernández K, Stange C, Handford M (2021) Abiotic stress in crop species: improving tolerance by applying plant metabolites. Plan Theory 10:186

Guerra D, Crosatti C, Khoshro HH, Mastrangelo AM, Mica E, Mazzucotelli E (2015) Post-transcriptional and post-translational regulations of drought and heat response in plants: a spider's web of mechanisms. Front Plant Sci 6:57

Hamid S, Ahmad I, Akhtar MJ, Iqbal MN, Shakir M, Tahir M, Zhu B (2021) *Bacillus subtilis* Y16 and biogas slurry enhanced potassium to sodium ratio and physiology of sunflower (*Helianthus annuus* L.) to mitigate salt stress. Environ Sci Pollut Res:1–11

Hasan MM, Ali MA, Soliman MH, Alqarawi AA, Abd-Allah EF, Fang XW (2020) Insights into 28-homobrassinolide (HBR)-mediated redox homeostasis, AsA–GSH cycle, and methylglyoxal detoxification in soybean under drought-induced oxidative stress. J Plant Interac 15:371–385

Hasanuzzaman M, Alhaithloul HAS, Parvin K, Bhuyan MHM, Tanveer M, Mohsin SM, Fujita M (2019) Polyamine action under metal/metalloid stress: regulation of biosynthesis, metabolism, and molecular interactions. Int J Mol Sci 20:3215

Hasanuzzaman M, Bhuyan MHM, Zulfiqar F, Raza A, Mohsin SM, Mahmud JA, Fotopoulos V (2020) Reactive oxygen species and antioxidant defense in plants under abiotic stress: revisiting the crucial role of a universal defense regulator. Antioxidants 9:681

Hezaveh TA, Pourakbar L, Rahmani F, Alipour H (2019) Interactive effects of salinity and ZnO nanoparticles on physiological and molecular parameters of rapeseed (*Brassica napus* L.). Commun Soil Sci Plant Anal 50:698–715

Hussain A, Rizwan M, Ali Q, Ali S (2019) Seed priming with silicon nanoparticles improved the biomass and yield while reduced the oxidative stress and cadmium concentration in wheat grains. Environ Sci Pollut Res 26:7579–7588

Hussain F, Hadi F, Rongliang Q (2021a) Effects of zinc oxide nanoparticles on antioxidants, chlorophyll contents, and proline in *Persicaria hydropiper* L. and its potential for Pb phytoremediation. Environ Sci Pollut Res:1–17

Hussain S, Hussain S, Ali B, Ren X, Chen X, Li Q, Ahmad N (2021b) Recent progress in understanding salinity tolerance in plants: story of Na+/K+ balance and beyond. Plant Physiol Biochem 160:239–256

Iftikhar A, Ali S, Yasmeen T, Arif MS, Zubair M, Rizwan M, Soliman MH (2019) Effect of gibberellic acid on growth, photosynthesis and antioxidant defense system of wheat under zinc oxide nanoparticle stress. Environ Pollut 254:113109

Irshad MA, Shakoor MB, Ali S, Nawaz R, Rizwan M (2019) Synthesis and application of titanium dioxide nanoparticles for removal of cadmium from wastewater: kinetic and equilibrium study. Water, Air, Soil Poll 230:1–10

Irshad MA, Nawaz R, ur Rehman MZ, Adrees M, Rizwan M, Ali S, Tasleem S (2021) Synthesis, characterization and advanced sustainable applications of titanium dioxide nanoparticles: a review. Ecotoxicol Environ Saf 212:111978

Karimi V, Karami E, Keshavarz M (2018) Climate change and agriculture: impacts and adaptive responses in Iran. J Integ Agr 17:1–15

Karimi V, Valizadeh N, Karami S, Bijani M (2021) Climate change and adaptation: recommendations for agricultural sector. In: Exploring synergies and trade-offs between climate change and the sustainable development goals. Springer, Singapore, pp 97–118

Khan MN, Siddiqui MH, Mohammad F, Naeem M, Khan MMA (2010) Calcium chloride and gibberellic acid protect linseed (*Linum usitatissimum* L.) from NaCl stress by inducing antioxidative defence system and osmoprotectant accumulation. Acta Physiol Plant 32:121–132

Khan MN, Mobin M, Abbas ZK, AlMutairi KA, Siddiqui ZH (2017) Role of nanomaterials in plants under challenging environments. Plant Physiol Biochem 110:194–209

Khan I, Awan SA, Raza MA, Rizwan M, Tariq R, Ali S, Huang L (2020a) Silver nanoparticles improved the plant growth and reduced the sodium and chlorine accumulation in pearl millet: a life cycle study. Environ Sci Pollut Res:1–13

Khan ZS, Rizwan M, Hafeez M, Ali S, Adrees M, Qayyum MF, Sarwar MA (2020b) Effects of silicon nanoparticles on growth and physiology of wheat in cadmium contaminated soil under different soil moisture levels. Environ Sci Pollut Res 27:4958–4968

Khan MN, AlSolami MA, Basahi RA, Siddiqui MH, Al-Huqail AA, Abbas ZK, Khan F (2020c) Nitric oxide is involved in nano-titanium dioxide-induced activation of antioxidant defense system and accumulation of osmolytes under water-deficit stress in *Vicia faba* L. Ecotoxicol Environ Saf 190:110152

Kirova E, Pecheva D, Simova-Stoilova L (2021) Drought response in winter wheat: protection fromoxidative stress and mutagenesis effect. Acta Physiol Plant 43:1–11
Koffler BE, Luschin-Ebengreuth N, Stabentheiner E, Müller M, Zechmann B (2014) Compartmentspecific response of antioxidants to drought stress in *Arabidopsis*. Plant Sci J 227:133–144
Kohli SK, Khanna K, Bhardwaj R, Abd-Allah EF, Ahmad P, Corpas FJ (2019) Assessment of subcellular ROS and NO metabolism in higher plants: multifunctional signaling molecules. Antioxidants (Basel) 12:641
Kour D, Rana KL, Yadav AN, Sheikh I, Kumar V, Dhaliwal HS, Saxena AK (2020) Amelioration ofdrought stress in foxtail millet (*Setaria italica* L.) by P-solubilizing drought-tolerant microbeswith multifarious plant growth promoting attributes. J Environ Sustain:1–12
Li X, Li Y, Ahammed GJ, Zhang XN, Ying L, Zhang L, Han WY (2019) RBOH1-dependent apoplastic H_2O_2 mediates epigallocatechin-3-gallate-induced abiotic stress tolerance in *Solanum lycopersicum* L. Environ Exp Bot 161:357–366
Li H, Shi J, Wang Z, Zhang W, Yang H (2020) H_2S pretreatment mitigates the alkaline salt stress on Malus hupehensis roots by regulating Na+/K+ homeostasis and oxidative stress. Plant Physiol Biochem 156:233–241
Lian J, Zhao L, Wu J, Xiong H, Bao Y, Zeb A, Liu W (2020) Foliar spray of TiO_2nanoparticles prevails over root application in reducing cd accumulation and mitigating cd-induced phytotoxicity in maize (*Zea mays* L.). Chemosphere 239:124794
Linh TM, Mai NC, Hoe PT, Lien LQ, Ban NK, Hien LTT, Van NT (2020) Metal-based nanoparticles enhance drought tolerance in soybean. J Nanomat:4056563. https://doi.org/10.1155/2020/4056563
Ma X, Geiser-Lee J, Deng Y, Kolmakov A (2010) Interactions between engineered nanoparticles (ENPs) and plants: phytotoxicity, uptake and accumulation. Sci Total Environ 408:3053–3061
Manzoor N, Ahmed T, Noman M, Shahid M, Nazir MM, Ali L, Wang G (2021) Iron oxide nanoparticles ameliorated the cadmium and salinity stresses in wheat plants, facilitating photosynthetic pigments and restricting cadmium uptake. Sci Total Environ 769:145221
Mauter MS, Zucker I, Perreault F, Werber JR, Kim JH, Elimelech M (2018) The role of nanotechnology in tackling global water challenges. Nat Sus 1:166–175
Misra N, Gupta AK (2005) Effect of salt stress on proline metabolism in two high yielding genotypes of green gram. Plant Sci 169:331–339
Mohammadi MHZ, Panahirad S, Navai A, Bahrami MK, Kulak M, Gohari G (2021) Cerium oxide nanoparticles (CeO_2-NPs) improve growth parameters and antioxidant defense system in Moldavian balm (*Dracocephalum moldavica* L.) under salinity stress. Plant Stress:100006
Natasha N, Shahid M, Khalid S, Bibi I, Naeem MA, Niazi NK, Rinklebe J (2021) Influence of biochar on trace element uptake, toxicity and detoxification in plants and associated health risks: a critical review. Crit Rev Environ Sci Technol:1–41
Noman M, Ahmed T, Hussain S, Niazi MBK, Shahid M, Song F (2020) Biogenic copper nanoparticles synthesized by using a copper-resistant strain *Shigella flexneri* SNT22 reduced the translocation of cadmium from soil to wheat plants. J HazMat 398:123175
Ogunkunle CO, Odulaja DA, Akande FO, Varun M, Vishwakarma V, Fatoba PO (2020) Cadmium toxicity in cowpea plant: effect of foliar intervention of nano-TiO_2on tissue cd bioaccumulation, stress enzymes and potential dietary health risk. J Biotechnol 310:54–61
Oraee A, Tehranifar A (2020) Evaluating the potential drought tolerance of pansy through its physiological and biochemical responses to drought and recovery periods. Sci Hortic 265:109225
Pakdel H, Hassani SB, Ghotbi-Ravandi AA, Bernard F (2020) Contrasting the expression pattern change of polyamine oxidase genes and photosynthetic efficiency of maize (*Zea mays* L.) genotypes under drought stress. J Biosci 45
Priyanka N, Geetha N, Manish T, Sahi SV, Venkatachalam P (2021) Zinc oxide nanocatalyst mediates cadmium and lead toxicity tolerance mechanism by differential regulation of photosynthetic machinery and antioxidant enzymes level in cotton seedlings. Toxicol Rep 8:295–302

Quadir M, Quillérou E, Nangia V, Murtaza G, Singh M, Thomas RJ, Noble AD (2014) Economics of salt-induced land degradation and restorantion. In Nat ResForum. Vol. 38, pp. 282–295
Ragab G, Saad-Allah K (2020) Seed priming with greenly synthesized sulfur nanoparticles enhances Antioxidative defense machinery and restricts oxidative injury under manganese stress in *Helianthus annuus* (L.) seedlings. J Plant Growth Regul:1–9
Raghib F, Naikoo MI, Khan FA, Alyemeni MN, Ahmad P (2020) Interaction of ZnO nanoparticle and AM fungi mitigates Pb toxicity in wheat by upregulating antioxidants and restricted uptake of Pb. J Biotechnol 323:254–263
Rajput VD, Minkina TM, Behal A, Sushkova SN, Mandzhieva S, Singh R, Movsesyan HS (2018) Effects of zinc-oxide nanoparticles on soil, plants, animals and soil organisms: a review. Environ Nanotechnol Monit Manag 9:76–84
Rashwan E, Alsohim AS, El-Gammaal A, Hafez Y, Abdelaal K (2020) Foliar application of Nano zinc-oxide can alleviate the harmful effects of water deficit on some flax cultivars under drought conditions. Fres Environ Bull 29:8889–8904
Raza A, Razzaq A, Mehmood SS, Zou X, Zhang X, Lv Y, Xu J (2019) Impact of climate change on crops adaptation and strategies to tackle its outcome: a review. Plan Theory 8:34
Rizwan M, Ali S, Ali B, Adrees M, Arshad M, Hussain A, ur Rehman MZ, Waris AA (2019) Zinc and iron oxide nanoparticles improved the plant growth and reduced the oxidative stress and cadmium concentration in wheat. Chemosphere 214:269–277
Sadiq Y, Zaid A, Khan MMA (2020) Adaptive physiological responses of plants under abiotic stresses: role of Phytohormones. In: Hasanuzzaman M (ed) Plant ecophysiology and adaptation under climate change: mechanisms and perspectives I. Springer, Singapore. https://doi.org/10.1007/978-981-15-2156-0_28
Salehi-Lisar SY, Bakhshayeshan-Agdam H (2020) Agronomic crop responses and tolerance to drought stress. In: Agronomic crops. Springer, Singapore, pp 63–91
Šamec D, Karalija E, Šola I, Vujčić Bok V, Salopek-Sondi B (2021) The role of polyphenols in abiotic stress response: the influence of molecular structure. Plan Theory 10:118
Sami F, Siddiqui H, Alam P, Hayat S (2021) Nitric oxide mitigates the salt-induced oxidative damage in mustard by up-regulating the activity of various enzymes. J Plant Growth Regul:1–24
Saradadevi R, Palta JA, Siddique KH (2017) ABA-mediated stomatal response in regulating water use during the development of terminal drought in wheat. Front Plant Sci 8:1251
Semida WM, Abdelkhalik A, Mohamed G, El-Mageed A, Taia A, El-Mageed A, Ali EF (2021) Foliar application of zinc oxide nanoparticles promotes drought stress tolerance in eggplant (*Solanum melongena* L.). Plan Theory 10:421
Shah AA, Aslam S, Akbar M, Ahmad A, Khan WU, Yasin NA, Ali S (2021) Combined effect of *Bacillus fortis* IAGS 223 and zinc oxide nanoparticles to alleviate cadmium phytotoxicity in *Cucumis melo*. Plant Physiol Biochem 158:1–12
Shaheen SM, Rinklebe J (2015) Impact of emerging and low cost alternative amendmentson the (im)mobilization and phytoavailability of cd and Pb in a contaminatedfloodplain soil. Ecol Eng 74:319–326
Shahid M (2021) Effect of soil amendments on trace element-mediated oxidative stress inplants: meta-analysis and mechanistic interpretations. J Haz Mat 407:124881
Shahid M, Dumat C, Khalid S, Schreck E, Xiong T, Niazi NK (2017) Foliarheavy metal uptake, toxicity and detoxification in plants: a comparison of foliar androot metal uptake. J Haz Mat 325:36–58
Shahid M, Niazi NK, Dumat C, Naidu R, Khalid S, Rahman MM, Bibi I (2018) A meta-analysis of the distribution, sources and health risks of arsenic-contaminated groundwater in Pakistan. Environ Poll 242:307–319
Sharifi P, Bidabadi SS, Zaid A, Latef AAHA (2021) Efficacy of multi-walled carbon nanotubes in regulating growth performance, total glutathione and redox state of *Calendula officinalis* L. cultivated on Pb and cd polluted soil. Ecotoxicol Environ Saf 213:112051

Sharma M, Gupta SK, Majumder B, Maurya VK, Deeba F, Alam A, Pandey V (2017) Salicylic acid mediated growth, physiological and proteomic responses in two wheat varieties under drought stress. J Proteome 163:28–51

Sheikhalipour M, Esmaielpour B, Behnamian M, Gohari G, Giglou MT, Vachova P, Skalicky M (2021) Chitosan–selenium nanoparticle (Cs–SeNP) foliar spray alleviates salt stress in bitter melon. Nano 11:684

Shelke DB, Nikalje GC, Chambhare MR, Zaware BN, Penna S, Nikam TD (2019) Na^{+} and cl^{-}induce differential physiological, biochemical responses and metabolite modulations in vitro in contrasting salt-tolerant soybean genotypes. 3 Biotech 9:1–15

Shin W, Siddikee MA, Joe MM, Benson A, Kim K, Selvakumar G, Sa T (2016) Halotolerant plant growth promoting bacteria mediated salinity stress amelioration in plants. Korean J Soil Sci Fert 49:355–367

Singh J, Lee BK (2016) Influence of nano-TiO_2particles on the bioaccumulation of cd in soybean plants (*Glycine max*): a possible mechanism for the removal of cd from the contaminated soil. J Environ Manag 170:88–96

Singh J, Dutta T, Kim KH, Rawat M, Samddar P, Kumar P (2018) 'Green'synthesis of metals and their oxide nanoparticles: applications for environmental remediation. J Nanobiotechnol 16:1–24

Singh P, Arif Y, Siddiqui H, Sami F, Zaidi R, Azam A, Hayat S (2021) Nanoparticles enhances the salinity toxicity tolerance in *Linum usitatissimum* L. by modulating the antioxidative enzymes, photosynthetic efficiency, redox status and cellular damage. Ecotoxicol Environ Saf 213:112020

Soliman M, Elkelish A, Souad T, Alhaithloul H, Farooq M (2020a) Brassinosteroid seed priming with nitrogen supplementation improves salt tolerance in soybean. Physiol Mol Biol Plants 26:501–511

Soliman MH, Alnusaire TS, Abdelbaky NF, Alayafi AA, Hasanuzzaman M, Rowezak MM, Elkelish A (2020b) *Trichoderma*-induced improvement in growth, photosynthetic pigments, proline, and glutathione levels in *Cucurbita pepo* seedlings under salt stress. Phyton 89(3):473

Soliman MH, Abdulmajeed AM, Alhaithloul H, Alharbi BM, El-Esawi MA, Hasanuzzaman M, Elkelish A (2020c) Saponin biopriming positively stimulates antioxidants defense, osmolytes metabolism and ionic status to confer salt stress tolerance in soybean. Acta Physiol Plant 42:1–13

Spanò C, Bottega S, Sorce C, Bartoli G, Castiglione MR (2019) TiO_2 nanoparticles may alleviate cadmium toxicity in co-treatment experiments on the model hydrophyte *Azolla filiculoides*. Environ Sci Pollut Res 26:29872–29882

Tanji KK (2002) Salinity in the soil environment. In: Salinity:environment-plants-molecules. Springer, Dordrecht, pp 21–51

Taran N, Storozhenko V, Svietlova N, Batsmanova L, Shvartau V, Kovalenko M (2017) Effect of zinc and copper nanoparticles on drought resistance of wheat seedlings. Nan Res Lett 12:1–6

Tarrahi R, Mahjouri S, Khataee A (2021) A review on in vivo and in vitro nanotoxicological studies in plants: a headlight for future targets. Ecotoxicol Environ Saf 208:111697

Tomar RS, Kataria S, Jajoo A (2021) Behind the scene: critical role of reactive oxygen species and reactive nitrogen species in salt stress tolerance. J Agr Crop Sci. https://doi.org/10.1111/jac.12490

Velmurugan A, Swarnam P, Subramani T, Meena B, Kaledhonkar MJ (2020) Water demand and salinity. In Desalination-challenges and opportunities. IntechOpen

Wani W, Masoodi KZ, Zaid A, Wani SH, Shah F, Meena VS, Mosa KA (2018) Engineering plants for heavy metal stress tolerance. Rend Lin Sci Fis e Nat 29:709–723

Yan N, Marschner P, Cao W, Zuo C, Qin W (2015) Influence of salinity and water content on soil microorganisms. Inter Soil Water Con Res 3:316–323

Ye Y, Cota-Ruiz K, Hernández-Viezcas JA, Valdés C, Medina-Velo IA, Turley RS, Gardea-Torresdey JL (2020) Manganese nanoparticles control salinity-modulated molecular responses in *Capsicum annuum* L. through priming: a sustainable approach for agriculture. ACS Sus Chem Eng 8:1427–1436

Zahedi SM, Hosseini MS, Daneshvar Hakimi Meybodi N, Peijnenburg W (2021) Itigation of the effect of drought on growth and yield of pomegranates by foliar spraying of different sizes of selenium nanoparticles. J Sci Food Agr. https://doi.org/10.1002/jsfa.11167

Zaid A, Wani SH (2019) Reactive oxygen species generation, scavenging and signaling in plant defense responses. In: Bioactive molecules in plant defense. Springer, Cham, pp 111–132

Zaid A, Asgher M, Wani IA, Wani SH (2020) Role of triacontanol in overcoming environmental stresses. Protective Chemical Agents in the Amelioration of Plant Abiotic Stress: Biochemical and Molecular Perspectives, 491–509

Zaid A, Ahmad B, Wani SH (2021) Medicinal and aromatic plants under abiotic stress: a cross-talk on Phytohormones' perspective. In: Aftab T, Hakeem KR (eds) Plant growth regulators. Springer, Cham. https://doi.org/10.1007/978-3-030-61153-8_5

Zhao C, Zhang H, Song C, Zhu JK, Shabala S (2020) Mechanisms of plant responses and adaptation to soil salinity. The Innovat 1:100017

Chapter 7
Mode of Action and Signaling of Nanoparticles to Alleviate Abiotic Stress in Crop Plants

Nazish, Babli, and Ajai Kumar Jaitly

Abstract Abiotic stress leads to a myriads of morphological, physiological, molecular, and biochemical modifications in plant and unfavourably impinge agricultural productivity, and global food security. Sustainable agriculture and food security requirements looking for the development of effective technologies to alleviate environmental stress. Nanotechnology is found to be a promising technology that could transform the agronomic sector with progressive and futuristic perspectives by ameliorating cultivation efficiency. Several metal and metal oxide based nanoparticles are executed as a pivotal tool to upgrade plant growth, development and productivity by combating against biotic and abiotic stresses. Applications of nanoparticles improved antioxidant potential of plants by enhancing the radical scavenging potential and antioxidant enzyme activities. A number of researches have been organized the study of mechanisms of nanoparticles action in order to fortifying plant growth and development. This chapter is a pursuit to explicate the mode of action and signalling pathway of nanoparticles in plants to combat and mitigate the abiotic stress. The relevant elucidation of physiological, biochemical and molecular mechanisms of nanoparticles in plants leads to better plant growth and yields under abiotic stress and also results in enhanced agricultural production.

Keywords Nanoparticles · Crop production · Abiotic stress · Crop plants · Antioxidant system

Nazish (✉) · A. K. Jaitly
Microbiology Lab Department of Plant Science, M.J.P.R.U Bareilly, Bareilly, Uttar Pradesh, India

Babli
Genetics Lab Department of Plant Science, M.J.P.R.U Bareilly, Bareilly, Uttar Pradesh, India

M. Faizan et al. (eds.), *Sustainable Agriculture Reviews 53*, Sustainable Agriculture Reviews 53, https://doi.org/10.1007/978-3-030-86876-5_7

7.1 Introduction

Plants being the sessile organism face abiotic stresses throughout their life cycle that adversely affects the plant growth and productivity moreover global food security. In response to abiotic stress, plant develops defence system at various levels by modifying their morphological, biochemical, physiological and molecular pathway. However, these are not enough to nullify the negative impacts of environmental stresses. Therefore, it is a major concern for scientist and researchers to mitigate the impacts of environmental stresses for the sake of enhancing the crop yield and productivity. According to FAO reports, the scientists have a crucial concern to increase 70% more food crop production (FAO 2009). Hence in this dynamic environmental framework there is an incessant requirement to venture the new research dimensions for removal of the technological challenges in context of yield barrier, efficiency of resource applications and eco-friendly technologies development. In past few year nanobiotechnology was found to occupy the privilege of being a promising tool to mitigate the limitations bound with abiotic and biotic stress to achieve a sustainable and reliable agriculture worldwide (Robinson et al. 2009). The major applications of nanotechnology could be exploited in agricultural sector, pathogen diagnostics, food processing industries, food engineering and packaging materials (Perez-de-Luque and Diego 2009).

Nanoparticles are microscopic molecules having dimensions of 1-100 nm. Beside their small size the nanoparticles may also have peculiar physio-chemical properties like large surface area, improved reactivity, diverse morphology and flexible pore size (Nel et al. 2006). Nanoparticles (NPs) possess lofty surface energy and good surface to volume ratio which enhances their reactivity and other biochemical activity, these unique features of NPs may manifest diverse behaviour and impact than their bulk counterparts (Dubchak et al. 2010). In the present scenario nanoparticles were found to have potential to enhance plant growth and development in the form of nano pesticides, herbicides, and nano fertilizers by loading their content in required amounts to targeted cellular organelles in plants. There is a substantial scope of nanotechnology in the cultivation sector; however promising applications of nanoparticles are still unravelled, especially their role and mode of action in plant development and growth (Manzer et al. 2015). Various metal and their oxide based nanoparticles are being studied to appraise their latent potential in plant growth and development, fortification of plants to condone biotic and abiotic stresses, enhancing production efficiency and embroilment in modulating the various processes of plants. Application of fertilizers in agriculture is a common practice that helps to spike the crop production and thereby fulfilled the increasing food demand. The excessive application of fertilizers was not completely utilized by plants because of leaching in soil, photo degradation, hydrolysis, and decomposition (Singh et al. 2015). Therefore, novel approaches needs to be developed with the help of nanotechnology. Which would increase the crop production and yield as well as prevent the loss of nutrient from fertilizers thus exacerbates their efficient accessibility to the plants. The applications of nano fertilizers or nano-encapsulated nutrients could be a potent tool in the discipline of sustainable agriculture. It also helps to bring down the soil toxicity induced by the accretion of chemical fertilizers and pesticides (Nair et al. 2010; De Rosa et al. 2010).

Plants do not have the ability to escape from environmental stress like salinity, chilling, drought, UV radiations etc. The stress response in plants was found to produce reactive oxygen species (ROS) causing an oxidative burst. Excessive production of ROS degrades the macromolecules and membrane lipids lead to the toxicity of cells (Yadav et al. 2014) and conquer plant growth (Khan et al. 2016). An antioxidant defense system was found in plants in order to mitigate the oxidative stress through the ROS scavenging (Khan and Khan 2017). The nanoparticles can enhance the antioxidative enzymes system by trapping the ROS (Wei and Wang 2013) whereas during abiotic stress, nanoparticles boost up the photosynthetic rate by conquering the osmotic and oxidative stress and rescuing the photosynthetic system (Qi et al. 2013; Siddiqui et al. 2014). Nanoparticles enter into the plants via size dependent mechanism and translocated to all growing tissues and germinating seeds. Although the potential applications of nanoparticles in agriculture sector is receiving global attention, but the accurate mechanism of nanoparticles in plants during stress conditions are still scanty. Therefore the present chapter is emphasises on the mode of action of nanoparticles during different abiotic stresses and their signalling in plants under stress conditions in order to achieve the increased agriculture production.

7.2 Plant Response During Abiotic Stress

The optimum abiotic environmental conditions are necessary for proper growth and development of plants. Any type of alteration from optimal environmental conditions which adversely affect the growth, activity and productivity of plants is referred as abiotic stress (Bray et al. 2000). Different abiotic stresses like drought, salinity, temperature, irradiation, water logging metal toxicity and nutrient deficiency affects the growth and development of plant which ultimately results in low yield and productivity (Emamverdian et al. 2015; Pasala et al. 2016). Plants found to have the defence mechanisms to overcome the abiotic stress conditions. These mechanisms lead the alteration of metabolic processes in the plant system in order to assist bio physicochemical progressions of the abiotic stress conditions (Bolton 2009; Massad et al. 2012; Mickelbart et al. 2015). Cellular responses of plants against stress include changes in the cell cycle, cell division and in cell wall architecture, leading to elevated stress tolerance of cells. Biochemically plants alter their metabolism to acclimatize environmental stresses through the production of osmoregulatory compounds like proline and glycine betaine. After the perception of a stress signal the molecular events take place with the genomic responses leading to tolerance has been studied extensively in recent years.

The initial prolonged salinity treatment disrupts the osmotic equilibrium and caused toxicity in the plant cell. However the consequence of ionic stress followed by osmotic stress results in the reduction of plant growth and development (Munns and Tester 2008). Plants retort the osmotic stress by mass production of inositol, sorbitol and glycerol as well as amino acids (betaine, proline, glycine, and taurine) which maintain the required osmolarity of the cells. During hypoxia conditions the unavailability of oxygen to the plants, cause energy depletion and low vigor. Plants found to alter their metabolic rate and adopt fermentation over starch metabolism to retain normal vigor level (Banti et al. 2013). In response to heavy metal toxicity, plants accumulate metal chelates, polyphosphates, and organic acids which results in sequestration of toxic metals in the plasma membrane.

Abiotic stress leads to the formation of ROS. Usually there is a balance found in production and elimination of ROS, however during abiotic stress this equilibrium gets disturbed because of increased ROS production (Fig. 7.1). The excess ROS production brings phytotoxicity through the alteration of protein structure and function. Oxygen radicals and hydrogen peroxide are originated in mitochondria due to the over reduction of electron transport chain. These ROS was also found to produce in chloroplast and peroxisomes (Davletova et al. 2005; Halliwell and Gutteridge 2000). These superoxides are converted into hydrogen peroxide with the help of enzyme superoxide dismutase (SOD) which is then broken down into water by peroxidase enzyme. The ROS could be damage the biomolecules i.e., carbohydrates, proteins, lipids and DNA, which ultimately results in cell death (Foyer and Noctor 2005) Fig 7.1.

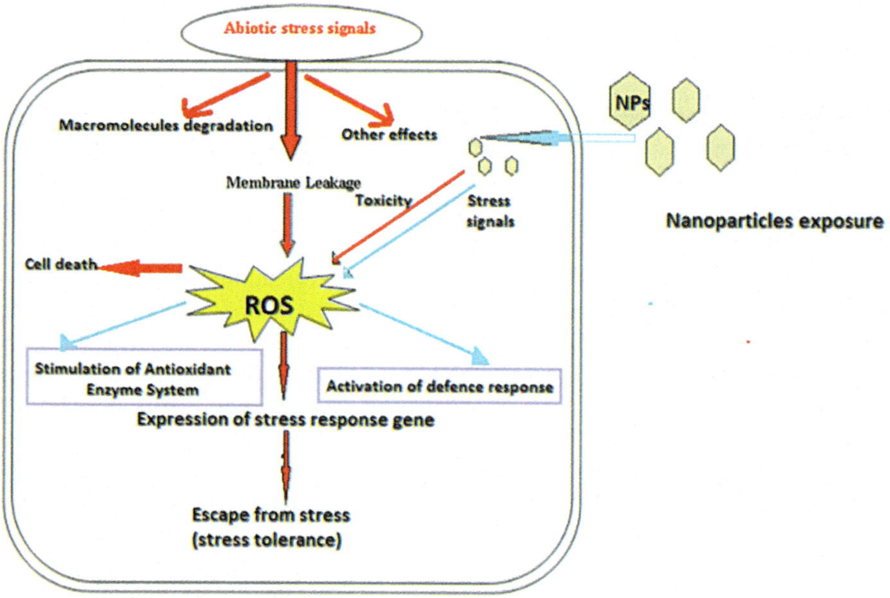

Fig. 7.1 Possible mechanism of Nanoparticles action during abiotic stress in plant cell. The response of plant and toxic effect of NPs cause cell death (Red arrows). The NPs acts as a stress signal and stimulate the antioxidant enzyme system for scavenging reactive oxygen species (ROS); induce the stress responsive gene which ultimately combats the plant to stress conditions (Blue arrow)

7.3 Abiotic Stresses and Mode of Nanoparticles Action

The initial response of plants to counter abiotic stresses leads to the transient escalation of cytoplasmic Ca^{+2} cascades which activates the intracellular second messengers (inositol, polyphosphate), ROS, abscisic acid, and increase in mitogen-activated protein kinase (MAPK) pathways (Baxter et al. 2013). The advanced level of stress response modulates the proteins involved in protection from cellular damage, and regulation of the stress-specific genes expression (Mahalingam and Fedoroff 2003). Secondary metabolites also play a key role in plants protection against abiotic stress by stabilizing the cell structure, signal transduction, protecting the photosystem from ROS and biosynthesis of polyamines (Oh et al. 2009). Oxidative stress triggers the generation of ROS, biosynthesis of enzymes, accumulation of phenylpropanoids, and management of gene expression in plant defence response (Minibayeva et al. 2009; Daudi et al. 2012).

Nanoparticles found to mediate their effect on plants growth and development in a concentration dependent manner (Laware and Raskar 2014). During their study on onion seedling they find out that TiO_2 nanoparticles elevate the activity of superoxide dismutase enzyme which was further enhanced with increasing the TiO_2 concentration. However its low concentration was required for seed germination and seedling growth in onion which get suppressed at higher concentration. A significant induction of hydrolytic enzyme (amylase), catalase and peroxidase was also observed here along with superoxide dismutase.

7.3.1 *Salt Stress*

Salinity is a profound concern to achieve sustainable crop production among the scientific community. More than 20% of cultivated land was found to experienced salinity stress worldwide. Salinity poses two major threats to plant growth i.e. osmotic stress and ionic stress (Flowers and Colmer 2008). Depletion of soil osmotic potential, increasing ionic toxicity and nutritional imbalance are some common insinuation of salinity stress experienced by plants (Ashraf 1994). High salt uptake by the plant leads to increase the osmotic potential of cytosol. Under the effect of saline condition, the sequestration of ions in the vacuole and formation of osmolytes (Munns, 2008) helps to maintain homeostasis in the cell. In majority of plant species proline and soluble sugars, is produced as a compatible osmoprotectant to resist the cells from adverse effects of salinity stress (Sairam and Tyagi 2004; Hokmabadi et al. 2005).

The application of silicon nanoparticles and silicon fertilizer was exhibited promising results on physiological and morphological features on vegetative state of basil (*Ocimum basilicum*) under salinity stress. The considerable increase in growth and development indices, chlorophyll content and proline level in basil was recorded under the influence of salinity when treated with silicon nanoparticles and silicon fertilizer (Kalteh et al. 2014). Similar consequences of SiO_2 nanoparticles have also been observed by (Haghighi et al. 2012) under salinity stress. The nanoparticles (SiO_2) have been found to drop the ionic toxicity caused by high Na^+ ion concentration leading to enhanced crop growth and yield under adverse conditions (Savvasd et al. 2009). The silica nanoparticles are exploited to mitigate salinity stress in plants to reduce Na^+ ion concentration, probably by reducing Na^+ ion uptake by plant tissues.

Multiwalled carbon nanotubes applied to broccoli under salinity stress found to induce water uptake and transportation by enhancing the net CO_2 assimilation and aquaporin transduction to alleviate the stress and enhancing growth (Martınez-Ballesta et al. 2016).

7.3.2 Drought Stress

Drought is also a major abiotic factor affecting plant growth and crop production. The application of nanoparticles was found to be an effective strategy among all the techniques used to alleviate drought stress. The applications of Si nanoparticles in different fractions enhance the plant tolerance toward drought stress. It was find out that hawthorn (*Crataegus sp.*) exhibited the increased drought resistance. The biochemical and physiological responses in seedlings of hawthorn vary for different concentration of Si nanoparticles at various level of dehydration stress from temperate to severe. The pre-treatment of Si nanoparticles indicates the productive changes on photosynthetic parameters, malondialdehyde (MDA), water content, leaf pigments, ion leakage, proline and carbohydrate content (Ashkavand et al. 2015). The metal-based NPs were found to support the drought tolerance of NP-treated plants in *Glycine max* (Linh et al. 2020) by triggering the expression of drought-associated gene. Pei et al. (2010) suggested that exposure to low concentration (1.0 mM) of sodium silicate can fairly alleviate the damaging effects of drought stress in wheat by improving the shoot growth, increasing the chlorophyll contents, and maintaining leaf water potential of plant. Radical length of germinating seeds has been increased by the application of Zn. Moreover higher concentration of Zn increases the seed viability especially for plants grown in Zn-deficient areas (Degenhardt and Gimmler 2000).

Application of PEG and silver nanoparticles (AgNPs) was found to increase germination rate and percentage, root length, fresh and dry weight of root in lentil (*Lens culinaris Medic*) seeds significantly (Hojjat 2016). Safflower cultivars exhibited drought stress mitigation on foliar application of iron-nanoparticles by increasing their biomass and oil content. Application of Fe nanoparticles augmented biomass production at different stages of flowering and granulation, when subjected to drought stress (Davar et al. 2014).

7.3.3 Heat Stress

The elevated temperature and excessive radiations are the limiting factors for growth and development of plant. The continuous exposure of plant to high temperature at severe level resulting in irretrievable loss of development and growth is referred as heat stress (Wahid 2007). Heat stress elevate the ROS production and generates oxidative stress leads to the membrane lipid degeneration and membrane ion leakage, followed by protein denaturation (Karuppanapandian et al. 2011) as well as reduced chlorophyll content and rate of photosynthesis. Se nanoparticles in small concentration were observed to minimize the effect of heat stress by enhancing hydration ability, chlorophyll content, and plant development (Haghighi et al. 2014). The anti oxidative properties towards the plants were also shown at low Se concentration whereas large concentration of Se nanoparticles could induce oxidative stress (Hasanuzzaman et al. 2013). Plants produced some heat shock protein (HSP) and molecular chaperones in response to heat stress. The HSPs activates some other proteins to conserve their constancy during stress conditions (Wahid 2007). Khodakovskaya et al. (2011) reported that multiwall carbon nanotubes were engaged in regulation of heat shock proteins expressions, like HSP90. The CeO_2 nanoparticles exposure to Maize plant causes overproduction of H_2O_2 and up regulation of HSP70 (Zhao et al. 2012). The application of silver nanoparticles (AgNPs) was observed to defend wheat plant from heat stress through the improvement of morphological growth (Iqbal et al. 2018).

7.3.4 Chilling Stress

Plants may develop physiological disarray while subjected to a low temperature i.e., non freezing temperature. The chilling stress thermodynamically lowers the kinetics of various physiological and metabolic processes in plants (Hussain et al. 2018). The major impact of chilling stress includes retarted metabolism, membrane viscosity i.e. distortion of permeability and ion leakage from the membrane affecting the germination and growth of plant (Suzuki et al. 2008).

Photosynthesis is an important physiological process of plant which shows high susceptibility towards the chilling stress. The photosystem of plants exposed to chilling stress found damaged by reducing chlorophyll content, CO_2 assimilation, transpiration rate and the degradation of enzyme Rubisco (Liu et al. 2012). Effect of nanoparticles on the photosystem is attributed to increased production of Rubisco enzyme, light immersion ability of chloroplast and inhibiting the ROS production (Giraldo et al. 2014). Application of TiO_2 nanoparticles enhances the expression of Rubisco, chlorophyll binding protein gene and activities of antioxidant enzyme, thereby improves the susceptibility of plant to chilling stress (Hasanpour et al. 2015).

7.3.5 *Heavy metal Stress*

Heavy metal toxicity is also a matter of great concern as it constraints crop production by increasing phytotoxicity to the plants which brings all detrimental manifestations. The exposure of plants to heavy metals triggers the ROS production, causing an oxidative burst in the cell which leads to the alteration of cell structure, reduction of membrane permeability, and protein degradation (Sharma et al. 2012). The metal toxicity could be counterbalanced by several biochemicals, physiological and molecular mechanisms of plants. As a defense mechanism plants produce metal-chelates, polyphosphates and organic acids which inhibit the heavy metal consumption, efflux of metal ions and activate the antioxidant enzymes system that scavenge ROS production (Ghori et al. 2019).

Nanoparticles enter into plant cells easily due to their small size and large surface area and retain more affinity for the heavy metals. Exposures to TiO_2 nanoparticles limit cadmium toxicity and enhance the photosynthesis rate as well as plant growth (Singh and Lee 2016). Zn nanoparticles were also play a crucial role in reducing Cd generated metal toxicity of plants (Baybordi 2005). Hydroxyapatite nanoparticles treatments in *Brassica juncea* have been found to reduced Cd toxicity (Li and Huang 2014). The supplement of Si nanoparticles in growth media alleviates chromium toxicity in pea (Tripathi et al. 2015). Similarly the application of gold ions on cow pea encourages the conversion of Au^{+3} to nontoxic gold nanoparticles with the help of phenolic compounds present in germinating seeds (Shabnam et al. 2014).

7.4 Nanoparticles Signalling During Abiotic Stress

The existing data on nanoparticles and their interactions with plants in the course of abiotic stress was indicates that, reactive oxygen species (ROS) production is a common response of plants to all type of stresses (Fig. 7.1). ROS play an important role in the form of stress signals to trigger the plant defence response thereby aggravates cellular damage (Dat et al. 2000). Various studies suggested that along with the induction of ROS production nanoparticles also stimulate the antioxidant enzyme activities for ROS scavenging (Ma et al. 2010; Simon et al. 2013). However, the extensive mechanism of nanoparticles operation and functioning is not well understood. Amalgamation of various approaches such as proteomics and genomic will make it possible to understand the mechanism of nanoparticles action in plants during abiotic stress conditions. The activities of various antioxidant enzymes i.e. ascorbate peroxidase, guaiacol peroxidase and catalase was enhanced through the treatment of silver nanoparticles in *Brassica juncea* that results in the reduction of ROS concentration.

The Gene expression analysis of *Arabidopsis* by RT-PCR has furnished advanced insights for the molecular mechanism of plant response to Ag-NPs.

The transcriptional response of *Arabidopsis* plants exposed to Ag-NPs was scrutinized through whole-genome cDNA expression microarrays (Kaveh et al. 2013) resulted in the up regulation of 286 genes, including the genes fundamentally associated with metal and oxidative stress (e.g., cytochrome P450-dependent oxidase, superoxide dismutase, vacuolar cation/proton exchanger and peroxidase), and suppression of 81 genes, including the genes involved in plant defense system and hormonal stimuli (e.g. auxin-regulated gene involved in organ size-ARGOS, ethylene signalling pathway and SAR against pathogens). mi RNAs have been found to regulate various metabolic processes in plants and animals, and also plays a pivotal role in plant defence from abiotic and biotic stresses (Frazier et al. 2014). The combined effect of nanoparticles and mi RNAs also reveals insight into the mechanism of nanoparticles under abiotic stress. It was observed that the application of TiO_2 and Al_2O_3 nanoparticles on tobacco plants triggers the up regulation of mi RNA expression as well as response to metal stress, though higher concentration of these nanoparticles caused wilting, reduced biomass, leaf sizes, and leaf counts (Burklew et al. 2012).

A signalling network switches the defense system in plants, which stimulate the molecular mechanism in response to a stress conditions. Ca^{+2} involved in signal transduction during various stress conditions serve as a second messenger. Induction of stress signals generates a cascade of Ca^{+2} to the cytosol through calcium ion channels, thus the level of Ca^{+2} in the cytosol increase, which recognized through a calcium ion-binding proteins (CBPs) causing modifications in gene expression and plant adaptation to stress conditions (Khan and Khan 2014; Singh and Husen 2019). It has been reported that nitric oxide (NO) is responsible for the increased cytosolic calcium ions concentration during biotic and abiotic stress conditions therefore calcium ions supports the synthesis of nitric oxide (Corpas et al. 2004). The Ag nanoparticles treatment on *Oryza sativa* roots asserted the involvement of nanoparticles responsive proteins in modulating and signalling of calcium ions, protein degeneration, cell wall formation, transcription, oxidative stress response pathway, cell division, and apoptosis (Mirzajani et al. 2014). They also suggested that nanoparticles imitate the Ca^{+2} and their binding with calcium-binding proteins (CBP) which stimulates a cascade of stress responsive genes.

The interaction of C60 nanocrystals initiated regulation of the Ca^{+}/calmodulin-dependent protein kinase II (Miao et al. 2014). Furthermore, the application of cadmium sulfide on *Arabidopsis thaliana* stimulates over expression of calcium-binding protein CML45 as well as calcium-dependent protein kinase 23 (Marmiroli et al. 2015). These calcium-binding proteins have been reported to control stress responses, and their over expression caused improved resistance in plants against several abiotic stress conditions (Boudsocq and Sheen 2013).

Nanoparticles also found to activates the nitrate reductase enzyme in plants, which enhanced the concentration of nitric oxide to modulate immune response in plants (Chandra et al. 2015). Contrarily, it has been find out that nitric oxide causes nanoparticles-induced toxicity and stimulates the antioxidant genes expression as well as suppresses the lipid oxidation and ROS generation (Chen et al. 2015). Syu et al. (2014) has been concludes that nanoparticles sustains an active antioxidant defense system at lower concentration that regulates the generation of ROS into a specific concentration that could be incapable of causing damage but adequate for signalling. However the complete mechanism of nanoparticle induced signal transduction in plants are under study but it has been find out that nanoparticles induced phytotoxicity have shown enhanced generation of ROS that act as toxic compounds as well as signalling molecules in plant cells. The different roles of ROS are precisely governed by their production and scavenging activity, disproportion in any of these activities may result in extreme production or diminished accessibility of ROS, which leads to oxidative stress or disruption of signalling respectively.

7.5 Conclusion

Nanoparticles enhance the stress tolerance of plants by stimulating the antioxidant enzymes production (SOD, CAT and POD) as well as the detoxification of ROS. The successful interaction of NPs with plants is mainly due to their small size, large surface area and intrinsic catalytic activity which makes them enable for easy penetration, high absorption and efficient catalytic reactions. The complete knowledge of mechanism of nanoparticles under stress conditions are still unknown and are under extensive study. However the available data indicates that nanoparticles trigger Ca^{+} dependent signals via the calcium binding proteins (CBPs) in cytosol. These signals induced the expression of stress response gene and thereby stimulating the defense mechanism of plant. The exact role and mode of action of NPs during abiotic stress needs further research at cellular and molecular levels.

References

Ashkavand P, Tabari M, Zarafshar M, Toma Skova I, Struve D (2015) Effect of SiO2 nanoparticles on drought resistance in hawthorn seedlings. Lesne Prace Badawcze/Forest Research Papers Grudzien 76(4):350–359

Ashraf M (1994) Organic substances responsible for salt tolerance in Eruca sativa. Biol Plant 36:255–259

Banti V, Giuntoli B, Gonzali S, Loreti E, Magneschi L, Novi G et al (2013) Low oxygen response mechanisms in green organisms. Int J Mol Sci 14:4734–4761

Baxter A, Mittler R, Suzuki N (2013) ROS as key players in plant stress signalling. J Exp Bot 65:1229–1240

Baybordi A (2005) Effect of zinc, iron, manganese and copper on wheat quality under salt stress conditions. J Water Soil 140:150–170

Bolton MV (2009) Primary metabolism and plant defence-fuel for the fire. Mol Plant-Microbe Interact 22:487–497

Boudsocq M, Sheen J (2013) CDPKs in immune and stress signalling. Trends Plant Sci 18:30–40

Bray AB, Bailey Serres J, Weretilnyk E (2000) Responses to abiotic stress. In: Buchanan BB, Gruissem W, Jones RL (eds) Biochemistry & molecular biology of plants. American Society of Plant Physiology, Rockville, pp 1158–1203

Burklew CE, Ashlock J, Winfrey WB, Zhang B (2012) Effects of aluminum oxide nanoparticles on the growth, development, and micro RNA expression of tobacco (*Nicotiana tabacum*). PLoS One 7:34783

Chandra S, Chakraborty N, Dasgupta A, Sarkar J, Panda K, Acharya K (2015) Chitosan nanoparticles: a positive modulator of innate immune responses in plants. Sci Rep 5:15195

Chen J, Liu X, Wang C, Yin, SS, Li XL, Hu WJ et al. (2015) Nitric oxide ameliorates zinc oxide nanoparticles-induced phytotoxicity

Corpas FJ, Barroso JB, Carreras A, Quiros M, Leon AM, Romero Puertas MC et al (2004) Cellular and subcellular localization endogenous nitric oxide in young and senescent pea plants. Plant Physiol 136:2722–2733

Dat J, Vandenabeele S, Vranova E, Van Montagu M, Inze D, Van Breusegem F (2000) Dual action of the active oxygen species during plant stress responses. Cell Mol Life Sci 57:779–795

Daudi A, Cheng Z, O'Brien JA, Mammarella N, Khan S, Ausubel FM et al (2012) The apoplastic oxidative burst peroxidase in Arabidopsis is a major component of pattern-triggered immunity. Plant Cell 24:275–287

Davar F, Zareii AR, Amir H (2014) Evaluation the effect of water stress and foliar application of Fe nanoparticles on yield, yield components and oil percentage of safflower (*Carthamus tinctorious* L.). Int J Adv Biol Biomed Res 4:1150–1159

Davletova S, Rizhsky L, Liang H, Shengqiang Z, Oliver DJ, Coutu J et al (2005) Cytosolic ascorbate peroxidase 1 is a central component of the reactive oxygen gene network of Arabidopsis. Plant Cell 17:268–281

De Rosa MC, Monreal C, Schnitzer M, Walsh R, Sultan Y (2010) Nanotechnology in fertilizers. Nat Nanotechnol 5:91

Degenhardt B, Gimmler H (2000) Cell wall adaptations to multiple environmental stresses in maize roots. J Exp Bot 51:595–603

Dubchak S, Ogar A, Mietelski JW, Turnau K (2010) Influence of silver and titanium knanoparticles on arbuscular mycorrhiza colonization and accumulation of radiocaesium in Helianthus annuus. Span J Agric Res 8:S103–S108

Emamverdian A, Ding Y, Mokhberdoran F, Xie Y (2015) Heavy metal stress and some mechanisms of plant defence response. Sci World J 756120

FAO (2009) High level expert forum-how to feed the world in 2050. Economic and Social Development, Food and Agricultural Organization of the United Nations, Rome

Flowers TJ, Colmer TD (2008) Salinity tolerance in halophytes. New Phytol 179:945–963

Foyer CH, Noctor G (2005) Oxidant and antioxidant signalling in plants: a re-evaluation of the concept of oxidative stress in a physiological context. Plant Cell Environ 28:1056–1071

Frazier TP, Burklew CE, Zhang B (2014) Titanium dioxide nanoparticles affect the growth and micro RNA expression of tobacco (*Nicotiana tabacum*). Funct Integr Genom 14:75–83

Ghori N-u-H, Ghori T, Hayat MQ, Imadi S, Gul A, Altay V, Ozturk M (2019) Heavy metal stress and responses in plants. Int J Environ Sci Technol 2019:1–22. https://doi.org/10.1007/s13762-019-02215-8

Giraldo JP, Landry MP, Faltermeier SM, Mc Nicholas TP, Iverson NM, Boghossian AA et al (2014) Plant nanobionics approach to augment photosynthesis and biochemical sensing. Nat Mater 13

Haghighi M, Afifipour Z, Mozafarian M (2012) The effect of N- Si on tomato seed germination under salinity levels. Int J Environ Sci 6:87–90

Haghighi M, Abolghasemi R, Teixeira da Silva JA (2014) Low and high temperature stress affect the growth characteristics of tomato in hydroponic culture with Se and nano-Se amendment. Sci Hortic 178:231–240

Halliwell B, Gutteridge JMC (2000) Free radicals in biology and medicine, 3rd edn. Oxford University Press, Oxford

Hasanpour H, Maali Amiri R, Zeinali H (2015) Effect of TiO2 nanoparticles on metabolic limitations to photosynthesis under cold in chickpea. Russ J Plant Physiol 62:779–787

Hasanuzzaman M, Nahar K, Fujita M (2013) Extreme temperature responses, oxidative stress and antioxidant defense in plants. In: Vahdati K, Leslie C (eds) Abiotic stress – plant responses and applications in agriculture. In Tech Open Access Publisher

Hojjat (2016) The effect of silver nanoparticle on lentil seed germination under drought stress. Int J Farm Alli Sci 5(3):208–212

Hokmabadi H, Arzani K, Grierson PF (2005) Growth, chemical composition, and carbon isotope discrimination of pistachio (*Pistacia vera* L.) rootstock seedlings in response to salinity. Aust J Agric Res 56:135–144

Hussain HA, Hussain S, Khaliq A, Ashraf U, Anjum SA, Men S, Wang L (2018) Chilling and drought stresses in crop plants: implications, cross talk, and potential management opportunities. Front Plant Sci 9:393. J Hazard Mater 297:173–182. https://doi.org/10.3389/fpls.2018.00393

Iqbal M, Raja N, Mashwani ZR, Wattoo F, Hussain M, Ejaz M, Saira H (2018) Assessment of silver nanoparticles exposure on physiological and biochemical changes and antioxidative defense system in wheat (Triticum aestivum L) under heat stress. IET Nanobiotechnol

Kalteh M, Zarrin TA, Shahram A, Maryam MA, Alireza FN (2014) Effect of silica nanoparticles on Basil (Ocimum basilicum) under salinity. Stress J Chem Health Risks 4(3):49–55

Karuppanapandian T, Wang HW, Prabakaran N, Jeyalakshmi K, Kwon M, Manoharan K et al (2011) 2, 4-dichlorophenoxyacetic acid-induced leaf senescence in mung bean (*Vigna radiate* L. Wilczek) and senescence inhibition by co-treatment with silver nanoparticles. Plant Physiol Biochemist 49:168–217

Kaveh R, Li YS, Ranjbar S, Tehrani R, Brueck CL, Aken BV (2013) Changes in Arabidopsis thaliana gene expression in response to silver nanoparticles and silver ions. Environ Sci Technol 47:10637e10644

Khan MIR, Khan NA (2014) Ethylene reverses photosynthetic inhibition by nickel and zinc in mustard through changes in PS II activity, photosynthetic-nitrogen use efficiency and antioxidant metabolism. Protoplasma 251:1007–1019

Khan MIR, Khan N (2017) Reactive oxygen species and antioxidant system in plants: role and regulation under abiotic stress. Springer, New York. 978-981-10-5254-5

Khan MIR, Khan NA, Masood A, Per TS, Asgher M (2016) Hydrogen peroxide alleviates nickel-inhibited photosynthetic responses through increase in use-efficiency of nitrogen and sulfur, and glutathione production in mustard. Front Plant Sci 7:44

Khodakovskaya MV, de Silva K, Nedosekin DA, Dervishi E, Biris AS, Shashkov EV et al (2011) Complex genetic, photothermal, and photoacoustic analysis of nanoparticle-plant interactions. Proc Natl Acad Sci U S A 108:1028–1033

Laware SL, Raskar S (2014) Effect of titanium dioxide nanoparticles on hydrolytic and antioxidant enzymes during seed germination in onion. Int J Curr Microbiol App Sci 3(7):749–760

Li Z, Huang J (2014) Effects of nanoparticle hydroxyapatite on growth and antioxidant system in pakchoi (*Brassica chinensis* L.) from cadmium-contaminated soil. J Nanomater 2014:1–7

Linh TM, Nguyen CM, Pham TH et al (2020) Metal based nanoparticles enhance drought tolerance in soybean. J Nanomater 2020., Article ID 4056563, 13 pages. https://doi.org/10.1155/2020/4056563

Liu YF, Qi MF, Li TL (2012) Photosynthesis, photo inhibition, and antioxidant system in tomato leaves stressed by low night temperature and their subsequent recovery. Plant Sci 196:8–17

Ma X, Geiser-Lee J, Deng Y, Kolmakov A (2010) Interactions between engineered nanoparticles (ENPs) and plants: phytotoxicity, uptake and accumulation. Sci Total Environ 408:3053–3061

Mahalingam R, Fedoroff N (2003) Stress response, cell death, and signalling: the many faces of reactive oxygen species. Physiol Plant 119:56–68

Manzer H, Siddiqui Mohamed H, Al-Whaibi Firoz M, Al-Khaishany MY (2015) Role of nanoparticles in plants. Nanotechnol Plant Sci:19–35

Marmiroli M, Imperiale D, Pagano L, Villani M, Zappettini A, Marmiroli N (2015) The proteomic response of *Arabidopsis thaliana* to cadmium sulphide quantum dots, and its correlation with the transcriptomic response. Front Plant Sci 6:1104

Martınez-Ballesta MC, Zapata L, Chalbi N, Carvajal M (2016) Multiwalled carbon nanotubes enter broccoli cells enhancing growth and water uptake of plants exposed to salinity. J Nanobiotechnol 14:42

Massad TJ, Dyer LA, Vega CG (2012) Cost of defence and a test of the carbon-nutrient balance and growth-differentiation balance hypotheses for two co-occurring classes of plant defence. PLoS One 7:7554

Miao Y, Xu J, Shen Y, Chen L, Bian Y, Hu Y et al (2014) Nanoparticle as signalling protein mimic: robust structural and functional modulation of CaMKII upon specific binding to fullerene C60 nanocrystals. ACS Nano 8:6131–6144

Mickelbart MV, Paul M, Hasegawa PM, Bailey-Serres J (2015) Genetic mechanisms of abiotic stress tolerance that translates to crop yield stability. Nat Rev Genet 16:237–251

Minibayeva F, Kolesnikov O, Chasov A, Beckett RP et al (2009) Wound-induced apoplastic peroxidase activities: their roles in the production and detoxification of reactive oxygen species. Plant Cell Environ 32:497–508

Mirzajani F, Askari H, Hamzelou S, Schober Y, Rhompp A, Ghassempour A et al (2014) Proteomics study of silver nanoparticles toxicity on Oryza sativa L. Ecotoxicol Environ Saf 108:335–339

Munns R, Tester M (2008) Mechanisms of salinity tolerance. AnnuRev Plant Biol 59:651–681

Nair R, Varghese SH, Nair BG, Maekawa T, Yoshida Y, Kumar DS (2010) Nanoparticulate material delivery to plants. Plant Sci 179:154–163

Nel A, Xia T, Madler L, Li N (2006) Toxic potential of materials at the nanolevel. Science 311:622–627

Oh MM, Trick HN, Rajashekar C (2009) Secondary metabolism and antioxidants are involved in environmental adaptation and stress tolerance in lettuce. J Plant Physiol 166:180–191

Pasala RK, Khan MIR, Minhas PS, Farooq MA, Sultana R, Per TS et al (2016) Can plant bioregulators minimize crop productivity losses caused by drought, heat and salinity stress? An integrated review. J Appl Bot Food 89

Pei ZF, Ming DF, Liu D, Wan GL, Geng XX, Gong HJ et al (2010) Silicon improves the tolerance to water-deficit stress induced by polyethylene glycol in wheat (*Triticum aestivum* L.) seedlings. J. Plant Growth Regul 29:106–115

Perez-de-Luque A, Diego R (2009) Nanotechnology for parasitic plant control. Pest Manag Sci 65:540–545

Qi M, Liu Y, Li T (2013) Nano-TiO2 improves the photosynthesis of tomato leaves under mild heat stress. Biol Trace Elem Res 156:323–328

Robinson DKR, Morrison M et al. (2009) Nanotechnology developments for the Agri food sector report

Sairam RK, Tyagi A (2004) Physiology and molecular biology of salinity stress tolerance in plants. Curr Sci 86(3):407–421

Savvasd G, Giotes D, Chatzieustratiou E, Bakea M, Patakioutad G (2009) Silicon supply in soilless cultivation of zucchini alleviates stress induced by salinity and powdery mildew infection. Environ Exp Bot 65:11–17

Shabnam N, Pardha-Saradhi P, Sharmila P (2014) Phenolics impart Au3þ-stress tolerance to cowpea by generating nanoparticles. PLoS One 9:85242

Sharma P, Jha AB, Dubey RS, Pessarakli M (2012) Reactive oxygen species, oxidative damage, and antioxidant defence mechanisms in plants under stressful conditions. J Bot 2012:1–26

Siddiqui MH, Al-Whaibi FM, Alsahli AA (2014) Nanosilicon dioxide mitigates the adverse effects of salt stress on *Cucurbita pepo* L. Environ Toxicol Chem 33:2429–2437

Simon DF, Domingos RF, Hauser C, Hutchins CM, Zerges W, Wilkinson KJ (2013) Transcriptome sequencing (RNA-seq) analysis of the effects of metal nanoparticle exposure on the transcriptome of *Chlamydomonas reinhardtii*. Appl Environ Microbiol 79:4774–4785

Singh S, Husen A (2019) Role of nanomaterials in the mitigation of abiotic stress in plants. In: Husen A, Iqbal M (eds) Nanomaterials and plant potential. Springer, Cham. https://doi.org/10.1007/978-3-030-05569-1-18

Singh J, Lee BK (2016) Influence of nano-TiO2 particles on the bioaccumulation of Cd in soybean plants (Glycine max): a possible mechanism for the removal of Cd from the contaminated soil. J Environ Manag 170:88–96

Singh A, Singh NB, Hussain I, Singh H, Singh SC (2015) Plant nanoparticle interaction: an approach to improve agricultural practices and plant productivity. Int J Pharm Sci Inv 4(8):25–40

Suzuki K, Nagasuga K, Okada M (2008) The chilling injury induced by high root temperature in the leaves of rice seedlings. Plant Cell Physiol 49:433–442

Syu YY, Hung JH, Chen JC, Chuang HW (2014) Impact of size and shape of silver nanoparticles on *Arabidopsis* plant growth and gene expression. Plant Physiol Biochem 83:57–64

Tripathi DK, Singh VP, Prasad SM, Chauhan DK, Dubey NK (2015) Silicon nanoparticles (Si Np) alleviate chromium (VI) phytotoxicity in *Pisum sativum* (L.) seedlings. Plant Physiol Biochemist 96:189–198

Wahid A (2007) Physiological implications of metabolites biosynthesis in net assimilation and heat stress tolerance of sugarcane (*Saccharum officinarum*) sprouts. J Plant Res 120:219–228

Wei H, Wang E (2013) Nanomaterials with enzyme-like characteristics (nanozymes): next-generation artificial enzymes. Chem Soc Rev 42:6060–6093

Yadav T, Mungray AA, Mungray AK (2014) Fabricated nanoparticles: current status and potential phytotoxic threats. In: Whitacre DM (ed) Reviews of environmental contamination and toxicology. Springer, Cham

Zhao L, Peng B, Hernandez-Viezcas JA, Rico C, Sun Y, Peralta Videa JR et al (2012) Stress response and tolerance of Zea mays to CeO2 nanoparticles: cross talk among H2O2, heat shock protein and lipid peroxidation. ACS Nano 6:9615–9622

Chapter 8
Impact of Nanoparticles and Nanoparticle-Coated Biomolecules to Ameliorate Salinity Stress in Plants with Special Reference to Physiological, Biochemical and Molecular Mechanism of Action

Akankhya Guru, Soumya Kumar Sahoo, Payel Saha, and Padmanabh Dwivedi

Abstract It is well known that salinity, classified among the major abiotic stresses, negatively influences the growth as well as yield of plants in the whole world. There is a serious concern among scientists to overcome crop losses due to salt stress. With the breakneck advancement in the context of nanotechnology, various engineered nanoparticles are being recognized and applied in plant science to safeguard plants from the devastating repercussions of salinity stress. Nanoparticles function as magic bullets owing to their unique physico-chemical properties. The incorporation of nanoparticles can potentially modulate physiological and biochemical approaches of plants which could possibly enhance salinity tolerance. To improve plants' adaptation to salinity, nanoparticles encapsulated with certain biomolecules and/or signaling agents might be a more effective approach. This chapter aims to deliver an update on the utilization of different classes of nanoparticles and their mechanisms of action to manage soil salinity in plants. Additionally, the impact of nano-encapsulation on salt-stressed plants at physio-biochemical levels will be discussed.

Keywords Nanoparticles · Salinity · Nano-encapsulation · Tolerance

A. Guru · P. Dwivedi (✉)
Department of Plant Physiology, Institute of Agricultural Sciences, Banaras Hindu University, Varanasi, India

S. K. Sahoo
Crop Physiology and Biochemistry Division, ICAR-NRRI, Cuttack, India

P. Saha
Department of Plant Physiology, Bidhan Chandra Krishi Vishwavidyalaya, Mohanpur, West Bengal, India

M. Faizan et al. (eds.), *Sustainable Agriculture Reviews 53*, Sustainable Agriculture Reviews 53, https://doi.org/10.1007/978-3-030-86876-5_8

8.1 Introduction

Plants endure various abiotic stresses, among which salinity stress is addressed as one of the leading constraints in crop establishment worldwide. On a global basis, about 800 million hectares of arable land have been affected due to soil salinity. Crop productivity losses due to salt stress are nearly 50% around the world (Zhao et al. 2020). Salinity stress evokes osmotic and ionic toxicity; ultimately causes cell death in plants, specifically in arid and semi-arid parts of the globe (Munns and Tester 2008). Consequently, salinity affects the whole plant physiology. To tailor circumstances of salt stress, plants have developed certain tolerance mechanisms such as uptake of toxic ions, ion exclusion, osmotic regulation, production of reactive oxygen species scavangers, photosynthetic electron transport, antioxidant defense, chlorophyll fluorescence, chlorophyll content, etc. (Tang et al. 2015). Recently, many pieces of research have focused on novel strategies to tackle the destructive aftereffects of salt stress in plants (Liang et al. 2018a, b).

In this modern age, 'Nanotechnology' is one of those branches of science, whose application is very vast. It is one of the most important concepts of ongoing discussion in the scientific community. 'Nanotechnology' is defined as the study and utilization of materials of small size ranging from 1 to 100 nm. It is one of the emerging techniques to improve plants' yield and growth performance under salinity stress (Duhan et al. 2017). Due to specific attributes like minimal size, large surface to volume ratio, and structural features, diverse kinds of nanoparticles such as gold, silver, zinc, carbon, cerium, titanium, etc., have been applied to protect plants against the devastating implications of abiotic stresses including salinity stress (Khan et al. 2017). Moreover, nanoparticles can form a coating around other substances that differ from both the substance (which is being coated) and the surrounding medium. These coatings can drastically alter the properties of substances, such as chemical reactivity, catalytic activity, suspension stability, etc., even though just a single molecule thick. Further, nanoparticles allow alterations in physiological and biochemical attributes in salt-treated plants via nano-encapsulation. Also, the application of nanoparticles alters gene expression profiles in several plants to minimize the reducing effects of salinity stress.

Therefore, this chapter discusses the role of various nanoparticles, nano-encapsulation in plants under salinity stress. Additionally, it briefly explains how nanoparticles affect the physio-biochemical procedures as well as stress-responsive genes in plants affected by salinity stress conditions.

8.2 Importance of Nanoparticles and Nanoparticle-Coated Biomolecules in Plants

Biological systems are highly complex. Plants are sessile organisms; hence they adjust their metabolism and physiology in response to the dynamic changes in their surroundings, including biotic and abiotic stresses. Highly coordinated metabolic as well as signaling networks are key to the physiological and metabolic changes that are required to continuously face these fluctuating situations. Furthermore, the system's important properties are frequently embodied in its dynamics. Changes in the network structure through epigenetic effects or mutation can cause changes in network dynamics, resulting in different physiological properties. The effects of these minute changes have resulted in massive developmental steps since the dawn of life. DNA and RNA containing hereditary information facilitated the transfer of these changes from generation after generation. In this way, the adaptability of the organism in its environment also developed further. In the entire life cycle, an organism faces several abiotic and biotic stresses, which reshapes the genetic configuration. Among the abiotic stresses; salinity, drought, cold stress, waterlogging, and heat stress are the predominant ones. The soil may be deficient in some of the essential nutrients and create a stressful environment for the plant. In brief, those stress-imposing factors which are directly related to environmental extremities and not associated with other biological entities are termed as abiotic stress factors. On the other hand, biotic stress factors are referred to as pathogens, pests, and toxins, which directly or indirectly hamper the normal life cycle of the plant. Among the biotic factors, unicellular organisms like plant pathogenic bacteria, viruses, protozoa, fungi, etc., and multicellular organisms like insect pests, rodents are important.

The plant has its own system of mitigating stress, and the process initiates with a network of complex signaling mechanisms. Whenever there is stress, the plant receives the signal from the environment through the receptors and transfers it throughout the system with a cascade of primary and secondary signaling reactions. Hormones like ABA, functional proteins like LEA, HSP, enzymes involving in the biosynthesis of osmoprotectants, and transcription factors like DREBs, NACs, and AREBs are an important part of these signaling mechanisms. Several genes also play a dynamic role in this entire event (Akpinar et al. 2012). On the other hand, molecular crosstalk between several growth hormones like jasmonic acid, salicylic acid, ethylene, ABA, auxins, etc. is central to the balance between growth and defensive responses in the plant (Yang et al. 2019). There are unique mechanisms for each type of stress. For example, under the influence of drought and salinity stress, ABA plays a pivotal role in signal transduction and closure of stomata, so that the moisture present inside the plant system doesn't escape. The plant also retains its moisture through the production of proline, glycine betaine, etc. which helps in osmotic regulation (Dikilitas et al. 2020). Similarly, in response to ROS (reactive oxygen species) plant involves different ROS scavenging enzymes like catalase, superoxide dismutase, peroxidase, etc. which protects the cell from oxidative damage (Tripathy and Oelmüller 2012).

Surely, plants are well equipped to mitigate mild to moderate stress via several pathways. But, both the economic, as well as biological yield of the plant, may get adversely affected under those situations. So, researchers are trying to find out various ways to enhance the productivity of plants under stressed conditions. It may be through the external supply of some compounds or stimuli; or internal regulation of genetic factors. Several agronomical approaches like soil amelioration, application of fertilizers, change in showing time etc. and biotechnological approaches like the introduction of the new gene, regulation of existing genes, RNA interference, etc. are very few examples of that. With the advancement of agricultural sciences and the introduction of nanotechnology in this field, the opportunities to produce higher yield and better-quality foods have widened up. Nanoparticles and biomolecules-coated-nanoparticles have produced promising results in drought stress, heavy metal toxicity, salinity stress, etc. Not only that, those were also used to improve the quality of soil, increase productivity, stimulate plant growth, and smart monitoring of crop plants. Following are few areas where nanoparticles and nanoparticle coated biomolecules play important roles in plant systems.

1. **Importance of nanoparticles in drought stress:**

There are numerous researches that have been conducted on mitigating drought stress with the help of nanotechnology. In maize, drought-induced damage to chloroplasts and mitochondria was successfully checked by using nano-ZnO. Drought stress causes subcellular ultrastructure modifications, as well as the accumulation of osmolytes and malondialdehyde in leaves. In plants, the application of nano-ZnO (100 mg L^{-1}) stimulates melatonin production and triggers the antioxidant enzyme defense mechanism. Changes were related to the upregulation of relative transcript abundance of APX, ASMT CAT, COMT, SNAT, TDC, Fe/Mn-SOD, and Cu/Zn SOD, induced by nano-ZnO particles (Sun et al. 2020). In rice, both drought stress and Cd toxicity were mitigated by applying IONPs (iron oxide nanoparticles) and HGNPs (hydrogel nanoparticles) simultaneously. Bio-fabrication of IONPs was done by using the RNT1 strain of Bacillus. On the other hand, HGNPs were synthesised artificially using chemicals followed by the authentication and characterization of both the nanoparticles by means of nanomaterials characterization methods (Ahmed et al. 2021). Correspondingly, in sorghum, foliar spray of nanoceria (Cerium oxide nanoparticles) @10 mg L^{-1} dramatically lowered the impacts of drought-stressed and improved pollen germination, rate of leaf carbon assimilation, and seed yield per plant; leading to higher grain yield (Djanaguiraman et al. 2018). Thus, nanoparticles perform an important function in dealing with the drought stress in the plant system.

2. **Importance of nanoparticles in salinity stress:**

Salinity stress is one of the foremost reasons of yield decrement in crops that are grown in saline soils. The presence of excess amount of salts creates a hypertonic environment to the plant root, which ultimately results in exosmosis of water from the plant tissue. Nanoparticles may play a vital role in mitigating stress-induced by saline soil. For example, silver nanoparticles (AgNP) can be used to check the effects of salinity stress in wheat, which is directly associated with proline metabolism, defense mechanism, and ion accumulation within the plant system (Wahid et al. 2020). Another study has shown that the application of TiO_2 nanoparticles @ 100 mg L^{-1} on *Dracocephalum moldavica* under salt stress not only reduced H_2O_2 concentration in the plant but also increased essential oil content significantly (up to 1.19%) (Gohari et al. 2020a). Similarly, in salt-treated grapes, iron nanoparticles and potassium silicate promoted an increase in membrane stability index and a decline in malondialdehyde content (Mozafari et al. 2017).

3. **Importance of nanoparticles in supporting plant growth and providing nutrients:**

Regular absorption of soil nutrients helps in the developmental process of plants. Soil is the provider of most of the essential nutrients, including macro and micronutrients. Soil deficient in such nutrients may adversely impact the life cycle and overall productivity of the plant. Hence, to achieve higher productivity, the application of fertilizer may be necessary. The development of nano-fertilizers is a fact of growing interest worldwide. Quite a few researchers have examined the infuences of various NPs on plant growth and phytotoxicity, for example, magnetite (Fe_3O_4) nanoparticles and plant growth (Shankramma et al. 2016), alumina, zinc, and zinc oxide on seed germination and root enlargement of radish, rape, corn, cucumber, and lettuce, sulfur nanoparticles on tomato, silver nanoparticles and wheat seedling growth (Salem et al. 2016), and zinc oxide in mungbean, etc. Wheat growth and yield can be boosted by soil application of silver nanoparticles (SNPs) @ 25 ppm. (Ghidan and Al Antary 2019).

4. **Importance of nanoparticles and nanoparticle-coated biomolecules in plant pest and disease management:**

Traditional laboratory methods of pathogen diagnosis, for instance, microscopic and cultural approaches, take time and necessitate complicated sample handling processes. Immunological and molecular approaches have progressed significantly, but they also have some problems associated with rapidity as well as signal strength. However, in today's agriculture systems, symptoms of infectious diseases can often be detected well before the actual onset of infection, with higher accuracy and rapidity. Nanotechnology has the ability to be used for pathogen diagnosis and removal because it can deal with certain problems created by viruses and other pathogens. Teams studying drug delivery mechanisms, nutrients, and probiotics have long dreamed of designing medication release systems, and nanotechnology now provides opportunities in this area (Kashyap et al. 2016). Different biological, physical, and chemical processes are often used in the agriculture and food industry to detect viral, or bacterial infections as well as chemical contamination. Nanotechnology has recently been used to detect contaminants, even at very low levels, thanks to the invention of nanoscale sensors. These sensors can also be used to quantify and track materials derived from bacterial metabolism and growth, allowing for bacterial and viral toxicity monitoring. For example, plant pathogens like *Xanthomonas axonopodis pv. vesicatoria* inducing bacterial spot in pepper and tomato, were effectively identified using fluorescent silica nanoparticles (FSNP) conjoined with antibody units (Yao et al. 2009). Similarly, copper oxide (CuO) nanomaterials and nanolayers were put to use to trace *Aspergillus niger* in growth medium (Etefagh et al. 2013)

5. **Importance of nanoparticles and nanoparticle-coated biomolecules in the formulation of pesticides with important functions:**

Some nanomaterial properties, like permeability, stiffness, crystallinity, solubility, thermal resilience, biodegradability, etc. have significantly improved the formulation and impact of nano pesticides, improving the dissemination and wettability of agricultural compositions while reducing organic solvent runoff. Controlled release kinetics are possible with nano-encapsulated pesticides. By blocking the early deterioration of active ingredients (AIs) under severe environmental conditions, nanoencapsulation can improve pest-control performance for longer periods of time (Kumar et al. 2019). For example, researchers have created a nanoparticle delivery system, containing γ-polyglutamic acid and chitosan for transportation of avermectin, a nematicide that controls pinewood nematode. The system has shown promising results with a 20% initial release of nematicide regardless of pH, followed by a 69.5% release at pH 8.5, in contrast to 60.4% at pH 7 and 57.5% at pH 5.5. Because of the electrostatic reciprocity between chitosan and the carboxyl groups of -polyglutamic acid, the nanocarrier set up demonstrated great stability at pH 5.5. This interaction gets affected due to high pH, which causes the release of avermectin in a controlled manner (Liang et al. 2018a, b).

6. **Other importance:**

Nanomaterials in agriculture are being used to limit the usage of plant protection chemicals, lessen fertilizer nutrient inputs, and boost outputs by better nutrient management strategies. The use of target-specific nanoparticles will minimize non-targeted plant tissue damage as well as the number of chemicals released to the environment. Plant breeding and genetic transformation have also been studied using nanotechnology-derived instruments and techniques. Metal and metal oxide nanoparticles of precisely defined sizes, weight, morphologies, and physicochemical characteristics, in particular, fluorescence, magnetic, and photocatalytic degradation effects, have brought a revolution in sensor development, degradation of agrochemicals, and soil remediation among nanomaterials. They also contributed to the advancement of new agricultural technologies. For example, selenium nanoparticles (SeNPs) based biosensors, with 35–40 nm size and face-centered cubic (FCC) structures are used for the investigation of heavy metal toxicity in agriculture (Ahmed et al. 2020). Similarly, Rice plants treated with a variety of carbon nanomaterials (nanotubes, C60, graphene) revealed that these materials can increase moisture content in seeds and can also be translocated to leaves (Nair et al. 2012). *Cicer arietinum* treated with water-soluble carbon nanotubes (wsCNTs) showed increased water absorption by carbon nanomaterials, which improved shoot, root, and branching growth rates (Tripathi et al. 2011).

8.3 Nanoparticles: Types and Synthesis

Silver Owing to better antimicrobial peculiarities, silver nanoparticles are considered as the most effective against microorganisms such as bacteria, viruses, and other eukaryotes (Rai et al. 2009). Among all, they are the most prevalent nanoparticles as they have manifold applications like antimicrobial agents, textile industries, treatment of water, sunscreen lotions, etc. (Sharma et al. 2009). As per the reports, they have been successfully synthesized by several plants namely *Azadirachta indica* (Shankar et al. 2004), *Capsicum annuum* (Bar et al. 2009), and *Carica papaya* (Jha and Prasad 2010).

Gold For immunochemical researches, gold nanoparticles (AuNPs) are utilized to recognize protein interactions. These particles are implemented as lab tracers in DNA fingerprinting to locate DNA in a sample. These are also employed for the identification of aminoglycoside antibiotics such as streptomycin, gentamycin, and neomycin. The application of gold nanorods leads to the detection of cancer stem cells, cancer diagnosis, and to identify various kinds of bacteria (Tomar and Garg 2013).

Alloy These nanoparticles possess structural features that vary from their bulk samples (Ceylan et al. 2006). Because of very high electrical conductivity (Junggwon et al. 2008), silver flakes are most commonly exploited. The properties of bimetallic alloy nanoparticles are governed by the set of two metals and display greater preferences over conventional metallic nanoparticles (Mohl et al. 2011).

Magnetic Magnetic nanoparticles, for example, Fe_3O_4 (magnetite) and Fe_2O_3 (maghemite) exhibit biocompatibility. They have been applied for the treatment of cancer (magnetic hyperthermia), sorting and manipulation of stem cells, drug delivery, DNA analysis, gene therapy, and magnetic resonance imaging (MRI) respectively (Fan et al. 2009).

The synthesis of nanoparticles occurs via chemical and biological methods. Unlike chemical methods including chemical reduction of metal ions, photoreduction in reverse micelles, thermal decomposition of organic solvents, etc., which have many adverse effects, nanoparticles synthesis through the biological mode is eco-friendly. Nanoparticles can be synthesized biologically through microorganisms, plants, or plant extracts.

8.4 Microorganisms for Nanoparticles Synthesis

The approach of utilizing microorganisms for synthesizing nanoparticles offers certain advantages: (1) It is a cost-effective tool, (2) It is free from harmful, toxic chemicals, and (3) It holds huge energy demand prescribed for physiochemical synthesis. Intracellular and extracellular synthesis of nanoparticles has been studied using microorganisms such as bacteria (actinomycetes), fungi, and yeasts. The intracellular process entails the movement of ions into microbial cells subjected to the availability of enzymes. On the other hand, extracellular synthesis entails the elimination of downstream processing steps required to recover nanomaterials in intracellular methods, numerous centrifugation and washing steps prescribed for the purification of nanoparticles, and others. Extracellular secretory constituents like metal-resistive genes, proteins, enzymes, peptides, cofactors, and organic substances act as reducing agents and provide natural capping for nanoparticle synthesis. Therefore, these help in preventing the assemblage of nanoparticles and ensuring stability for a longer period.

Bacteria Principally, enzymes help in nanoparticle synthesis through bacteria (Zhang et al. 2011). For instance, silver nanoparticle was synthesized using nitrate reductase enzyme in *Bacillus licheniformis*. Bacteria, including *Pseudomonas deceptionensis* (Jo et al. 2015), *Weissella oryzae* (Singh et al. 2015f), *Bacillus methylotrophicus* (Wang et al. 2015), *Brevibacterium frigoritolerans* (Singh et al. 2015d) and *Bhargavaea indica* (Singh et al. 2015c, e), have been utilized for the synthesizing silver and gold nanoparticles. Metal nanoparticle synthesis has been possible using *Bacillus, Pseudomonas, Klebsiella, Escherichia, Enterobacter, Aeromonas, Corynebacterium, Lactobacillus, Pseudomonas, Weissella, Rhodobacter, Rhodococcus, Brevibacterium, Streptomyces, Trichoderma, Desulfovibrio, Sargassum, Shewanella, Plectonemaboryanum, Rhodopseudomonas, Pyrobaculum*, and others (Li et al. 2011).

Fungi To achieve an easy and stable nanoparticle synthesis, mycosynthesis is a comprehensible approach. Several fungi characterized by chief metabolic substances with more bioaccumulation capacity and basic downstream processing can be easily cultured to synthesize low-cost nanoparticles effectively (Alghuthaymi et al. 2015). Fungi possess a higher ability to tolerate and uptake metals in comparison to bacteria, thereby metal salts bind with the wall of fungal biomass for producing high yielded nanoparticles (Castro-Longoria et al. 2011). There are three possible mechanisms for mycosynthesis of metallic nanoparticles i.e., nitrate reductase action, electron shuttle quinones, or duo. Reductase enzymes derived from fungi like *Penicillium* species and *Fusarium oxysporum*, α-NADPH-dependent reductases, and nitrate reductase participate actively in nanoparticle synthesis (Alghuthaymi et al. 2015).

Actinomycetes Although actinomycetes mediated nanoparticles are stable and have considerable biocidal activities against different pathogens, nanoparticle synthesis using actinomycetes has not been well investigated (Golinska et al. 2014). A study illustrated that reductase enzyme from *Streptomyces* species promoted the reduction of metal salts while synthesizing silver, zinc, and copper nanoparticles using *Streptomyces* (Karthik et al. 2014).

Yeast In parallel to other microorganisms, yeasts play a considerable role in producing nanoparticles extracellularly on a huge scale (Waghmare et al. 2015). Examples – intracellular synthesis of lead nanoparticles by *Rhodosporidium diobovatum* (Seshadri et al. 2011), extracellular synthesis of silver and gold nanoparticles by means of Extremophilic yeast (Mourato et al. 2011).

Viruses The synthesis of nanoparticles via viruses is also possible. Viruses enable the synthesis of nanowires with working components that have wide applications in battery electrodes, photovoltaic devices, and supercapacitors (Nam et al. 2006).

Despite that, microorganisms-mediated synthesis involves specific problems including slow synthesis, minimum productivity, need for downstream processing to recover nanoparticles, complicated steps involving microbial sampling, isolation, culturing, and maintenance.

8.5 Steps Involved in the Microorganisms-Mediated Synthesis of Nanoparticles

Extracellular synthesis

- culturing of microorganisms for 1–2 days in a rotating shaker in accordance with ideal temperature, pH, medium factors, etc.
- centrifugation of the culture to exclude biomass
- addition of a filter-sterilized metal salt solution to the obtained supernatant followed by incubation
- monitoring of nanoparticle synthesis by noticing the colour change in the culture medium
- centrifugation of the reaction mixture at varying speeds post-incubation so as to remove any large particle or medium components
- final centrifugation of the nanoparticles at elevated speed or with a density gradient
- washing of nanoparticles in water/ethanol/methanol and collection of nanoparticles in a bottom pellet fashion

Intracellular synthesis

- culturing of microorganisms
- collection of biomass by centrifugation
- washing with distilled water
- dissolving in sterile water with filter-sterilized metal salt solution
- monitoring the reaction mixture for a color change
- removal of biomass by ultrasonication, washing and centrifugation to initiate cell wall breakage and release of nanoparticles
- further centrifugation, washing, and collection of the synthesized nanoparticles

8.6 Biosynthesis of Nanoparticles Using Plants

Phytonanoparticle synthesis is a very simple, rapid, eco-friendly, and cost-effective strategy. Besides, it has several benefits in terms of biocompatibility, adaptability, applications in the medical sector. Water, a universal solvent is used as a reducing agent for synthesizing plant-based nanoparticles (Noruzi 2015). Nanoparticles can be synthesized from plants using easily available plant extracts which are non-toxic in nature. Reports suggested that medicinal plants can be utilized as resources to synthesize nanoparticles, for instance, root and leaf extracts of *Panax ginseng* have been exploited for the synthesis of silver and gold nanoparticles (Singh et al. 2015b). Furthermore, numerous plant organs (leaves, stem, roots, fruits) and their extracts are utilized for metal nanoparticle synthesis (Table 8.1). Nevertheless, the appropriate mechanism and the ingredients accountable for the synthesis of plant-derived nanoparticles are still unexplored. It has been envisaged that amino acids, proteins, organic acids, vitamins, secondary metabolites (flavonoids, alkaloids, polyphenols, terpenoids, heterocyclic compounds, and polysaccharides) have far-reaching contributions in reducing metal salts and also they serve as capping and stabilizing medium for biosynthesized nanoparticles (Duan et al. 2015). For example, gold nanoparticles could be synthesized with the help of proteins and polyamines of *Corallina officinalis* extract (El-kassas et al. 2014). It has been highlighted that silver and gold nanoparticles were synthesized and stabilized by biomolecule adherence in leaf extract of *Murraya koenigii* (Philip et al. 2011). Different plants have different mechanisms for the synthesis of nanoparticles. For instance, emodin in xerophytes assists in silver nanoparticle synthesis; cyperoquinone, dietchequinone, and remirin in mesophytes are worthwhile for the synthesis of metal nanoparticles (Baker et al. 2013). Eugenol, the principal terpenoid of *Cinnamomum zeylanisum*, takes part in synthesizing gold and silver nanoparticles (Makarov et al. 2014).

Table 8.1 Synthesis of nanoparticles from plants/plant extracts

Plants	Plant extract	Type of nanoparticle	References
Azadirachta indica	Leaves	Silver	Poopathi et al. (2015)
Panax ginseng	Root	Silver and gold	Singh et al. (2015b)
Ginkgo biloba	Leaves	Copper	Nasrollahzadeh et al. (2015)
Euphorbia prostrata	Leaves	Silver and titanium dioxide (TiO2)	Zahir et al. (2015)
Catharanthus roseus	Leaves	Palladium	Kalaiselvi et al. (2015)
Artocarpus gomezianus	Fruit	Zinc	Suresh et al. (2015)
Lawsonia inermis	Leaves	Iron	Naseem et al. (2015)
Musa sp.	Peel	Cadmium sulfide	Zhou et al. (2014)

Steps involved in plant-mediated nanoparticle synthesis:

washing of plant parts such as leaf, root, bark, etc. using distilled water
cutting into pieces of small size and boiling for extraction
purification of the extract by filtration and centrifugation
incubation of the reaction mixture for reducing the metal salt
monitoring for a color change
collection of synthesized nanoparticles

8.7 Application of Nanoparticles in Salt Stress Management

Different types of nanoparticles play pivotal roles in imparting salinity tolerance to plants. These are briefly discussed in this section.

1. **Zinc (Zn) Nanoparticles**

Zinc is an essential micronutrient for crops. It performs diverse functions in the cell. Moreover, zinc is also associated with DNA transcription, inter-and intra-cellular signaling (Caldelas and Weiss 2017). It is believed that zinc performs a primary role in plants susceptible to abiotic stress conditions including salinity (Sofy et al. 2020). Nowadays, zinc oxide (ZnO) nanoparticles are being applied commercially with an expectation to serve as a boon for the agricultural sector. It has been suggested that Zn nanoparticles can be a good remedy for abating the unfavorable effects of environmental stresses in plants (Caldelas and Weiss 2017). Studies emphasized that plants treated with Zn-based nanoparticles showed better morpho-physiological traits under control (Mahmoud et al. 2020) and salinity stress (Alabdallah and Alzahrani 2020; Farouk and Al-Amri 2019). Application of ZnO nanoparticles enhanced photosynthetic pigments, antioxidative enzymes (CAT, SOD) performance, but decreased proline and total soluble sugar content (Alabdallah and Alzahrani 2020). It was found that the employment of Zn nanoparticles to salt-stressed canola plants reduced the salt-triggered harsh impacts via enhancing osmolyte biosynthesis, ion regulation, and antioxidant system (Farouk and Al-Amri 2019). While studying salinity stress in potato plants, it was found that the combined soil incorporation of zeolite, zinc, boron and silicon nanoparticles improved water retention capacity, nutrient use efficiency, antioxidative enzymatic profiles, photosynthetic, growth, and yield (Mahmoud et al. 2019). Further, Zn nanoparticles produce positive effects when applied in low doses, while high concentrations lead to toxicity even under normal conditions (Molnar et al. 2020). Recently a study in rapeseed plants showed that the exogenous application of ZnO-nanoparticles caused a reduction in ion leakage and improvement in Hill reaction and henceforth regulated the expression of stress response genes, for instance, *ARP* expression was upregulated while that of *MYC, SKRD2*, and *MPK4* were downregulated (Hezaveh et al. 2019). Therefore, from these pieces of literature, it is quite evident Zn nanoparticles ameliorate salinity stress, however, future research must be emphasized on molecular impacts to completely interprete the mode of actions of these nanoparticles under salinity stress.

2. **Silver (Ag) nanoparticles**

Because of their antifungal and antibacterial properties, Ag nanoparticles are being used in agriculture. It is known that Ag nanoparticles improve multiple growth traits such as germination, growth via modulating various physio-chemical characteristics of crops (Soliman et al. 2020; Mohamed et al. 2017). Treatment of *Pennisetum glaucum* seeds with Ag nanoparticles lowered oxidative damage under saline conditions as a result of higher antioxidant enzyme activities (Khan et al. 2020). Additionally, Ag nanoparticles suppressed the Na+/K+ ratio in leaves but enhanced flavonoids and phenolic composition (Khan et al. 2020). During an experiment in wheat, it was demonstrated that salt-stressed wheat seedlings, when treated with Ag nanoparticles, showed an improvement in POD activity, total soluble sugar, and hence an overall growth of the plants was observed (Mohamed et al. 2017). According to reports, germination and grain yield in wheat are influenced under saline regimes by Ag nanoparticles-induced- modulation of photosynthetic efficiency and growth regulators because 6-benzylaminopurine, 1-naphthalene acetic acid, and indole-3-butyric acid levels were boosted, while abscisic acid (ABA) content was lowered (Abou-Zeid and Ismail 2018). There are scanty reports on Ag nanoparticles to minimize the deteriorating effects of salt stress in plants. Thus, future investigations should be focused on annotating their involvement in salinity stress management at physiological, biochemical, and molecular stages. Since Ag nanoparticles are highly toxic, they should be used carefully (de Souza et al. 2019; Tortella et al. 2020). Thus, it is necessary to understand how Ag nanoparticles stimulate the growth of plants, and in what way they can put forth threat to the environment.

3. **Iron oxide (Fe_2O_3) nanoparticles**

Iron is an indispensable micronutrient for plants. It cooperates in various cellular processes such as photosynthesis, respiration, chlorophyll biosynthesis (Kim and Guerinot 2007). It is also associated with the synthesis of proteins related to cellular metabolism, oxygen balance and transport, DNA repair and thereby promoting the overall productivity of crops (Tripathi et al. 2018). Iron is known to intervene salinity tolerance supported by up-regulation of antioxidative enzymes in plants (Singh and Bhatla 2016). For instance, foliar application of nano-Fe_2O_3 in combination with salicylic acid (SA) restored the physiological performance of ajowan plants under saline conditions. The authors observed that the plants treated with combined SA and Fe_2O_3 nanoparticles showed enhancement in K^+ uptake, K^+/Na^+ ratio, Fe content, endogenous salicylic acid level, antioxidant enzyme activities, and key osmolytes. These alterations led to the improvement in membrane stability index, leaf water content, photosynthetic pigments, growth of root and shoot, and ultimately seed yield of salinity stressed plants (Abdoli et al. 2020). Similarly, in *Helianthus annus* plants, Fe nanoparticles increased the activities of catalase, peroxidase, polyphenol oxidase and hence alleviated salt stress effects (Torabian et al.

2018). However, there is limited information on the particular metabolic pathways regulated by these nanoparticles.

4. **Silicon dioxide (SiO_2) nanoparticles**

Though silicon is surely not an essential mineral nutrient, it contributes to several metabolic pathways associated with different environmental stresses including salinity in plants (Vaculík et al. 2020). It has been noted that supplementation of silicon to plants raises plant water status by withstanding water loss during salinity (Abdelaal et al. 2020). Furthermore, the application of silicon to salt-stressed plants increases photosynthesis, dry matter production, vegetative growth, k^+ deposition; decreases Na^+ and Cl^- accumulation in the shoot (Hurtado et al. 2020).

Nano forms of silicon influence crop growth and yield in response to salt stress. For instance, SiO_2 nanoparticles enhance the rate of germination and growth in wheat seedlings under salinity (Mushtaq et al. 2019). These nanoparticles mediate tolerance against salinity stress in soybean seedlings via increasing k^+ content in leaves, stimulating antioxidative enzyme profiles while decreasing leaf Na^+ concentration, lipid peroxidation, and ROS generation (Farhangi-Abriz and Torabian 2018). Application of nano SiO_2 to strawberry plants retained epicuticular wax structure, leaf pigments (chlorophyll and carotenoid) but reduced proline accumulation in comparison to the salt-stressed plants without nano supplementation (Avestan et al. 2019). In tomato, plants treated with silicon nanoparticles performed well by maintaining glutathione reductase level, chlorophyll concentration, phenylalanine ammonia-lyase activity relative to untreated plants under saline conditions (Pinedo-Guerrero et al. 2020).

From the above studies, it can be inferred that silicon nanoparticles improve plant responses against harmful consequences of salinity stress by regulating various physiological and biochemical parameters. But, their application at the gene level is yet to be understood and further research should be done indicating how silicon nanoparticles affect stress-responsive genes in crops.

5. **Manganese (Mn) nanoparticles**

Mn, like other micronutrients, is critically required for plant metabolism. It functions as a cofactor for few enzymes like Mn superoxide dismutase, oxalate oxidase, and Mn-protein in PS II (Eaton 2015. Mn nanoparticles have the capacity to mitigate salt stress through nano-priming. For example, in capsicum plants, priming with Mn nanoparticles strengthened root elongation and ameliorated salinity stress in the process of germination. Further, nano-forms of Mn assisted in controlling oxidative stress in capsicum seedlings (Ye et al. 2020). Foliar application of Mn nanoparticles boosted the antioxidant defense system, increased leaf photosynthetic pigments, net photosynthesis, and maintained biomass of cucumber plants and thereby counteracting unfavorable effects of salinity stress (Lu et al. 2020). In another study, supplementation of Mn nanoparticles improved membrane stability index, chlorophyll concentrations, and nitrate reductase activity in mung bean plants under salt stress (Shahi and Srivastava 2018). The contribution of Mn nanomaterials in influencing various physiological as well as biochemical attributes in salt-stressed plants is still needed to be explored. So, researchers should focus on this domain in the future so as to facilitate effective amelioration of salinity stress conditions.

6. **Copper (Cu) nanoparticles**

Cu, an essential micronutrient, is broadly distributed in plant tissues and is concerned with various physiological mechanisms (Chibber et al. 2013). It performs a chief function in ethylene perception, photosynthesis, respiration, carbon, and nitrogen metabolisms (Iqbal et al. 2018). Furthermore, Cu is a crucial component of specific enzymes associated with redox reactions (Lwalaba et al. 2020). Investigations demonstrated that Cu imparted tolerance against salinity stress by reducing the production of harmful ROS and increasing the accumulation of osmoprotectants as well as amino acids in plants, for example, maize (Iqbal et al. 2018). It has been examined that foliar application of Cu nanoparticles to tomato plants lowered the damaging impacts of salinity stress by ameliorating the Na^+/K^+ ratio and growth performance. In addition, plants with Cu nanoparticles treatment showed a higher percentage of phenol and vitamin C in leaves; increased glutathione peroxidase content in fruits as compared to untreated tomato plants (Pérez-Labrada et al. 2019). The application of Cu NPs to salt-treated tomato plants upregulated the expression profile of jasmonic acid and SOD genes, which contributed to the alleviation of oxidative and ionic stresses (Hernández-Hernández et al. 2018a, b). This study proposed that the supplementation of Cu nanoparticles could efficiently promote salt tolerance via triggering the antioxidant defense mechanism and through the octadecanoid pathway of jasmonates. Thus, the utilization of Cu nanoparticles is a feasible way to manage salinity stress and enhance crop growth.

7. **Cerium (Ce) Nanoparticles**

Nano-forms of 'Ce' are mostly used in cosmetics, semiconductor, pharmaceutical, drug delivery, and optical markets (Hussain et al. 2019). Application of lower doses of Ce-based nanoparticles enhances physiological and biochemical traits in plants grown under unstressed conditions (Salehi et al. 2018). It is reported that poor concentrations of Ce nanoparticles increased the rate of photosynthesis and overall growth performance of wheat plants while high levels of Ce nanoparticles had negative influences on physio-biochemical variables (Abbas et al. 2020). Ce nanoparticles are known to have the ability to relieve abiotic stresses such as salinity (Rossi et al. 2019). For instance, under salinity stress, the supplementation of Cerium oxide (CeO_2) nanoparticles to Brassica plants pruned the barriers of root apoplast which enabled more transportation of Na+ to shoots and low Na+ accumulation in plant roots. These alterations in transport and fluxes of Na+ led to the improved physiological activity of the plants (Rossi et al. 2017). In another study, the morpho-physiological, biochemical, and molecular mechanisms were found to be strongly associated with Ce nanoparticles mediated seed priming in salt-stressed cotton plants (An et al. 2020). The authors noticed an enhancement in biomass and growth, on the other hand, root transcripts had differential expression responding to seed priming with nanomaterials of Ce. Furthermore, plants treated with Ce nanoparticles showed low levels of ROS. It was concluded by the authors that seed priming with Ce nanoparticles prompted salinity tolerance which was associated with ion homeostasis, ROS mechanisms, and Ca^{2+} signaling pathways in plants. Despite that, the availability of research on the impact of Ce nanoparticles in mitigating salt stress in crop plants is scarce. Future studies should be undertaken to scrutinize the molecular mechanism of action of Ce nanoparticles for combating salinity stress conditions in plants.

8. **Titanium dioxide nanoparticles**

Titanium is a transition element. Since 0.33% of the earth's outer layer is composed of titanium, it is categorized as the ninth abundant metal (Buettner and Valentine 2012). It improves the performance of plants by regulating enzyme activities, enhancing chlorophyll content and photosynthesis (Carbajal-VáZquez et al. 2020). Specifically, titanium plays an imperative role in making plants resistant to handle salinity stress (Lyu et al. 2017). There are three distinct mechanisms of application of TiO_2 nanoparticles in plants. The first mechanism is associated with alterations in ROS signaling via TiO_2 nanoparticles; the second mechanism describes the improvement of TiO_2 nanoparticles-induced-nitrogen metabolism in plants by converting atmospheric nitrogen to nitrate in presence of UV or sun's irradiation and the third mechanism deals with the size, shape, and surface features of TiO_2 nanoparticles, that modulates its accessibility to the plants. The combination of these three attributes determines the inclusive functioning of TiO_2 nanoparticles in plants. In leguminous crops such as broad beans, the supplementation of 0·01% TiO_2 nanoparticles significantly promoted the activities of antioxidative enzymes and profiles of total soluble sugars, proline, and amino acids in plants subjected to salt stress against plants with salinity treatment alone (Abdel Latef et al. 2018). Hence, the upregulated activities of antioxidant enzymes attributed to the detected drop in the levels of hydrogen peroxide and malondialdehyde, while higher amounts of proline and various metabolites were found to be responsible for the osmoprotection, unanimously leading to the significant advancement in the growth of plants under salinity. Moreover, the positive effects mediated by TiO_2 nanoparticles were dependent on the concentration, for example, 0·01% TiO2 nanoparticle was highly effective, contrastingly 0·02% exhibited an intermediate response and 0·03% was futile under both normal and stressed environments (Abdel Latef et al. 2018).

In another study, the effect of TiO_2 nanoparticles was evaluated on salt-affected *Dracocephalum moldavica* plants. The findings revealed that these nanoparticles improved some enzyme activities, which consequently promoted plant growth under salinity. Further, the levels of oil like z-citral, geraniol, geranyl acetate, and geranial were enhanced in response to the application of TiO_2 nanoparticles under salt stress (Gohari et al. 2020a, b).

Recent pieces of literature evidenced the utilization of TiO_2 nanoparticles in reducing the negative effects of plants under salinity stress conditions. Furthermore, it is needed to know the doses of application of TiO_2 nanoparticles and to elucidate the mechanism of TiO_2 nanoparticles-triggered metabolic processes in plants grown under salinity.

9. **Potassium (K) nanoparticles**

Potassium, an essential macronutrient, plays a substantial role in the growth and metabolism of plants. It helps to alleviate unfavorable environmental stresses in plants. Exogenously-added potassium has been reported to elevate salinity tolerance in crops (Xu et al. 2020). For instance, in alfalfa plants, the application of potassium-based nanoparticles (K_2SO_4) improved plants' responses to salt stress via reducing leakage of electrolytes, increasing proline content and antioxidative enzyme activities, thereby overcoming the unfavorable impacts of salinity stress (El-Sharkawy et al. 2017). Further, the role of potassium nanoparticles was evaluated in wheat plants. While supplementing potassium nanoparticles as the foliar spray in salt-treated wheat plants, it was observed that ROS level decreased significantly with a concomitant increase in morphological characteristics, leaf pigment concentration, total carbohydrates, total phenols, proline content, antioxidative enzyme activities than that of the potassium nanoparticles-untreated plants under salt stress (Jan et al. 2017).

There exists no further recent information on potassium-mediated nanomaterials in combating salinity stress in crop plants. Hence, researchers should show more attention to evaluate the effects of various concentrations of nanoforms of potassium on several crops for unraveling the morpho-physiological, biochemical, and molecular traits regulated by them.

10. **Sulphur (S) nanoparticles**

Sulphur is regarded as a major macroelement that participates in plant growth and developmental processes. It is required for chlorophyll functioning and synthesis of chief proteins (Duncan et al. 2018). The efficiency of sulphur nanoparticles in mitigating salinity stress in crops has been reported. An experiment demonstrated that the lettuce plants supplemented with green synthesized sulphur nanoparticles exhibited increased compatible solutes, total phenols, total soluble sugars, total flavonoids, anthocyanin, tannins, etc. under saline conditions while control plants showed a significant reduction in these traits (Najafi et al. 2020). Studies on the interaction between sulphur nanoparticles and salinity stress are rare. Therefore, plant scientists should focus on the application of these nanoparticles to foster our knowledge on how these nanoparticles would benefit plants with respect to salt stress management. Also, the doses and type of application of these nanoparticles should be clearly understood for obtaining sound results in plants.

11. **Carbon (C) nanoparticles**

There are heterogeneous forms of 'C' nanoparticles such as graphenes, nanotubes, and fullerenes (Zaytseva and Neumann 2016). The impact of carbon nanomaterials depends upon the dose, size, and solubility of the supplemented nanoparticles (Husen and Siddiqi 2014). It has been reported that carbon nanomaterials could help in limiting the disadvantageous effects of salinity stress and henceforth, enriching the quality as well as yield of crops under stressed conditions (Khan et al. 2017). For instance, the application of water-soluble carbon nanomaterials to lettuce seedlings improved germination under salinity stress (Baz et al. 2020). In addition to that, the elongation of primary roots was inhibited whereas the lateral roots showed enhanced growth in plants with carbon nanomaterials treatment. Further, the authors observed an improvement in the chlorophyll content of salt-stressed lettuce plants in response to the treatment of carbon nanoparticles (Baz et al. 2020). Treatment with multi-walled carbon nanotubes resulted in an increased rate of water uptake through improved aquaporin transport and photosynthesis in broccoli plants under salt stress (Martínez-Ballesta et al. 2016). Additionally, carbon nanotubes promoted fluctuations in the composition of lipids and the permeability of root plasmalemma under saline regimes and thus, mediating salinity tolerance. Similarly, in rapeseed seedlings, the supplementation of carbon nanotubes increased the activity of nitrate reductase, nitric oxide production; re-established ion and redox instability followed by the suppression in the production of thiobarbituric acid and reduction in Na+/K+ ratio under salinity stress conditions (Zhao et al. 2019). Moreover, carbon nanotubes affect plants at the molecular level via altering transcripts such as Na^+/H^+ exchanger 1 (NHX1) and K^+ transporter 1 (KT1), salt overly sensitive 1 (SOS1) genes, and genes related to the antioxidant defense system (Zhao et al. 2019). In another study, low concentrations of carbon nanotubes ameliorated salinity effects in *Ocimum basilicum* plants by stimulating photosynthetic pigments and increasing antioxidative enzyme activities and phenolics content, in contrast to high levels of carbon nanotubes which showed phytotoxicity as proved through biochemical and epifluorescence microscopy observations (Gohari et al. 2020a, b). The application of polyhydroxy fullerene nanomaterials enhanced antioxidative enzyme activities, chlorophyll content, phosphorous and potassium concentrations, total soluble sugars in wheat seedlings, thereby inducing salinity tolerance (Shafiq et al. 2019).

Although carbon-based nanoparticles have created their own position in the realm of abiotic stress management, exploration in light of the effects of carbon nanodots, nanobeads, nano-diamonds, and nano-fibers is limited. Therefore, future studies are critically required in this arena to develop a clear concept on the mode of action of these carbon nanomaterials concerning the remediation of salt stress effects in plants (Table 8.2).

8.8 Role of Nano-Encapsulation in Mitigating Salinity Stress

Currently, nano-encapsulation is one of the main approaches for enhancing salinity tolerance in plants. For instance, encapsulation of the NO (signaling molecule) donor i.e., S-nitroso-mercaptosuccinic acid into chitosan-based nanoparticles resulted in high photosystem II activity and chlorophyll content in salt-stressed maize plants and hence, eliminated the deleterious outcomes of salinity stress (Oliveira et al. 2016). Moreover, S-nitrosothiol content was increased in leaves of maize plants treated with S-nitroso-mercaptosuccinic acid encapsulated chitosan nanoparticles under salinity stress (Oliveira et al. 2016). Researchers evaluated the effects of polyvinylpyrrolidone (PVP) coated silver nanoparticles on Bitter Vetch

Table 8.2 Effects of nanoparticles on plants grown under salinity stress

Nanoparticle and its form	Concentration of application	Mode of application	Plants	Effects	References
Zinc – ZnO	10 mg/l	Foliar spray	Okra	Increased photosynthetic pigments, decreased total soluble sugars, proline	Alabdallah and Alzahrani (2020)
Silver – Ag	0, 10, 20, and 30 mM	Seed priming	Pearl millet	Enhanced relative water content, minimized oxidative damage, decreased Na+/K+ ratio	Khan et al. (2020)
Silicon – SiO_2	50, 100 mg/l	Foliar application	Strawberry	Maintained epicuticular wax structure, leaf pigments, decreased proline	Avestan et al. (2019)
Copper – Cu	250 mg/l	Foliar spray	Tomato	Improved antioxidant defense machinery, vitamin C, glutathione level	Pérez-Labrada et al. (2019)
Iron – Fe_2O_3	3 mM	Foliar spray	Ajowan	Enhanced K^+/Na^+ ratio, K^+ uptake	Abdoli et al. (2020)
Titanium – TiO_2	50, 100, 200 mg/l	Nutrition solution	Moldavian balm	Upregulated antioxidative enzyme activities	Gohari et al. (2020a)
Cerium – CeO_2	500 mg Kg^{-1} dry sand		Canola	Pruned the barriers of root apoplast which enabled more transportation of Na+ to shoots and low Na+ accumulation in plant roots	Rossi et al. (2017)
Potassium-K_2SO_4	1/4, 1/8, and 1/10 of the 'K' levels in full strength Hoagland's solution (235 mg/l)		Alfalfa	Reduced leakage of electrolytes increased proline content	El-Sharkawy et al. (2017)
Manganese	0.1, 0.5 and 1 mg/l	Seed priming	Capsicum	Increased root growth	Ye et al. 2020
Carbon	20 mg/l	Seedling treatment	Rapeseeds	Re-established redox balance and ion homeostasis	Zhao et al. (2019)

plants exposed to salinity stress (Hojjat et al. 2019). It was stated that the application of lower concentrations of PVP-silver nanoparticles boosted seedling vigor index and rate of germination in the plants grown under high salinity conditions in relation to the control samples. Possibly, rapid germination was due to the silver nanoparticles mediated formation of new pores on the seed coat while penetrating which in turn facilitated the movement of water into the seed. Another reason could be the entry of silver nanoparticles into the seed with the help of the cracks present over the seed surface and as a result, the enzymes activated in the early phase and hence enhancing the germination speed (Hojjat et al. 2019). It is reported that chitosan, a natural polymer is being used for nanoencapsulation due to its various properties such as bio-degradability, toxic-free, biocompatibility, antibacterial, etc. (Kashyap et al. 2015). Copper nanoparticles encapsulated with chitosan enhanced the quality of tomato fruits and induced salinity tolerance in those plants via upregulating the activities of antioxidative enzymes (catalase, peroxidase, superoxide dismutase, glutathione peroxidase, ascorbate peroxidase) and suppressing Na^+ accumulation in the leaves (Hernǻndez-Hernǻndez et al. 2018a, b).

Research regarding the effects of nano-encapsulation on salinity stressed plants is very limited. Furthermore, whether the nano-encapsulation strategy has any impact on stress-related genes is quite unknown. Future investigations should be focused on exploring the molecular aspects of nano-encapsulation in order to combat the ill effects of salt stress in crop plants. Furthermore, a large number of active compounds should be tested to evaluate their efficiency for encapsulation into nanoparticles in plants.

8.9 Mechanism of Action of Nanoparticles to Ameliorate Salt Stress

Salt stress is one of the abiotic stresses, which in particular, is concerned with ionic toxicity and interrupts ionic homeostasis in plants (Tanveer and Shabala 2018). In this stress, Na^+ and Cl^- ions accumulate significantly in plant cells resulting in ionic toxicity (Tanveer and Shah 2017). During salt stress, plants generally efflux K^+ out of mesophyll cells and accumulate Na^+ ions in the cytosol (Shabala and Cuin 2008). Salt stress decreases the chlorophyll content (Yasar et al. 2008) which may be due to the chlorophyll degrading enzyme, chlorophyllase (Nazarbeygi et al. 2011) in fennel plants (Rahimi et al. 2012), Faba Bean (Azooz et al. 2013) and rice (Djanaguiraman and Ramadass 2004). Proline is also accumulated under salt stress which may be due to the degradation of proteins to form amino acids, including proline (Ashraf and Foolad 2007) or due to proline biosynthesis related enzymes or due to decreased proline oxidation (Ahmad and Jhon 2005). Salinity stress has a superior role in the formation of reactive oxygen species (ROS) which results in cell toxicity, membrane dysfunction, and cell death (Chookhampaeng 2011). Plants have evolved enzymatic and non-enzymatic mechanisms to scavenge ROS

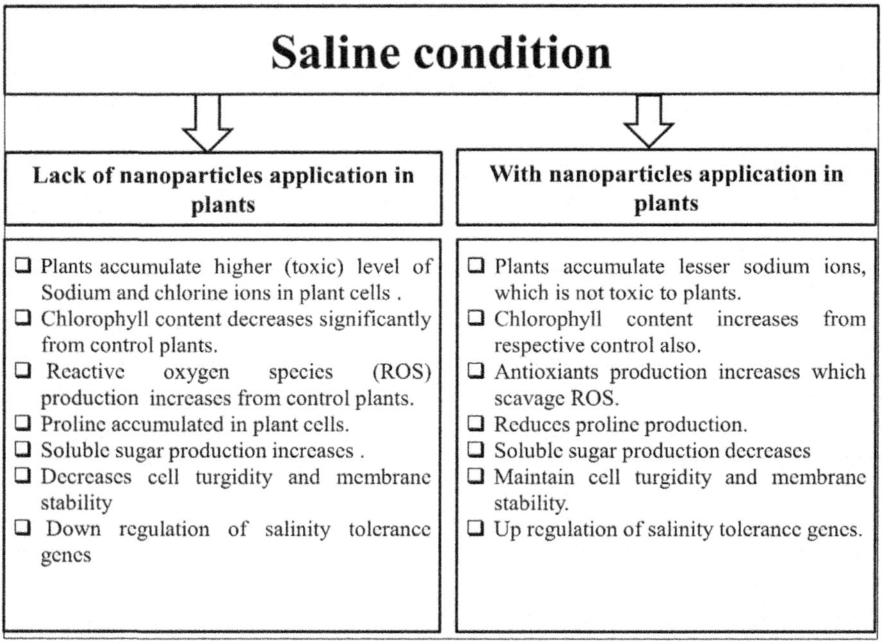

Fig. 8.1 An overview of the responses of the plants treated with and without nanoparticles under salinity stress

(Hassanein et al. 2012). Antioxidant enzymes like SOD, POD, and APX are increased in *Zea mays* (Hassanein et al. 2009), pea (Farag 2009), Faba Bean (Qados and Moftah 2015) in response to salinity stress. As soluble sugar is highly responsible for the turgidity of cells and maintains membrane stabilization by behaving as ROS scavengers (Hossain et al. 2013), their production increases during salt stress conditions.

It has been noticed that the application of suitable nanoparticles would efficiently ameliorate the critical effects of salt stress by modulating germination rate, antioxidant defense system, leaf turgor, photosynthetic rate, and carbon assimilation process (Haghighi and Pessarakli 2013; Qados 2015). Si, Zn, Ag, and some other metallic nanoparticles are being used efficiently in different crops to repair the effects of salinity stress by reducing oxidative and ionic damage. Here, in this section, attempts have been taken to uncover the mechanisms of how the nanoparticles are contributing towards salt tolerance in crop plants (Fig. 8.1).

ZnO-NPs treatment helps in the development of photosynthetic systems (Singh et al. 2015a; Govorov and Carmeli 2007). Alabdallah and Alzahrani (2020) in Lady's finger and Venkatachalam et al. (2017) in cotton found that ZnO-NPs treated plants enhanced chlorophyll a, b, total chlorophyll and carotenoid content under the salt stress. Zn also plays a critical role in the synthesis of protochlorophyllide (an intermediate compound of the chlorophyll biosynthesis) and has a role in chloroplast development and the repair of photosystem II (Hänsch and Mendel 2009;

Salama et al. 2019). Under salt stress conditions proline content, total soluble sugar, antioxidant activity were higher than corresponding controls, but exposing ZnO-NPs treated plants to salt stress conditions directed a reduction in proline content, total soluble sugar content whereas, antioxidant activity increased as compared to their respective controls.

It has been found that Si nano-particles were also predominantly used for the growth improvement of several species i.e., maize (Suriyaprabha et al. 2012), rice (Debnath et al. 2011) and *V. faba* (Qados 2015) plants. In Faba bean, Si nano-particles treatment showed an improvement in growth and germination parameters (Qados and Moftah 2015). When Faba Bean was treated with nano silicon and subjected to salt stress, then the plants were suitable for overcoming oxidative damage and protecting themselves from numerous injurious effects of salinity by escalating the antioxidant enzyme content, which lessened the assimilation of Na^+ ions during salt stress conditions. The salt tolerance in Faba bean was also clearly predicted from the increase in photosynthetic pigments and carbohydrates content in plants. Application of Si nano-particles to the salt-stressed plant leaves showed a decrease in electrolyte leakage (EL), lipid peroxidation (LP), and increased in membrane stability index (MSI) as compared to the leaves without Si nano-particles treatment and that of control indicating that Si nano-particles have a constructive impact on the cell membrane so as to enhance tolerance against salt stress. Si-nanoparticles activate the antioxidant enzyme system, as a result of which the membrane gets stabilized. Lower EL and higher MSI have also been noticed in salt-tolerant varieties of rice (Tijen and Ismail 2005) and sugarcane (Gomathi and Rakkiyapan 2011) grown under salt stress conditions.

The decline in chlorophyll content of salt-treated Faba bean plants is linked with the decrease in enzyme activities and high proline content, which is dependable upon the indication that nitrogen might be redirected to the accumulation of proline rather than chlorophyll. Further, it was reported that silicon increased proline (considered as key osmoticum) content in stressed plant tissues (Crusciol et al. 2009). However, it is also reported that the total proline was enhanced appreciably in banana plants with the increases in the doses of nanoparticles (Helaly et al. 2014). In saline exposed plants, Si nanoparticles protected their chlorophyll content probably due to higher antioxidant enzyme with preservation of leaf chlorophyll from degradation and stabilized the integrity of chloroplast membrane (Siddiqui and Al-Whaibi 2014). Nano silicon applications triggered a noticeable increase in soluble sugar content and promoted sugar formation by eliminating the inhibitory effect (Siddiqui and Al-Whaibi 2014). Furthermore, it was also reported that a strong correlation has been existing between sugar accumulation and salt tolerance (Siddiqui and Al-Whaibi 2014). The role of silicon on soluble sugar biosynthesis in salt stress conditions is crucial as it acts as a main osmotic compound in several glycophytes (Hassanein et al. 2012). Increasing antioxidant enzyme activities (CAT and SOD) and establishing antioxidant metabolism are the efficient means to improve salinity tolerance in plants (Mao et al. 2004). Nanosilicate significantly increases SOD, POD, APX, and CAT in plants and ameliorates the salinity effect. Liang et al. (2007) described the role of nano silicon in increasing SOD activity in salt-stressed barley leaves and increasing SOD, GPX, CAT, and GR activity in salt-stressed barley roots. It is known that salinity-affected plants suffered from the high concentration of Na^+ ions but the application of Si resulted in the decrease in Na+ ions concentration and increase in P and K/Na ratio in plants. A similar result was also found in the shoot of Faba Beans (Hellal et al. 2012). The expression profiles of the tomato plant treated with nano Si under salt stress showed that out of 14 genes four genes namely AREB, TAS14, NCED3, and CRK1, were upregulated, and six genes, RBOH1, APX2, MAPK2, ERF5, MAPK3, and DDF2, were down-regulated. This showed the feasible involvement of nano Si in the plant's response to stress, by alternating gene expression for salinity tolerance in plants (Nabil and Motaweh 2015). In addition to this, X-ray analysis showed unique silica-filled cells in the leaf margin nearer to hydathodes and around trichome bases along the margins of the leaf (Frantz et al. 2008). It has been considered that deposition of leaf silica leads to leaf strengthening, transpiration reduction, and increased resistance for biotic and abiotic stresses (Postek 1981).

The application of AgNPs is also one of the efficient strategies for mitigating oxidative damage due to salt stress in plants (Younes and Nassef 2015). The non-toxic nature and chemical stability of these compounds regarded as 'biocompatible precursors' are contributing towards the overall development in plants (Castro-González et al. 2019). As compared to control, under salt stress proline oxidase (POX) activity was reduced and glutamyl kinase and proline contents were increased, but when Ag Nanoparticles were applied individually or combined with NaCl, the POX activity decreased, and the glutamyl kinase (GK) activity and proline content in wheat plants were increased as compared to their respective control and NaCl treated plants. Like other nanoparticles, AgNPs also reversed the negative impacts of salt stress and promoted germination, plant growth, and development by enhancing enzymatic and non-enzymatic antioxidants by creating a buffer system in the plant cell (Hasanuzzaman et al. 2019). Ag nanoparticles treated plants have the ability to exclude toxic amounts of Na^+ and translocate K^+ into root and shoot (Almeida et al. 2017). AgNPs-mediated increased proline content coupled with up-regulation in the activity of GK, and repression in the activity of POX justifies the salt tolerance of the wheat plant. Foliar application of AgNPs promotes NiR, NR activities and also increases nitrogen uptake and assimilation under salt stress conditions. In addition to this, nitrogen is an observable variable that is straightforwardly associated with the chlorophyll content (Bojovoic and Markovic 2009). Moreover, it is also found that AgNPs increase chlorophyll content. The increase in chlorophyll content shows AgNPs-mediated enhanced photosynthetic activity (Kataria et al. 2019). However, ABA content declines in response to AgNPs under salinity stress.

8.10 Conclusion and Future Perspectives

Salt stress in plants is inevitable under current climatic conditions. Nanotechnology is widely used in different sectors but the application of nanotechnology and nanoparticles is still in the juvenile phase in agriculture and crop improvement programs. Two forms of nanoparticles i.e. pure metals and metal oxides of Zn, Si, Ag, Ce, Ti, etc. are efficient to ameliorate the ionic and oxidative effects of the salinity stress in plants. The application of nanoparticles showed promising results in improving salt tolerance by alternating several physiological and biochemical processes relevant to growth and development in plants. Therefore, it is much more essential to develop a fundamental understanding regarding the interactions of NPs with plants at cellular as well as molecular levels. Further, it has been seen that there is a toxic effect of nanoparticles on plants. The useful and harmful effects of nanoparticles depend upon several considerations such as their concentration, surface charge, surface area, and exposure regime. Forthcoming research of nanoparticles for salt stress amelioration must be targeted to explore the mechanisms of how nanoparticle modulate the cellular metabolisms or molecular pathways, again the researchers must focus on eliminating toxic effects and enhancing beneficial effects of nanoparticles which will promote salt stress tolerance in plants. Moreover, research on nano-encapsulation of biomolecules, bioactive compounds, and other active substances and their impacts on different salt-affected plants should be the priority. Furthermore, green nanotechnology should be encouraged in a variety of agricultural crops to make the plants resistant to salinity stress conditions.

References

Abbas Q, Liu G, Yousaf B, Ali MU et al (2020) Biochar-assisted transformation of engineered-cerium oxide nanoparticles: effect on wheat growth, photosynthetic traits and cerium accumulation. Ecotoxicol Environ Saf 187:109845

Abdel Latef AAH, Srivastava AK, El-sadek MSA, Kordrostami M, Tran LSP (2018) Titanium dioxide nanoparticles improve growth and enhance tolerance of broad bean plants under saline soil conditions. Land Degrad Dev 29(4):1065–1073

Abdelaal KA, Mazrou YS, Hafez YM (2020) Silicon foliar application mitigates salt stress in sweet pepper plants by enhancing water status, photosynthesis, antioxidant enzyme activity and fruit yield. Plan Theory 9:733

Abdoli S, Ghassemi-Golezani K, Alizadeh-Salteh S (2020) Responses of ajowan (Trachyspermum ammi L.) to exogenous salicylic acid and iron oxide nanoparticles under salt stress. Environ Sci Pollut Res 27:36939–36953

Abou-Zeid H, Ismail G (2018) The role of priming with biosynthesized silver nanoparticles in the response of *Triticum aestivum* L to salt stress. Egypt J Bot 58:73–85

Ahmad P, Jhon RJ (2005) Effect of salt stress on growth and biochemical parameters of Pisum sativum L. (Einfluss von Salzstress auf Wachstum und biochemische Parameter von Pisum sativum L.). Arch Agron Soil Sci 51(6):665–672

Ahmed F, Dwivedi S, Shaalan NM, Kumar S, Arshi N, Alshoaibi A, Husain FM (2020) Development of selenium nanoparticle based agriculture sensor for heavy metal toxicity detection. Agriculture 10(12):610. https://doi.org/10.3390/agriculture10120610

Ahmed T, Noman M, Manzoor N, Shahid M, Abdullah M, Ali L, Wang G, Hashem A, Al-Arjani ABF, Alqarawi AA, Abd_Allah EF, Li B (2021) Nanoparticle-based amelioration of drought stress and cadmium toxicity in rice via triggering the stress responsive genetic mechanisms and nutrient acquisition. Ecotoxicol Environ Saf 209:111829. https://doi.org/10.1016/j.ecoenv.2020.111829

Akpinar BA, Avsar B, Lucas SJ, Budak H (2012) Plant abiotic stress signaling. Plant Signal Behav 7(11):1450–1455. https://doi.org/10.4161/psb.21894

Alabdallah NM, Alzahrani HS (2020) The potential mitigation effect of ZnO nanoparticles on (*Abelmoschus esculentus* L. Moench) metabolism under salt stress conditions. Saudi J Biol Sci 27(11):3132–3137

Alghuthaymi MA et al (2015) Myconanoparticles: synthesis and their role in phytopathogens management. Biotechnol Biotechnol Equip 29:221–236

Almeida DM, Oliveira MM, Saibo NJ (2017) Regulation of Na+ and K+ homeostasis in plants: towards improved salt stress tolerance in crop plants. Gen Mol Biol 40:326–345

An J, Hu P, Li F, Wu H, Shen Y, White JC et al (2020) Emerging investigator series: molecular mechanisms of plant salinity stress tolerance improvement by seed priming with cerium oxide nanoparticles. Environ Sci Nano 7:2214–2228

Ashraf M, Foolad MRJE (2007) Roles of glycine betaine and proline in improving plant abiotic stress resistance. Environ Exp Bot 59(2):206–216

Avestan S, Ghasemnezhad M, Esfahani M, Byrt CS (2019) Application of nanosilicon dioxide improves salt stress tolerance in strawberry plants. Agronomy 9:246

Azooz MM, Alzahrani AM, Youssef MM (2013) The potential role of seed priming with ascorbic acid and nicotinamide and their interactions to enhance salt tolerance in broad bean (*Vicia faba* L.). Aust J Crop Sci 7:2091–2100

Baker S et al (2013) Plants: emerging as nanofactories towards facile route in synthesis of nanoparticles. Bioimpacts 3:111–117

Bar H, Bhui DK, Sahoo GP, Sarkar P, De SP, Misra A (2009) Green synthesis of silver nanoparticles using latex of Jatropha curcas. Colloids Surf A Physicochem Eng Asp 339:134–139

Baz H, Creech M, Chen J, Gong H, Bradford K, Huo H (2020) Water-soluble carbon nanoparticles improve seed germination and post-germination growth of lettuce under salinity stress. Agronomy 10(8):1192

Bojovoic B, Markovic A (2009) Correlation between nitrogen and chlorophyll in wheat (Triticumaestivum L.). Kragujevac J Sci 31:69–74

Buettner KM, Valentine AM (2012) Bioinorganic chemistry of titanium. Chem Rev 112:1863–1881

Caldelas C, Weiss DJ (2017) Zinc homeostasis and isotopic fractionation in plants: a review. Plant Soil 41:17–46

Carbajal-VáZquez VH, GóMez-Merino FC, Herrera-Corredor JA, Contreras-Oliva A, Alcantar-Gonzalez G, Trejo-TéLlez LI (2020) Effect of titanium foliar applications on tomato fruits from plants grown under salt stress conditions. Not Bot Horti Agrobot Cluj-Napoca 48:924–937

Castro-González CG, Sánchez-Segura L, Gómez-Merino FC, Bello-Bello JJ (2019) Exposure of qwstevia (Stevia rebaudiana B.) to silver nanoparticles in vitro: transport and accumulation. Sci Rep 9:10372

Castro-Longoria E et al (2011) Biosynthesis of silver, gold and bimetallic nanoparticles using the filamentous fungus *Neurospora crassa*. Colloids Surf B Biointerfaces 83:42–48

Ceylan A, Jastrzembski K, Shah SI (2006) Enhanced solubility Ag-Cu nanoparticles and their thermal transport properties. Metall Mater Trans A 37(7):2033–2038

Chibber S, Ansari SA, Satar R (2013) New vision to CuO, ZnO, and TiO_2 nanoparticles: their outcome and effects. Journal of Nanoparticle Research 15(4):1–13

Chookhampaeng S (2011) The effect of salt stress on growth, chlorophyll content proline content and antioxidative enzymes of pepper (*Capsicum annuum* L.) seedling. Eur J Sci Res 49:103–109

Crusciol CAC, Pulz AL, Lemos LB, Soratto RP, Lima GPP (2009) Effects of silicon and drought stress on tuber yield and leaf biochemical characteristics in potato. Crop Sci 49:949–954

de Souza TAJ, Souza LRR, Franchi LP (2019) Silver nanoparticles: an integrated view of green synthesis methods, transformation in the environment, and toxicity. Ecotoxicol Environ Saf 171:691–700

Debnath N, Das S, Seth D, Chandra R, Bhattacharya SC, Goswami A (2011) Entomotoxic effect of silica nanoparticles against *Sitophilus oryzae* (L.). J Pest Sci 84:99–105

Dikilitas M, Simsek E, Roychoudhury A (2020) Role of proline and glycine betaine in overcoming abiotic stresses. In: Protective chemical agents in the amelioration of plant abiotic stress. Hoboken, Wiley, pp 1–23. https://doi.org/10.1002/9781119552154.ch1

Djanaguiraman M, Ramadass R (2004) Effect of salinity on chlorophyll content of rice genotypes. Agric Sci Dig 24:178–181

Djanaguiraman M, Nair R, Giraldo JP, Prasad PVV (2018) Cerium oxide nanoparticles decrease drought-induced oxidative damage in sorghum leading to higher photosynthesis and grain yield. ACS omega 3(10):14406–14416

Duan H et al. (2015) Green chemistry for nanoparticle synthesis. Chem Soc Rev 44: 5778–5792

Duhan JS et al (2017) Nanotechnology: the new perspective in precision agriculture. Biotechnol Rep 15:11–23

Duncan E, O'Sullivan C, Roper M, Biggs J, Peoples M (2018) Influence of co-application of nitrogen with phosphorus, potassium and Sulphur on the apparent efficiency of nitrogen fertilizer use, grain yield and protein content of wheat: review. Field Crop Res 226:56–65

Eaton ET (2015) Manganese. In: Barker Allen V, Pilbeam DJ (eds) Handbook of plant nutrition. CRC Press, Boca Raton, pp 427–485

El-Kassas HY et al (2014) Cytotoxic activity of biosynthesized gold nanoparticles with an extract of the red seaweed *Corallina officinalis* on the MCF-7 human breast cancer cell line. Asian Pac J Cancer Prev 15:4311–4317

El-Sharkawy MS, El-Beshsbeshy TR, Mahmoud EK, Abdelkader NI, Al-Shal RM, Missaoui AM (2017) Response of alfalfa under salt stress to the application of potassium sulfate nanoparticles. Am J Plant Sci 8:1751–1773

Etefagh R, Azhir E, Shahtahmasebi N (2013) Synthesis of CuO nanoparticles and fabrication of nanostructural layer biosensors for detecting Aspergillus niger fungi. Sci Iran 20(3):1055–1058

Fan TX, Chow SK, Zhang D (2009) Biomorphic mineralization: from biology to materials. Prog Mater Sci 54(5):542–659

Farag AA (2009) Increasing tolerance of *Vigna sinensis* L. to salt stress using an organic acid and a polyamine. M.Sc. thesis. Ain Shams University, Cairo

Farhangi-Abriz S, Torabian S (2018) Nano-silicon alters antioxidant activities of soybean seedlings under salt toxicity. Protoplasma 255:953–962

Farouk S, Al-Amri SM (2019) Exogenous zinc forms counteract NaCl-induced damage by regulating the antioxidant system, osmotic adjustment substances, and ions in canola (Brassica napus L. Cv. Pactol). Plants J Soil Sci Plant Nutr 19:887–899

Frantz JM, Locke JC, Datnoff L, Omer M, Widrig A, Sturtz D, Horst L, Krause CR (2008) Detection, distribution, and quantification of silicon in floricultural crops utilizing three distinct analytical methods. Commun Soil Sci Plant Anal 39:2734–2751

Ghidan AY, Al Antary TM (2019) Applications of nanotechnology in agriculture. Appl Nanobiotechnol. https://doi.org/10.5772/intechopen.88390

Gohari G, Mohammadi A, Akbari A, Panahirad S, Dadpour MR, Fotopoulos V, Kimura S (2020a) Titanium dioxide nanoparticles (TiO_2 NPs) promote growth and ameliorate salinity stress effects on essential oil profile and biochemical attributes of *Dracocephalum moldavica*. Sci Rep 10:1–14. https://doi.org/10.1038/s41598-020-57794-1

Gohari G, Safai F, Panahirad S, Akbari A, Rasouli F, Dadpour MR, Fotopoulos V (2020b) Modified multiwall carbon nanotubes display either phytotoxic or growth promoting and stress protecting activity in *Ocimum basilicum* L. in a concentration-dependent manner. Chemosphere 249:126171

Golinska P et al (2014) Biogenic synthesis of metal nanoparticles from actinomycetes: biomedical applications and cytotoxicity. Appl Microbiol Biotechnol 98:8083–8097

Gomathi R, Rakkiyapan P (2011) Comparative lipid peroxidation, leaf membrane thermostability, and antioxidant system in four sugarcane genotypes differing in salt tolerance. Inter J Plant Physiol Biochem 3:67–74

Govorov AO, Carmeli I (2007) Hybrid structures composed of photosynthetic system and metal nanoparticles: plasmon enhancement effect. Nano Lett 7(3):620–625

Haghighi M, Pessarakli M (2013) Influence of silicon and nano-silicon on salinity tolerance of cherry tomatoes (Solanum lycopersicum L.) at early growth stage. Sci Hortic 161:111–117

Hänsch R, Mendel RR (2009) Physiological functions of mineral micronutrients (Cu, Zn, Mn, Fe, Ni, Mo, B, Cl). Curr Opin Plant Biol 12(3):259–266

Hasanuzzaman M, Bhuyan MHM, Anee TI, Parvin K, Nahar K, Mahmud JA, Fujita M (2019) Regulation of ascorbate-glutathione pathway in mitigating oxidative damage in plants under abiotic stress. Antioxidants 8:384

Hassanein RA, Hassanein AA, Haider AS, Hashem HA (2009) Improving salt tolerance of *Zea mays* L. plant by presoaking their grains in glycine betaine. Aust J Basic Appl Sci 3:928–942

Hassanein RA, Hashem HA, Khalil RR (2012) Stigmasterol treatment increases salt stress tolerance of faba bean plants by enhancing antioxidant systems. Plant Osmics J 5:476–485

Helaly MN, El-Metwally MA, El-Hoseiny H, Omar SA, El-Sheery NI (2014) Effect of nanoparticles on biological contamination of in vitro cultures and organogenic regeneration of banana. Aust J Crop Sci 8:612–624

Hellal FA, Abdelhameid M, Abo-Basha DM, Zewainy RM (2012) Alleviation of the adverse effects of soil salinity stress by foliar application of silicon on faba bean (*Vica faba* L.). J Appl Sci Res 8:4428–4433

Hernández-Hernández H, González-Morales S, Benavides-Mendoza A, Ortega-Ortiz H, Cadenas-Pliego G, Juárez-Maldonado A (2018a) Effects of chitosan–PVA and Cu nanoparticles on the growth and antioxidant capacity of tomato under saline stress. Molecules 23(1):178

Hernández-Hernández H, Júarez-Maldonado A, Benavides-Mendoza A, Ortega-Ortiz H, Cadenas-Pliego G, Sánchez-Aspeytia D, González-Morales S (2018b) Chitosan-PVA and copper nanoparticles improve growth and overexpress the SOD and JA genes in tomato plants under salt stress. Agronomy 8:175

Hezaveh TA, Pourakbar L, Rahmani F, Alipour H (2019) Interactive effects of salinity and ZnO nanoparticles on physiological and molecular parameters of rapeseed (*Brassica napus* L.) Commun. Soil Sci Plant Anal 50:698–715

Hojjat SS, Mozumder C, Bora T, Hornyak GL (2019) Polyvinylpyrrolidone-coated silver nanoparticle mitigation of salinity on germination and seedling parameters of bitter vetch (Vicia ervilia L.) plants. Nanotechnol Russ 14(11):582–587

Hossain MA, Mostofa MG, Fujita MJ (2013) Cross protection by cold-shock to salinity and drought stress-induced oxidative stress in mustard (*Brassica campestris* L.) seedlings. Mol Plant Breed 4:50–70

Hurtado AC, Chiconato DA, de Mello PR et al (2020) Different methods of silicon application attenuate salt stress in sorghum and sunflower by modifying the antioxidative defense mechanism. Ecotoxicol Environ Saf 203:110964

Husen A, Siddiqi KS (2014) Carbon and fullerene nanomaterials in plant system. J Nanobiotechnol 12(1):1–10

Hussain I, Singh A, Singh NB, Singh P (2019) Plant-nanoceria interaction: toxicity, accumulation, translocation and biotransformation. South Afr J Bot 121:239–247

Iqbal MN, Rasheed R, Ashraf MY, Ashraf MA, Hussain I (2018) Exogenously applied zinc and copper mitigate salinity effect in maize (*Zea mays* L.) by improving key physiological and biochemical attributes. Environ Sci Pollut Res 25:23883–23896

Jan AU, Hadi F, Nawaz MA, Rahman K (2017) Potassium and zinc increase tolerance to salt stress in wheat (*Triticum aestivum* L.). Plant Physiol Biochem 116:139–149

Jha AK, Prasad K (2010) Green synthesis of silver nanoparticles using Cycas leaf. Int J Green Nanotechnol Phys Chem 1:110–117

Jo JH et al. (2015) Pseudomonas deceptionensis DC5-mediated synthesis of extracellular silver nanoparticles. Artif Cells Nanomed Biotechnol. Published online July 31, 2015. https://doi.org/10.3109/21691401.2015.1068792

Junggwon Y, Kyoungah C, Byoungjun P, Ho-Chul K, Byeong-Kwon J, Sangsig KJ (2008) Optical heating of ink-jet printable Ag and Ag–Cu nanoparticles. J Appl Phys 47:5070

Kalaiselvi A et al (2015) Synthesis and characterization of palladium nanoparticles using *Catharanthus roseus* leaf extract and its application in the photo-catalytic degradation. Spectrochim Acta A Mol Biomol Spectrosc 135:116–119

Karthik L et al (2014) Streptomyces sp. LK3 mediated synthesis of silver nanoparticles and its biomedical application. Bioprocess Biosyst Eng 37:261–267

Kashyap PL, Xiang X, Heiden P (2015) Chitosan nanoparticle based delivery systems for sustainable agriculture. Int J Biol Macromol 77:36–51

Kashyap PL, Rai P, Sharma S, Chakdar H, Kumar S, Pandiyan K, Srivastava AK (2016) Nanotechnology for the detection and diagnosis of plant pathogens. Sustain Agric Rev:253–276. https://doi.org/10.1007/978-3-319-39306-3_8

Kataria S, Jain M, Rastogi A, Zivcak M, Brestic M, Liu S, Tripathi DK (2019) Role of nanoparticles on photosynthesis: avenues and applications. In: Nanomaterials in plants, algae, and microorganisms. Academic Press, Cambridge, MA, pp 103–127

Khan MN, Mobin M, Abbas ZK, Al Mutairi KA, Siddiqui ZH (2017) Role of nanomaterials in plants under challenging environments. Plant Physiol Biochem 110:194–209

Khan I, Raza MA, Awan SA, Shah GA et al (2020) Amelioration of salt induced toxicity in pearl millet by seed priming with silver nanoparticles (AgNPs): the oxidative damage, antioxidant enzymes and ions uptake are major determinants of salt tolerant capacity. Plant Physiol Biochem 156:221–232

Kim SA, Guerinot ML (2007) Mining iron: iron uptake and transport in plants. FEBS Lett 581:2273–2280

Kumar S, Nehra M, Dilbaghi N, Marrazza G, Hassan AA, Kim KH (2019) Nano-based smart pesticide formulations: emerging opportunities for agriculture. J Control Release 294:131–153. https://doi.org/10.1016/j.jconrel.2018.12.01

Li X et al (2011) Biosynthesis of nanoparticles by microorganisms and their applications. J Nanomater 16:270974

Liang Y, Sun W, Zhu YG, Christie P (2007) Mechanisms of silicon mediated alleviation of abiotic stresses in higher plants: a review. Environ Poll 147:422–428

Liang W, Ma X, Wan P, Liu L (2018a) Plant salt-tolerance mechanism: a review. Biochem Biophys Res Commun 495(1):286–291

Liang W, Yu A, Wang G, Zheng F, Jia J, Xu H (2018b) Chitosan-based nanoparticles of avermectin to control pine wood nematodes. Int J Biol Macromol 112:258–263. https://doi.org/10.1016/j.ijbiomac.2018.01.174

Lu L, Huang M, Huang Y, Corvini PFX, Ji R, Zhao L (2020) Mn_3O_4 nanozymes boost endogenous antioxidant metabolites in cucumber (Cucumis sativus) plant and enhance resistance to salinity stress. Environ Sci Nano 7(6):1692–1703

Lwalaba JLW, Louis LT, Zvobgo G, Richmond MEA et al (2020) Physiological and molecular mechanisms of cobalt and copper interaction in causing phytotoxicity to two barley genotypes differing in Co tolerance. Ecotoxicol Environ Saf 187:109866

Lyu S, Wei X, Chen J, Wang C, Wang X, Pan D (2017) Titanium as beneficial element for crop production. Front Plant Sci 8:597

Mahmoud AWM, Abdelaziz SM, El-mogy MM, Abdeldaym EA (2019) Effect of foliar zno and feo nanoparticles application on growth and nutritional quality of red radish and assessment of their accumulation on human health. Agriculture 65(1):16–29

Mahmoud AWM, Abdeldaym EA, Abdelaziz SM, El-Sawy MB, Mottaleb SA (2020) Synergetic effects of zinc, boron, silicon, and zeolite nanoparticles on confer tolerance in potato plants subjected to salinity. Agronomy 10:19

Makarov VV et al (2014) "Green" nanotechnologies: synthesis of metal nanoparticles using plants. Acta Nat 6:35–44

Mao G, Xu X, Xu ZJ (2004) Advances in physiological and biochemical research of salt tolerance in plant. Chin J Eco-Agic 12(1):43–46

Martínez-Ballesta MC, Zapata L, Chalbi N et al (2016) Multi-walled carbon nanotubes enter broccoli cells enhancing growth and water uptake of plants exposed to salinity. J Nanobiotechnol 14:42

Mohamed AKS, Qayyum MF, Abdel-Hadi AM, Rehman RA, Ali S, Rizwan M (2017) Interactive effect of salinity and silver nanoparticles on photosynthetic and biochemical parameters of wheat. Arch Agron Soil Sci 63:1736–1747

Mohl M, Dobo D, Kukovecz A, Konya Z, Kordas K, Wei J, Vajtai R, Ajayan PM (2011) J Phys Chem C 115:9403

Molnar A, Papp M, Kovacs DZ, Belteky P et al (2020) Nitro-oxidative signalling induced by chemically synthetized zinc oxide nanoparticles (ZnO NPs) in Brassica species. Chemosphere 251:126419

Mourato A et al (2011) Biosynthesis of crystalline silver and gold nanoparticles by extremophilic yeasts. Bioinorg Chem Appl 2011:546074

Mozafari A, Ghadakchi Asl A, Ghaderi N (2017) Grape response to salinity stress and role of iron nanoparticle and potassium silicate to mitigate salt induced damage under in vitro conditions. Physiol Mol Biol Plants 24(1):25–35. https://doi.org/10.1007/s12298-017-0488-x

Munns R, Tester M (2008) Mechanisms of salinity tolerance. Annu Rev Plant Biol 59:651–681

Mushtaq A, Rizwan S, Jamil N, Ishtiaq T, Irfan S, Ismail T, Malghani MN, Shahwani MN (2019) Influence of silicon sources and controlled release fertilizer on the growth of wheat cultivars of Balochistan under salt stress. Pak J Bot 51:1561–1567

Nabil M, Motaweh HA (2015) Silica nanoparticles preparation using alkali etching process. In Applied Mechanics and Materials. Trans Tech Publications Ltd. 749:155–158)

Nair R, Mohamed MS, Gao W, Maekawa T, Yoshida Y, Ajayan PM, Kumar DS (2012) Effect of carbon nanomaterials on the germination and growth of rice plants. J Nanosci Nanotechnol 12(3):2212–2220. https://doi.org/10.1166/jnn.2012.5775

Najafi S, Razavi SM, Khoshkam M, Asadi A (2020) Effects of green synthesis of sulfur nanoparticles from *Cinnamomum zeylanicum* barks on physiological and biochemical factors of Lettuce (*Lactuca sativa*). Physiol Mol Biol Plants 26:1–12

Nam KT et al (2006) Virus-enabled synthesis and assembly of nanowires for lithium ion battery electrodes. Science 312:885–888

Naseem T et al (2015) Antibacterial activity of green synthesis of iron nanoparticles using *Lawsonia inermis* and *Gardenia jasminoides* leaves extract. J Chem 2015:912342

Nasrollahzadeh M et al (2015) Green synthesis of copper nano-particles using Ginkgo biloba L. leaf extract and their catalytic activity for the Huisgen [3+2] cycloaddition of azides and alkynes at room temperature. J Colloid Interface Sci 457:141–147

Nazarbeygi E, Yazdi HL, Naseri R, Soleimani R (2011) The effects of different levels of salinity on proline and A-, B- chlorophylls in canola. Am-Eurasian J Agric Environ Sci 10:70–74

Noruzi M (2015) Biosynthesis of gold nanoparticles using plant extracts. Bioprocess Biosyst Eng 38:1–14

Oliveira HC, Gomes BC, Pelegrino MT, Seabra AB (2016) Nitric oxide-releasing chitosan nanoparticles alleviate the effects of salt stress in maize plants. Nitric Oxide 61:10–19

Pérez-Labrada F, López-Vargas ER, Ortega-Ortiz H, Cadenas-Pliego G, Benavides-Mendoza A, Júarez-Maldonado A (2019) Responses of tomato plants under saline stress to foliar application of copper nanoparticles. Plan Theory 8(6):151

Philip D et al (2011) Murraya Koenigii leaf-assisted rapid green synthesis of silver and gold nanoparticles. Spectrochim Acta A Mol Biomol Spectrosc 78:899–904

Pinedo-Guerrero ZH, Cadenas-Pliego G, Ortega-Ortiz H, González-Morales S, Benavides-Mendoza A, Valdes-Reyna J, Juárez-Maldonado A (2020) Form of silica improves yield, fruit quality and antioxidant defense system of tomato plants under salt stress. Agriculture 10:367

Poopathi S et al (2015) Synthesis of silver nanoparticles from *Azadirachta indica*–a most effective method for mosquito control. Environ Sci Pollut Res Int 22:2956–2963

Postek MT (1981) The occurrence of silica in the leaves of *Magnolia grandiflora* L. Bot Gaz 142:124–134

Qados AMA (2015) Mechanism of nanosilicon-mediated alleviation of salinity stress in faba bean (*Vicia faba* L.) plants. Journal of Experimental Agriculture International, 78–95

Qados AMA, Moftah AE (2015) Influence of silicon and nano-silicon on germination, growth and yield of faba bean (*Vicia faba* L.) under salt stress conditions. Journal of Experimental Agriculture International, 509–524

Rahimi R, Mohammakhani A, Roohi V, Armand N (2012) Effects of salt stress and silicon nutrition on cholorophyll content, yield and yield components in fennel (*Foeniculum vulgar* Mill.). Int J Agric Crop Sci 4:1591–1595

Rai M, Yadav A, Gade A (2009) Silver nanoparticles as a new generation of antimicrobials. Biotechnol Adv 27:76–83

Rossi L, Zhang W, Ma X (2017) Cerium oxide nanoparticles alter the salt stress tolerance of Brassica napus L. by modifying the formation of root apoplastic barriers. Environ Pollut 229:132–138

Rossi L, Bagheri M, Zhang W, Chen Z, Burken JG, Ma X (2019) Using artificial neural network to investigate physiological changes and cerium oxide nanoparticles and cadmium uptake by Brassica napus plants. Environ Pollut 246:381–389

Salama DM, Osman SA, Abd El-Aziz ME, Abd Elwahed MSA, Shaaban EA (2019) Effect of zinc oxide nanoparticles on the growth, genomic DNA, production, and the quality of common dry bean (Phaseolus vulgaris). Biocatal Agric Biotechnol 18:101083

Salehi H, Chehregani A, Lucini L, Majd A, Gholami M (2018) Morphological, proteomic and metabolomic insight into the effect of cerium dioxide nanoparticles to *Phaseolus vulgaris* L. under soil or foliar application. Sci Total Environ 616:1540–1551

Salem NM, Albanna LS, Abdeen AO, Ibrahim QI, Awwad AM (2016) Sulfur nanoparticles improves root and shoot growth of tomato. J Agric Sci 8(4):179

Seshadri S et al (2011) Green synthesis of lead sulfide nanoparticles by the lead resistant marine yeast, *Rhodosporidium diobovatum*. Biotechnol Prog 27:1464–1469

Shabala S, Cuin TA (2008) Potassium transport and plant salt tolerance. Physiol Plant 133(4):651–669

Shafiq F, Iqbal M, Ali M, Ashraf MA (2019) Seed pre-treatment with polyhydroxy fullerene nanoparticles confer salt tolerance in wheat through upregulation of H_2O_2 neutralizing enzymes and phosphorus uptake. J Soil Sci Plant Nutr 19:734–742

Shahi S, Srivastava M (2018) Influence of foliar application of manganese on growth, pigment content, and nitrate reductase activity of *Vigna radiata* (L.) R. Wilczek under salinity. J Plant Nutr 41:1397–1404

Shankar SS, Rai A, Ankamwar B, Singh A, Ahmad A, Sastry M (2004) Biological synthesis of triangular gold nanoprisms. Nat Mater 3:482–488

Shankramma K, Yallappa S, Shivanna MB, Manjanna J (2016) Fe_2O_3 magnetic nanoparticles to enhance S. lycopersicum (tomato) plant growth and their biomineralization. Appl Nanosci 6(7):983–990

Sharma VK, Yngard RA, Lin Y (2009) Silver nanoparticles: green synthesis and their antimicrobial activities. Adv Colloid Interf Sci 145(1–2):83–96

Siddiqui MH, Al-Whaibi MH (2014) Role of nano-SiO2 in germination of tomato (*Lycopersicum esculentum* Mill.) seeds. Saudi J Biol Sci 21:13–17

Singh N, Bhatla SC (2016) Nitric oxide and iron modulate heme oxygenase activity as a long distance signaling response to salt stress in sunflower seedling cotyledons. Nitric Oxide 53:54–64

Singh A, Singh N, Hussain I, Singh H, Singh SJ (2015a) Plant-nanoparticle interaction: an approach to improve agricultural practices and plant productivity. Int J Pharm Sci Invent 4(8):25–40
Singh P et al (2015b) A strategic approach for rapid synthesis of gold and silver nanoparticles by Panax ginseng leaves. Artif Cells Nanomed Biotechnol 44(8):1–9
Singh P et al (2015c) Biosynthesis of anisotropic silver nanoparticles by Bhargavaea indica and their synergistic effect with antibiotics against pathogenic microorganisms. J Nanomater 1–10
Singh P et al (2015d) Biosynthesis, characterization, and antimicrobial applications of silver nanoparticles. Int J Nanomedicine 10:2567–2577
Singh P et al. (2015e) Microbial synthesis of flower-shaped gold nanoparticles. Artif Cells Nanomed Biotechnol. Published online May 6, 2015. https://doi.org/10.3109/21691401.2015.1041640
Singh P et al. (2015f) Weissella oryzae DC6-facilitated green synthesis of silver nanoparticles and their antimicrobial potential. Artif Cells Nanomed Biotechnol. Published online July 27 https://doi.org/10.3109/21691401.2015.1064937
Sofy MR, Elhindi KM, Farouk S, Alotaibi MA (2020) Zinc and paclobutrazol mediated regulation of growth, upregulating antioxidant aptitude and plant productivity of pea plants under salinity. Plan Theory 9:1197
Soliman M, Qari SH, Abu-Elsaoud A et al (2020) Rapid green synthesis of silver nanoparticles from blue gum augment growth and performance of maize, fenugreek, and onion by modulating plants cellular antioxidant machinery and genes expression. Acta Physiol Plant 42:148
Sun L, Song F, Guo J, Zhu X, Liu S, Liu F, Li X (2020) Nano-ZnO-induced drought tolerance is associated with melatonin synthesis and metabolism in maize. Int J Mol Sci 21(3):782. https://doi.org/10.3390/ijms21030782
Suresh D et al. (2015) *Artocarpus gomezianus* aided green synthesis of ZnO nanoparticles: luminescence, photocatalytic and antioxidant properties. Spectrochim Acta A Mol Biomol Spectrosc 141:128–134
Suriyaprabha R, Karunakaran G, Yuvakkumar R, Rajendran V, Kannan N (2012) Silica nanoparticles for increased silica availability in maize (*Zea mays* L) seeds under hydroponic conditions. Curr Nanosci 8:1–7
Tang X, Mu X, Shao H (2015) Global plant-responding mechanisms to salt stress: physiological and molecular levels and implications in biotechnology. Crit Rev Biotechnol 35:425–437
Tanveer M, Shabala S (2018) Targeting redox regulatory mechanisms for salinity stress tolerance in crops, salinity responses and tolerance in plants, vol 1. Springer, Cham, pp 213–234
Tanveer M, Shah AN (2017) An insight into salt stress tolerance mechanisms of Chenopodium album. Environ Sci Pollut Res 24:16531–16535
Tijen D, Ismail T (2005) Comparative lipid peroxidation, antioxidant defense systems and proline content in roots of two rice cultivars differing in salt tolerance. Environ Exp Bot 53:247–257
Tomar A, Garg G (2013) Short review on application of gold nanoparticles. Global Journal of Pharmacology 7(1):34–38
Torabian S, Farhangi-Abriz S, Zahedi M (2018) Efficacy of FeSO4 nano formulations on osmolytes and antioxidative enzymes of sunflower under salt stress. Indian J Plant Physiol 23:305–315
Tortella GR, Rubilar O, Duran N, Diez MC, Martínez M, Parada J, Seabra AB (2020) Silver nanoparticles: toxicity in model organisms as an overview of its hazard for human health and the environment. J Hazard Mater 390:121974
Tripathi S, Sonkar SK, Sarkar S (2011) Growth stimulation of gram (Cicer arietinum) plant by water soluble carbon nanotubes. Nanoscale 3(3):1176. https://doi.org/10.1039/c0nr00722f
Tripathi DK, Singh S, Gaur S, Singh S, Yadav V, Liu S, Dubey NK (2018) Acquisition and homeostasis of iron in higher plants and their probable role in abiotic stress tolerance. Front Environ Sci 5:86
Tripathy BC, Oelmüller R (2012) Reactive oxygen species generation and signaling in plants. Plant Signal Behav 7(12):1621–1633. https://doi.org/10.4161/psb.22455
Vaculík M, Lukčová Z, Bokor B, Martinka M, Tripathi DK, Lux A (2020) Alleviation mechanisms of metal (loid) stress in plants by silicon: a review. J Exp Bot. https://doi.org/10.1093/jxb/eraa288

Venkatachalam P, Priyanka N, Manikandan K, Ganeshbabu I, Indiraarulselvi P, Geetha N, Muralikrishna K, Bhattacharya RC, Tiwari M, Sharma N, Sahi SV (2017) Enhanced plant growth promoting role of phycomolecules coated zinc oxide nanoparticles with P supplementation in cotton (Gossypium hirsutum L.). plant Physiol. Biochemist 110:118–127

Waghmare SR et al (2015) Ecofriendly production of silver nanoparticles using *Candida utilis* and its mechanistic action against pathogenic microorganisms. 3 Biotech 5:33–38

Wahid I, Kumari S, Ahmad R, Hussain SJ, Alamri S, Siddiqui MH, Khan MIR (2020) Silver nanoparticle regulates salt tolerance in wheat through changes in ABA concentration, ion homeostasis, and defense systems. Biomol Ther 10(11):1506. https://doi.org/10.3390/biom10111506

Wang C et al (2015) Green synthesis of silver nanoparticles by Bacillus methylotrophicus, and their antimicrobial activity. Artif Cells Nanomed Biotechnol. Published online March 6. https://doi.org/10.3109/21691401.2015.1011805

Xu X, Du X, Wang F, Sha J et al (2020) Effects of potassium levels on plant growth, accumulation and distribution of carbon, and nitrate metabolism in apple dwarf rootstock seedlings. Front Plant Sci 11:904

Yang J, Duan G, Li C, Liu L, Han G, Zhang Y, Wang C (2019) The crosstalks between jasmonic acid and other plant hormone signaling highlight the involvement of jasmonic acid as a Core component in plant response to biotic and abiotic stresses. Front Plant Sci 10. https://doi.org/10.3389/fpls.2019.01349

Yao KS, Li SJ, Tzeng KC, Cheng TC, Chang CY, Chiu CY, Liao CY, Hsu JJ, Lin ZP (2009) Fluorescence silica nanoprobe as a biomarker for rapid detection of plant pathogens. Adv Mater Res 79–82:513–516. https://doi.org/10.4028/www.scientific.net/amr.79-82.513

Yasar F, Ellialtioglu S, Yildiz K (2008) Effect of salt stress on antioxidant defense systems, lipid peroxidation, and chlorophyll content in green bean. Russ J Plant Physiol 55:782–786

Ye Y, Cota-Ruiz K, Hernandez-Viezcas JA, Valdes C, Medina-Velo IA, Turley RS, Peralta-Videa JR, Gardea-Torresdey JL (2020) Manganese nanoparticles control salinity-modulated molecular responses in *Capsicum annuum* L. through priming: a sustainable approach for agriculture. ACS Sustain Chem Eng 8:1427–1436

Younes N, Nassef D (2015) Effect of silver nanoparticles on salt tolerance of tomato transplants (solanum Lycopersicum, mill.). Assiut J Agric Sci 46(6):76–85

Zahir AA et al (2015) Green synthesis of silver and titanium dioxide nanoparticles using Euphorbia prostrata extract shows shift from apoptosis to G0/G1 arrest followed by necrotic cell death in *Leishmania donovani*. Antimicrob. Agents Chemother 59:4782–4799

Zaytseva O, Neumann G (2016) Carbon nanomaterials: production, impact on plant development, agricultural and environmental applications. Chem Biol Technol Agric 3(1):1–26

Zhang X et al (2011) Synthesis of nanoparticles by microorganisms and their application in enhancing microbiological reaction rates. Chemosphere 82:489–494

Zhao G, Zhao Y, Lou W, Su J, Wei S, Yang X, Wang R, Guan R, Pu H, Shen W (2019) Nitrate reductase-dependent nitric oxide is crucial for multi-walled carbon nanotube-induced plant tolerance against salinity. Nanoscale 11:10511–10523

Zhao C, Zhang H, Song C, Zhu JK, Shabala S (2020) Mechanisms of plant responses and adaptation to soil salinity. Innovations 1:100017

Zhou GJ et al (2014) Biosynthesis of CdS nanoparticles in banana peel extract. J Nanosci Nanotechnol 14:4437–4442

Chapter 9
Effect of Carbon Nanotubes on Abiotic Stress Response in Plants: An Overview

Mohammad Faizan, Anjuman Hussain, Anayat Rasool Mir, Vishnu D. Rajput, Tatiana Minkina, and Shamsul Hayat

Abstract In the period of climate change, abiotic stresses are propitious menace to agriculture, restricting the production of crops by inflicting primary effects like osmotic and ionic stress, along with other impacts like oxidative stress, disturbance in hormonal balance, and nutrient imbalance. However, increasing evidences reveal that augmentation of nanoparticles to plants can notably reduce detrimental effects caused by several severe environmental conditions and therefore, modulate several mechanisms in plants. Several types of nanoparticles and nanofertilizers have unveiled propitious corroboration regarding abiotic stress management. Carbon nanotubes (CNTs) are eminent members of the nanomaterial family. Because of the exceptional physical, chemical, and mechanical properties, CNTs are demonstrated to be efficacious means in the plant science field. CNTs were commonly considered to lead to important biotechnological and agricultural applications which are still far from experimental realization. Various studies have shown capability of CNTs to cross diverse plant cell blockades. These investigations, also, evaluated the harmful effects of these nanomaterials. Whilst several types of nanoparticles have shown to enact physiological processes in plants, carbon nanotubes experienced specific interest. Recent investigations have revealed CNTs to be chemically captured into plant tracheary elements. This ought to start off studies in the fields of plant defense. This chapter emphasizes on the effect of carbon nanotubes on plants under various abiotic stresses and the possible use of these distinctive nanomaterials in crop management.

M. Faizan (✉)
Collaborative Innovation Center of Sustainable Forestry in Southern China, College of Forest Science, Nanjing Forestry University, Nanjing, China

A. Hussain · A. R. Mir · S. Hayat
Plant Physiology Lab Department of Botany, Aligarh Muslim University, Aligarh, India

V. D. Rajput · T. Minkina
Academy of Biology and Biotechnology, Southern Federal University, Rostov-on-Don, Russia

M. Faizan et al. (eds.), *Sustainable Agriculture Reviews 53*, Sustainable Agriculture Reviews 53, https://doi.org/10.1007/978-3-030-86876-5_9

Keywords Abiotic stress · Agriculture · Carbon nanotubes · Hormonal balance · Nutrient imbalance

9.1 Introduction

Abiotic stresses including drought, salinity, extremely low or high temperatures, heavy metal, soil alkalinity or acidity and nutrient deficiency etc. limit plant development. Majority of the crop plants show sensitivity to abiotic stresses. First responses of plants exposed to stress conditions include restricted vegetative growth and reproductive development (Boscaiu et al. 2008). Agricultural productivity is greatly affected by climatic changes as it regulates the development of plants by influencing variations in temperature and precipitation levels and carbon dioxide. However, a remarkable reduction in crop production occurs by elements other than global warming like salinity, water deficiency, floods, and nutrient deficiency. Hence, for the enhanced food production improved crop varieties by genetic engineering and efficacious initiatives for nutrient management and irrigation practices will be needed for meeting the needs of expanding population. For agriculture under abiotic stresses, nanotechnology appears to be emerging solution (Hatami et al. 2016). In recent years, the importance of nanobiotechnology in the field of agriculture has earned attraction of the researchers. The formation of materials to nano level regulates their biological, chemical properties as well as their catalytic attributes (Manzer et al. 2015). The atomic agglomeration with dimension (at least one) between 1 and 100 nm are referred to as nanoparticles (Ball 2002). Nanoparticles are specified by characteristic physical and chemical features differing from their bulk forms. Nanoparticles can be employed for enhancing crop yield under abiotic stress (Tamer et al. 2018). Because of the unparallel physicochemical properties of nanoparticles or nanomaterials, nanotechnology has found great novel applications in the agriculture sector. In the agricultural sector, nanotechnology has found use in fertilization sector (Mani and Mondal 2016; Monreal et al. 2016; Panpatte et al. 2016), plant protection (Saharan and Pal 2016), food sector (Oprea and Grumezescu 2017) precision farming (Chhipa and Joshi 2016), remediation of terrestrial environments (Patra et al. 2016; Gil-Díaz et al. 2017), etc. Of all the nanoparticles, carbon nanotubes (CNTs) are getting more consideration because of their attractive arrangement, shape, size and distinctive physical, chemical, and biological properties (Bianco et al. 2005; Serag et al. 2015). CNTs are cylindrical structures having a diameter of several nanometers and comprise rolled graphene sheets. Recently, attempts have been devoted to divulge utilization of carbon nanotubes due to their remarkable physicochemical properties (Gao et al. 2006; Kam et al. 2005, 2006). During previous years, there has been wide concern in employing nanoparticles to plants for agricultural and crop production (Torney et al. 2007; Khodakovskaya et al. 2009, 2012; Serag et al. 2011, 2012; Husen and Siddiqi 2014). Indeed, utilization of carbon nanomaterials in crop regulation turns out to be more pressing with regards to expanding population and exhausting resources. Researchers have established that application of carbon nanotubes to plant seeds can enhance the germination percentage and thus can increase seedling growth. These observations could bring about critical improvements in the productivity of valuable crops such as maize and tomato, by enhanced biomass of the plants. But, clashing studies on the safety of carbon nanotubes have been laid out. These controversial findings require clarification to avoid confusion to the public (Liu et al. 2007, 2008; Singh et al. 2006; Lacerda et al. 2008). On the bases of their structure, CNTs are divided into two categories: single-walled nanotubes (SW) and multi-walled nanotubes (MW). Single-walled carbon nanotubes comprise of one rolled sheet of grapheme with diameter close to 1 nm, while multi-walled carbon nanotubes consist of several sheets of graphenes flapped into concentric cylinders in which diameter can reach several tenths of nanometers. Because of the distinctive properties of CNTs, they are capable of penetrating the cell wall and membrane of cells and also provide a suitable delivery system of chemicals to cells. The single-walled-CNTs (SWCNTs) function as nanotransporters for transfer of DNA and dye molecules into plants cells (Srinivasan and Saraswathi 2010). Moreover it has been reported that that multi-walled-CNTs (MWCNTs) possess ability to regulate the seed germination and plant growth, and also function as a delivery system of DNA and chemicals to plants cells. MWCTs stimulate water and nutrients uptake that increase the seed germination and plant growth and development (Villagarcia et al. 2012; Tiwari et al. 2014). The addition of MWCNTs sterile agar medium accelerated seed germination of three important crops (barley, soybean, corn) because of ability to penetrate the seed coats (Lahiani et al. 2013). It has also been reported that MWCNTs modulate the expression of several genes encoding various types of water channel proteins in soybean, corn and barley seeds coat. In tomato, hybrid Bt cotton, *Brassica juncea*, *Phaseocclus mungo* and rice maximum germination rate was observed upon the application of MWCNTs (Morla et al. 2011; Nalwade and Neharkar 2013).

9.2 Physiological Impacts of Carbon Nanotubes on Plants

CNTs play a role as modulators of seed germination and development and are capable of changing morphology and physiology of plant cells. It has been reported that CNTs can alter morphological and physiological attributes of plant cells (Pourkhaloee et al. 2011; Lahiani et al. 2013) and are believed to modulate seedling and plant growth (Khodakovskaya et al. 2012; Haghighi and da Silva 2014).

9.2.1 Effects of CNT on Seed Germination

Seed germination is considered as the initial stage of life cycle of plants which is prone to environmental stress, and successful organization of seedlings is needed to develop plants with desirable capacity for growth and productivity. An increment in the rate of seed germination was noticed with CNT. Boulmaali and Zafour

Hadj-Ziane (Boulmaali and Hadj-Ziane 2017) observed that germination of *Phaseolus vulgaris* seeds was favored in the presence of CNT. In rice the lower concentration of CNTs were found to promote seed germination (Jiang et al. 2012). Application of multiwalled carbon nanotubes (MWCNT) on tomato seeds exhibited an increase in the seed germination (Khodakovskaya et al. 2009). Germinated tomato seeds also showed a considerable enhancement in the vegetative biomass. Water is the fundamental component for seed germination. Imbibition of water depends on the presence of water and on the permeability of the seed coat. It is ascertained that the seeds manifested with CNT absorbed a higher concentration of moisture in comparison to untreated seeds. This could be described by the penetration of CNT through the seed coat by making large number of pores, thereby permitting a higher content of water uptake into the seeds (Khodakovskaya et al. 2009). Hence, seeds soaked in water having CNT enhanced uptake of water and seed germination. Similarly, Morla et al. (2011) revealed an increase in percentage of seed germination and seedling growth by MWCNT at 40 μg/ml concentration in *Lycopersicum esculentum.* Likewise, in *Medicago sativa* and *Triticum aestivum*, 10 nm MWCNTs increased the growth of seedling and enlarged roots (Miralles et al. 2012). Additionally, Lahiani et al. (2013) demonstrated increased growth and germination rate in *Hordeum vulgare*, *Zea mays*, and *Glycine max*, from 25 to 100 μg/mL levels of multiwalled carbon nanotubes. Incorporation of single walled carbon nanotubes (SWCNTs: 50–800 μg/mL) alleviated the water scarcity up to moderate levels, by stimulating germination and growth attributes of *Hyoscya musniger* (Hatami et al. 2017).

9.2.2 Photosynthetic Rate Effects

Photosynthesis is the central process for plants on earth which converts light energy to chemical energy. Plants transform only 2–4% of the accessible radiation energy into new plant growth (Kirschbaum 2011). Currently, researchers are attempting to amend this low productivity of plants by controlling strategies and gene manipulations. To stimulate photosynthesis and turbocharged crops, scientists are working with an indispensable enzyme for photosynthesis, Rubisco, to catalyze the integration of carbon dioxide into biological compounds. Recently, new tobacco plants were grown by substituting the Rubisco gene in tobacco plant, with two genes of cyanobacterium *Synechococcus elongates*; the new plants were photosynthetically more efficient that native plants (Lin et al. 2014). In nanobiotechnology domain, researchers want to produce bionic plants that can have higher photosynthesis efficiency. CNTs have ability to associate with plant tissue to increase biological activities and also elevate light-related reactions in the chloroplast and promote light harvesting and affect the properties such as biochemical detection of various solutes (Iijima and Ichihashi 1993; Giraldo et al. 2014). CNT are capable of gathering more light energy via plant chloroplast while elevating electron transport rate, and eventually the rate of photosynthesis (Giraldo et al. 2014). It has also been revealed that CNTs absorb the comprehensive range of light wavelength (UV, visible and near infrared), exceeding chloroplast pigments and SWCNT—chloroplast transforms this absorbed light into exciton and enhance photosynthetic rate by transporting large number of electrons (Giraldo et al. 2014; Wong et al. 2017). MWCNT also boost the water uptake capability, total biomass and yield of plants (Husen and Siddiqi 2014; Joshi et al. 2018). Joshi et al. described that in rice, concentration of chlorophyll was more in CNT-supplemented plant, which was directly correlated to the growth of the plant and leaves. Concurrently, the photosynthetic efficiency was also enhanced which promoted the growth and yield. It was revealed that electron transport rate increased with the supplementation of CNT to the chloroplast, as sample supplemented with the CNTs captured more light photons and transferred this light energy to enhance the rate of electron transport, which eventually increased the photosynthetic rate (Giraldo et al. 2014; Deng et al. 2017).

9.3 Impact of Carbon Nanotubes on ROS and Antioxidant System of the Plants

Nanomaterials of diverse chemical nature like fullerenes, CNT, and metal oxides have been demonstrated to stimulate oxidative damage by generating reactive oxygen species (Vallyathan and Shi 1997). The chief characteristics implicated in NP-stimulated ROS comprise (i) prooxidant groups on the reactive surface of NP; (ii) functional redox cycling on the surface of NP due to transition metal-based NP; and (iii) particle-cell interactions (Risom et al. 2005; Knaapen et al. 2004). One of the commonly revealed harmful outcomes for CNT is the generation of ROS which can be either defensive or deleterious during biological interactions. CNT-induced ROS can directly cause oxidative stress in proximity or in the cell or could appear more indirectly due to the impacts of internalized CNT on mitochondrial respiration (Xia et al. 2008) or in decreasing antioxidant species inside the cell (Park et al. 2008). Begum et al. (2014) observed that application of MWNCTs increased the presence of ROS in red spinach leaf. Hatami (2017) observed treatment of MWCNTs caused an increment in stress response in pumpkin plants. There was an increment in the oxidative stress indices peroxide and malondialdehyde (Hatami 2017). Moreover, antioxidant enzymes, such as superoxide dismutase, catalase, and peroxidase were observed to decline (Hatami 2017). Oxidative stress indices and antioxidant enzymes act as substitute for plant stress. It is likely that a loss or reduced level of antioxidant enzymes can lead to enhanced levels of oxidative stress. In response to CNT exposure the level of ROS increased rice, but however no differentiation in the generation of ROS noticed in the presence of CNTs in wheat or rapeseed, indicating that stress responses to CNTs vary depending on plant species. Similar to CNT exposure, the level of ROS enhanced in onion after the treatment of fullerol (C60 (OH)), which could impact cell vitality and may impair the cell wall (Husen and Siddiqi 2014). Ghorbanpour and Hadian (2015) reported that the exposure of *Satureja khuzestanica* to MWCNTs activate some enzymatic antioxidants. Smirnova et al. (2012) revealed that MWCNTs, increased peroxidase activity and reduced the production of H_2O_2 in *Onobrychis* seedlings. It is likely that decreased level of H_2O_2 by MWCNTs occurred because of increased peroxidase activity in MWCNT-treated carrot leaf tissue. Since H_2O_2 accumulation can prompt senescence in plants (Hung and Kao 2004), decreased levels of H_2O_2 by MWCNTs might be able to delay senescence in carrot plants.

9.3.1 Carbon Nanotubes and Drought Stress

Water deficiency is the most significant environmental challenge that restricts plant growth and brings about extreme yield losses in comparison to other abiotic stress factors leading to acute food shortage. Nanomaterials are regarded as essential tools which are employed to repress current and future challenges in agricultural crop production. Currently there is not a comprehensive review regarding the possible function of nanomaterials in reducing the drought-induced deleterious impacts in crop plants. NMs have been observed to generally decrease drought-incited osmotic stress by accretion of osmolytes that cause osmotic adjustment and upgrade plant water status. Earlier reports demonstrated promising role of SWCNTs with diameter of 1–3 nm, and MWCNTs with diameter of 5–40 nm (Zaytseva and Neumann 2016) in mitigating cell damages by drought stress. Supplementation of SWCNTs at lower level (50 and 100 g mL^{-1}) to water stressed *Hyoscyamus niger* seedlings reduced the water scarcity (up to moderate levels only) by enhancing seed imbibitions and amylase stimulation (a key germination enzyme in starch hydrolysis process) and initiation of metabolic processes. However, elevated dose of SWCNTs (400 and 800 g mL^{-1}) exhibited negative impacts on seed germination and seedling functioning, enhanced the H_2O_2, MDA and ELI, and changed antioxidant enzyme activities (multiple biomarkers) and specific metabolite contents under control and all drought levels applied (Hatami et al. 2017). Similarly, foliar application of MWCNTs at 50 mg/L to Salvia mirzayanii plants enhanced membrane stability index, chlorophyll index, total phenolics, and antioxidant capacity in presence of moderate drought stress (Chegini et al. 2017). Seed nano-priming with MWCNTs at 30 mg/L resulted in the maximum seedling growth of Caucasian alder (Alnus subcordata) under drought stress (from −2 to −10 bars) (Rahimi et al. 2016). Additionally this level of MWCNTs was only enough to enhance seedling growth of Dodonaea viscosa (L.) Jacq. (Hopbush) under stress free conditions, while 50–100 mg/L were needed to increase growth under drought stress (Yousefi et al. 2017). By elevating the uptake of water, the levels of MWCNTs (500–1000 mg/L) stimulate drought and salinity tolerance in barley (Karami and Sepehri 2017). Contrarily the different concentrations of MWCNTs (125–1000 μg/mL) had detrimental effects on seed germination and seedling growth of cucumber under both PEG-prompted stress and normal growing conditions. These ill impacts are result of oxidative damage due to inactivation of several cellular antioxidant enzymes (Hatami 2017). In maize seedling, nano-priming under drought stress was effectual on germination rate and percentage, contents of chlorophyll a, b, and total chlorophyll and carotenoid. Supplementation of 25 mg/L CTNs at −1 Mpa drought stress proved efficacious in increasing germination percentage of maize seed. Also, 50 mg/L CTNs application enhanced photosynthetic pigmentation at −1 Mpa drought stress. Typically, priming with CTNs enhanced seed germination traits and photosynthetic pigmentation of maize seedlings under drought stress conditions (Shahriari et al. 2019).

9.3.2 Carbon Nanotubes and Salinity Stress

In the changing climatic conditions, salt stress is a rising menance to agriculture, decreasing production of crops through its osmotic and ionic effect, generation of oxidative stress, hormonal homeostasis and nutrient imbalance. Contrarily, production areas are flourishing in the regions where salinity stress occurs because of the exorbitant pressure for satisfying food security targets to meet needs of the expanding population. It has been revealed that application of nanoparticles to plants can remarkably diminish the negative impacts due to several intolerant conditions like salt stress, and therefore, modulate the adaptive mechanisms in plants. Different kinds of NPs and nanofertilizers have shown a favorable proof so far regarding salt stress control (Zulfiqar and Ashraf 2021). The supplementation of carbon nanomaterials in agriculture exhibited distinctive ability towards enhancing crop production under non-stress and stressed conditions (Martínez-Ballesta et al. 2016; Baz et al. 2020), even though cytotoxic effects of these materials have also been revealed. Martínez-Ballesta et al. (2016) supplementation of multi-walled carbon nanotubes (MWCNTs) to potential vegetable crop broccoli under saline conditions, improved the net photosynthetic rate and water uptake. These authors also reported that MWCNTs also enhanced aquaporin transduction, ensued increased water uptake, thus mediating salinity tolerance (Martínez-Ballesta et al. 2016). In other study, treatment of MWCNTs to salt treated rapeseed (*Brassica napus* L.) seedlings stimulated escalation of nitrate reductase dependent NO biosynthesis, re-establishment of ion and redox imbalance indicated by the reduction in ROS over generation, decrease in thiobarbituric acid production and Na^+/K^+ ratio (Zhao et al. 2019). furthermore, molecular studies revealed that the aforesaid impacts were associated to MWCNTs stimulated regulation in Na^+/H^+ exchanger 1 (*NHX1*) and K^+ transporter 1 (*KT1*) transcripts, antioxidant defense system genes, and salt overly sensitive 1 (*SOS1*) genes (Zhao et al. 2019). Recently, Gohari et al. (Gohari et al. 2020a, b) have noticed that low levels of MWCNTs resulted in salt amelioration in *Ocimum basilicum* plants through enhanced photosynthetic pigments and induced both enzymatic (i.e. APX, CAT and GP) and non-enzymatic components (i.e. phenolic content) of the antioxidant defense system, however, high levels of MWCNTs exhibited phytotoxic impacts as established by biochemical and fluorescence microscopy studies. Pandey et al. (2018) observed that the supplementation of graphene and CNTs enhanced the rate of germination of switch grass seeds and resulted in an early germination of sorghum seeds under salt stress. They suggested that carbon based nanomaterials remarkably decreased symptoms of salt stress inflicted by the application of NaCl into the growth medium. With the use of ion selective electrode, these authors demonstrated that the level of Na^+ ions in NaCl solution can be remarkably reduced by adding CNTs to the salt solution. Data established the promising role of carbon based nanomaterials as plant growth regulators for non-food crops and manifested the role of carbon based nanomaterials in the defence of plants against salt stress (Pandey et al. 2018). Under salinity stress, carbon nanotubes increased number and length of roots and the number and dry weight of shoots in *Satureja rechingeri*. By increasing carbon nanotubes, the contents of caffeic acid and rosmarinic acid were enhanced, while caffeic acid and rosmarinic acid levels were remarkably reduced in presence of different NaCl concentrations. This finding showed carbon nanotubes diminish the stress condition.

9.3.3 Carbon Nanotubes and Other Abiotic Stresses

Besides the abiotic stresses mentioned above carbon nanotubes are also employed to overcome various other abiotic stresses. Under paraquat (MV) toxicity, the exposure of Arabidopsis to MWCNT induced the relative electron transport rate and the effective photochemical quantum yield of PSII value in comparison to control by around 12% and lateral root formation up to nearly fourfold in comparison to the control. The defensive role of MWCNT on MV toxicity on the root surface area could be elucidated by the magnitude of MV adsorption on MWCNT and was associated to activation of photosynthesis, antioxidant protection and number and area of lateral roots which in turn helped nutrient assimilation (Fan et al. 2018). In *Boehmeria nivea* (L.) Gaudich (ramie) seedlings the application of 500 mg kg^{-1} MWCNTs elevated the accretion and translocation of Cd in seedlings and mitigated the toxicity induced by Cd by triggering plant growth, diminishing oxidative stress, enhancing specific antioxidant level and stimulating antioxidant enzyme activities (Gong et al. 2019). Sun et al. (2020) in *Scenedesmus obliquus* showed that CNTs, at a level of 5 mg/L, stimulated its growth and upgraded photosynthetic regulation by enhancing exciton trap efficiency and quantum yield for electron transport. Supplementation of CNTs seemed to reduce the ill effects of Cu, Cd or Zn on these microalgae which were demonstrated by enhanced growth, increased total chlorophyll content and photosynthetic indices. Amplification of photosynthesis and intercession of metal uptake by CNTs, have a significant role in the impacts of CNTs on metal toxicity (Sun et al. 2020). In Cd stressed maize, the MWCNTs treatment augmented root and shoot fresh mass and antioxidant enzyme activities, comprising peroxidase (POD), superoxide dismutase (SOD), and catalase (CAT) activities, and decreased the malondialdehyde (MDA) content in presence of Cd. However, the toxic impacts were also exhibited by higher concentration of MWCNTs suggesting a possible risk caused by MWCNTs. Application of MWCNTs diminished the Cd toxicity in growing maize plants; however, the effect of MWCNTs were concentration dependent (Chen et al. 2021). MWCNTs at 500 mg kg^{-1} resulted in a notable enhancement in the accretion of heavy metal (loid)s in *Solanum nigrum* (18.29% for Cd and 32.47% for As) and reduced co-contamination stimulated toxicity, by triggering the growth of plants, activating antioxidant enzymatic activities, and enhancing micronutrient content. As bio-concentration factor was reduced by 15.31–28.08% by the supplementation of MWCNTs, thereby mitigating phytotoxicity. Besides, availability of Cd and As were decreased in rhizosphere soils, and the remarkable decrease (16.29% for Cd and 8.19% for As) were shown in 500 mg kg^{-1} MWCNTs application. These findings signified that appropriate concentration of MWCNTs can increase remediation efficiency (Chen et al. 2021). Similarly, foliar-application of MWCNT up to 250 mg L^{-1} to pot marigold plants not only reduced Pb and Cd-induced toxicity by decreasing oxidative damage and improving both enzymatic and non-enzymatic antioxidant defense system but also boosted the phytoremediation property of plants by increasing the accumulation of both Pb and Cd from the soil (Sharifi et al. 2021).

9.4 Conclusion and Future Perspectives

In this chapter we outlined the possible role of CNTs under normal conditions and in presence of abiotic stress. CNTs increase the quality and quantity of crops significantly, because at specific.

Concentrations they alter the morphology, the physiology, and even the genetic constitutions of the plant, eventually, enhancing biomass. So, it can be useful in increasing the biomass of agricultural fields at global scale in the presence of changing environment.

References

Ball P (2002) Natural strategies for the molecular engineer. Nanotechnology 13:15

Baz H, Creech M, Chen J, Gong H, Bradford K, Huo H (2020) Water-soluble carbon nanoparticles improve seed germination and post-germination growth of lettuce under salinity stress. Agronomy 10:1192

Begum P, Ikhtiari R, Fugetsu B (2014) Potential impact of multi-walled carbon nanotubes exposure to the seedling stage of selected plant species. Nanomaterials 4:203–221

Bianco A, Hoebeke J, Kostarelos K, Prato M, Partidos CD (2005) Carbon nanotubes: on the road to deliver. Curr Drug Deliv 2:253–259

Boscaiu M, Lull C, Lidon A, Bautista I, Donat P, Mayoral O, Vicente O (2008) Plant responses to abiotic stress in their natural habitats. Bull Univ Agric Sci Vet Med Cluj-Napoca Hortic 65:53–58

Boulmaali M, Hadj-Ziane AZ (2017) Impact of carbon nanotubes on the germination of the *Phaseolus Vulgaris* seeds. In: Euro-Mediterranean conference for environmental integration. Springer, Cham, pp 391–393

Chegini E, Ghorbanpour M, Hatami M, Taghizadeh M (2017) Effect of multi-walled carbon nanotubes on physiological traits, phenolic contents and antioxidant capacity of *Salvia mirzayanii* Rech. f.& Esfand under drought stress. J Med Plants 2:191–207

Chen J, Zeng X, Yang W, Xie H, Ashraf U, Mo Z, Liu J, Li G, Li W (2021) Seed priming with multiwall carbon nanotubes (MWCNTs) modulates seed germination and early growth of maize under cadmium (Cd) toxicity. J Soil Sci Plant Nutr 21:1–13

Chhipa H, Joshi P (2016) Nanofertilisers, nanopesticides and nanosensors in agriculture. In: Nanoscience in food and agriculture 1. Springer, Cham, pp 247–282

Deng Y, Petersen EJ, Challis KE, Rabb SA, Holbrook RD, Ranville JF, Nelson BC, Xing B (2017) Multiple method analysis of TiO_2 nanoparticle uptake in rice (*Oryza sativa* L.) plants. Environ Sci Technol 51:10615–10623

Fan X, Xu J, Lavoie M, Peijnenburg WJ, Zhu Y, Lu T, Fu Z, Zhu T, Qian H (2018) Multiwall carbon nanotubes modulate paraquat toxicity in *Arabidopsis thaliana*. Environ Pollut 233:633–641

Gao L, Nie L, Wang T, Qin Y, Guo Z, Yang D, Yan X (2006) Carbon nanotube delivery of the GFP gene into mammalian cells. Chembiochem 7:239–242

Ghorbanpour M, Hadian J (2015) Multi-walled carbon nanotubes stimulate callus induction, secondary metabolites biosynthesis and antioxidant capacity in medicinal plant *Satureja khuzestanica* grown in vitro. Carbon 94:749–759

Gil-Díaz M, Alonso J, Rodríguez-Valdés E, Gallego JR, Lobo MC (2017) Comparing different commercial zero valent iron nanoparticles to immobilize As and Hg in brownfield soil. Sci Total Environ 584:1324–1332

Giraldo JP, Landry MP, Faltermeier SM, McNicholas TP, Iverson NM, Boghossian AA, Strano MS (2014) Plant nanobionics approach to augment photosynthesis and biochemical sensing. Nat Mater 13:400–408

Gohari G, Mohammadi A, Akbari A, Panahirad S, Dadpour MR, Fotopoulos V, Kimura S (2020a) Titanium dioxide nanoparticles (TiO_2 NPs) promote growth and ameliorate salinity stress effects on essential oil profile and biochemical attributes of *Dracocephalum moldavica*. Sci Rep 10:1–4

Gohari G, Safai F, Panahirad S, Akbari A, Rasouli F, Dadpour MR, Fotopoulos V (2020b) Modified multiwall carbon nanotubes display either phytotoxic or growth promoting and stress protecting activity in *Ocimum basilicum* L in a concentration-dependent manner. Chemosphere 249:126171

Gong X, Huang D, Liu Y, Zeng G, Wang R, Xu P, Zhang C, Cheng M, Xue W, Chen S (2019) Roles of multiwall carbon nanotubes in phytoremediation: cadmium uptake and oxidative burst in *Boehmeria nivea* (L.) Gaudich. Environ Sci Nano 6:851–862

Haghighi M, da Silva JA (2014) The effect of carbon nanotubes on the seed germination and seedling growth of four vegetable species. J Crop Sci Biotechnol 17:201–208

Hatami M (2017) Toxicity assessment of multi-walled carbon nanotubes on *Cucurbita pepo* L. under well-watered and water-stressed conditions. Ecotoxicol Environ Saf 142:274–283

Hatami M, Kariman K, Ghorbanpour M (2016) Engineered nanomaterial-mediated changes in the metabolism of terrestrial plants. Sci Total Environ 571:275–291

Hatami M, Hadian J, Ghorbanpour M (2017) Mechanisms underlying toxicity and stimulatory role of single-walled carbon nanotubes in *Hyoscyamus niger* during drought stress simulated by polyethylene glycol. J Hazard Mater 324:306–320

Hung KT, Kao CH (2004) Hydrogen peroxide is necessary for abscisic acid-induced senescence of rice leaves. J Plant Physiol 161:1347–1357

Husen A, Siddiqi KS (2014) Plants and microbes assisted selenium nanoparticles: characterization and application. J Nanobiotechnol 12:28

Iijima S, Ichihashi T (1993) Single-shell carbon nanotubes of 1-nm diameter. Nature 363:603–605

Jiang Y, Hua Z, Zhao Y, Liu Q, Wang F, Zhang Q (2012) The effect of carbon nanotubes on rice seed germination and root growth. In: Proceedings of the 2012 international conference on applied biotechnology (ICAB 2012) 2014. Springer, Berlin, Heidelberg, pp 1207–1212

Joshi A, Kaur S, Dharamvir K, Nayyar H, Verma G (2018) Multi-walled carbon nanotubes applied through seed-priming influence early germination, root hair, growth and yield of bread wheat (*Triticum aestivum* L.). J Sci Food Agric 98:3148–3160

Kam NW, Liu Z, Dai H (2005) Functionalization of carbon nanotubes via cleavable disulfide bonds for efficient intracellular delivery of siRNA and potent gene silencing. J Am Chem Soc 127:12492–12493

Kam NW, Liu Z, Dai H (2006) Carbon nanotubes as intracellular transporters for proteins and DNA: an investigation of the uptake mechanism and pathway. Angew Chem Int Ed 45:577–581

Karami A, Sepehri A (2017) Multiwalled carbon nanotubes and nitric oxide modulate the germination and early seedling growth of barley under drought and salinity. Agric Conspec Sci 82:331–339

Khodakovskaya M, Dervishi E, Mahmood M, Xu Y, Li Z, Watanabe F, Biris AS (2009) Carbon nanotubes are able to penetrate plant seed coat and dramatically affect seed germination and plant growth. ACS Nano 3:3221–3227

Khodakovskaya MV, De Silva K, Biris AS, Dervishi E, Villagarcia H (2012) Carbon nanotubes induce growth enhancement of tobacco cells. ACS Nano 6:2128–2135

Kirschbaum MU (2011) Does enhanced photosynthesis enhance growth? Lessons learned from CO_2 enrichment studies. Plant Physiol 155:117–124

Knaapen AM, Borm PJA, Albrecht C, Schins RPF (2004) Inhaled particles and lung cancer, part A: mechanisms. Int J Cancer 109:799–809

Lacerda L, Soundararajan A, Singh R, Pastorin G, Al-Jamal KT, Turton J, Frederik P, Herrero MA, Li S, Bao A, Emfietzoglou D (2008) Dynamic imaging of functionalized multi-walled carbon nanotube systemic circulation and urinary excretion. Adv Mater 20:225–230

Lahiani MH, Dervishi E, Chen J, Nima Z, Gaume A, Biris AS, Khodakovskaya MV (2013) Impact of carbon nanotube exposure to seeds of valuable crops. ACS Appl Mater Interfaces 5:7965–7973

Lin MT, Occhialini A, Andralojc PJ, Parry MA, Hanson MR (2014) A faster rubisco with potential to increase photosynthesis in crops. Nature 513:547–550

Liu Z, Sun X, Nakayama N, Dai H (2007) Supramolecular chemistry on water-soluble carbon nanotubes for drug loading and delivery. ACS Nano 1:56

Liu Z, Davis C, Cai W, He L, Chen X, Dai H (2008) Circulation and long-term fate of functionalized, biocompatible single-walled carbon nanotubes in mice probed by Raman spectroscopy. Proc Natl Acad Sci U S A 105:1410–1415

Mani PK, Mondal S (2016) Agri-nanotechniques for plant availability of nutrients. In: Plant nanotechnology. Springer, Cham, pp 263–303

Manzer S, Mohamed A, Mohammad F, Mutahhar A (2015) Role of nanoparticles in plants. In: Nanotechnology and Plant Science. Springer, Cham, pp 19–35

Martínez-Ballesta MC, Zapata L, Chalbi N, Carvajal M (2016) Multiwalled carbon nanotubes enter broccoli cells enhancing growth and water uptake of plants exposed to salinity. J Nanobiotechnol 14:1–4

Miralles P, Johnson E, Church TL, Harris AT (2012) Multiwalled carbon nanotubes in alfalfa and wheat: toxicology and uptake. J R Soc Interface 9:3514–3527

Monreal CM, DeRosa M, Mallubhotla SC, Bindraban PS, Dimkpa C (2016) Nanotechnologies for increasing the crop use efficiency of fertilizer-micronutrients. Biol Fertil Soils 52:423–437

Morla S, Rao CR, Chakrapani R (2011) Factors affecting seed germination and seedling growth of tomato plants cultured *in vitro* conditions. J Chem Biol Phys Sci 1:328

Nalwade AR, Neharkar SB (2013) Carbon nanotubes enhance the growth and yield of hybrid Bt cotton Var. ACH-177-2. Int J Adv Sci Technol 3:840–846

Oprea AE, Grumezescu AM (eds) (2017) Nanotechnology applications in food—flavor, stability, nutrition and safety. Academic, Elsevier

Pandey K, Lahiani MH, Hicks VK, Hudson MK, Green MJ, Khodakovskaya M (2018) Effects of carbon-based nanomaterials on seed germination, biomass accumulation and salt stress response of bioenergy crops. PLoS One 13:e0202274

Panpatte DG, Jhala YK, Shelat HN, Vyas RV (2016) Nanoparticles: the next generation technology for sustainable agriculture. In: Microbial inoculants in sustainable agricultural productivity. Springer, New Delhi, pp 289–300

Park E, Choi J, Park Y, Park K (2008) Oxidative stress induced by cerium oxide nanoparticles in cultured BEAS-2B cells. Toxicology 245:90–100

Patra AK, Adhikari T, Bhardwaj AK (2016) Enhancing crop productivity in salt-affected environments by stimulating soil biological processes and remediation using nanotechnology. In: Innovative saline agriculture. Springer, New Delhi, pp 83–103

Pourkhaloee A, Haghighi M, Saharkhiz MJ, Jouzi H, Doroodmand MM (2011) Investigation on the effects of carbon nanotubes (CNTs) on seed germination and seedling growth of salvia (*Salvia microsiphon*), pepper (*Capsicum annum*) and tall fescue (*Festuca arundinacea*). J Seed Technol 33:155–160

Rahimi D, Kartoolinejad D, Nourmohammadi K, Naghdi R (2016) Increasing drought resistance of *Alnus subcordata* CA Mey. seeds using a nano priming technique with multi-walled carbon nanotubes. J For Sci 62:269–278

Risom L, Møller P, Loft S (2005) Oxidative stress-induced DNA damage by particulate air pollution. Mutat Res 592:119–137

Saharan V, Pal A (2016) Current and future prospects of chitosan-based nanomaterials in plant protection and growth. In: Chitosan based nanomaterials in plant growth and protection. Springer, New Delhi, pp 43–48

Serag MF, Kaji N, Gaillard C, Okamoto Y, Terasaka K, Jabasini M, Tokeshi M, Mizukami H, Bianco A, Baba Y (2011) Trafficking and subcellular localization of multiwalled carbon nanotubes in plant cells. ACS Nano 5:493–499

Serag MF, Kaji N, Tokeshi M, Baba Y (2012) Introducing carbon nanotubes into living walled plant cells through cellulase-induced nanoholes. RSC Adv 2:398–400

Serag MF, Kaji N, Tokeshi M, Baba Y (2015) Carbon nanotubes and modern nanoagriculture. In: Nanotechnology and plant sciences. Springer, Cham, pp 183–201

Shahriari A, Omidi H, Mohammadi H, Mohammadi A, Ahmadi K (2019) The effect of carbon nanotubes seed priming on germination and photosynthetic pigmentation of maize hybrids under drought stress. Agroecol J 15:1–2

Sharifi P, Bidabadi SS, Zaid A, Latef AAHA (2021) Efficacy of multi-walled carbon nanotubes in regulating growth performance, total glutathione and redox state of *Calendula officinalis* L. cultivated on Pb and Cd polluted soil. Ecotoxicol Environ Saf 213:112051

Singh R, Pantarotto D, Lacerda L, Pastorin G, Klumpp C, Prato M, Bianco A, Kostarelos K (2006) Tissue biodistribution and blood clearance rates of intravenously administered carbon nanotube radiotracers. Proc Natl Acad Sci 103:3357–3362

Smirnova E, Gusev A, Zaytseva O, Sheina O, Tkachev A, Kuznetsova E, Lazareva E, Onishchenko G, Feofanov A, Kirpichnikov M (2012) Uptake and accumulation of multiwalled carbon nanotubes change the morphometric and biochemical characteristics of *Onobrychis arenaria* seedlings. Front Chem Sci Eng 6:132–138

Srinivasan C, Saraswathi R (2010) Nano-agriculture-carbon nanotubes enhance tomato seed germination and plant growth. Curr Sci 99:274–275

Sun C, Li W, Xu Y, Hu N, Ma J, Cao W, Sun S, Hu C, Zhao Y, Huang Q (2020) Effects of carbon nanotubes on the toxicities of copper, cadmium and zinc toward the freshwater microalgae *Scenedesmus obliquus*. Aquatic Toxicol 224:105504

Tamer TM, Abou-Taleb WM, Roston GD, Mohyeldin MS, Omer AM, Khalifa RE, Hafez AM (2018) Formation of zinc oxide nanoparticles using alginate as a template for purification of wastewater. Environ Nanotechnol Monit Manag 10:112–121

Tiwari DK, Dasgupta-Schubert N, Cendejas LV, Villegas J, Montoya LC, García SB (2014) Interfacing carbon nanotubes (CNT) with plants: enhancement of growth, water and ionic nutrient uptake in maize (*Zea mays*) and implications for nanoagriculture. Appl Nanosci 4:577–591

Torney F, Trewyn BG, Lin VSY, Wang K (2007) Mesoporous silica nanoparticles deliver DNA and chemicals into plants. Nat Nanotechnol 2:295–300

Vallyathan V, Shi X (1997) The role of oxygen free radicals in occupational and environmental lung diseases. Environ Health Perspect 105:165–177

Villagarcia H, Dervishi E, de Silva K, Biris AS, Khodakovskaya MV (2012) Surface chemistry of carbon nanotubes impacts the growth and expression of water channel protein in tomato plants. Small 8:2328–2334

Wong MH, Giraldo JP, Kwak SY, Koman VB, Sinclair R, Lew TTS, Strano MS (2017) Nitroaromatic detection and infrared communication from wild-type plants using plant nanobionics. Nat Mater 16:264–272

Xia T, Kovochich M, Liong M, Zink JI, Nel AE (2008) Cationic polystyrene nanosphere toxicity depends on cellspecific endocytic and mitochondrial injury pathways. ACS Nano 2:85–96

Yousefi S, Kartoolinejad D, Naghdi R (2017) Effects of priming with multi-walled carbon nanotubes on seed physiological characteristics of hopbush (*Dodonaea viscosa* L.) under drought stress. Int J Environ Stud 74:528–539

Zaytseva O, Neumann G (2016) Carbon nanomaterials: production, impact on plant development, agricultural and environmental applications. Chem Biol Technol Agr 3:17

Zhao Z, Wang W, Li C, Zhang Y, Yu T, Wu R, Zhao J, Liu Z, Liu J, Yu H (2019) Reactive oxygen species–activatable liposomes regulating hypoxic tumor microenvironment for synergistic photo/chemodynamic therapies. Adv Funct Mater 29:1905013

Zulfiqar F, Ashraf M (2021) Nanoparticles potentially mediate salt stress tolerance in plants. Plant Physiol Biochem 160:257–268

Chapter 10
Responses of Crop Plants Under Nanoparticles Supply in Alleviating Biotic and Abiotic Stresses

Sameer H. Qari, Awatif M. Abdulmajeed, Taghreed S. Alnusaire, and Mona H. Soliman

Abstract Crop plants are continuously exposed to abiotic and biotic constraints like salt, drought, heat, heavy metal, ultraviolet, fungi, bacteria, pathogens etc. Plants respond to them by altering their physiology and metabolism. The response is initiated at molecular, cellular and at whole plant level. Plants undergo alterations in growth, photosynthesis, yield and quality attributes when exposed to stresses. Nevertheless, these complex processes are modulated by the application of different kinds of nanoparticles (NPs). In the current times, nanotechnology, as a broad interdisciplinary area of research, finds potential in agriculture regarding plant disease management, pathogen detection and imparting stress tolerance. Nanoparticles at particular concentrations control growth, morpho-physiology and yield attributes under stress conditions. In the present chapter, we attempt to discuss various response alterations initiated by crop plants towards abiotic and biotic stresses, vis-à-vis controlling nature of various NPs doses in protecting these processes under these stress conditions.

Keywords Nanoparticles · Abiotic stress · Biotic stress · Crop plants

---

S. H. Qari
Biology Department, Aljumum University College, Umm Al – Qura University, Makkah, Saudi Arabia
e-mail: shqari@uqu.edu.sa

A. M. Abdulmajeed
Biology Department, Faculty of Science, University of Tabuk, Umluj, Saudi Arabia

T. S. Alnusaire
Biology Department, College of Science, Jouf University, Sakaka, Saudi Arabia

M. H. Soliman (✉)
Botany and Microbiology Department, Faculty of Science, Cairo University, Giza, Egypt

Biology Department, Faculty of Science, Taibah University, Yanbu, Kingdom of Saudi Arabia
e-mail: hmona@sci.cu.edu.eg

M. Faizan et al. (eds.), *Sustainable Agriculture Reviews 53*, Sustainable Agriculture Reviews 53, https://doi.org/10.1007/978-3-030-86876-5_10

10.1 Introduction

In current times due to the unsettling increase in the growth of population together with changing climate, a bunch of enhancement in crop productivity of agriculture is the need of the hour. As plants are sessile creatures and are continuously exposed to a variety of environmental pressures like biotic and abiotic that act as simultaneous confronting factors to decrease the crop productivity (Rejeb et al. 2014; Alhaithloul et al. 2020; Zaid et al. 2020, 2021). In the current decade, biotic and abiotic stresses are currently being realized as one of the most potential threats not only to plants but also to human existence that jointly results in the broad consequences to human health as well as to crop agricultural system (Roberts and Mattoo 2018; Alhaithloul et al. 2020). A conspicuous decrease in crop yield and as well as fertility of soil together with the over-accumulation of the toxic trace elements in the rhizospheric and their transfer to the subsequent food chains are threatening human health, thus leading to severe health as well as ecological issues (Verstraeten et al. 2008). It has been estimated that 70% increase in the agricultural productivity has to be achieved in order to feed 2.3 billion human populations by 2050 (Tilman et al. 2011). Therefore, in order to meet these challenges, biotic and abiotic stress tolerance in crop plants is of paramount importance. Nevertheless, the underlying stress tolerance mechanisms are more complex as compared to animals and are governed by multigenes in diverse crop plants (Qin et al. 2011). It is thus this complex nature of polygenic trait of crop plants, scientists have begun to look for alternative mechanistic tools in this regard. Thus, sophisticated and potent techniques are needed to meet these desired aims. Now-a-days, much attention is laid on the synthesis of diverse nano-products having size 1–100 nm that have potential advantage but risk associated in different industries, like food, medicine, and agriculture sector. An estimate suggests that during 2014–2019, the worldwide application of nanoparticles (NPs) would increase from 225,060 metric tons to approximately 585,000 metric tons (Rajput et al. 2018). The nano-products have remarkable properties, like small size ranging 1–100 nm diameters, specific surface properties, wettability, melting point, electrical and thermal conductivity, light absorption, high reactivity, large surface areas, and show strong binding interactions with other biomolecules and structures (Asgari-Targhi et al. 2018; Jeevanandam et al. 2018). Nevertheless, NPs show interaction with the plant at different hierarchical levels. NPs affect uptake, translocation, accumulation and transformation that occur within the plant and these interactions can have positive, negative or neutral impacts depending upon the species involved.

Among the various approaches, engineering of crop plants by NPs supply have proven a very good and effective way and has yielded prominent results so far. NPs have gained great interests because of their physicochemical properties and biological activities as compared to their bulk parent materials. Thus in the present chapter, we attempt to describe the responses of various crop plants under NPs supply, modulation of gene expression and the alleviating role induced by different NPs under biotic and abiotic stresses. A tabulated view of responses of various plants under various NP supply are presented in Table 10.1.

Table 10.1 Representing biotic and abiotic stress condition, nanoparticle application and the corresponding response of plants

Nanoparticle	Abiotic or Biotic stress	Effects	References
ZnO-NPs (50 mg/L)	Cadmium (0.4, 0.6 or 0.8 mM)	Cd-increased oxidative burst in tomato, the antioxidant defense activities and proline content, reduced carbonic anhydrase (CA) and nitrate reductase (NR) activities but the follow-up treatment with ZnO-NPs significantly incremented the plant growth traits, chlorophyll content, leaf gas exchange attributes, protein content, activities of NR and CA, stomatal aperture and reduced the malondialdehyde and ROS levels	Faizan et al. (2020)
TiO_2 (50 mg/L)	2 to 20 mg/L of $CdCl_2$	The TiO_2 NPs application induced a higher translocation of Cd to the aerial portion in *Azolla filiculoides*. The high dose of Cd in leaves did not cause damage to the photosynthetic machinery. The NPs boosted antioxidant apparatus and decrease H_2O_2 content, but did not suppress the TBARS level	Spanò et al. (2019)
MgO (0, 25, 50 or 100 ppm)	*Meloidogyne incognita*	The MgO-NPs application reduced nematode fecundity, decreased number and size of gall cells; enhanced plant growth parameters, chlorophyll, carotenoid, seed protein, and root and shoot N contents	Tauseef et al. (2021)
Fe_2O_3-NPs (100, 200 and 400 mg/L)	As (0.5, 1 and 2 μM)	The As application induced oxidative stress, enhanced total antioxidant capacity, SOD and CAT activity, but declined GPOX activity and lowered root oxidisability. The Fe_2O_3-NPs application reduced As toxicity by reducing As availability in *Vigna radiata*	Shabnam et al. (2019)
TiO_2 (50, 100, 150, and 200 ppm)	Salt (1.5 and 0.5 M and 0.5 and 0.17 M)	The TiO_2 decreased the specific growth rate, chlorophyll, and photosynthesis in *Dunaliella salina* and *Dunaliella tertiolecta* The toxic effect of NPs decreased at higher doses of salt stress	Ghazaei and Shariati (2020)
Se/SiO_2 (50 and 100 mg/L)	Drought stress (30%, 60%, and 100% field capacity)	The praying of NPs was found to improve the growth and yield parameters of strawberry plants grown under drought stress. The plants grown under Se/SiO_2 (100 mg/L) had more photosynthetic pigments, carbohydrate and proline, relative water content, membrane stability index (MSI) and water use efficiency (WUE), increased activity of CAT, APX, GPX and SOD and decreased levels of stress biomarkers of strawberry plants	Zahedi et al. (2020)

(continued)

Table 10.1 (continued)

Nanoparticle	Abiotic or Biotic stress	Effects	References
TiO_2 (0, 50, 100 and 200 mg/L)	Salinity (0, 50 and 100 mM)	The imposition of salt stress decreased the agronomic traits in *Dracocephalum moldavica* L. The application of TiO_2-NPs improved all agronomic traits, antioxidant enzyme activity, significantly lowered ROS concentration and caused highest essential oil content. It was found that the highest amounts of secondary metabolites were obtained by 100 mg/L TiO_2-NP-application under control conditions	Gohari et al. (2020)
SiO_2 (100 μM or 500 μM)	Pb (50, 250, 500, 1000, or 1500 μM)	At first the antioxidant enzyme activity increased at low levels of Pb and then decreased in *Pleioblastus pygmaeus*. The application of SiO_2-NPs increased SOD, CAT, GR and phenylalanine ammonia-lyase (PAL) activities under Pb stress. The NPs application lowered the Pb-induced oxidative stress which resulted in increased plant photosynthesis and growth under Pb toxicity. The 500 μM SiO_2 NPs treatment showed more pronounced effects than 100 μM in countering Pb toxicity	Emamverdian et al. (2020)
FeO (25, 50 and 100 mg/kg)	Salt and Cd stress	The application of 100 mg/kg the FeO-NPs stimulated and reprogrammed the morpho-physiological state of wheat plants caused a significant increment in the N, P and K^+ but reduced the toxic salt ions in the wheat grain. Under Cd stress, application of the FeO-NPs reduced Cd uptake in the wheat plants	Manzoor et al. (2021)
ZnO (100 mg/L)	Cd and water limited stress	Both drought and Cd stress decreased plant dry weight, chlorophyll contents but increased oxidative stress in *Triticum aestivum* L. The application of NPs improved the plants biomass, chlorophyll contents by minimizing the oxidative stress and boosting antioxidants enzymatic activities in stressed plants and restricting the uptake of Cd	Bashir et al. (2021)

10.2 Modulation of Gene Expression by Nanoparticle Supply

Various kinds of NPs are known to modulate expression of various genes in diverse crop plants. When a plant faces stress, perception of stress stimulus elevates the level of cytosolic Ca^{2+} concentration ([Ca^{2+}] cyt). The Ca^{2+} binding protein (CaBP) present sense the elevated [Ca^{2+}] cyt. It is known that the binding of Ca^{2+} activates the CaBP that directly binds to the promoter regions of specific genes and either induces or represses their expression (Tuteja and Mahajan 2007). The over-expression of CaBPs by the NPs supply has been reported earlier (Marmiroli et al. 2015). In addition, NPs also bind to CaBPs (Miao et al. 2014); the transcription activator then trigger downstream signalling and finally the expression of genes related to stress that and activates the defense system of plants (Delk et al. 2005; Boudsocq and Sheen 2013). The supply of various forms of NPs also modulates the expression of genes of plants' vital components. Khattab et al. (2021) evaluated the roles of Si and Si-NPs in improving thermo-tolerance of wheat photosynthesis by studying the up-regulation of *PsbH*, *PsbB* and *PsbD* genes encoding the core proteins of photosystem II (PSII) under heat stress (45 °C, 5 h). The results revealed that at the molecular level, RT-PCR analysis showed that K_2SiO_3 (1.5 mM) and SiO2NPs (1.66 mM) mitigated the negative impacts and stimulated the over-expression of *PsbH*, *PsbB*, and *PsbD* genes to a considerable extent. Under salt stress, it was molecularly revealed that multi-walled carbon nanotube (MWCNT) induced the alteration in Na^+/H^+ exchanger 1 (*NHX1*) and K^+ transporter 1 (*KT1*) transcripts, antioxidant defense system genes, and salt overly sensitive 1 (*SOS1*) genes which confer tolerance to salt stress (Zhao et al. 2019). Chandrakar et al. (2020) studied the influence of Carbon dots (C-dots) in enhancing tolerance to As stress by analysing the defense-related gene expression in *Cicer arietinum* L. The As treatment in tested plants resulted in the up-regulation of expressions of *NADPH* oxidase and defense-related genes in *Cicer arietinum* L. It was however, revealed that exogenous application of C-dots, enhanced the expressions of defense-related genes. In *Catharanthus roseus* (L.), the vital role of chitosan NPs (C-NPs) in enhancing drought (100% and 50% of field capacity [FC], respectively) stress tolerance through modulation of gene expression was studied (Ali et al. 2021). It was found that drought stress increased the plant alkaloid content, of which application of C-NPs has an additive effect. It was further revealed that under drought stress conditions, the maximum alkaloid content was engaged with induced gene expression of strictosidine synthase (*STR*), deacetylvindoline-4-O-acetyltransferase (*DAT*), peroxidase 1 (*PRX1*) and geissoschizine synthase (*GS*). The application of C-NPs caused their additive gene expression of alkaloid biosynthesis. Hedayati et al. (2020) applied SiO_2-NPs (0, 25, 50, 100 and 200 mg/L) in improving tropane alkaloid production and studied the changes in gene expression levels in two species of *Hyoscyamus*. The SiO_2-NPs (100 and 200 mg/L), at 24 h of treatment maximally elicited the total phenol and flavonoid accumulation, the 100 mg/L dose induced the highest amount of hyoscyamine, and scopolamine accumulation as compared to non-treated roots as revealed by High-performance liquid chromatography (HPLC) technique. Furthermore the semi-quantitative RT-PCR analysis showed the highest level of *pmt* (putrescine *N*9 methyltransferase) and *h6h* (hyoscyamine 6 < *beta* > _ hydroxylase) gene expression. In order to study the effect of engineered magnetic NPs (M-NP) (from 125 to 1000 mg/L) in enhancing chlorophyll content and growth of barley plants, the induction of photosystem genes was studied (Tombuloglu et al. 2020). Two magnetic elements (Co and Nd) were doped into technologically important Fe_2O_3 NPs by utilizing a sonochemical synthetic approach. It was found that M-NPs enhanced the germination rate, tissue growth, biomass, chlorophyll (a, b), and carotenoids contents. The highest growth enhancement at 125 or 250 mg/L treatments was noticed but the higher doses diminished the growth traits. The gene expressions of photosystem marker genes (*BCA*, *psbA*, and *psaA*) were found to be over-expressed in M-NPs treated plants as compared to non-treated control in a proportional basis as increasing M-NPs doses, indicated a positive correlation with the photosynthetic machinery. Mosa et al. (2018) in *Cucumis sativus* plants, studied the impacts of Cu-NPs in inducing the changes in gene expression of superoxide dismutase. It was found that Cu-Zn (SOD) gene expression was induced by Cu-NP treatment.

10.3 Effect of Nanoparticles Under Biotic Stress Conditions

The use of NPs in biotic stress conditions has become more common as technology advances. The various NPs have been well documented for their profound counteracting role against diverse categories of biotic stresses and are now-a-days being utilized in controlling of various plant pathogens. In addition, NPs mediated controlling various plant pathogens are approaches that are more environment-friendly as compared to other chemically synthesized fungicides (Jo et al. 2009). Thus, the use of various NPs in controlling the pathogenic agricultural-microbes is cost-effective as well as eco-friendly approach (Kumar and Yadav 2009; Swamy and Prasad 2012). The effect of *Bacillus* sp. and Ag-NPs on *Zea mays* was evaluated (Kumar et al. 2020). The leaf and flower extracts of *Tagetes erecta* was used for the synthesis of Ag-NPs, whereas *Bacillus* sp. was isolated from rhizosphere of spinach plant. By using ultra centrifugation, UV–Vis spectroscopy and high resolution transmission electron microscopy (HRTEM), the average particles size of Ag-NPs was recorded approximately 60 nm. The results revealed that almost all the potential isolates of PGPR were able to produce indole acetic acid, ammonia and hydrogen cyanide, solubilised tricalcium phosphate and potentially restricted the growth of *Macrophomina phaseolina* in vitro but the isolate LPR2 was adjudged as best performing. By conducting 16S rRNA gene sequence, this isolate; LPR2 was characterized as *Bacillus cereus* LPR2. It was found that the combination of *Bacillus cereus* LPR2 and Ag-NPs enhanced the plant growth and LPR2 strongly inhibited the growth of deleterious fungal pathogen. Ina recent study, Lopez-Lima et al. (2021) studied the role of Cu-NPs in tomato for the effective management of *Fusarium* wilt and in promoting tomato plant growth. The antifungal activity of Cu-NPs at different concentrations (0.1, 0.25, 0.5, 0.75, and 1.0 mg/mL) against *Fusarium oxysporum* f. sp. *lycopersici* (FOL) was studied. The results showed that a strong inhibitory effect on mycelial FOL growth was observed with 0.5 mg/mL Cu-NPs treatment. It was further found that a significant reduction in the symptoms of *Fusarium* wilt was observed with NPs supply. In addition, Cu-NPs treatments were found to increase the growth and chlorophyll content (from 19.3% to 28.6%) of tomato plants. Khan and Siddiqui (2021) evaluated the role of ZnO-NPs (100 and 200 mg/ L) in the management of disease complex of *Beta vulgaris* L. caused by *Pectobacterium betavasculorum* (Pb),

Meloidogyne incognita (Mi) and *Rhizoctonia solani* (Rs). In one experiment, the results showed that treatment with ZnO-NPs inhibited the hatching and induced mortality of Mi and also inhibited the growth of Pb and Rs pathogens. Simultaneously, in pot experiments, it was found that ZnO-NPs treatments as foliar spray and seed priming to tested plants infected pathogens and more efficiently improved plant dry mass and physiological and biochemical parameters. The foliar spray of ZnO-NPs at 200 mg/L to plants was best as compared to seed priming and infected pathogens more and resulted in the greatest increase in plant dry mass, and physiological and biochemical parameters of beetroot. Both the treatment methods viz-seed priming and foliar spray of ZnO-NPs caused a reduction in disease indices, nematode population, and root galling. By conducting inductively coupled plasma mass spectrometry (ICP-MS) analysis, their results showed that ZnO-NPs were accumulated in shoots and roots of both infected and uninfected plants. Tauseef et al. (2021) studied the role of MgO-NPs in suppressing the effects of Mi, infecting *Vigna unguiculata* L. for the improvement in growth and physiology of plants. The MgO-NPs were synthesized by sol-gel method and characterized by transmission and scanning electron microscopy (TEM, SEM), UV-Vis spectroscopy, X-ray diffraction (XRD) and Fourier transform infrared spectroscopy (FTIR) techniques. The MgO-NPs were used in different concentrations (0, 25, 50 or 100 ppm) and by various modes of applications (root dip, soil drench or foliar spray) on cowpea plants infected with (Mi). The SEM images of Mi showed that treatment with MgO-NPs (50 and 100 ppm) exhibited indentations, roughness and distortions in the cuticular surface as compared to control juveniles of Mi. Further 100 ppm dose of MgO-NPs as root dip application; reduced the fecundity of nematodes, decreased number and reduce size of galls; enhanced plant growth, chlorophyll, carotenoids, seed protein, and root and shoot nitrogen contents. In yet another study under greenhouse conditions, Khan and Siddiqui (2020) employed the use of silicon dioxide nanoparticles (SiO_2-NPs) as seed priming and foliar spray in two concentrations (100 and 200 mg/L) for the management of *Meloidogyne incognita, Pectobacterium betavasculorum* and *Rhizoctonia solani* disease complex of beetroot. The results showed that plants grown with SiO_2-NPs primed seeds and inoculated with pathogens were found to be more beneficial than the foliar spray in enhancing dry weights of shoot and root, chlorophyll content, chlorophyll fluorescence traits and the activities of antioxidant enzymes. It was found that priming seeds with SiO_2-NPs at 200 mg/L resulted in the highest improvements in dry weight of shoot and root, chlorophyll content, chlorophyll fluorescence characters and boosted the activities of defense enzymes followed by seed priming with 100 mg/L, foliar spray of 200 mg/L and foliar spray of 100 mg/L. Both methods of seed priming as well as foliar spray resulted in a reduction in root galling, nematode multiplication and disease indices. Treatment with 200 mg/L SiO_2-NPs resulted in the highest reduction in root galling, nematode multiplication and disease indices. In a recent study, Danish et al. (2021) studied the ameliorative role of green synthesized Ag-NPs (25, 50, and 100 ppm) to mitigate stress induced by *Meloidogyne incognita* in *Trachyspermum ammi* (L.) plants. Various techniques viz.,

UV – visible spectrophotometer, scanning electron microscopy (SEM), transmission electron microscopy (TEM), and energy-dispersive X-ray analyzer (EDX) were used for characterization of Ag-NPs from *Senna siamea*. The results depicted that 50 ppm Ag-NPs treatment before inoculation of *M. incognita* showed maximum and significant ($p \leq 0.05$) increment in plant growth traits, biochemical characteristics, and activities of peroxidise (POX), catalase (CAT), superoxide dismutase (SOD), and ascorbate peroxidase (APX) with respect to control plants. However, it was noticed that the maximum reduction in the number of galls, egg masses, and root-knot indices was observed in plants in the order 100 > 50 ppm Ag-NPs treatment before nematode inoculation as compared to the inoculated plants. Nevertheless, transverse section in *T. ammi* roots showed a considerable lignification when treated with 50 ppm Ag-NPs. Khan and Siddiqui (2018) applied two concentrations (100 and 200 ppm) of ZnO-NPs for the management of *Ralstonia solanacearum*, *Phomopsis vexans* and *M. incognita* induced disease complex of eggplant. It was observed that the highest increase in plant growth traits, chlorophyll and carotenoid contents was in the order of spray>seed>soil inoculation of NP treatments. It was however, found that plants sprayed with 200 ppm NPs resulted in higher increase in plant growth characters, and higher reduction in galling and nematode multiplication. The SEM analysis showed that ZnO-NPs induced an adverse effect on cell wall/body wall of tested pathogens in addition, the disease index was also found to be reduced to one in plants treated with 200 ppm ZnO-NPs. In a study conducted on tomato plants, the applicability of ZnO-NPs in protecting the bacterial speck pathogen, caused by *Pseudomonas syringae* pv. tomato DC3000 (Pst) was evaluated (Elsharkawy et al. 2020). The results depicted that ZnO-NPs showed significant direct antibacterial activity against *Pseudomonas syringae* pv. tomato, caused significant reductions in disease severity index and proliferation of bacteria as compared to non-treated tomato plants. The ZnO-NPs treated plants showed higher defense related enzyme activity and the regulatory and defense genes like *LePR-1a* and lipoxygenase (*LOX*), that were involved in the salicylic acid and jasmonic acid signalling pathways were also maximally expressed. He et al. (2011) studied the antifungal activity of ZnO-NPs (0, 3, 6 and 12 mmol/l) against two pathogenic fungi viz-*Botrytis cinerea* and *Penicillium expansum* by using traditional microbiological plating, SEM, and Raman spectroscopy techniques. The results revealed that the application of ZnO-NPs more than 3 mmol/l caused a significant inhibition in the growth of *B. cinerea* and *P. expansum*. The ZnO-NPs inhibited cellular functions of *B. cinerea*, which resulted in the deformation of hyphae of fungi and prevented the development of conidiophores and conidia of *P. expansum*, as a result of which the death of fungal hyphae was observed. Siddiqui et al. (2019a) tested the efficacy of TiO_2 and ZnO-NPs at 0.25 and 0.50 mL/L for controlling bacterial diseases, and growth and physiological changes of beetroot plants. It was found that inoculating plants with *P. betavasculorum*, *X. campestris* pv. *beticola*, and *P. syringae* pv. *aptata* caused a reduction in plant growth traits, chlorophyll and carotenoid contents, but increased the activities of SOD, CAT, APX, PAL, and contents of GSH, proline, H_2O_2, and MDA. The application of ZnO-NPs excelled over TiO_2-NPs supply. The disease indices were reduced considerably by 0.50 mL/L ZnO-NPs application as compared to TiO_2-NP. In yet another study in carrot plants, Siddiqui et al. (2019b) evaluated the effects of graphene oxide (GO) and ZnO-NPs application at two concentrations, i.e., 0.05 mg ml^{-1} and 0.10 mg ml^{-1} on growth traits, chlorophyll, carotenoids, proline contents and disease complex. The plants grown with GO or ZnO-NPs with or without pathogens resulted in a significant increase in plant growth, chlorophyll, carotenoid and proline contents. It was observed that ZnO-NPs application showed more prominent effects than GO-NPs in increasing plant growth, chlorophyll, carotenoid and proline contents. The NPs application inhibited the galling and multiplication of *M. javanica* with ZnO-NPs (0.10 mg ml^{-1}) causing maximum reduction as compared to GO NPs. The disease indices were also found to be reduced when plants were treated with either GO or ZnO-NPs treatments. Thus it is clear from the above discussion that various forms of NPs confer biotic tolerance in different crop plants.

10.4 Alleviative Effect of Different Nanoparticles Under Abiotic Stress Conditions

The use of various forms of NPs has recently emerged as potential signalling agents in diverse crop plants to ameliorate abiotic stresses by modulating cascades of plant processes in one or the other way. On the other hand, various abiotic stress like drought, salt, heavy metal toxicity, ultraviolet radiations (UV) decrease the crop productivity (Ahmad et al. 2019; Zaid et al. 2020, 2021). Thus in order to increase resilience, various forms and doses of NPs are being applied to increase the crop productivity. In a recent study, in order to decrease to toxicity of Cu (100 mg/kg), Faizan et al. (2021) applied the combined applications of ZnO-NPs (foliar) and 24-epibrassinolide (EBL; root dipping) in *Solanum lycopersicum* plants. The plantlets of tomato were submerged in 10^{-8} M of EBL for 2 h, and the Cu was applied at 30 days after sowing. The results revealed that Cu-stress decreased photosynthesis by 17.30%, stomatal conductance by 18.10%, and plant height by 19.70%, and nitrate reductase (NR) activity by 19.20%, but causes significant increments in MDA by 29.40%, superoxide radical content by 22.30% and H_2O_2 by 26.20%. It was however; observed that ZnO-NPs and EBL application improved the photosynthetic activity, stomatal aperture, growth traits, cell viability and antioxidant defense system to ameliorate the Cu induced toxicity. In salinity (250 mM NaCl) stressed safflower plants, the combined application of ZnO-NPs (17 mg/L) and a biofertilizer (BF) induces salt stress resistance by regulating the homeostasis of ions and the activity of antioxidant defence enzymes (Yasmin et al. 2021). Noman et al. (2021) applied biogenic Cu-NPs produced from *Klebsiella pneumoniae* strain *NST2* to ameliorate salt stress toxicity in maize plants. The bacterial strain *NST2*, was genetically identified as *Klebsiella pneumoniae*. The biogenic synthesis of Cu-NPs was confirmed by UV spectroscopy and the material characterization by Fourier transform infrared spectroscopy, X-ray diffractometer, SEM and TEM techniques. It was found that 100 mg/kg Cu-NPs increased the maize root and shoot length by 43.52% and 44.06%, fresh weight by 46.05% and 51.82%, and dry weight 47.69% and 30.63%. Ahmad et al. (2020) applied ZnO-NPs to alleviate As toxicity in soybean plants. The As stress was initiated by applying AsIII, salt sodium arsenite ($NaAsO_2$,) to 60-day-old soybean plants. After 2 weeks of As treatment, three different (0, 50 and 100 mg/L) ZnO-NPs concentrations were sprayed onto the plant foliage. The results showed that As content in plant root and shoot declined with the application of ZnO–NPs supply. In contrast, the shoot and root length, net photosynthetic rate, transpiration, stomatal conductance, photochemical yield which were decremented by As were found to show a considerable increase by ZnO-NPs application. A conspicuous increment in the enzymes of the ascorbate–glutathione cycle including SOD, CAT, APX and GR were observed with the application of ZnO–NPs to the As-stressed plants. Hussain et al. (2021) studied the effects of ZnO-NPs (5, 10, 15, and 20 mg/L) on antioxidants, chlorophyll contents, and proline content in *Persicaria hydropiper* L. grown in Pb-polluted media. It was shown that Pb significantly decreased that seedling growth but caused a significant increase in Pb accumulation in roots, stem and leaves of tested plants. The application of ZnO-NPs alleviated Pb-induced oxidative stress by promoting plant growth, improved chlorophyll and carotenoid contents and enhanced production of free proline, phenolics, flavonoids, and modulation of antioxidative defense enzymes. In contrast the higher concentration of ZnO-NPs (20 mg/L) caused suppression in plant growth, Pb accumulation, secondary metabolites, and antioxidative enzyme activities. In cotton seedlings, ZnO-NPs mediated the Cd and Pb toxicity tolerance (Priyanka et al. 2021). It was revealed that ZnO-NPs under Cd and Pb treatments significantly promoted the shoot, root growth, biomass which were decreased by Cd and Pb exposed treatments alone. The application of ZnO-NPs increased the level of chlorophyll *a*, *b* and carotenoid contents but decreased the MDA contents. The Cd and Pb significantly increased the antioxidant defense enzymes viz., SOD, CAT, POX and APX which were further up-regulated under ZnO-NPs treatments in tested seedlings. It was further observed by the Random amplified polymorphic DNA (RAPD) fingerprinting analysis that no genomic changes/alterations were noticed in when seedlings were applied by co-exposure of ZnO-NPs and Pb and Cd. Rai-Kalal, and Jajoo (2021) studied the priming effect of ZnO-NPs (10 mg/L) in improving wheat germination and photosynthetic performance. The primed wheat seeds with ZnO-NPs showed an increase in seed germination and vigour index, seed water uptake and α-amylase activity, chlorophyll *a*, chlorophyll *b*, total chlorophyll content and chlorophyll *a* fluorescence traits. Moreover, the application of ZnO-NPs increased the water splitting complex at donor side of PSII (Fv/Fo), the numbers of active reaction centres (RC) per molecule of chlorophyll, efficiency of excitation energy trapping (TR) and electron transport (ET) from active RC and the activities of POD, CAT, and SOD but decreased the ROS content. Under drought stressed conditions, Ibrahim, and Neamah (2021) studied the effect of SA (20 mg/L) and Zno-NPs (0, 1.0, 2.0, and 3.0 mg/L) on physiological traits, antioxidant enzyme activity and phenolic compounds in *Cucurbita Pepo* L. The results depicted that fresh, dry weight were increased significantly. The phenolic compounds were also increased after SA and ZnO-NPs treatment under drought stress. Cui et al. (2017) prepared three different sizes (19, 48 and 202 nm) of Si-NPs to alleviate Cd stress in rice by unravelling the molecular mechanisms. The results revealed that that application of three doses of Si-NPs caused a substantial enhancement of live cells by 95.4%, 78.6% and 66.2%, respectively. In the presence of high concentration of Cd, Si-NPs prevented the dramatic damage and severe structural changes as remained nearly intact under exposure to the Si-NPs.With Si-NPs treatment; the noninvasive microtest technology revealed that the doses of Si-NPs (19 nm, 48 nm and 202 nm) decreased the rate of Cd^{2+} influx by 15.7-, 11.1- and 4.6-fold. In addition the gene expression studies showed that Si-NPs inhibited Cd uptake and transport (*OsLCT1 and OsNramp5*), but enhanced Cd transport into vacuole (*OsHMA3*) and Si uptake (*OsLsi1*) suggesting the ameliorative role of Si-NPs under Cd stress. In a pot experiment run at botanical garden of Government College University, Faisalabad, Hussain et al. (2019) applied different levels of Si-NPs (0, 300, 600, 900, 1200 mg/L) for 24 h as seed priming in reducing Cd concentration in wheat plants. It was depicted that Si-NPs increased the wheat growth traits, chlorophyll contents but diminished the oxidative stress and Cd concentrations by 10–52% in shoot, 11–60% in roots, and by 12–75% in wheat grains. In yet another study, Ali et al. (2019) applied different Si-NPs doses (0, 300, 600, 900 and 1200 mg/kg) as soil and foliar treatments in *Triticum aestivum* L. in increasing the growth attributes and reducing the Cd content in wheat. The results showed that soil and foliar Si-NPs application improved the dry biomass of shoots, roots, spikes and grains, leaf gas exchange traits, chlorophyll *a and b* contents, but significantly decreased electrolyte leakage and Cd content in shoots, roots, and grains. Nevertheless an increment in antioxidant enzyme activities was also observed. In order to counteract the Cd and Pb toxicity in rice, Hussain et al. (2020) applied Se and Si-NPs as foliage application. The results showed that Se and Si-NPs (5, 10 and 20 mg/L) caused a significant reduction in Cd and Pb contents and the combined application of Se and Si-NPs was more effective than individual. The effect of foliar applications of Si and TiO_2-NPs (0, 5, 10, 15, and 20 mg/L) on growth, oxidative stress, and Cd accumulation in *Oryza sativa* plants was evaluated (Rizwan et al. 2019). The results demonstrated that the foliar NPs application diminished the Cd accumulation in a dose dependent manner. The plant photosynthesis and gas exchange traits were increased by NPs application. The NPs application also decreased the oxidative stress parameters (MDA and EL); but boosted the activity of antioxidant enzymes. The ration of Cd shoot-to-root also registered a decreasing trend with increasing doses NPs application. In order to study the potentiality of single-walled carbon nanohorns (SWCNHs) and ZnO-NPs by comparative physiological and metabolomics analysis in *Sophora alopecuroides* under salt stress tolerance, Wan et al. (2020) found that SWCNHs incremented root length, root fresh weight and leaf soluble sugar content, while ZnO-NPs treatment increased the root fresh weight, leaf dry weight and soluble sugar content. The application of SWCNHs result an increase the PSII activity, total protein content in leaves and roots, leaf soluble sugar content and Cu content in the leaves under salt stress. Simultaneously, under salt stress, ZnO-NPs supply increased plantlet height, root fresh weight, leaf total protein content, soluble sugar contents of leaves and roots, leaf Zn and root Cu content. Furthermore as revealed by the metabolome analysis both SWCNHs and ZnO-NPs under salt stress modulated the metabolism of the carbon/nitrogen by promoting glycolysis and the Kreb cycle to generate energy and incrementing unsaturated fatty acids levels for maintaining the integrity of membranes. It was shown that ZnO-NPs triggered the accumulation of metabolites in leaves and roots by 4.83-fold and 3.7-fold more than those in SWCNH-treated plants under salt stress. Under drought stress conditions, the soybean plants grown with Fe, Cu, Co, and ZnO-NPs were analyzed (Linh et al. 2020). The results demonstrated that Fe, Cu, Co, and ZnO-NPs led to an improvement in shoot and root morphology and drought stress tolerance by enhancing the significant expressions of drought stress tolerance related marker genes under water deficit conditions. Kardavan Ghabel and Karamian (2020) evaluated the effects of TiO_2-NPs (0, 2 and 5 ppm) and spermine (at 0 and 1 mM) under cold stress (4 °C) in a factorial and completely randomized design experiment with three replications. The results depicted that all the growth parameters photosynthetic pigments were found to be decreased under cold stress. The application of TiO_2-NPs and spermine decreased the MDA and H_2O_2 contents but increased phenolics, total protein and osmolytes. El-Saadony et al. (2021) applied Se-NPs synthesized by *Lactobacillus acidophilus* ML14 in controlling the diseases of crown root and rot in *Triticum aestivum* L. induced by *Fusarium culmorum* and *Fusarium graminearum* in improving the yield under concurrent drought and heat stress conditions. The obtained results showed that Se-NPs (100 μg/mL) significantly scavenged radicals and caused an inhibition in the growth of fungi, enhanced plant growth, grain quantity and quality, photosynthetic pigment contents and gas exchange parameters. The potentiality of plant growth-promoting rhizobacteria (PGPR) and Si-NPs in curtailing the adversely impacts of water deficit and salt stress on *Zea mays* L. growth and productivity was studied (Hafez et al. 2021). It was found that the application of PGPR under salt stress improved the activity of dehydrogenase and alkaline phosphatase and soil physicochemical characteristics which improves yield-related traits and maize productivity. The foliar spraying of Si-NP under salt-affected soil was found to decline the oxidative stress traits (reduced MDA and EL) by boosting the activities of CAT, SOD, and POD, increased K^+/Na^+ ratio, net photosynthetic rate, relative water content, photosynthetic pigment contents, and stomatal conductance, along with less Na^+ content. The combined treatment of Si-NPs with PGPR induced greater improvement in maize productivity and the contents of N, P, and K.

10.5 Conclusion

It is clear from the above discussion that various doses and forms of NPs alleviate the biotic and abiotic stress in various crop plants. The application of NPs modulates various plant processes including photosynthesis, respiration and antioxidant defense systems of plants. It is also known that NPs modulate the expression of various genes involved in conferring stress tolerance in plants by controlling principal plant processes.

References

Ahmad B, Zaid A, Sadiq Y, Bashir S, Wani SH (2019) Role of selective exogenous elicitors in plant responses to abiotic stress tolerance. In: Plant abiotic stress tolerance. Springer, Cham, pp 273–290

Ahmad P, Alyemeni MN, Al-Huqail AA, Alqahtani MA, Wijaya L, Ashraf M et al (2020) Zinc oxide nanoparticles application alleviates arsenic (As) toxicity in soybean plants by restricting the uptake of As and modulating key biochemical attributes, antioxidant enzymes, ascorbate-glutathione cycle and glyoxalase system. Plan Theory 9(7):825

Alhaithloul HAS, Abu-Elsaoud AM, Soliman MH (2020) Abiotic stress tolerance in crop plants: role of phytohormones. In: Abiotic stress in plants. IntechOpen, Rijeka

Ali S, Rizwan M, Hussain A, ur Rehman MZ, Ali B, Yousaf B et al (2019) Silicon nanoparticles enhanced the growth and reduced the cadmium accumulation in grains of wheat (Triticum aestivum L.). Plant Physiol Biochem 140:1–8

Ali EF, El-Shehawi AM, Ibrahim OHM, Abdul-Hafeez EY, Moussa MM, Hassan FAS (2021) A vital role of chitosan nanoparticles in improvisation the drought stress tolerance in *Catharanthus roseus* (L.) through biochemical and gene expression modulation. Plant Physiol Biochem 161:166–175

Asgari-Targhi G, Iranbakhsh A, Ardebili ZO (2018) Potential benefts and phytotoxicity of bulk and nano-chitosan on the growth, morphogenesis, physiology, and micropropagation of Capsicum annuum. Plant Physiol Biochem 127:393–402

Bashir A, ur Rehman MZ, Hussaini KM, Adrees M, Qayyum MF, Sayal AU et al (2021) Combined use of zinc nanoparticles and co-composted biochar enhanced wheat growth and decreased Cd concentration in grains under Cd and drought stress: a field study. Environ Technol Innov 23:101518

Boudsocq M, Sheen J (2013) CDPKs in immune and stress signaling. Trends Plant Sci 18(1):30–40

Chandrakar V, Yadu B, Korram J, Satnami ML, Dubey A, Kumar M, Keshavkant S (2020) Carbon dot induces tolerance to arsenic by regulating arsenic uptake, reactive oxygen species detoxification and defense-related gene expression in Cicer arietinum L. Plant Physiol Biochem 156:78–86

Cui J, Liu T, Li F, Yi J, Liu C, Yu H (2017) Silica nanoparticles alleviate cadmium toxicity in rice cells: mechanisms and size effects. Environ Pollut 228:363–369

Danish M, Altaf M, Robab MI, Shahid M, Manoharadas S, Hussain SA, Shaikh H (2021) Green synthesized silver nanoparticles mitigate biotic stress induced by Meloidogyne incognita in Trachyspermum ammi (L.) by improving growth, biochemical, and antioxidant enzyme activities. ACS Omega 6:11389–11403

Delk NA, Johnson KA, Chowdhury NI, Braam J (2005) CML24, regulated in expression by diverse stimuli, encodes a potential Ca^{2+} sensor that functions in responses to abscisic acid, daylength, and ion stress. Plant Physiol 139(1):240–253

El-Saadony MT, Saad AM, Najjar AA, Alzahrani SO, Alkhatib FM, Shafi ME, Hassan MA (2021) nanoparticles in controlling Triticum aestivum L. crown root and rot diseases induced by fusarium species and improve yield under drought and heat stress. Saudi J of Biol Sci. https://doi.org/10.1016/j.sjbs.2021.04.043

Elsharkawy M, Derbalah A, Hamza A, El-Shaer A (2020) Zinc oxide nanostructures as a control strategy of bacterial speck of tomato caused by *Pseudomonas syringae* in Egypt. Environ Sci Pollut Res 27(16):19049–19057

Emamverdian A, Ding Y, Mokhberdoran F, Xie Y, Zheng X, Wang Y (2020) Silicon dioxide nanoparticles improve plant growth by enhancing antioxidant enzyme capacity in bamboo (Pleioblastus pygmaeus) under lead toxicity. Trees 34(2):469–481

Faizan M, Faraz A, Mir AR, Hayat S (2020) Role of zinc oxide nanoparticles in countering negative effects generated by cadmium in *Lycopersicon esculentum*. J Plant Growth Regulat 40:1–15

Faizan M, Bhat JA, Noureldeen A, Ahmad P, Yu F (2021) Zinc oxide nanoparticles and 24-epibrassinolide alleviates Cu toxicity in tomato by regulating ROS scavenging, stomatal movement and photosynthesis. Ecotoxicol Environ Saf 218:112293

Ghazaei F, Shariati M (2020) Effects of titanium nanoparticles on the photosynthesis, respiration, and physiological parameters in Dunaliella salina and Dunaliella tertiolecta. Protoplasma 257(1):75–88

Gohari G, Mohammadi A, Akbari A, Panahirad S, Dadpour MR, Fotopoulos V, Kimura S (2020) Titanium dioxide nanoparticles (TiO 2 NPs) promote growth and ameliorate salinity stress effects on essential oil profile and biochemical attributes of Dracocephalum moldavica. Sci Rep 10(1):1–14

Hafez EM, Osman HS, Gowayed SM, Okasha SA, Omara AED, Sami R et al (2021) Minimizing the adversely impacts of water deficit and soil salinity on maize growth and productivity in response to the application of plant growth-promoting rhizobacteria and silica nanoparticles. Agronomy 11(4):676

He L, Liu Y, Mustapha A, Lin M (2011) Antifungal activity of zinc oxide nanoparticles against Botrytis cinerea and Penicillium expansum. Microbiol Res 166(3):207–215

Hedayati A, Hosseini B, Palazon J, Maleki R (2020) Improved tropane alkaloid production and changes in gene expression in hairy root cultures of two Hyoscyamus species elicited by silicon dioxide nanoparticles. Plant Physiol Biochem 155:416–428

Hussain A, Rizwan M, Ali Q, Ali S (2019) Seed priming with silicon nanoparticles improved the biomass and yield while reduced the oxidative stress and cadmium concentration in wheat grains. Environ Sci Pollut Res 26(8):7579–7588

Hussain B, Lin Q, Hamid Y, Sanaullah M, Di L, Khan MB et al (2020) Foliage application of selenium and silicon nanoparticles alleviates Cd and Pb toxicity in rice (Oryza sativa L.). Sci Total Environ 712:136497

Hussain F, Hadi F, Rongliang Q (2021) Effects of zinc oxide nanoparticles on antioxidants, chlorophyll contents, and proline in Persicaria hydropiper L. and its potential for Pb phytoremediation. Environ Sci Pollut Res 28:1–17

Ibrahim LD, Neamah SI (2021) Influnce of salicylic acid and Zno-Nps on physiological triats, antioxidant enzymes and phenolic compounds production in a Cucurbita pepo L. callus culture under normal and drought stress. Ann Rom Soc Cell Biol 25:4056–4068

Jeevanandam J, Barhoum A, Chan YS, Dufresne A, Danquah MK (2018) Review on nanoparticles and nanostructured materials: history, sources, toxicity and regulations. Beilstein J Nanotechnol 9:1050–1074

Jo YK, Kim BH, Jung G (2009) Antifungal activity of silver ions and nanoparticles on phytopathogenic fungi. Plant Dis 93(10):1037–1043

Kardavan Ghabel V, Karamian R (2020) Effects of TiO2 nanoparticles and spermine on antioxidant responses of Glycyrrhiza glabra L. to cold stress. Acta Bot Croat 79(2):137–114

Khan M, Siddiqui ZA (2018) Zinc oxide nanoparticles for the management of Ralstonia solanacearum, Phomopsis vexans and Meloidogyne incognita incited disease complex of eggplant. Indian Phytopathol 71(3):355–364

Khan MR, Siddiqui ZA (2020) Use of silicon dioxide nanoparticles for the management of *Meloidogyne incognita*, *Pectobacterium betavasculorum* and *Rhizoctonia solani* disease complex of beetroot (Beta vulgaris L.). Sci Hortic 265:109211

Khan MR, Siddiqui ZA (2021) Role of zinc oxide nanoparticles in the management of disease complex of beetroot (Beta vulgaris L.) caused by Pectobacterium betavasculorum, Meloidogyne incognita and Rhizoctonia solani. Hortic Environ Biotechnol 62(2):225–241

Khattab H, Alatawi A, Abdulmajeed A, Emam M, Hassan H (2021) Roles of Si and SiNPs in improving thermotolerance of wheat photosynthetic machinery via upregulation of PsbH, PsbB and PsbD genes encoding PSII core proteins. Horticulturae 7(2):16

Kumar V, Yadav SK (2009) Plant-mediated synthesis of silver and gold nanoparticles and their applications. J Chem Technol Biotechnol 84:151–157

Kumar P, Pahal V, Gupta A, Vadhan R, Chandra H, Dubey RC (2020) Effect of silver nanoparticles and Bacillus cereus LPR2 on the growth of Zea mays. Sci Rep 10(1):1–10
Linh TM, Mai NC, Hoe PT, Lien LQ, Ban NK, Hien LTT, Van NT (2020) Metal-based nanoparticles enhance drought tolerance in soybean. J Nanomater. https://doi.org/10.1155/2020/4056563
Lopez-Lima D, Mtz-Enriquez AI, Carrión G, Basurto-Cereceda S, Pariona N (2021) The bifunctional role of copper nanoparticles in tomato: effective treatment for fusarium wilt and plant growth promoter. Sci Hortic 277:109810
Manzoor N, Ahmed T, Noman M, Shahid M, Nazir MM, Ali L et al (2021) Iron oxide nanoparticles ameliorated the cadmium and salinity stresses in wheat plants, facilitating photosynthetic pigments and restricting cadmium uptake. Sci Total Environ 769:145221
Marmiroli M, Imperiale D, Pagano L, Villani M, Zappettini A, Marmiroli N (2015) The proteomic response of Arabidopsis thaliana to cadmium sulfide quantum dots, and its correlation with the transcriptomic response. Front Plant Sci 6:1104
Miao Y, Xu J, Shen Y, Chen L, Bian Y, Hu Y et al (2014) Nanoparticle as signaling protein mimic: robust structural and functional modulation of CaMKII upon specific binding to fullerene C60 nanocrystals. ACS Nano 8(6):6131–6144
Mosa KA, El-Naggar M, Ramamoorthy K, Alawadhi H, Elnaggar A, Wartanian S, Ibrahim E, Hani H (2018) Copper nanoparticles induced genotoxicty, oxidative stress, and changes in superoxide dismutase (SOD) gene expression in cucumber (Cucumis sativus) plants. Frontiers in plant science, 9:872
Noman M, Ahmed T, Shahid M, Niazi MBK, Qasim M, Kouadri F et al (2021) Biogenic copper nanoparticles produced by using the Klebsiella pneumoniae strain NST2 curtailed salt stress effects in maize by modulating the cellular oxidative repair mechanisms. Ecotoxicol Environ Saf 217:112264
Priyanka N, Geetha N, Manish T, Sahi SV, Venkatachalam P (2021) Zinc oxide nanocatalyst mediates cadmium and lead toxicity tolerance mechanism by differential regulation of photosynthetic machinery and antioxidant enzymes level in cotton seedlings. Toxicol Rep 8:295–302
Qin F, Shinozaki K, Yamaguchi-Shinozaki K (2011) Achievements and challenges in understanding plant abiotic stress responses and tolerance. Plant Cell Physiol 52:1569–1582
Rai-Kalal P, Jajoo A (2021) Priming with zinc oxide nanoparticles improve germination and photosynthetic performance in wheat. Plant Physiol Biochem 160:341–351
Rajput VD, Minkina TM, Behal A, Sushkova SN, Mandzhieva S, Singh R, Gorovtsov A, Tsitsuashvili VS, Purvis WO, Ghazaryan KA, Movsesyan HS (2018) Effects of zinc-oxide nanoparticles on soil, plants, animals and soil organisms: a review. Environ Nanotechnol Monitor Manag 9:76–84
Rejeb IB, Pastor V, Mauch-Mani B (2014) Plant responses to simultaneous biotic and abiotic stress: molecular mechanisms. Plan Theory 3(4):458–475
Rizwan M, Ali S, ur Rehman MZ, Malik S, Adrees M, Qayyum MF et al (2019) Effect of foliar applications of silicon and titanium dioxide nanoparticles on growth, oxidative stress, and cadmium accumulation by rice (Oryza sativa). Acta Physiol Plant 41(3):1–12
Roberts DP, Mattoo AK (2018) Sustainable agriculture—enhancing environmental benefits, food nutritional quality and building crop resilience to abiotic and biotic stresses. Agriculture 8(1):8
Shabnam N, Kim M, Kim H (2019) Iron (III) oxide nanoparticles alleviate arsenic induced stunting in vigna radiata. Ecotoxicol Environ Saf 183:109496
Siddiqui ZA, Khan MR, Abd_Allah EF, Parveen A (2019a) Titanium dioxide and zinc oxide nanoparticles affect some bacterial diseases, and growth and physiological changes of beetroot. Int J Veg Sci 25(5):409–430
Siddiqui ZA, Parveen A, Ahmad L, Hashem A (2019b) Effects of graphene oxide and zinc oxide nanoparticles on growth, chlorophyll, carotenoids, proline contents and diseases of carrot. Sci Hortic 249:374–382
Spanò C, Bottega S, Sorce C, Bartoli G, Castiglione MR (2019) TiO 2 nanoparticles may alleviate cadmium toxicity in co-treatment experiments on the model hydrophyte Azolla filiculoides. Environ Sci Pollut Res 26(29):29872–29882

Swamy VS, Prasad R (2012) Green synthesis of silver nanoparticles from the leaf extract of *Santalum album* and its antimicrobial activity. J Optoelectron Biomed Mater 4(3):53–59

Tauseef A, Khalilullah A, Uddin I (2021) Role of MgO nanoparticles in the suppression of Meloidogyne incognita, infecting cowpea and improvement in plant growth and physiology. Exp Parasitol 220:108045

Tilman D, Balzer C, Hill J, Befort BL (2011) Global food demand and the sustainable intensification of agriculture. Proc Natl Acad Sci 108:20260–20264

Tombuloglu H, Slimani Y, Tombuloglu G, Alshammari T, Almessiere M, Korkmaz AD, Samia ACS (2020) Engineered magnetic nanoparticles enhance chlorophyll content and growth of barley through the induction of photosystem genes. Environ Sci Pollut Res 27(27):34311–34321

Tuteja N, Mahajan S (2007) Calcium signaling network in plants: an overview. Plant Signal Behav 2(2):79–85

Verstraeten S, Aimo L, Oteiza P (2008) Aluminium and lead: molecular mechanisms of brain toxicity. Arch Toxicol 82:789–802

Wan J, Wang R, Bai H, Wang Y, Xu J (2020) Comparative physiological and metabolomics analysis reveals that single-walled carbon nanohorns and ZnO nanoparticles affect salt tolerance in Sophora alopecuroides. Environ Sci Nano 7(10):2968–2981

Yasmin H, Mazher J, Azmat A, Nosheen A, Naz R, Hassan MN et al (2021) Combined application of zinc oxide nanoparticles and biofertilizer to induce salt resistance in safflower by regulating ion homeostasis and antioxidant defence responses. Ecotoxicol Environ Saf 218:112262

Zahedi SM, Moharrami F, Sarikhani S, Padervand M (2020) Selenium and silica nanostructure-based recovery of strawberry plants subjected to drought stress. Sci Rep 10(1):1–18

Zaid A, Ahmad B, Jaleel H, Wani SH, Hasanuzzaman M (2020) A critical review on iron toxicity and tolerance in plants: role of exogenous phytoprotectants. In: Plant Micronutrients. Springer, Cham, pp 83–99

Zaid A, Ahmad B, Wani SH (2021) Medicinal and aromatic plants under abiotic stress: A cross-talk on phytohormones' perspective. In: Aftab T, Hakeem KR (eds) Plant growth regulators. Springer, Cham. https://doi.org/10.1007/978-3-030-61153-8_5

Zhao G, Zhao Y, Lou W, Su J, Wei S, Yang X et al (2019) Nitrate reductase-dependent nitric oxide is crucial for multi-walled carbon nanotube-induced plant tolerance against salinity. Nanoscale 11(21):10511–10523

Chapter 11
Nanotechnological Approaches for Efficient Delivery of Plant Ingredients

Maysoun M. Saleh, Amira S. Mahmoud, Heba S. Abbas, Farrag F. B. Abu-Ellail, Muddukrishnaiah Kotakonda, and Khaled F. M. Salem

Abstract Nanotechnology is a multidisciplinary study field. Through numerous nanotechnology research, tremendous efforts have recently been undertaken to boost the yield of agricultural output. Excessive use of pesticides and chemical fertilizers was a result of the Green revolution, which resulted in the loss of agricultural soil biodiversity and the development of resistance to pathogens, whether diseases or pests. Only nanoparticles or nanochips can give ingredients to plants via nanoparticles or improved biosensors for precision cultivation. Fertilizers and pesticides, whether traditional nano-encapsulated weeds or insecticides, aid the slow and long-term release of nutrients and agricultural pesticides, giving plants a precise amount. Nanotechnology-based plant viral disease detection tools are also gaining in popu-

---

M. M. Saleh (✉)
Genetic Resources Department, General Commission for Scientific Agricultural Research (GCSAR), Damascus, Syria
e-mail: mzainsamasaleh@gcsar.gov.sy

A. S. Mahmoud
Department of Environmental Studies, Institute of Graduate Studies and Research, Alexandria University, Alexandria, Egypt

H. S. Abbas
National Organization for Drug Control and research, Egyptian Drug Authority, Cairo, Egypt

F. F. B. Abu-Ellail
Department of Breeding and Genetics, Sugar Crops Research Institute, Agricultural Research Center (ARC), Giza, Egypt

M. Kotakonda
Department of Technology, Anna University, Chennai, India

Centre for Biotechnology and Phyto Pharmacognosy Research, Coimbatore, India

K. F. M. Salem
Department of Plant Biotechnology, Genetic Engineering and Biotechnology Research Institute (GEBRI), University of Sadat City, Sadat City, Egypt

Department of Biology, College of Science and Humanitarian Studies, Shaqra University, Qwaieah, Saudi Arabia
e-mail: khaled.salem@gebri.usc.edu.eg; khaledfathi@su.edu.sa

M. Faizan et al. (eds.), *Sustainable Agriculture Reviews 53*, Sustainable Agriculture Reviews 53, https://doi.org/10.1007/978-3-030-86876-5_11

larity, which is useful for the rapid and early detection of viral infections. This chapter presents an overview of the importance of nanotechnology in agriculture, uptake and translocation system, nano-based for ingredients delivery, silica-based nanosystem for gene delivery, nano-based for fertilizers and pesticides delivery.

Keywords Agriculture · Crop improvement · Ingredients delivery · Fertilizers delivery · Nanotechnology · Nanoparticle uptake · Pesticides delivery · Gene delivery

11.1 Introduction

11.1.1 Importance of Nanotechnology in Agriculture

Nanotechnology is a cutting-edge field of study that enables the manufacture of a diverse range of materials, including particulate materials with at least one dimension of fewer than 100 nanometers (nm). Nanomaterials (NMs) are chemical substances or materials produced and used on a very small scale, usually 1–100 nm in at least one dimension (Gleiter 2000). We are living in the 'Nano Age,' a period in which nanomaterials pervade every aspect of human life, nanomaterials are already in, on, and around us. whether it is in the cosmetics we use, the textiles we wear, the machines we use, the food we consume (Kim et al. 2017). Nanotechnology has impacted every aspect of life and every scientific field, including biology, chemistry, physics, engineering, and medicine. Nanoparticles, nanocoating, nanosheets and nanoclusters are examples of nanomaterials (Pokropivny and Skorokhod 2007).

There are several methods to produce Nanomaterials as Condensation, attrition, chemical precipitation, ion implantation, pyrolysis, and hydrothermal synthesis (Pokropivny and Skorokhod 2007). Because of tunable chemical, physical, and mechanical properties of nanomaterials (NMs) becoming increasingly important in technological applications like water treatment plants, oil refineries, petrochemical industries, manufacturing processes, catalytic processes, buildings and construction materials, diagnostics and drug delivery (Saleh 2020). Also, nanomaterials are used in a wide variety of commercial products in agriculture as fertilizers, herbicides, pesticides, fungicides, and nano sensors. This will aid in meeting potential agricultural demands by improving crop quality and yield, reducing contamination caused by biochemicals and protecting crops from environmental stresses.

Agricultural nanomaterials have an important role in reducing chemical distribution, reducing fertilizer nitrogen loss, and increasing productivity through pest and nutrient management. Nanotechnology aims to enhance the agricultural and food sectors by developing new nanotools for disease diagnosis and improving the ability of plants to absorb nutrients, among other things (Prasad et al. 2017). Agriculture's potential will be enhanced by nanobiotechnology. Feedstocks for industrial processes are harvested. In the meantime, tropical Rubber, cocoa, coffee, and other agricultural products Cotton, as well as the small-scale farmers who cultivate it, will prosper. in a modern nano economy, they are quaint and insignificant "Flexible matter" is a term used to describe material with industrial properties. Nanoparticles can be tweaked to create different effects (Manjunatha et al. 2016).

The positive applications (hormesis) of plant–NP interaction outcomes are depicted in Fig. 11.1 (Rai et al. 2018). The broad applications of plant–NP interactions in environmental remediation and bioenergy or biofuel production were listed in this diagram (Babadi et al. 2016; Dash et al. 2014; Makarov et al. 2014; Sharma et al. 2007). Because of their non-toxic nature due to the stabilizing and capping effects of plant metabolites (Daisy and Saipriya 2012). Photosynthesized NPs have a lot of applications in nanomedicine and biomedicine. The graphic also depicts the use of nanoparticles in agriculture, including nano fertilizers, nano pesticides, nano priming and water or other resource management via smart delivery systems (Jayaseelan et al. 2011; Kah and Hofmann 2014; Naderi and Danesh-Shahraki 2013; Parisi et al. 2015; Salah et al. 2015). Crops in the agro-ecosystem are sustained by photosynthesized NPs, which help the agriculture industry achieve environmental sustainability (El-Ramady et al. 2016; Shanmugam et al. 2016; Zheng et al. 2005). Biogenic NPs, such as Ag NPs are also employed in food packaging due to their antibacterial properties, which prevent food spoiling (Ahmed et al. 2016; Fayaz et al. 2009). In addition, biosensors that take advantage of plant–NP interactions have applications in a variety of fields (Fig. 11.1) (Li et al. 2013; Zhang

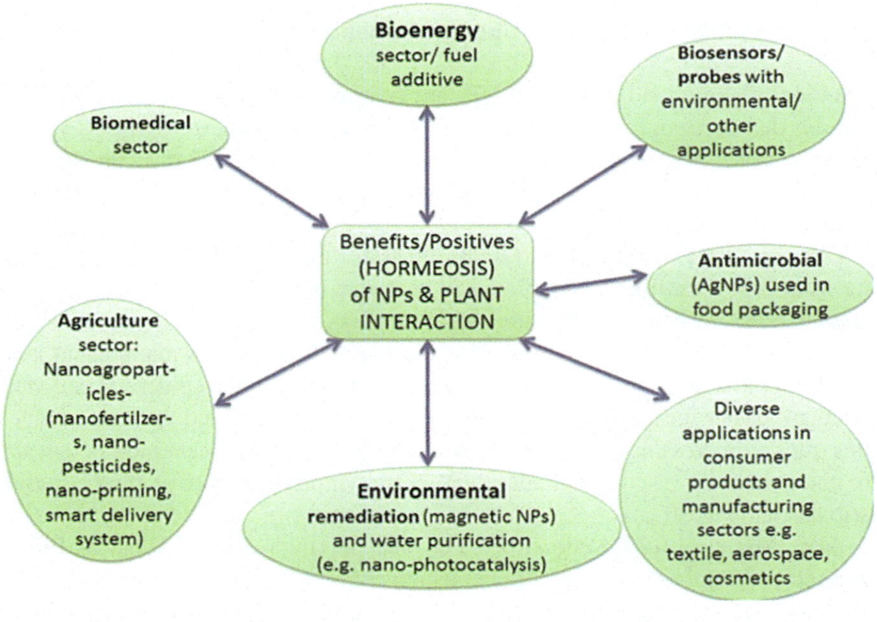

Fig. 11.1 Plant and NP interactions and the resulting applications in diverse energy, environmental, and agricultural sectors. (Source: Rai et al. 2018)

et al. 2012). Nonetheless, the widespread use of NPs/NMs has raised worries about the environment and human health.

11.1.2 Uptake and Translocation System

Plants provide a potential route for nanoparticles to enter the environment and serve as a major source of bioaccumulation in the food chain. Two decades ago, research into the uptake of engineered nanoparticles (ENMs) by plants was motivated primarily by an interest in their potential as gene transfer vectors (Fisk and Dandekar 1993; Torney et al. 2007). Ali et al. 2021 studied the effects of different types of nanoparticles (NPs) on plants to see whether they could improve seed germination, biomass, or grain yield. The plant resistance to biotic and abiotic stresses was also improved by the NPs. The biological functions of the plant are determined by events at the molecular level. However, despite the effect of the nanoparticles, little progress has been made at the molecular stage. The biological functions of the plant are determined by events at the molecular level. However, little progress has been made at the molecular level, which is a crucial step in determining possible mechanisms and plant effects. As a result, it's critical to comprehend plants' underlying mechanisms and responses to nanoparticles, as well as gene expression changes. By 2050, the planet will need to produce 50% more food to meet the needs of 9 billion people.

11.1.2.1 Plants' Nanoparticle Uptake Mechanisms

The bioavailability and toxicity of NPs are determined by a series of bio/transformations that occur in the soil. After communicating with plant roots, the NPs translocate to aerial portions and accumulate in cellular or subcellular organelles. The first step in bioaccumulation is the adsorption of NPs from the soil by plant roots (Nair et al. 2010).

Du et al. (2011) and Lin et al. (2008) illustrated that small NPs (diameters ranging from 3 to 5 nm) have been found to permeate plant roots via several ways as osmotic pressure, capillary pressures or direct passage via root epidermal cells. The root cell wall epidermal cells are semipermeable, with tiny pores that prevent large NPs from passing through. Some NPs caused new pores in the epidermal cell wall, making it easier for them to enter. NPs are apoplastically carried through extracellular gaps after crossing cell walls, allowing the xylem to flow linear motion upward. To enter the core vascular cylinder, NPs must symplastically cross the Casparian strip barrier (Fig. 11.2).

Also, Mei et al. (2021) described Per- and polyfluoroalkyl substances PFASs (per and polyfluoroalkyl l compounds) which common in the environment, but can still harm ecosystems and human health. In the bioaccumulation of PFASs, the soil-plant system plays a critical role. Figure 11.3 depicts the uptake and accumulation of PFASs chemicals that can flow from the soil solution to the vascular tissues of

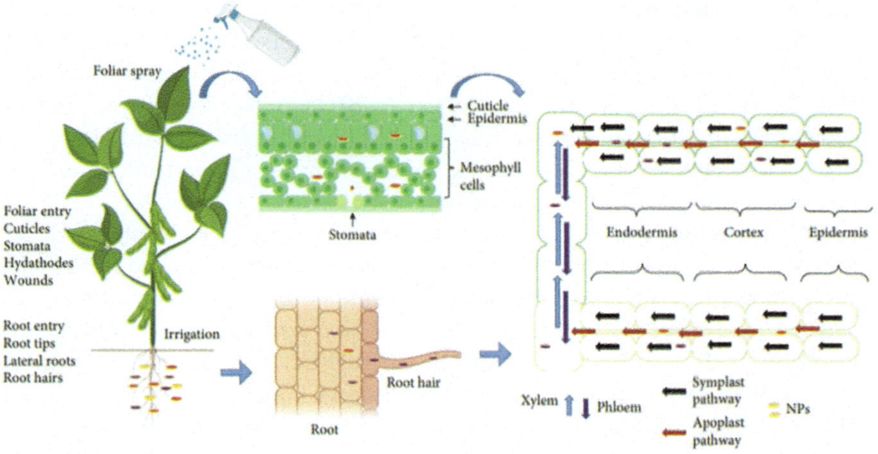

Fig. 11.2 A diagram depicting nanoparticle uptake through various routes and their translocation pathways in various plant sections. (Source: Ali et al. 2021)

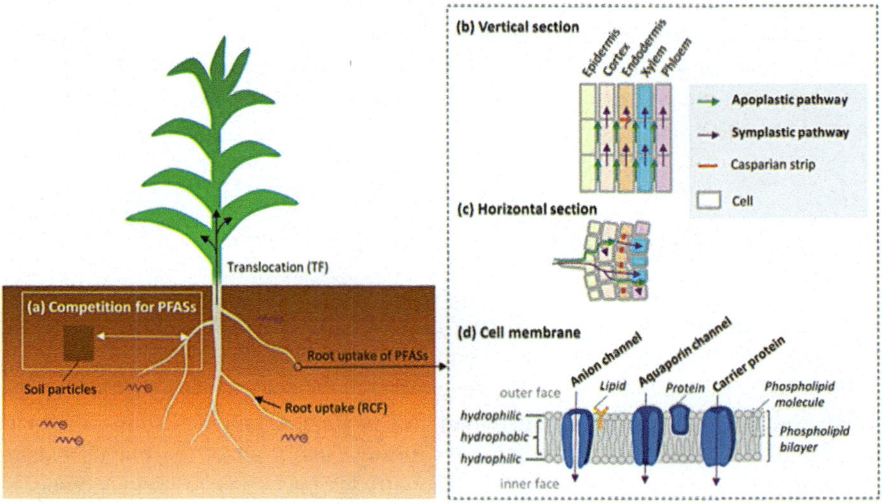

Fig. 11.3 Possible uptake and accumulation processes of Per- and polyfluoroalkyl substances (PFASs) by a plant grown in the soil environment (Modified from Blaine et al. 2014; Wang et al. 2020a, b)

plant roots via apoplastic (cell to cell) pathways, symplastic (plasmodesmata to plasmodesmata) pathways, or transmembrane (cell to the cell membrane) pathways (Miller et al. 2016).

When nanoparticles come into contact with plants, they enter the plant cell wall and the cell membrane of the root epidermis, triggering a complex series of events that allow them to invade the plant vascular bundle (xylem) and more simply, transfer to the stele, from which they can be transferred to leaves. Furthermore, NPs pass via apertures or holes in the cell membrane to traverse the unbroken cell membrane, indicating that nanomaterial absorption is size-specific (Banerjee et al. 2019; Ma et al. 2010; Rico et al. 2011; Tripathi et al. 2017). Before adhering to the stele, nanoparticles must be mixed inactively over the endodermis' apoplast (Judy et al. 2012). Aslani et al. (2014) identified the xylem as the most important carrier in the dissemination and transfer of nanoparticles.

Furthermore, Hu et al. (2021) investigated the effects of the chemical structure of organophosphate esters (OPEs), plant cultivar and copper (Cu) on OPE uptake and translocation by plants. OPE bioaccumulation differed between plant varieties. They were primarily enhanced in carrots, with maize having the lowest quantities. OPEs with electron-ring substituents (EROPEs) have a larger root uptake potential than OPEs with open-chain substituents (OC-OPEs), possibly due to ER-OPE stronger sorption onto root charged surfaces. The increased noncovalent interactions with the electron-rich structure of ER-OPEs explained this. The inclusion of Cu minimized the significant difference in root ability to absorb OC-OPEs and ER-OPEs. The interaction of Cu ions with the electron-rich structure of ER-OPEs reduced the sorption of ER-OPEs on the root surface, which was explained. OPEs + Cu, implying that hydrophobicity plays a substantial role in OPEs acropetal transfer. The findings will help us better understand OPE uptake and translocation by plant cultivars, as well as how the chemical structure of OPEs and Cu affects the process, resulting in better ecological risk assessments of OPEs in the food chain.

11.1.3 Barriers of Plant Delivery System

The plant cell wall establishes a barrier that prevents any foreign agent, including nanoparticles, from easily entering the plant system. The pore diameter of the cell wall, which ranges from 5 to 20nm, determines the sieving capabilities within the plant system. As a result, only nanoparticles or their aggregates with a diameter smaller than the pore diameter can pass through and contact the plant's plasma membrane (Navarro et al. 2008). The nanoparticles may cling to various cytoplasmic organelles, obstructing metabolic processes at that location (Jia et al. 2005). Nanoparticles on leaf surfaces penetrate through stomatal apertures and then translocate to numerous plant organs and tissues (Uzu et al. 2010). However, increasing the concentration of nanoparticles on the photosynthetic surface produces foliar heating, which leads to changes in gas exchange due to stomatal obstruction, resulting in changes in many plant physiological processes and cellular functions (Banerjee et al. 2019; Da Silva et al. 2006).

While nanomaterials can enter living plant tissues, their accumulation in the environment and efficiency as smart delivery systems in living plants have consequences. It is crucial to figure out if plants can eat intact NPs and transport them to other parts of the plant. NPs may penetrate plant tissue through root or above-ground tissues, as well as through the root junction (Chaudhary et al. 2018; Duhan et al. 2017; Verma et al. 2018).

Agriculture was made possible by technological advances, as is well recognized. Agriculture, on the other hand, allows for more population growth. People come into contact with one another, and they share various experiences in the production of agricultural products, life experiences, and other things, resulting in a greater variety of foodstuffs to choose from scale-up to delivery at field scale. At the field level, data are scarce on delivery techniques. Soil and foliar applications currently in use Conventional agrochemicals approach yields less than half of the results. agrochemicals achieving their objectives that is the root, leaf or target pest.

Increasing absorption efficiency and allowing targeting to certain plant cell compartments and organelles, such as chloroplasts, mitochondria and the nucleus may be possible by adjusting charge and size or coating nanomaterials with guiding biomolecules. This will improve plant protection and nutrition delivery while also conserving embodied resources (such as raw materials and processing energy and water) that are currently lost in the agrochemicals. Additionally, it may be able to reduce ammonia, nitrogen dioxide, and the greenhouse gas N_2O emissions while increasing nitrogen delivery efficiency to the crops. Highly efficient foliar distribution and retention could be achieved with formulations that improve leaf adherence and precise spraying. In the lab, nanomaterials developed for soil applications have been explored, but more research is needed to determine the amounts required for a favourable response. Most nanomaterials for crop growth have too high an economic and embodied resource cost to be viable in the field at current soil application rates, as employed for conventional agrochemicals. Technology readiness and overcoming challenges to implementing nanotechnology-enabled plant agriculture on a long-term basis (Hofmann et al. 2020).

A greater mechanistic understanding of how applied nanoparticles are taken up from a leaf-applied suspension might hasten the development of nano-enabled herbicides and fertilizers and their commercial penetration. High material production costs may still stymie foliar application deployment. Kocide 3000 is a common copper-based nanomaterial fungicide applied as an aqueous suspension onto plant leaves. There are just a handful of other field-scale trials where experimental nanomaterials are sprayed onto plants at the moment (Anusuya et al. 2015; Prasad et al. 2012). Additionally, Hydrophilic particles can pass through the apertures of stomata in leaves, according to Eichert and Goldbach (2008), if their diameters are more than or equal to 40 nm. Furthermore, when designed nanoparticles are given leaves, they collect in the stomata rather than the vascular bundle and are then distributed to other areas via phloem.

11.2 Nano-Based for Ingredients Delivery

Smart delivery system depending on nanomaterials seems to be a promise solution to avoid the huge lose of nutrients and micronutrients or immune ingredients from fertilizers or pesticides in plants (Arias-Estévez et al. 2008) (Fig. 11.4), since it has the capability to assure the effective deliverance of specific ingredients to the targeted cells in plants mainly because of the proper size of nano-materials (Nair et al. 2010). Fraceto et al. (2016) mentioned that nanomaterials can save time in delivering several ingredients in plants.

11.2.1 Nutrients

Depending on the science of nanotechnology, several manufactured applications for the precise delivery of ingredients such as nutrients were used in plants (Alam et al. 2016; Sertova 2015; Dhewa 2015; NRC 2008) (Fig. 11.5). Gogos et al. (2012) declared that the active delivery of nutrients can be achieved by using nanomaterials. According to Boverhof et al. (2015), one dimension of the nanomaterials at least, is measured by the nano-meter scale, which is appropriate to make NMs play the role of agro-chemicals carriers such as pesticides or fertilizers to deliver nutrients in plants resulting in enhancing yield and immune system in plants. One example of nutrients delivery system, is the nanoparticles and nano-scale fertilizers (De Rosa et al. 2010), as the outside nanomaterials surface layer of nano-scale fertilizers

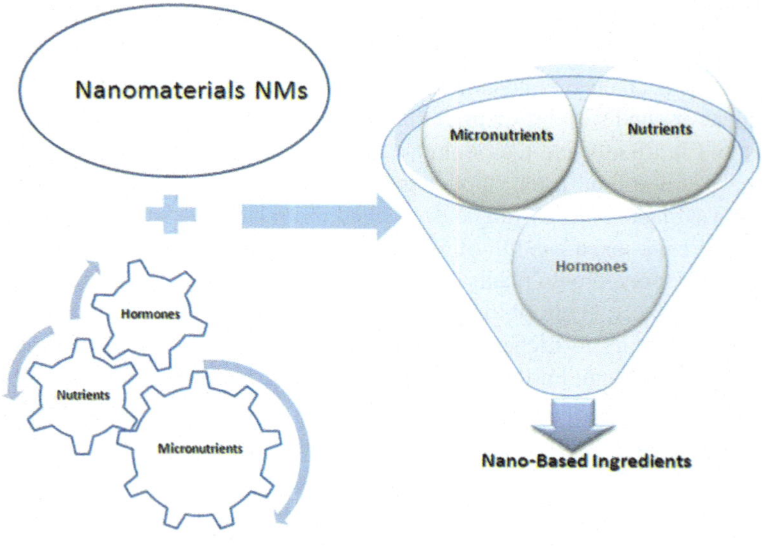

Fig. 11.4 Illustration of the nano-based -formation used in the ingredients delivery system (Constructed by M.M. Saleh)

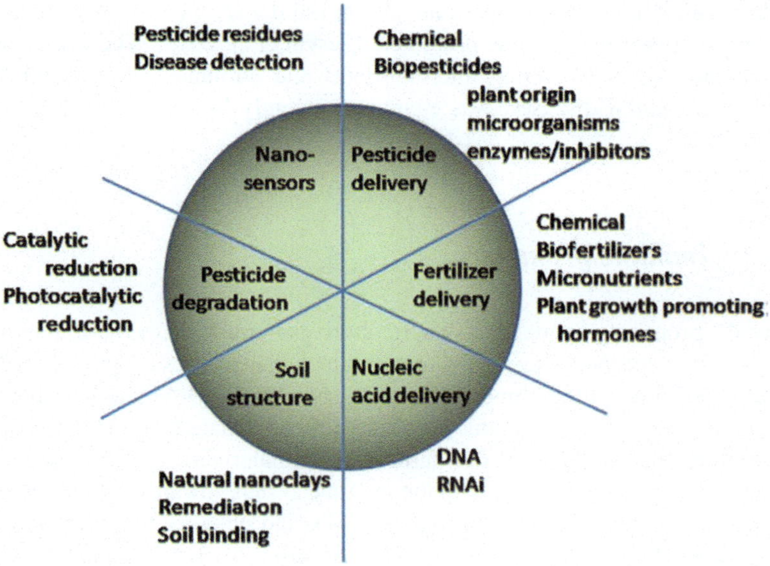

Fig. 11.5 Applications of nano-biotechnology in plant protection and nutrition (Ghormade et al. 2011)

assure the powerful holding of the effective materials which assist in the deliverance of nutrients and micronutrients in plants.

Tarafdar et al. (2012) mentioned that scientists were working extensively concerning the smart delivery system to create supplementation of almost all needed vital nutrients in proper quantities as nano-complex compounds. Boehm et al. (2003) declared about applying the smart delivery systems in order to control the various effects of nutrients deliverance in plants. In addition, many scientists confirmed the role of smart delivery system depending on NMs in enhancing the accessibility of nutrients to plants (Rai and Ingle 2012) and manages their release within time (Kah et al. 2013). The applying of nano-based biosensors guarantee is one method of the smart delivery system that improves the exactly deliverance of nutrients (Solanki et al. 2015), within the perfect time according to plant demand (Liu and Lal 2015).

11.2.2 Micronutrients

Each of iron, zinc, boron, molybdenum, copper, manganese and chlorine which are all micronutrients, had an important role in promoting crops growth, but unfortunately, some micronutrients were depleted from the arable lands since the green revolution, in addition to the reduction of micronutrients accessibility after applying some agricultural procedures (Alloway, 2008).

Nanomaterials, plays an important role in the deliverance of micronutrients in plants and activate the immune promoters (Servin et al. 2015). The important task of nanomaterials in delivering micronutrients and immune factors to inhibit the infection of several diseases, was examined recently by many researchers (Guo et al. 2018).

11.2.3 Immune promoters

One of the promising approaches in agriculture, is to improve the natural immunity system towards various diseases in crops and to enhance crops adaptation to the rising pathogens due to the climate change (Servin et al. 2015) and to achieve this goal, more studies must be conducted concerning nanomaterials role in the delivery system of immune stimulators. Furthermore, nanomaterials emphasize the nutrients deliverance in plants and strengthen the immune system towards unfavorable biotic and abiotic stresses (Kah and Hofmann 2014), and enhance the quantitative and quality parameters of crops yield (Chhipa and Joshi 2016; Mukhopadhyay 2014). Saharan and Pal (2016) included that the treatment of salicylic acid and chitosan combination promoted the activities of defense and antioxidant enzymes which improved the final plant yield because this combination assured the slow deliverance of the salicylic acid, their results confirmed the role of nanomaterials in raising the effectiveness of immune promoter delivery system. Kumaraswamy et al. (2018) reported the usage of the biopolymer chitosan as a nano-capsulation agent for the agro-chemicals deliverance, mainly due to its biodegradable and harmless characteristics which enhance the immune promoters in plants. In maize, to overcome *Fusarium verticillioides* that causes stalk rot, Kumaraswamy et al. (2019) improved a combination between salicylic acid and chitosan nanoparticles which inhibited the growth of mycelia up to 100% and for that it considered as an effective antifungal compound. Thakur and Sohal (2013) declared that some of the immune promoters in plant like salicylic acid and harpin had caught a wide attention in the field of crops disease management since they were eco-friendly (Kumar 2014), but their low bioactivity affected negatively their efficiency.

11.2.4 Hormones

Many researchers like De La Torre-Roche et al. (2013) confirmed the efficient role of nanomaterials in the management of nutrients and hormones delivery for the maintenance of growth and protection of crops (Pérez-de-Luque and Rubiales 2009). For instance, Chitosan NPs according to Thakur et al. (2018) and Abdel-Aziz et al. (2016) positively affected the delivery of the 1-naphthylacetic acid which is a growth hormone (Saharan et al. 2013; Silva et al., 2011).

11.3 Silica-Based Nanosystem for Gene Delivery

In addition to traditional medication administration, mesoporous silica nanoparticles can also be used as a carrier in gene transfection. Carriers are widely recognized to serve an essential role in gene transfer, as bare nucleic acids demonstrate minimal cell membrane penetration (Kim et al. 2011). Two principal mechanisms are in place for the delivery of genes: viral and non-viral. The more effective viral methods are subject to important safety issues, including immunogenicity, recombination with the gene and non-specific issues (Marshall 2001; Thomas et al. 2003). Non-viral systems have been extensively investigated in recent years including cationic chemicals, recombinant proteins and polymeric and inorganic nanoparticles (Arıs and Villaverde 2004; De Smedt et al. 2000; Panyam and Labhasetwar 2012; Zhang et al. 2016). Cationic compounds, however, are generally linked to severe toxicity and recombinant proteins exhibit a low price-performance ratio (Lv et al. 2006).

Mesoporous silica nanoparticles are very desirable because of their unique characteristics of diverse materials. They are therefore regarded as a successful vehicle for the transfer of genes to improve cellular uptake and transfection efficiency (Zhou et al. 2018).

11.3.1 Surface Modification of Mesoporous Silica Nano particulates for Gene Delivery

11.3.1.1 Amination alteration

Untreated mesoporous silica nanoparticles generally have a negative charge because of the ionized surface silanol groups, which decreases binding to nucleic acids that are loaded negatively, for example, DNA. Silica nanoparticles are therefore frequently changed by processes such as amination-alteration, livered cations coding and cationic polymer functionalizing to express overall positive charges. Using these amended mesoporous Silica nanoparticles, improved electrostatic interactions with nuclear acids and gene loading. Modifying aminotriethoxysilane (TAP) or amino propyl trimethoxysilane (APTMS) is a simple and popular attempt to boost the genetic load capacity of the mesoporous silica nanoparticles (Ganguly and Ganguli 2013; Zheng et al. 2013). The positive link between plasmid-DNA adsorption (pDNA) and the degree of amination was also examined by Yang et al. (2012).

11.3.1.2 Metal Cations

In addition to facilitating mesoporous silica nanoparticles mediated gene transfer, metal cations have been employed to increase interaction between DNA and silica surface. The effects of metal counter ions on gene adsorption were studied by Solberg and Landry (2006) and discovered that Mg^{+2} had a greater attraction with DNA vs. Na^{+} or Ca^{+2}. In comparison with the presence of the amino group, however, DNA appears to attach less firmly to mesoporous silica nanoparticles via metal cations.

11.3.1.3 Cationic Polymers

Cationic polymers can bind and provide high transfection efficiency genes of mesoporous silica nanoparticles. Polyamidoamine dendrimer-grafted mesoporous silica nanoparticles were successfully used by Radu et al. (2004) in delivering DNA plasmid. Chen et al (2009) described the first method of using polyamidoamine decorated Silica NPs to deliver Dox and BCL-2 siRNA concurrently into multidrug-resistant cancer cells. The BCL-2 siRNA was shown to dramatically silence the mRNA of BCL-2 and successfully lower the resistance to non-pump, improving Dox anti-cancer activity.

The proton sponge effect is another efficient means of encouraging gene transfection of mesoporous silica NPs. This method is intended to assist the escape of polyethyleneimine decorated mesoporous silica NPs formulations from the endosomes or lyposomes. Xia et al. (2009) discovered that these formulations had excellent DNA and siRNA binding affinity as well as surprise high cell transfection efficiency up to 70%. In addition, polyethyleneimine may be combined with other compounds to regulate gene release before an attachment to mesoporous silica NPs (Shen et al. 2014).

Poly-L-lysine polymers are often employed for gene transfer, as they can transport big DNA and infiltrate low-immunogenic cell membranes. In order to create regulated release behavior, Poly-L-lysine may be destroyed by enzymes (Wang et al. 2009). Also, Zhu et al. (2011) combined poly-L-lysine polymer and mesoporous silica NPs to create an enzyme-controlled system that could concurrently regulate the release of drugs and genes. Natural poly-L-arginine may be biocompatible and less harmful than the manufactured polycationic polymers such as polyamidoamine and polyethyleneimine. In Kar et al. (2013) literature, the simple silica NPs synthesis of poly-L-arginine was proposed and the efficacy of transfection with plasmid DNA was shown to be up to 60%.

Finally, the positive charges of these materials might lead to significant electrostatic interactions with the cell membrane that is negatively charged, which might contribute to an increase in cellular absorption and cell toxicity. Therefore, to balance transfection effectiveness and toxicity of gene delivery of the modified silica-based nanosystem, it is necessary to manage the quantity of cationic polymer utilized (Zhou et al. 2018).

11.3.1.4 Magnetic Silica Nanosphere for Gene Delivery

Magnetic nanoparticles have also been employed frequently to distribute vehicles efficiently for targeted organs or tissue and even to enable magnetic reaction imagery. Gao et al. (2009) developed poly-L-lysine functional magnetic silica nanospheres with massive mesopores (13–24 nm). This platform has shown robust DNA adsorption and an effective cellular supply capacity correspondingly for miRNA. In contrast to commercial reagent Polymag™, Yiú et al. (2010) developed Polyethyleneimine-Fe_3O_4-MCM-48 nanocomposite with a 4 times better transfection efficiency. Zhang et al. (2012) constructed a functional, multipurpose fluorescent-magnetic polyethylene platform with mesoporous silicone that concurrently fulfilled the fluorescence tracking and magnetically controlled siRNA delivery.

11.3.2 Carbon Nanotubes for Gene Delivery

It should be mentioned that to make carbon nanotubes compatible with the biological environment, carbon nanotubes do not have adequate functional groups for covalent binding. The carbon nanotubes' surface area, therefore, has to be changed to provide adequate functional groups by using several synthetic chemical techniques that allow the covalent attachment of appropriate solubilizing, bioactive or biocompatible moieties. In terms of their reactivity, i.e. sidewalls and tips, researchers classically have classified carbon nanotubes into two areas. Tips have been demonstrated to be more likely than sidewalls to be modified. It is obvious that by modifying the carbon nanotubes surface, its capacity to bind genetic materials may be designed and the cell membrane may be traversed to transfer the genetic material to the cell (Dovbeshko et al. 2003).

Biomolecules are intracellularly disseminated based on their ability to pass cell membrane and entree subcellular media. Furthermore, a non-viral cargo fits the requirements for gene therapy as a powerful vector such as high payload sums, intracellular incorporation, targeted administration and minimal toxicity. For example, DNA, RNA, oligonucleotides, gene therapy aptamers and a large-scale tool box containing functionalized carbon nanotubes have been used by scientists for the active transfer of biomolecules in plants and mammalian cells (Kaboudin et al. 2018; Panwar et al. 2019; Verma et al. 2018). Also, Pantarotto et al. (2004) demonstrate the widest usage of ammonium functional carbon nanotubes to create a plasmid-DNA complex by ionic bonding that effectively transmits plasmid-DNA intracellularly.

Carboxylated single-walled carbon nanotubes (SWCNTs) were used by Nia et al. (2017) for the conjugation of polymers of different diameters of DNA and polyethyleneimine (PEI). The SWCNT-COOH-PEI has demonstrated enhanced cell survival and efficiency in transfection than the polyethyleneimine polymer alone. Similarly, the carboxylated multi-walled carbon nanotubes (MWCNTs) employed by Versiani and coll (2017) and Calegari et al. (2016) as a protein transporter to stimulate an immunological reaction in mice as well as to incorporate the dengue VERO virus cells with cytoplasm and cells nuclei transfection.

11.3.3 Carbon Nanotubes and Plant Biotechnology

The standard agricultural methods of plant biotechnology have been revised to tackle the rising need for food and energy. Recent progress in plant genetic engineering has led to improved agricultural yields and environmental stress tolerance. Due to the complicated cell absorption of a hard cell wall, transportation of biomolecule into plant cells is crucial. Traditional gene transfer methods are subject to constraints such as harm to tissues, inefficiency in targeting specific cell organelles and low cellular penetration (Cunningham et al. 2018)

Transfer of nanomaterials to genetic material into mammalian cells has already been researched and significant tractions were also achieved among plant cell biologists (Wang et al. 2019; Zhang et al. 2019). The high aspect ratio of single-wall carbon nanotubes has functionalized with PEI to retain DNA and discharge it electrostatically inside a weakly basic medium. Electron microscopy demonstrated the transportation of DNA-PEI-carbon nanotubes into the mature leaf cells; NIR fluorescence imaging demonstrated that this freight carrier was seven hundred times more effective than plasmid DNA adsorbed on the single-walled carbon nanotubes (Demirer et al. 2019). In other research work, GFP-encoding plasmid DNA was successfully transferred into intact tobacco root cells by arginine-functionalized single-walled carbon nanotubes (Golestanipour et al. 2018). Also, Burlaka et al. (2015) demonstrated the internalization of pGreen 0029, a form of plasmid DNA, into the plant cell walls of *N. tabacum* using non-covalent functionalized carbon nanotubes. At the stated concentrations, the modified callus and leaf explants of *N. tabacum* produced considerable sprout regeneration with minimal toxicity. Even though the mechanism of carbon nanotubes absorption is unknown, the surface functionalization and size of the nanotube may play a crucial role in determining the mechanistic pathway (Zhang et al. 2018) Perepelytsina et al. (2018) demonstrated that doxorubicin multiwalled carbon nanotubes internalization may be affected by the variations in zeta potential caused by nanotube surface functionalization. Compared to free doxorubicin, it was shown to be less cytotoxic.

11.3.4 Micro RNA Delivery in Crop Protection

In the field of biology, a large number of well-organized macromolecular structures and nanomachines were constructed for specific biological tasks. Due to its unique qualities of DNA, RNA and proteins at nano-size, its interesting maquillage has fostered several biometric designs that can serve as a platform to design nano constructions and instruments. The concept was born longtime ago by constructing nanostructures utilizing DNA that led to information bang in the well-known field of DNA nanotechnology (Seeman 2010).

Because of the RNA incredible diversity in form and function, they have recently piqued the curiosity of scientists. Several therapeutic aids, including aptamers, ribozymes, small interfering RNA (siRNA), and miRNA, can be attached to the packaging RNA (pRNA) molecule. These findings have paved the road for RNA nanotechnology to create revolutionary therapeutic ways for the treatment and diagnostics of numerous types of tumors, viral infections, and genetic illnesses in people, animals, and plants (Shu et al. 2013).

The miRNA is an endogenous non-coding RNA that has recently been discovered to have a significant part in regulatory processes responsible for gene expression in all entities. The miRNA is normally 18–23 nucleotides long and plays a crucial function in post-transcriptional regulation. Although their synthesis and transcription occur in the nucleus, these are energetic and dynamic in the cytoplasm. In a sick condition, the miRNA profile variations, making miRNAs an excellent target for pharmacological treatment using RNA nanotechnology. The restore of the down-regulated miRNA or the restriction of the over-expressed miRNA to return miRNA to the normal condition is the basis of miRNA-based illness monitoring. The miRNA therapeutic potential was initially established by the finding that miR-15 and miR-16 are down-regulated by the progress of B-cell leukemia (Calin et al. 2002). These little non-coding RNAs have since gained interest not only in biomedical research and medicine progress but also in the protection of crops since modification of cellular miRNAs can regulate the degree of expression of essential genes.

The miRNA delivery approach based on RNA nano particulates in plants may be used to aim biotic and abiotic stresses that are not effectively tackled. RNA nanotechnology can change agricultural enhancement programs that represent a step towards global food security. The artificial microRNA (amiRNA) technology can target natural miRNA pioneers to produce tiny RNAs that guide gene silencing in floras (Alvarez et al. 2006; Parizotto et al. 2004; Schwab et al. 2006). Oligonucleotide alterations that resemble the intact secondary structure of natural miRNA precursors have been shown to result in targeted gene silence (Qu et al. 2012). The amiRNA approach was initially applied in human cell lines, then in Arabidopsis by Parizotto et al. (2004). For the usage in agriculture, the practice of amiRNA short sequences is preferable to long hairpin mediated silencing because of declined of-target influences on host genes and a lower hazard of binding to non-target pathogens. Plant resistance to numerous infections has lately been accomplished by the expression of amiRNA in transgenic Arabidopsis plants, enabling resistance to turnip yellow mosaic virus and turnip mosaic virus. Niu et al. (2006) created it utilizing amiRP69^{159} and amiR-Hc-Pro^{159}. AmiRNA is also employed for the development of transgenic Arabidopsis plants that provide resistance against cucumber mosaic virus-based on Arabidopsis pre miRNA159a (Ai et al. 2011). Also, the resistance against wheat streak mosaic virus was advanced in wheat by rice miR359 to create amiRNA pioneers. The transgenic lines exhibited three resistant forms, i.e. whole resistance, originally immune but resistance was discontinuous and the early resistance followed by the recovery of the plant (Fahim et al. 2012). Table 11.1 lists endogenous genes silenced in several plant species utilizing the amiRNA method.

11.4 Nano Based for Fertilizers Delivery

Nanoscience is a new technology that has made innovative advances in electronics, electricity, remediation, transportation, space technology, and biological sciences. Nanotechnology has infinite importance in plant biology and medicine, including drug and gene delivery, biosensing, diagnostics, and tissue engineering (Oberdörster et al. 2005; Solanki et al. 2015). The term nano originates from a Greek word that means dwarf. The term nano refers to a billionth of a meter. Nanoparticles are the dimension of fewer than 100 nanometers. Nanoparticles are highly applicable due

Table 11.1 Artificial micro RNA applied to resist disease in plants (Source: Chaudhary et al. 2018)

Plant species	Disease	miRNA/Pre-miR backbone	References
Arabidopsis	Turnip yellow mosaic virus and Turnip mosaic virus	amiR-P69^{159} and amiR-Hc-Pro^{159}	Niu et al. (2006)
Tobacco	Cucumber mosaic virus	amiR-2b	Qu et al. (2007)
Tobacco	Cassava brown streak viru	amiR-159a	Wagaba et al. (2016)
Tobacco	Potato virus Y	amiR-HC-Pro	Simón-Mateo et al. (2006)
Grapes	grapevine fan leaf virus	amiRCP-2	Jelly et al. (2012)
Tomato	Viral infection	amiR-2a/b	Zhang et al. (2011)
Tobacco	Potato virus Y and potato virus X	amiR-Hc-Pro^{159a} amiR-Hc-Pro^{167b} amiR-Hc-Pro^{171a}	Ai et al. (2011)
Arabidopsis	Cucumber mosaic virus	amiR-159a	Duan et al. (2008)
Cotton	Cotton leaf curl virus	Pre-miR-169a	Ali et al. (2013)
Wheat	Wheat streak mosaic virus	Pre-miR395	Fahim et al. (2012)
Arabidopsis	Water melon silver mottle virus	Pre-miR-159a	Kung et al. (2012)
Tomato	Tomato leaf curl virus	amiR-AV1-3	Van et al. (2013)
Arabidopsis	Turnip mosaic virus	amiR^{159}-P69	Lin et al. (2009)
Rice	Rice stripe virus and Rice black-streaked swarf virus	Osa-pre-miR528	Kis et al. (2016)
Barely	Wheat dwarf virus	huv-pre-miR171	Kis et al. (2016)
Tobacco	Tomato spotted wilt virus	Pre-miR-159a	Mitter et al. (2016)

to their high surface area to volume ratio, nanometer regime, and specific properties. By combining science and engineering, nanotechnology offers a new interdisciplinary venture into agriculture and food sciences. It can significantly contribute to agricultural science, opening up new avenues for solving various agrarian problems. Nanocomposites could be used in farming to detect contaminants, plant diseases, pests and pathogens; deliver pesticides, fertilizers, minerals and gene products in a regulated manner; and serve as nano architects in the forming and binding soil structure (Ghormade et al. 2011; Thakkar et al. 2010).

There are numerous compositions for nanoparticles; metal oxide, ceramic, silicate, magnetic, quantum dot, lipid, polymer and dendrimer. The formulation of nanomaterials has a significant impact on the biological application. Polymer conjugated nanocomposites are used in agrochemicals due to their prolonged and control release properties. Metal conjugated nanoparticles had various natural properties like fluorescence and photocatalytic degradation (Huynh et al. 2020). A huge rising population over the past decade has contributed to improved agricultural productivity to meet the increased food requirements of billions of people in developing countries. There is a widespread deficiency of nutrients in the soil, which means significant losses for farmers and reductions in nutritional quality and grain quantity available for humans and livestock. To nano, famulated fertilizer application may be beneficial to increase agriculture production. However, plant nutrients present in the bulk chemical forms are not as accessible to plants as they could be. Nano fertilizer use typically accounts for less than half of the chemical fertilizer applied to crop plants.

In the current agricultural industry, the extensive use of fertilizers is necessary to maintain agricultural practices (Kah et al. 2019). There is a solid expectation that this percentage of fertilizers will increase in the coming years to a quantity that could provide the world with 9.6 billion people's daily caloric needs by 2050 (Diatta et al. 2020).

Synthetic chemical fertilizers are used for optimal performance and the development of crops (Guo et al. 2018). Still, current agricultural practices are not particularly successful in increasing plant nutrient uptake, nutrient use efficiency and crop productivity. For the most part, synthetic fertilizers used in extensive agriculture have low-efficiency values for nutrient use. The intensive use of synthetic fertilizers is caused by a low nutrient use efficiency, resulting in more crops requiring artificial fertilizers. On the other hand, intensive application of synthetic fertilizers over the long term can result in severe environmental problems such as air pollution, soil degradation, water eutrophication and water contamination. Additionally, too much synthetic fertilizer use increases the costs of producing it, decreases farmers' profit margins, and further contributes to mass food shortages in third-world countries (Seleiman et al. 2020).

As a result, increasing food demand while also protecting natural and environmental resources is possible utilizing modern approaches that could enhance the supply of food while also managing natural resources (Shang et al. 2019). In a few studies, it has been hypothesized that the potential of nanotechnology to augment the current synthetic framework utilized in modern agriculture systems may be able to enhance the effectiveness of novel agrochemicals and provide new solutions for environmental and agricultural issues (Prasad et al. 2017). Due to this, recent years have seen a rise in interest in studying the use of nanoparticles among agricultural researchers. There is a scientific curiosity for producing novel sources of fertilizer to increase fertilizer usage efficiency in this context. From an organic farming perspective, nanoscience can develop new and innovative fertilizers, including nano fertilizers, to enhance global food production to feed the rising global population (Ruffo Roberto et al. 2019).

11.4.1 Nano fertilizer

Plants need various nutrients for growth and development, some in higher concentrations (known as macronutrients) and others in lower concentrations (known as micronutrients). Nitrogen (N), phosphorus (P), potassium (K), calcium (Ca), magnesium (Mg), and sulphur (S) are the macronutrients (S). Carbon (C), hydrogen (H), and oxygen (O) are also macronutrients, but they are found in significant amounts in the atmosphere and are assimilated directly by plants. Boron (B), chlorine (Cl), cobalt (Co), copper (Cu), iron (Fe), manganese (Mn), molybdenum (Mo), nickel (Ni), and zinc make up the group of micronutrients (Zn) (Finck et al. 1992).

Fertilizer is defined as any organic or inorganic material of biological origin added to soil to supply essential nutrients for average plant growth and development. Fertilizers are now needed to achieve high levels of crop yields. Nitrogen stands out among the macronutrients that are commonly used in crop soil management. It is required for the composition of various fertilizers and is applied in large quantities. Ammonium salts or nitrates, and urea are examples of nitrogen compounds. Urea is the most popular nitrogen source, accounting for nearly half of all nitrogen-based fertilizer demand worldwide. It's commonly used because of its high nitrogen content and ease of use, whether as dry granules or aqueous solutions (Chien et al. 2009).

The availability of essential elements has a significant impact on crop nutrition and yield. Several long-term field studies have shown that nutrient production from commercial fertilizers accounts for 30 to 50% of crop yield (Chhipa et al. 2017). Given the benefits of nanomaterials, these nutrients could be delivered in nanosized forms to increase release and efficiency of usage, resulting in greater plant crop improvement with fewer environmental impacts. Nano fertilizers are nanomaterials that provide one or more essential nutrients to plants directly (Fig. 11.6) (Liu et al. 2015).

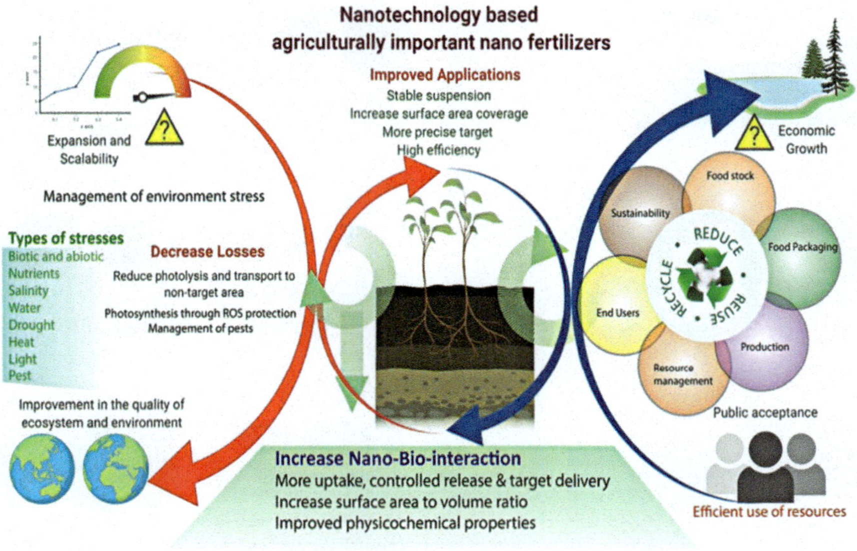

Fig. 11.6 Agriculture-relevant nano fertilizers based on nanotechnology improve agricultural sustainability, performance, and reduce environmental stress. (Source: López-Moreno et al. 2018).

11.4.2 Nano Fertiliser Formulations

11.4.2.1 Chemical-Based Nano Fertilizers Formulations

Nanomaterials for nano fertilizers can be made in various ways, including top-down, bottom-up and biological methods. The top-down approach is focused on reducing the size of bulk materials to nanoscale well-organized assemblies. Top-down is a material-milling-based physical process. The lack of control over nanoparticle size and a higher quantity of impurities are two drawbacks of this method. The bottom-up approach starts with chemical reactions at the atomic or molecular level to create nanoparticles. This method better controls particle size and eliminates impurities since it is a chemically regulated synthetic process (Fig. 11.7) (Pradhan et al. 2017; Sekhon et al. 2014).

11.4.2.2 Biological Based Nano Biofertilizers Formulations

Nanotechnology has emerged as an advanced technology for agriculture and environment service providers, with a wide range of applications in ecological sustainability. Biosynthesized nanoparticles are biocompatible, highly reproducible, and are simple to scale up (Singh et al. 2016). The biological synthesis of nanoparticles can be divided into two categories: intracellular and extracellular synthesis.

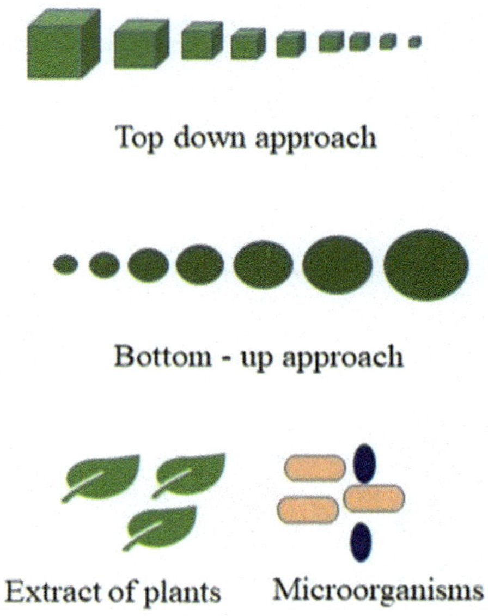

Fig. 11.7 Different approaches for the synthesis of nano fertilizers (Constructed by: M. Kotakonda)

Intracellular synthesis takes place within the biomass, fungi, bacteria and other organisms (Hulkoti et al. 2014). In contrast, extracellular synthesis takes place outside the organism and is assisted by various biomolecules and extracellular metabolites (Li et al. 2018). Because of the higher surface tension of the nanocoating, the nanoparticles encapsulate nutrients in a nanoemulsion, which ensures a stronghold of nutrients on the plant surface and has excellent potential for improving crop performance (Iavicoli et al. 2017). Soil enhancement goods such as nano zeolites and nano clays help efficiently release and preserve water and nutrients (Jahangirian et al. 2020; Sanzari et al. 2019).

Spraying conventional fertilizers on crops is the most popular method of application. However, the final concentration of fertilizers reaching the plant is one of the significant factors determining the mode of application (Tsuji et al. 2001). Owing to chemical leaching, drift, runoff, evaporation, hydrolysis by soil moisture, and photolytic and microbial degradation, very little concentration reaches the intended site in practice (Boehm et al. 2003). The nitrogen, phosphorus and potassium content of applied fertilizers has been confirmed to be lost in the atmosphere and unable to enter the plant, resulting in long-term and economic losses. These issues also resulted in the overuse of fertilizer, which harms the soil's natural nutrient balance. The widespread use of chemicals as fertilizers and pesticides has resulted in contamination that has harmed normal flora and fauna. Excessive use of fertilizers causes insect tolerance, decreases soil microflora, reduces nitrogen fixation, leads to pesticide bioaccumulation and destroys bird habitat. As a result, optimizing chemical fertilization to meet crop nutrient requirements while minimizing the risk of environmental contamination is critical (Fig. 11.8) (Miransari et al. 2011).

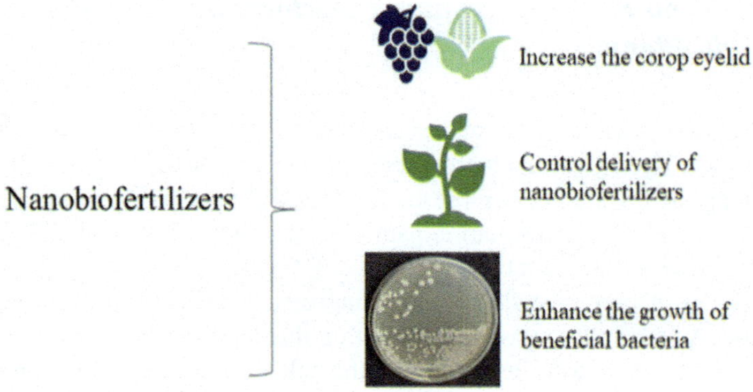

Fig. 11.8 Application of nano-biofertilizers (Constructed by: M. Kotakonda)

Table 11.2 Applications of nano fertilizers vs traditional fertilizers (Source: Solanki et al. 2015)

Characteristics	Modern methods	Nano-fertilizers
Controlled fertilizer released	Difficult to control and excess release can damage the ecological balance	Fertilizer release and rate of action handled easily
The time of nutrient release	Nutrient release the time of spelling	Nano formulations can control nutrient release
The efficiency of nutrient uptake	Larger particle difficult to absorption	Nano fertilizers increase the efficacy
Mineral micronutrient Solubility and dispersion	Poor bioavailability due to particle size and shape	Effective bioavailability
Mode of action	Slow mode of effect	Fast effect and efficacy

Nanoscience has made it possible to investigate nanostructured materials as fertilizer carriers for the development of innovative fertilizers as new facilities to improve nutrient use and lower environmental emission costs. A nano-fertilizer is a commodity that provides nutrients to plants in the nanometre range. Nanomaterial surface coatings on fertilizer particles keep the substance in place more tightly than traditional surfaces due to higher surface tension, allowing for more regulated release (DeRosa et al. 2010). The application of nanotechnology in agriculture includes delivering agrochemical materials such as fertilizer that provide macro-and micronutrients to plants. Nano-fertilizers demonstrate regulated agrochemical release, site-specific distribution, reduced toxicity, and improved nutrient utilization of delivered fertilizers. Because of their small scale, high mobility, and low toxicity, nanoparticles have a high surface area to volume ratio, high solubility, and precise targeting (Table 11.2) (Brady et al. 1996).

11.4.3 Nano Fertilizer Uptake, Translocation and Fate in Plants

The absorption, translocation, and accumulation of nanoparticles depend on the plant species, age, growth environment, physicochemical property. Rico et al. (2011) suggested a graphical representation of different nanoparticle absorption, translocation and biotransformation pathways, as well as potential cellular uptake modes in a plant system (Rico et al. 2011).

The cell wall pore size determines nanoparticle penetration through the cell membrane. As a result, nanostructures with a diameter smaller than the plant cell wall's pore size may easily pass through the cell wall and meet the plasma membrane (Fleischer et al. 1999). Functionalized nanoparticles enable the enlargement of pore size or new cell wall pore induction. Several studies have looked at how nanoparticles are taken into plant cells by binding to carrier proteins through aquaporin and ion channels. Nanoparticles can also be transported into plants through membrane transporters when they form complexes with them (Fernández et al. 2009). Nanomaterials can transfer apo plastically upon entering cells. They can be relocated from one cell to another through plasmodesmata. Nanomaterials in the cytoplasm interact with various cytoplasmic organelles and disrupt the cell's metabolic reactions (Nair et al. 2010).

11.4.4 Nano-Fertilizers for Abiotic and Biotic Stress Tolerance

Biological and ecological stresses are significant constraints to crop production, affecting plant growth, development and substantial threat to global food security. Drought, flooding, heat, hail, salinity, heavy metals and mineral deficiencies are some of the most common abiotic stresses that affect crop growth, yield and quality. Different forms of insect pests and diseases, on the other hand, are biotic stresses that reduce crop yields (Ashkavand et al. 2016). It is critical to identify and use new approaches to address current yield-limiting factors and improve resource use performance (Srivastava et al. 2004). Numerous studies have shown that using nano formulations to increase the levels of plant antioxidant compounds may effectively reduce the adverse effects of different environmental stresses (Farhangi-Abriz et al. 2018; Wu et al. 2015; Zulfiqar et al. 2019).

11.4.5 Nanofertilizers Limitations

In the field of sustainable farming, recent advancements have certainly seen the effective use of nano fertilizers to improve crop yields. Even so, the intentional use of this technology in farming methods may have several unforeseen and irreversible consequences (Kah 2015). Nanomaterials are extremely responsive due to their small size and high surface area. These materials' reactivity and variability are also a source of concern. Given the expected benefits, it is necessary to investigate the feasibility and suitability of these new smart fertilizers (Singh et al. 2021). Indeed, concerns about their transport, toxicity and bioavailability, as well as unintended environmental consequences from exposure to biological systems, restrict their use in sustainable agriculture and horticulture. Indeed, the use of nano fertilizers made from nanomaterials has sparked serious concerns about food safety, public health and food security (Table 11.3) (López-Moreno et al. 2018).

11.5 Nano-Based for Pesticides Delivery

Nanotechnology is still in its early stages of development when it comes to pesticide delivery. This method attempts to limit the use of traditional pesticides in an indiscriminate manner while also ensuring their safety. This critical evaluation looked into the potential of nanotechnology, particularly the nanoencapsulation method for pesticide administration (Gajbhiye et al. 2009). A complete analysis of various nanoencapsulation materials and processes, as well as the effectiveness of application and existing new research trends, was also conducted. To develop a nano encapsulated pesticide formulation with increased solubility, permeability and stability, as well as slow-release qualities. These attributes are primarily achieved by preventing the encapsulated active ingredients from premature degradation or extending the active ingredients' pest control effectiveness. (Torney 2009). The wrapping of several chemicals within another material at varying sizes in the nano-range is known as nanoencapsulation. The internal phase, core substance, or fill refers to the enclosed substance, such as insecticides. The external phase, shell, coating, or membrane of encapsulating material, such as nanocapsules, is referred to as the external phase, shell, coating, or membrane. In order to improve their physical qualities and restrict pesticide use, commercial pesticides and biocides have been encapsulated using nanoparticles. The development of pesticide-loaded or entrapped particles with a diameter in the nanometer range is known as nanoencapsulation (Nair et al. 2010).

Table 11.3 Advantages and limitations of nano fertilizers (Source: Sekhon 2014)

Nano fertilizers	
Advantages	Limitations
Reduce the rate of nutrient loss	Environmental effect
Regulation of nutrient delivery	High reactivity
Nutrient bioavailability increases	Farmers are concerned about their safety
The formulation as per applications	Consumers are concerned about their safety

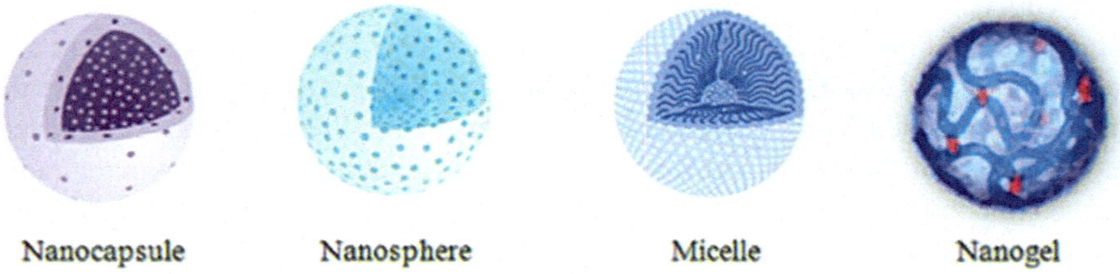

Fig. 11.9 Polymeric nanoencapsulated insecticides in various morphological shapes (adapted from Perlatti et al. 2013)

11.5.1 Polymer-Based Encapsulation

In a wide range of applications, polymers and polymeric materials are used. Polymer nano-composites, which are made up of a polymer matrix with nanoparticles or nano-fillers scattered throughout, are commonly used to encapsulate active ingredients. Amphiphilic block copolymers have recently piqued researcher interest due to their ability to form various types of nanoparticles as well as their low cost. Based on the number of blocks, copolymers are known as bi-block or tri-block copolymers. Polyethylene glycol, poly-caprolactone, chitosan, sodium alginate, and other synthetic and natural polymers, as well as block copolymers, have all been utilized to encapsulate a wide range of insecticides. Nanocapsules contain active chemicals and are made of natural polymers such as ketosan, gelatin, or polymers (Fig. 11.9) (Ghormade et al. 2011).

11.5.1.1 Nanocapsules

Nanocapsules are vesicular structures made up of a polymeric membrane encasing the active substances as an inner liquid center at the nanoscale level. The nanocapsule structure consists of a core-shell configuration, with the shell being a polymeric membrane or coating. The active ingredients are dissolved in the inner liquid center in most cases. Pesticide formulations or a polymeric matrix can be added to the inner core, and the polymeric shell can absorb active chemicals. In this way, the active compounds are encased by nanocapsules that form spontaneously during nanocapsule formation (Yang et al. 2009). Insecticides and biocides are effectively encapsulated in nanocapsules made from a variety of polymers. A polymer such as polyethylene glycol was used as the shell material in the production of nanocapsules. Using a melt-dispersion technique, developed round-shaped polyethylene glycol nanocapsules filled with garlic essential oil. Similarly, monomers including methyl methacrylate and styrene are suitable for the emulsion polymerization phase in nanocapsule production. Grillo et al. (2012) discovered that the emulsion polymerization process is ideal for producing nanocapsules.

11.5.1.2 Nanospheres

Nanospheres are uniformly distributed and embedded in the polymeric matrix to form the active nano-carrier structure. Although nanocapsules synthesis methods can be used to create this nanoparticle, polymerization techniques such as emulsion or interfacial polymerization are crucial in the fabrication of this nanomaterial. A schematic diagram of nanosphere synthesis. Pesticide encapsulation has been demonstrated using polymers such as PCL, which can encapsulate pesticides in nanoparticles. Forim et al. (2012), for example, used this polymer to encapsulate azadirachtin, resulting in nanocapsules and nanospheres with average diameters of 150 and 200 nm, respectively. Boehm et al. (2000) used this polymer to encapsulate active substances with an average diameter of 200–250 nm and generate nanospheres using nanoprecipitation in a previous study. The addition of different surfactants did not affect the loading efficiency, but they were shown to be efficient in stabilizing the nanospheres holding suspensions over two months (Trivedi et al. 2010).

11.5.1.3 Micelles

Micelles are excellent bioactive nano-carriers for encapsulating insecticides, particularly water-insoluble insecticides. Micelles are made up of substances such as amphiphilic block copolymers, polymers and surfactants. In an aqueous solution, the material amphiphilic properties cause them to self-assemble into spherical micelles, with the hydrophilic ends serving as the outer shell and the hydrophobic ends serving as the nucleus (Zhang et al. 2013). When it comes to water-soluble copolymers, there are two methods for making micelles. The first method is direct dissolution, which entails simply adding copolymers to an aqueous solution at their essential micelle concentration (CMC). Hydrophobic insecticides can become stuck in the center of micelles during the manufacturing process. The second method is the film casting method, which involves creating copolymer films containing pesticides and then solubilizing them using different methods (Letchford et al. 2007).

11.5.1.4 Nanogels

Nanogels are aqueous dispersions of hydrogel particles made up of physically or chemically cross-linked polymer networks that are nanoscale in size. Nanogels have demonstrated their utility as carriers of active compounds in a drug delivery system. This is considered remarkable for conventional medicinal nano-carriers due to their high loading power, high stability and reactivity to environmental conditions such as ionic strength, pH and temperature (Kabanov and Vinogradov 2009). Nanogels are formed by the regulated aggregation of interacting polymers in aqueous conditions due to their self-assembly properties. Water-soluble or hydrophilic polymers may have hydrophobic properties applied to them, allowing them to interact electrostatically and/or form hydrogen bonds with one another. The nano gel stays swollen because there is enough water in them, making it easier to load active molecule spontaneously through electrostatic, and/or hydrophobic interactions between active chemicals and the polymer matrix. The active chemicals are encapsulated in the polymer matrix, resulting in stable nano gel particles a group of people who work together to solve problems (Daoud-Mahammed et al. 2007).

11.5.2 Lipid NMS-based encapsulation

Lipid-based nanomaterials have been identified as possible bioactive substance delivery systems with improved encapsulation efficiency and low toxicity. Lipid-based nanomaterials have a lot of potential for encapsulating active ingredients that are hydrophilic, hydrophobic, or lipophilic. The efficacy of lipid-based nanoparticles, nanoliposomes and solid lipid nanoparticles in encapsulating pesticide active ingredients has already been demonstrated (Tamjidi et al. 2013).

11.5.2.1 Nanoliposomes

Nanoliposomes are a nanometric version of liposomes with similar chemical, structural and thermodynamic properties. Nanoliposomes, among other things, are thought to be very good at encapsulating and transporting bioactive substances to biological, biochemical, pharmacological, and agricultural goals (Mozafari 2010). Nanoliposomes have more surface area, greater solubility and better bioavailability of active compounds than liposomes. They also aid in the development of the controlled release system, allowing for more accurate encapsulation material targeting (Ziaee et al. 2014). Nanoliposomes are nanoscale vesicles with a watery interior made up of a bilayer lipid. Colloidal structures are formed when lipids, most commonly phospholipids, are arranged in an aqueous solution. The amphiphilic nature of phospholipids is critical for the growth of vesicle structures. When phospholipids are placed in aqueous media, they clump together to shield the hydrophobic tails (acyl chain) from water molecules, while the hydrophilic heads stay in close contact with the aqueous process. These aggregated phospholipids rearrange and form nanocapsules that resemble bilayer vesicles or nanoliposomes when given enough energy. Other molecules, such as sterols, may cause significant changes in the properties of nanoliposome bilayers. Cholesterol is the most common sterol used in the production of lipid vesicles. Although it has no part in the creation of the bilayer structure, it can be impregnated into the phospholipid membrane at extremely high concentrations. Nanoliposomes are versatile in terms of encapsulation, allowing them to encapsulate both hydrophilic (water-soluble) and hydrophobic (lipid-soluble) molecules. Hydrophilic molecules are trapped in the inner center of vesicles with aqueous media during vesicle formation, while lipophilic molecules are incorporated into liposomal bilayers.

11.5.2.2 Solid Lipid Nanoparticles (SLNs)

In 1991, solid lipid nanoparticles (SLNs) were introduced as a promising colloidal carrier material in a controlled delivery method. In colloidal systems, SLN has been stated to be a superior carrier material to other nano-carrier materials such as polymeric nanoparticles, liposomes, nanoemulsion and nanosuspension, according to the literature (Saupe and Rades 2006). SLNs are said to have all of the advantages of other colloidal carriers while having none of the disadvantages in terms of physical and chemical storage stability, toxicity, loading capability, production size, target-oriented release qualities, and viability. Surfactants stabilized semi-crystalline or crystalline solid lipid spherical nanostructure compounds. Triglycerides, partial glycerides, fatty acids, steroids and waxes are all part of the lipid. The type of lipid used to make SLN determines the emulsifiers or surfactants are needed to stabilize the aqueous lipid dispersion. Various emulsifiers with varying charges and molecular weights may be used, but phospholipids, such as soybean or egg lecithin, can be one of the better options (Potta et al. 2011). The crystallization process of lipids plays an effective role in stabilizing SLN suspensions and the crystallization process of lipids in nanoparticles is different compared to bulk lipid material. There were just a few experiments where pesticide active compounds were successfully loaded onto SLNs, for example, used the high-pressure homogenization approach to prepare SLN-based pesticide formulations. The ecological pesticide Artemisia arborescent L essential oils were loaded into lipid as Compritol 888 ATO and stabilized by surfactants such as Poloxamer 188 or Miranol Ultra C32.110.

11.5.3 Clay NMS-based Encapsulation

Clays are well-known for their applications in agriculture, environment and they contribute to the health of humans and other living things. Nanoclays are widely regarded as cost-effective materials with enormous potential for the development of multifunctional nano-carrier materials and their applications in the life and material sciences (Choy et al. 2007). Clay materials and layered double hydroxides (LDHs) have similar effects in agricultural applications. Strong electrostatic interactions are created between brucite-type sheets and anions due to higher layer charge densities, preventing LDHs from swelling and resulting in a more stable particle size than clay minerals. Furthermore, numerous studies on clay nanoparticles have revealed that functionalizing clay nanoparticles with various polymers and surfactants is required to alter the electrostatic interactions between chemical loading and the clay particles. Because of their positive charge, anionic compounds can be intercalated into the layers of LDHs (Lee et al. 2003).

11.5.3.1 Clay Nanomaterials

Nanoclays are fine-grained minerals with sheet-like structures that belong to a larger group of minerals known as naturally occurring aluminum silicates or hydrous silicates. Phyllosilicates are sheet-structured hydrous silicates that are essential because they help to construct the structure of individual clay minerals. The phyllosilicates formed a layered structure made up of two different types of sheets: tetrahedral silicate and octahedral aluminum. The arrangement of these sheets in the layer determines the diversity of clay minerals (Majeed et al. 2013). Clays can be classified as 1:1 type of clay (kaolinite), 2:1 type of clay (montmorillonite) or 2:1:1 type of clay (kaolinite) (chlorite). A 1:1 clay layer is made up of one tetrahedral sheet and one octahedral sheet, whereas a 2:1 clay layer is made up of two tetrahedral sheets and one octahedral sheet, forming a sandwich-like structure. When one octahedral sheet is added to the 2:1 type of clay, the 2:1:1 kind of clay is created. Clay and clay minerals can be considered nanomaterials of geological and pedological origins since the layers of primary clay particles are in the nanometer range, even though their length ranges in the millimeter range (Yuan et al. 2007). Clay materials can interact with organic compounds in a variety of ways, including cation exchange, ligand exchange, hydrophobic interaction, hydrogen bonding, protonation, cation bridging and water sorption. Several factors influence active compound intercalation, including exchangeable cations, interlayer distance and the presence of water molecules between the layers. Furthermore, clay minerals can be altered in a variety of ways (Annabi-Bergaya 2008).

11.5.3.2 Layered Double Hydroxides (LDHs)

Layered double hydroxides (LDHs) are anionic lamellar compounds made up of positively charged brucite-type layers of mixed metal hydroxides found in nature and synthetically. Their positively charged layer structure has a variety of stacking defects, resulting in a variety of polytypes with a wide range of chemical compositions. The hydroxide ions create endless two-dimensional nanosheets with a thickness of roughly one nanometer and a lateral size varying from a sub-micrometer to several tens of micrometers by connecting. The metal cations occupy the central position in the shared octahedral, while the hydroxide ions stay at the vertices (Fig. 11.10). LDHs are multifunctional nanomaterials with a variety of appealing characteristics, including tunable attributes in layer charge density, particle size and shape. They're also biocompatible and have low toxicity (Bi et al. 2014). The use of LDHs in the production of controlled/slow-release pesticide formulations is a relatively new phenomenon. Their excellent buffering capacity, water retention ability, acid-neutralizing potentials and affinity for common carbonate ions may make it easier to formulate complete and controlled release intercalated pesticide formulations. LDHs are good candidates for pesticide administration because of their soil compatibility and they act as a matrix for encapsulating or entrapping the pesticide. They used three alternative approaches to intercalate anionic herbicides

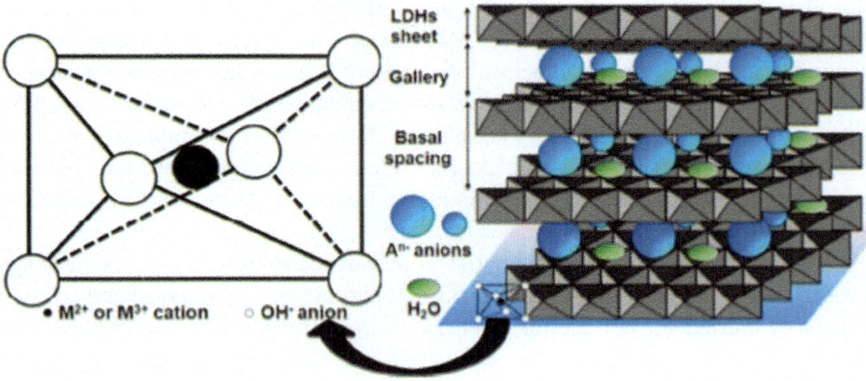

Fig. 11.10 A schematic representation of the structure of layered double hydroxides is shown (Bi et al., 2014)

2,4-D, MCPA, and picloram into MgAl-LDH: direct synthesis (SD) by co-precipitation, regeneration technique (RE) and ion-exchange approach (IE) (Cardoso et al. 2006).

11.5.4 Others Encapsulation (Starch—etc)

Other nanomaterials, such as poly (citric acid) grafted carbon nanotube, have the potential to be effective pesticide carriers in addition to the nanoencapsulation materials described above. On this subject, nanoencapsulation materials used for medicine delivery and bioactive food additives have shown that they can also be utilized as pesticide carriers. Polymersomes, nanostructured lipid carriers, dendrimers, inorganic porous materials such as porous nano-$CaCO_3$, nano zeolite and others are examples of potential nanoencapsulation materials (Lee and Feijen 2012; Soussan et al. 2009).

11.6 Conclusion and Prospects

Nanotechnology keeps a close eye on one of the essential agricultural control processes, owing to its small size. In agriculture, nanoparticles aim to reduce the number of chemicals spread, reduce nutrient losses in fertilization, and boost output through insect and nutrient management. Nanotechnology can benefit the agriculture and food industries by developing revolutionary nanotools for disease management, nutrient absorption capacity and other applications.

Soft materials, such as nano-gel formulations and conjugated nano polymers, must be extra investigated as promising options for developing new ways for sustainable release of macromolecules and editing the genome of plants, based on cumulative evidence published in the cell. The polymeric nature of hydrogel-based nanoparticles has evident compensations in medication administration because of their safety profile, storage efficiency for loading and excellent product prevention from deterioration. In cell, nanomaterials have been used to accomplish active molecule delivery that is regulated by external inputs. Before polymeric nanoparticles are widely used in agriculture, a thorough investigation of manufacturing scalability and cost-effectiveness is required.

Nano formulations provide potential as part of intelligent crop production systems in organic farming since they have small particle sizes, a delayed and constant release of nutrients. They are ideal for use in modern agriculture because of their promising properties. Agriculture production and resilience to biotic and abiotic stresses can both benefit from the usage of nanoformulations. As a result, the application of nanoformulation in agriculture cannot be overlooked. Nano formulation may reduce fertilizer losses due to evaporation, leakage and spent energy through creation by allowing active ingredients to be delivered more efficiently, increasing nutrient uptake values.

Additionally, seed coverings with nanoformulations and nanosensors may reduce agricultural production costs and environmental concerns. On the other hand, nanoformulations release nutrients slower than nano fertilizers and they can dramatically boost crop produces and excellence attributes, according to a study. Because of the significant improvements in crop development, physiological and biochemical features, work and quality. The great application of nanoformulations is healthier and more favored than soil treatment, especially in intelligent agriculture. Before marketing nanoformulations on a commercial scale, nano-formulation's scientific-based and practical use must be thoroughly investigated.

Despite significant advances in plant genetics, delivering DNA molecules and enzymes for genome editing remains a considerable barrier. New plant genetics strategies based on nanoparticles clustered regularly intermittent polynomial repetitions CRISPR technology, similar to those used in other natural processes, could lead to game-changing innovation.

The use of nanomaterials in plants is minimal. Recent findings and current applications suggest that more research is needed to improve the formulation and biological use of nanomaterials for agriculture importance and better understand the method of herbal uptake and enhance the management of agriculture. Surprisingly, advantages must be expanded to include previously unknown components of plant physiology. Nanobiosensors for real-time detection of secondary metabolites could improve plant growth and relationship with the environment, especially in constrained environment growth situations.

Finally, supporting the initiation of different methods for developing beneficial nanoparticles could help overcome the delay in plant nanotechnology. Joint collaborative endeavors combining complementary professional capabilities such as plant ecologists, genetic engineers, chemical scientists and biochemist's engineers may reveal new possibilities in agri-nanotechnology to achieve this goal. Future research should concentrate on the safety, bioavailability and toxicity of various nanoformulations utilized in agriculture. To enhance agricultural production using biosynthesized, nano-biofertilizers and nano compositions should also be investigated. morning

References

Abdel-Aziz HM, Hasaneen MN, Omer AM (2016) Nano chitosan-NPK fertilizer enhances the growth and productivity of wheat plants grown in sandy soil. Spanish J Agric Res 14:0902

Ahmed S, Ahmad M, Swami BL et al (2016) A review on plants extract mediated synthesis of silver nanoparticles for antimicrobial applications: a green expertise. J Adv Res 7:17–28

Ai T, Zhang L, Gao Z et al (2011) Highly efficient virus resistance mediated by artificial microRNAs that target the suppressor of PVX and PVY in plants. Plant Biol 13:304–316

Alam AR, Ismat F, Sayeed K et al (2016) Applciation of nanotechnology in agri-food and food science. World J Pharm Sci 4(7):45–54

Ali I, Amin I, Briddon RW et al (2013) Artificial microRNA-mediated resistance against the monopartite begomovirus cotton leaf curl Burewala virus. Virol J 10:231

Ali S, Mehmood A, Khan N (2021) Uptake, translocation, and consequences of nanomaterials on plant growth and stress adaptation. J Nanomater 2021:6677616. https://doi.org/10.1155/2021/6677616

Alloway B (2008) In: Alloway BJ (ed) Micronutrient deficiencies in global crop production. Springer

Alvarez JP, Pekker I, Goldshmidt A et al (2006) Endogenous and synthetic microRNAs stimulate simultaneous, efficient, and localized regulation of multiple targets in diverse species. Plant Cell 18:1134–1151

Annabi-Bergaya F (2008) Layered clay minerals. Basic research and innovative composite applications. Micropor Mesopor Mat 107:141–148

Anusuya S, Sathiyabama M (2015) Protection of turmeric plants from rhizome rot disease under field conditions by β-D-glucan nanoparticle. Int J Biol Macromol 77:9–14

Arias-Estévez M, López-Periago E, Martínez-Carballo E et al (2008) The mobility and degradation of pesticides in soils and the pollution of groundwater resources. Agric Ecosyst Environ 123(4):247–260

Arıs A, Villaverde A (2004) Modular protein engineering for non-viral gene therapy. Trends Biotechnol 22:371–377

Ashkavand P, Tabari M, Zarafshar M et al (2016) Effect of SiO_2 nanoparticles on drought resistance in hawthorn seedlings. Forest Research Papers 76(4):350–359

Aslani F, Bagheri S, Muhd Julkapli N et al (2014) Effects of engineered nanomaterials on plants growth: an overview. Sci World J

Babadi AA, Bagheri S, Hamid SBA (2016) Progress on implantable biofuel cell: nanocarbon functionalization for enzyme immobilization enhancement. Biosens Bioelectron 79:850–860

Banerjee K, Pramanik P, Maity A et al (2019) Methods of using nanomaterials to plant systems and their delivery to plants (Mode of entry, uptake, translocation, accumulation, biotransformation and barriers). Advances in Phytonanotechnology, Academic Press, pp 123–152

Bi X, Zhang H, Dou L (2014) Layered double hydroxide-based nanocarriers for drug delivery. Pharmaceutics 6:298–332

Blaine AC, Rich CD, Sedlacko EM (2014) Perfluoroalkyl acid distribution in various plant compartments of edible crops grown in biosolids-amended soils. Environ Sci Technol 48:7858–7865

Boehm ALLR, Zerrouk R, Fessi H (2000) Poly epsilon-caprolactone nanoparticles containing a poorly soluble pesticide: formulation and stability study. J Microencapsul 17:195–205
Boehm AL, Martinon I, Zerrouk R et al (2003) Nanoprecipitation technique for the encapsulation of agrochemical active ingredients. J Microencapsul 20(4):433–441
Boverhof DR, Bramante CM, Butala JH et al (2015) Comparative assessment of nanomaterial definitions and safety evaluation considerations. Regul Toxicol Pharmacol 73:137–150. https://doi.org/10.1016/j.yrtph.2015.06.001
Brady NC, Weil RR (1996) Organisms and ecology of the soil. In: Brady NC, Weil RR (eds) The nature and properties of soils, 13th edn. Prentice-Hall, New Jersey, pp 328–360
Burlaka OM, Pirko YV, Yemets AI et al (2015) Plant genetic transformation using carbon nanotubes for DNA delivery. Cytol Genet 49(6):349–357
Calegari LP, Dias RS, de Oliveira MD et al (2016) Multi-walled carbon nanotubes increase antibody-producing B cells in mice immunized with a tetravalent vaccine candidate for dengue virus. J Nanobiotechnol 14:61
Calin GA, Dumitru CD, Shimizu M et al (2002) Nonlinear partial differential equations and applications: frequent deletions and down-regulation of micro-RNA genes miR15 and miR16 at 13q14 in chronic lymphocytic leukemia. Proc Natl Acad Sci 99:15524–15529
Cardoso LP, Celis R, Cornejo J et al (2006) Layered double hydroxides as supports for the slow release of acid herbicides. J Agric Food Chem 54:5968–5975
Chaudhary V, Jangra S, Yadav NR (2018) Nanotechnology based approaches for detection and delivery of microRNA in healthcare and crop protection. J Nanobiotechnol 16:40. https://doi.org/10.1186/s12951-018-0368-8
Chen AM, Zhang M, Wei D et al (2009) Co-delivery of doxorubicin and Bcl-2 siRNA by mesoporous silica nanoparticles enhances the efficacy of chemotherapy in multi drug resistant cancer cells. Small 5:2673–2677
Chhipa H (2017) Nanofertilizers and nanopesticides for agriculture. Environ Chem Lett 15:15–22
Chhipa H, Joshi P (2016) Nanofertilisers, Nanopesticides and Nanosensors in Agriculture. In: Ranjan S, Dasgupta N, Lichtfouse E (eds) Nanoscience in Food and Agriculture 1, v. 20. Springer International Publishing, Cham, pp 247–282
Chien S, Prochnow L, Cantarella H (2009) Recent developments of fertilizer production and use to improve nutrient efficiency and minimize environmental impacts. Adv Agron 102:267–322
Choy JH, Choi SJ, Oh JM et al (2007) Clay minerals and layered double hydroxides for novel biological applications. Appl Clay Sci 36:122–132
Cunningham FJ, Goh NS, Demirer GS et al (2018) Nanoparticle-Mediated Delivery towards Advancing Plant Genetic Engineering. Trends Biotechnol 36(9):882–897. https://doi.org/10.1016/j.tibtech.2018.03.009
Da Silva LC, Oliva MA, Azevedo AA et al (2006) Responses of resting a plant species to pollution from an iron pelletization factory. Water Air Soil Pollut 175(1-4):241–256
Daisy P, Saipriya K (2012) Biochemical analysis of Cassia fistula aqueous extract and phytochemically synthesized gold nanoparticles as hypoglycemic treatment for diabetes mellitus. Int J Nanomedicine 7:1189
Daoud-Mahammed S, Couvreur P, Gref R (2007) Novel self-assembling nanogels: 2003 Stability and lyophilisation studies. Int J Pharm 332:185–191
Dash SS, Majumdar R, Sikder AK et al (2014) Saracaindica bark extract mediated green synthesis of polyshaped gold nanoparticles and its application in catalytic reduction. Appl Nanosci 4:485–490
De La Torre-Roche R, Hawthorne J, Deng Y et al (2013) Multiwalled carbon nanotubes and C60 fullerenes differentially impact the accumulation of weathered pesticides in four agricultural plants. Environ Sci Technol 47:12539–12547
De Rosa MR, Monreal C, Schnitzer M et al (2010) Nanotechnology in fertilizers. *Nat Nanotechnol J* 5:91–96
De Smedt SC, Demeester J, Hennink WE (2000) Cationic polymer based gene delivery systems. Pharm Res 17:113–126

Demirer GS, Zhang H, Matos JL et al (2019) High aspect ratio nanomaterials enable delivery of functional genetic material without DNA integration in mature plants. Nat Nanotechnol 14(5):456–464. https://doi.org/10.1038/s41565-019-0382-5

Dhewa T (2015) Nanotechnology applications in agri-food: an update. Octa J Environ Res 3(2):204–211

Diatta AA, Thomason WE, Abaye O et al (2020) Assessment of nitrogen fixation by mungbean genotypes in different soil textures using 15 N natural abundance method. J Soil Sci Plant Nutr 20(4):2230–2240

Dovbeshko GI, Repnytska OP, Obraztsova ED et al (2003) Study of DNA interaction with carbon nanotubes. Semiconduct Phys Quantum Electron & Optoelectron 6(1):105–108

Du W, Sun Y, Ji R et al (2011) TiO_2 and ZnO nanoparticles negatively affect wheat growth and soil enzyme activities in agricultural soil. J Environ Monit 13(4):822–828

Duan CG, Wang CH, Fang RX et al (2008) Artifcial microRNAs highly accessible to targets confer efcient virus resistance in plants. J Virol 82:11084–11095

Duhan JS, Kumar R, Kumar N et al (2017) Nanotechnology: the new perspective in precision agriculture. Biotechnology Reports 15:11–23

Eichert T, Goldbach HE (2008) Equivalent pore radii of hydrophilic foliar uptake routes in stomatous and astomatous leaf surfaces–further evidence for a stomatal pathway. Physiol Plant 132(4):491–502

El-Ramady H, Abdalla N, Taha HS et al (2016) Selenium and nano-selenium in plant nutrition. Environ Chem Lett 14:123–147

Fahim M, Millar AA, Wood CC et al (2012) Resistance to Wheat streak mosaic virus generated by expression of an artificial polycistronic microRNA in wheat. Plant Biotechnol J 10:50–63

Farhangi-Abriz S, Torabian S (2018) Nano-silicon alters antioxidant activities of soybean seedlings under salt toxicity. Protoplasma 255(3):953–962

Fayaz AM, Balaji K, Girilal M et al (2009) Mycobased synthesis of silver nanoparticles and their incorporation into sodium alginate films for vegetable and fruit preservation. J Agric Food Chem 57:6246–6252

Fernández V, Eichert T (2009) Uptake of hydrophilic solutes through plant leaves: current state of knowledge and perspectives of foliar fertilization. Crit Rev Plant Sci 28(1-2):36–68

Finck A (1992) World fertilizer use manual. International Fertilizer Industry Association (IFA), Paris, 632 pp (Glossary of Soil Science Terms 2008) Soil Science Society of America, SSSA, 92. pp

Fisk HJ, Dandekar AM (1993) The introduction and expression of transgenes in plants. Sci Hortic 55:5–36

Fleischer A, O'Neill MA, Ehwald R (1999) The pore size of non-graminaceous plant cell walls is rapidly decreased by Borate Ester cross-linking of the pectic polysaccharide Rhamnogalacturonan II. Plant Physiol 121(3):829–838

Forim MR, Silva MFDGFD, Fernandes JB (2012) Secondary metabolism as a measurement of efficacy of botanical extracts: the use of Azadirachtaindica (Neem) as a model, insecticides. In Advances in Integrated Pest Management, Dr. Farzana Perveen (Ed.), In Tech, 2012. https://doi.org/10.5772/27961.

Fraceto et al (2016) mentioned that nanomaterials can save time in delivering several ingredients in plants.

Gajbhiye M, Kesharwani J, Ingle A et al (2009) Fungus mediated synthesis of silver nanoparticles and its activity against pathogenic fungi in combination of fluconazole. Nanomedicine 5(4):282–286

Ganguly A, Ganguli AK (2013) Anisotropic silica mesostructures for DNA encapsulation. Bull Mater Sci 36:329–332

Gao F, Botella P, Corma A et al (2009) Monodispersed mesoporous silica nanoparticles with very large pores for enhanced adsorption and release of DNA. J Phys Chem B 113:1796–1804

Ghormade V, Deshpande MV, Paknikar KM (2011) Perspectives for nano-biotechnology enabled protection and nutrition of plants. Biotechnol Adv 29:792–803

Gleiter H (2000) Nanostructured materials: basic concepts and microstructure. Acta Mater 48:1–29

Gogos A, Knauer K, Bucheli TD (2012) Nanomaterials in plant protection and fertilization: current state, foreseen applications, and research priorities. J Agric Food Chem 60(39):9781–9792

Golestanipour A, Nikkhah M, Aalami A et al (2018) Gene delivery to tobacco root cells with single-walled carbon nanotubes and cell-penetrating fusogenic peptides. Mol Biotechnol 60(12):863–878. https://doi.org/10.1007/s12033-018-0120-5. PMID: 30203379

Grillo R, dos Santos NZP, Maruyama CR et al (2012) Poly(ε-caprolactone) nanocapsules as carrier systems for herbicides: Physico chemical characterization and genotoxicity evaluation. J Hazard Mater 231–232:1–9

Guo H, White JC, Wang Z et al (2018) Nano-enabled fertilizers to control the release and use efficiency of nutrients. Curr Opinion Environ Sci Health 6:77–83

Hofmann T, Lowry GV, Ghoshal S et al (2020) Technology readiness and overcoming barriers to sustainably implement nanotechnology-enabled plant agriculture. Nature Food 1(7):416–425

Hu B, Jiang L, Zheng Q et al (2021) Uptake and translocation of organophosphate esters by plants: impacts of chemical structure, plant cultivar and copper. Environ Int 155:106591

Hulkoti NI, Taranath TC (2014) Biosynthesis of nanoparticles using microbes- a review. Colloids and surfaces. B Biointerfaces 121:474–483

Huynh KH, Pham XH, Kim J et al (2020) Synthesis, properties, and biological applications of metallic alloy nanoparticles. Int J Mol Sci 21(14):5174

Iavicoli I, Leso V, Beezhold DH et al (2017) Nanotechnology in agriculture: Opportunities, toxicological implications, and occupational risks. Toxicol Appl Pharmacol 329:96–111

Jahangirian H, Rafiee-Moghaddam R, Jahangirian N et al (2020) Green synthesis of zeolite/Fe_2O_3 nanocomposites: toxicity & cell proliferation assays and application as a smart iron nanofertilizer. Int J Nanomedicine 15:1005–1020

Jayaseelan C, Rahuman AA, Rajakumar G et al (2011) Synthesis of pediculocidal and larvicidal silver nanoparticles by leaf extract from heartleaf moonseed plant, Tinosporacordifolia Miers. Parasitol Res 109:185–194

Jelly NS, Schellenbaum P, Walter B et al (2012) Transient expression of artificial microRNAs targeting Grapevine fanleaf virus and evidence for RNA silencing in grapevine somatic embryos. Transgenic Res 21:1319–1327. https://doi.org/10.1007/s11248-012-9611-5.205

Jia G, Wang H, Yan L et al (2005) Cytotoxicity of carbon nanomaterials: single-wall nanotube, multi-wall nanotube, and fullerene. Environ Sci Technol 39(5):1378–1383

Judy JD, Unrine JM, Rao W et al (2012) Bioavailability of gold nanomaterials to plants: importance of particle size and surface coating. Environ Sci Technol 46(15):8467–8474

Kabanov AV, Vinogradov SV (2009) Nanogels as pharmaceutical carriers: finite 1992 networks of infinite capabilities. Angew Chem Int 48:5418–5429

Kaboudin B, Saghatchi F, Kazemi F et al (2018) A novel magnetic carbon nanotubes functionalized with pyridine groups: synthesis, characterization and their application as an efficient carrier for plasmid DNA and Aptamer. Chemistry Select 3(24):6743–6749. https://doi.org/10.1002/slct.201800708

Kah M (2015) Nanopesticides and Nanofertilizers: Emerging Contaminants or Opportunities for Risk Mitigation? Front Chem 3:64

Kah M, Hofmann T (2014) Nanopesticide research: current trends and future priorities. Environ Int 63:224–235

Kah M, Beulke S, Tiede K et al (2013) Nanopesticides: state of knowledge, environmental fate, and exposure modeling. Crit Rev Environ Sci Technol 43(16):1823–1867

Kah M, Tufenkji N, White JC (2019) Nano-enabled strategies to enhance crop nutrition and protection. Nat Nanotechnol 14(6):532–540

Kar M, Tiwari N, Tiwari M et al (2013) Poly-L-arginine grafted silica mesoporous nanoparticles for enhanced cellular uptake and their application in DNA delivery and controlled drug release. Part Part Syst Charact 30:166–179

Kim MH, Na HK, Kim YK et al (2011) Facile synthesis of monodispersed mesoporous silica nanoparticles with ultra large pores and their application in gene delivery. ACS Nano 5:3568–3576

Kim D, Gopal J, Sivanesan I (2017) Nanomaterials in plant tissue culture: the disclosed and undisclosed. J RSC Adv 7:36492–36505

Kis A, Tholt G, Ivanics M et al (2016) Polycistronicartifcial miRNA-mediated resistance to Wheat dwarf virus in barley is highly efficient at low temperature. Mol Plant Pathol 17:427–437

Kumar D (2014) Salicylic acid signaling in disease resistance. Plant Sci 228:127–134

Kumaraswamy RV, Kumari S, Choudhary RC et al (2018) Engineered chitosan based nanomaterials: bioactivities, mechanisms and perspectives in plant protection and perspectives in plant protection and growth. Int J Biol Macromol 113:494–506

Kumaraswamy RV, Kumari S, Choudhary RC et al (2019) Salicylic acid functionalized chitosan nanoparticle: a sustainable biostimulant for plant. Int J Biol Macromol 123:59–69

Kung YJ, Lin SS, Huang YL et al (2012) Multiple artificial microRNAs targeting conserved motifs of the replicase gene confer robust transgenic resistance to negative-sense single-stranded RNA plant virus. Mol Plant Pathol 3:303–317

Lee JS, Feijen J (2012) Polymersomes for drug delivery: design, formation and characterization. J Control Release 161:473–483

Lee WF, Fu YT (2003) Effect of montmorillonite on the swelling 2147 behavior and drug release behavior of nanocomposite hydrogels. J Appl Polym Sci 89:3652–3660

Letchford K, Burt H (2007) A review of the formation and classification of amphiphilic block copolymer nanoparticulate structures: micelles, nanospheres, nanocapsules and polymersomes. Eur J Pharm Biopharm 65:259–269

Li J, Kuang D, Feng Y et al (2013) Green synthesis of silver nanoparticles–graphene oxide nanocomposite and its application in electrochemical sensing of tryptophan. Biosens Bioelectron 42:198–206

Li J, Tian B, Li T et al (2018) Biosynthesis of Au, Ag and Au-Ag bimetallic nanoparticles using protein extracts of Deinococcus radiodurans and evaluation of their cytotoxicity. Int J Nanomedicine 13:1411–1424

Lin D, Xing B (2008) Root uptake and phytotoxicity of ZnO nanoparticles. Environ Sci Technol 42(15):5580–5585

Lin SS, Wu HW, Elena SF et al (2009) Molecular evolution of a viral non-coding sequence under the selective pressure of amiRNA-mediated silencing. PLoSPathog 5:e1000312. https://doi.org/10.1371/journal.ppat.1000312

Liu R, Lal R (2015) Potentials of engineered nanoparticles as fertilizers for increasing agronomic productions. Sci Total Environ 514:131–139

López-Moreno ML, Cassé C, Correa-Torres SN (2018) Engineered nanomaterials interactions with living plants: benefits, hazards and regulatory policies. Curr Opin Environ Sci Heal 6:36–41

Lv H, Zhang S, Wang B et al (2006) Toxicity of cationic lipids and cationic polymers in gene delivery. J Control Release 114:100–109

Ma X, Geiser-Lee J, Deng Y et al (2010) Interactions between engineered nanoparticles (ENPs) and plants: phytotoxicity, uptake and accumulation. Sci Total Environ 408(16):3053–3061

Majeed K, Jawaid M, Hassan A et al (2013) Potential materials for food packaging from nanoclay/natural fibres filled hybrid composites. Mater Des 46:391–410

Makarov VV, Makarova SS, Love AJ et al (2014) Biosynthesis of stable iron oxide nanoparticles in aqueous extracts of Hordeum vulgare and Rumexacetosa plants. Langmuir 30:5982–5988

Manjunatha SB, Biradar DP, Aladakatti YR (2016) Nanotechnology and its applications in agriculture: A review. J Farm Sci 29(1):1–13

Marshall E (2001) Viral vectors still pack surprises. Science 294:1640

Mei W, Sun H, Song M et al (2021) Per-and polyfluoroalkyl substances (PFASs) in the soil–plant system: Sorption, root uptake, and translocation. Environ Int 156:106642

Miller EL, Nason SL, Karthikeyan KG et al (2016) Root uptake of pharmaceuticals and personal care product ingredients. Environ Sci Technol 50:525–541

Miransari M (2011) Soil microbes and plant fertilization. Appl Microbiol Biotechnol 92(5):875–885
Mitter N, Zhai Y, Bai AX et al (2016) Evaluation and identification of candidate genes for artifcial microRNA-mediated resistance to tomato spotted wilt virus. Virus Res 211:151–158
Mozafari M (2010) Nanoliposomes: preparation and analysis. In: Liposomes, vol 605. Springer, pp 29–50
Mukhopadhyay SS (2014) Nanotechnology in agriculture: prospects and constraints. Nanotechnol Sci Appl 63
Naderi M, Danesh-Shahraki A (2013) Nanofertilizers and their roles in sustainable agriculture. Int J Agric Crop Sci 5:2229
Nair R, Varghese SH, Nair BG et al (2010) Nanoparticulate material delivery to plants. Plant Sci 179(3):154–163
Navarro E, Baun A, Behra R et al (2008) Environmental behavior and ecotoxicity of engineered nanoparticles to algae, plants, and fungi. Ecotoxicology 17(5):372–386
Nia AH, Eshghi H, Abnous K et al (2017) The intracellular delivery of plasmid DNA using cationic reducible carbon nanotube – disulfide conjugates of polyethylenimine. Eur J Pharm Sci 100:176–186. https://doi.org/10.1016/j.ejps.2017.01.014
Niu QW, Lin SS, Reyes JL et al (2006) Expression of artifcial microRNAs in transgenic Arabidopsis thaliana confers virus resistance. Nat Biotechnol 24:1420–1428
NRC National Research Council (2008) Emerging technologies to benefit farmers in sub-Saharan Africa and South Asia, National Academic Press (Eds.), Washington, DC. United States of America
Oberdörster G, Oberdörster E, Oberdörster J (2005) Nanotoxicology: an emerging discipline evolving from studies of ultrafine particles. Environ Health Perspect 113(7):823–839
Pantarotto D, Singh R, McCarthy D et al (2004) Functionalized Carbon Nanotubes for Plasmid DNA Gene Delivery. Angew Chem Int Ed Engl 43:5242–5246
Panwar N, Soehartono AM, Chan KK et al (2019) Nanocarbons for biology and medicine: sensing, imaging, and drug delivery. Chem Rev 119(16):9559–9656
Panyam J, Labhasetwar V (2012) Biodegradable nanoparticles for drug and gene delivery to cells and tissue. Adv Drug Deliv Rev 64:61–71
Parisi C, Vigani M, Rodríguez-Cerezo (2015) Agricultural nanotechnologies: what are the current possibilities? Nano Today 10:124–127
Parizotto EA, Dunoyer P, Rahm N et al (2004) In vivo investigation of the transcription, processing, endonucleolytic activity, and functional relevance of the spatial distribution of a plant miRNA. Genes Dev 18:2237–2242
Perepelytsina OM, Ugnivenko AP, Dobrydnev AV et al (2018) Influence of carbon nanotubes and its derivatives on tumor cells in vitro and biochemical parameters, cellular blood composition In vivo. Nanoscale Res Lett 13(1):286. https://doi.org/10.1186/s11671-018-2689-9
Pérez-de-Luque A, Rubiales D (2009) Nanotechnology for parasitic plant control. Pest Manag Sci 65:540–545
Perlatti B, Bergo PLDS, Silva MFDGFD et al (2013) Polymeric nanoparticle-based insecticides: a controlled release purpose for agrochemicals. In insecticides – development of safer and more effective technologies, Prof. Stanislav Trdan (Ed.), INTECH Open Access Publisher, 523–550
Pokropivny VV, Skorokhod VV (2007) Classification of nanostructures by dimensionality and concept of surface forms engineering in nanomaterial science. Mater Sci Eng C 27:990–993
Potta SG, Minemi S, Nukala RK et al (2011) Preparation and characterization of ibuprofen solid lipid nanoparticles with enhanced solubility. J Microencapsul 28:74–81
Pradhan S, Mailapalli DR (2017) Interaction of engineered nanoparticles with the agri-environment. J Agric Food Chem 65(38):8279–8294
Prasad TNVKV, Sudhakar P, Sreenivasulu Y et al (2012) Effect of nanoscale zinc oxide particles on the germination, growth and yield of peanut. J Plant Nutr 35(6):905–927
Prasad R, Bhattacharyya A, Nguyen QD (2017) Nanotechnology in sustainable agriculture: recent developments, challenges, and perspectives. Front Microbiol 8:1014

Qu J, Ye J, Fang R (2007) Artifcial microRNA-mediated virus resistance in plants. J Virol 81:6690–6699

Qu J, Ye J, Fang R (2012) Artifcial microRNAs for plant virus resistance. Methods Mol Biol 894:209–222

Radu DR, Lai CY, Jeftinija K et al (2004) A polyamidoamine dendrimer-capped mesoporous silica nanospherebased gene transfection reagent. J Am Chem Soc 126:13216–13217

Rai M, Ingle A (2012) Role of nanotechnology in agriculture with special reference to management of insect pests. Appl Microbiol Biotechnol 94(2):287–293

Rai PK, Kumar V, Lee S et al (2018) Nanoparticle-plant interaction: Implications in energy, environment, and agriculture. Environ Int 119:1–19

Rico CM, Majumdar S, Duarte-Gardea M, Peralta-Videa JR, Gardea-Torresdey JL (2011) Interaction of nanoparticles with edible plants and their possible implications in the food chain. J Agric Food Chem 59(8):3485–3498

Roberto SR, Youssef K, Hashim AF, Ippolito A (2019) Nanomaterials as alternative control means against postharvest diseases in fruit crops. Nanomaterials-Basel 9(12):1752. https://doi.org/10.3390/nano9121752

Saharan V, Pal A (2016) Chitosan based nanomaterials in plant growth and protection. Springer, India

Saharan V, Mehrotra A, Khatik R, Rawal P, Sharma SS, Pal A (2013) Synthesis of chitosan-based nanoparticles and their in vitro evaluation against phytopathogenic fungi. Int J Biol Macromol 62:677–683. https://doi.org/10.1016/j.ijbiomac.2013.10.012. Epub 2013 Oct 16

Salah SM, Yajing G, Dongdong C, Jie L, Aamir N, Qijuan H, Weimin H, Mingyu N, Jin H (2015) Seed priming with polyethylene glycol regulating the physiological and molecular mechanism in rice (Oryza sativa L.) under nano-ZnO stress. Sci Rep 5:14278

Saleh TA (2020) Nanomaterials: Classification, properties, and environmental toxicities. Environ Technol Innov 101067

Sanzari I, Leone A, Ambrosone A (2019) Nanotechnology in plant science: to make a long story short. Front Bioeng Biotechnol 7:120

Saupe A, Rades T (2006) Solid lipid nanoparticles. In: Nanocarrier Technologies. Springer, pp 41–50

Schwab R, Ossowski S, Riester M, Warthmann N, Weigel D (2006) Highly specifc gene silencing by artifcial microRNAs in Arabidopsis. Plant Cell Online 18:1121–1133

Seeman NC (2010) Nanomaterials based on DNA. Annu Rev Biochem 79:65–87. https://doi.org/10.1146/annurev-biochem-060308-102244

Sekhon BS (2014) Nanotechnology in agri-food production: an overview. Nanotechnol Sci Applic 7:31–53

Seleiman MF, Almutairi KF, Alotaibi M, Shami A, Alhammad BA, Battaglia ML (2020) Nano-fertilization as an emerging fertilization technique: why can modern agriculture benefit from its use? Plants (Basel, Switzerland) 10(1):2

Sertova N (2015) Application of nanotechnology in detection of mycotoxin and in agricultural sector. J Central Eur Agric 16(2):117–130

Servin A, Elmer W, Mukherjee A et al (2015) A review of the use of engineered nanomaterials to suppress plant disease and enhance crop yield. J Nanopart Res 17:92

Shang Y, Hasan MK, Ahammed GJ, Li M, Yin H, Zhou J (2019) Applications of nanotechnology in plant growth and crop protection: a review. Molecules (Basel, Switzerland) 24(14):2558

Shanmugam C, Sivasubramanian G, Parthasarathi B, Baskaran K, Balachander R, Parameswaran V (2016) Antimicrobial, free radical scavenging activities and catalytic oxidation of benzyl alcohol by nano-silver synthesized from the leaf extract of Aristolochiaindica L.: a promenade towards sustainability. Appl Nanosci 6:711–723

Sharma NC, Sahi SV, Nath S, Parsons JG, Gardea-Torresde JL, Pal T (2007) Synthesis of plant-mediated gold nanoparticles and catalytic role of biomatrix-embedded nanomaterials. Environ Sci Technol 41:5137–5142

Shen J, Kim HC, Su H, Wang F, Wolfram J, Kirui D et al (2014) Cyclodextrin and polyethylenimine functionalized mesoporous silica nanoparticles for delivery of siRNA cancer therapeutics. Theranostics 4:487–497

Shu Y, Shu D, Haque F, Guo P (2013) Fabrication of pRNA nanoparticles to deliver therapeutic RNAs and bioactive compounds into tumor cells. Nat Protoc 8:1635–1659

Silva MDS, Cocenza DS, Grillo R et al (2011) Paraquat-loaded alginate/chitosan nanoparticles: preparation, characterization and soil sorption studies. J Hazard Mater 190:366–374

Simón-Mateo C, García JA (2006) MicroRNA-guided processing impairs Plum pox virus replication, but the virus readily evolves to escape this silencing mechanism. J Virol 80:2429–2436

Singh P, Kim YJ, Zhang D et al (2016) Biological synthesis of nanoparticles from plants and microorganisms. Trends Biotechnol 34(7):588–599

Singh H, Sharma A, Bhardwaj SK et al (2021) Recent advances in the applications of nano-agrochemicals for sustainable agricultural development. Environmental Science. Process & Impact 23(2):213–239

Solanki P, Bhargava A, Chhipa H et al (2015) Nano-fertilizers and their smart delivery system. In: Rai M, Ribeiro C, Mattoso L, Duran N (eds) Nanotechnologies in food and agriculture. Springer International Publishing, Cham, pp 81–101

Solberg SM, Landry CC (2006) Adsorption of DNA into mesoporous silica. J Phys Chem B 110:15261–15268

Soussan E, Cassel S, Blanzat M (2009) Drug delivery by soft matter: matrix and vesicular carriers. Angew Chem Int Ed 48:274–288

Srivastava A, Srivastava ON, Talapatra S et al (2004) Carbon nanotube filters. Nat Mater 3(9):610–614

Tamjidi F, Shahedi M, Varshosaz J et al (2013) Nanostructured lipid carriers (NLC): a potential delivery system for bioactive food molecules. Innov Food Sci Emerg Technol 19:29–43

Tarafdar JC, Raliya R, Rathore I (2012) Microbial synthesis of phosphorus nanoparticles from Tricalcium phosphate using Aspergillus tubingensis TFR-5. *J Bionanoscience* 6:84–89

Thakkar KN, Mhatre SS, Parikh RY (2010) Biological synthesis of metallic nanoparticles. Nanomed Nanotechnol Biol Med 6(2):257–262

Thakur M, Sohal BS (2013) Role of elicitors in inducing resistance in plants against pathogen infection: a review. ISRN Biochemistry 762412

Thakur S, Thakur S, Kumar R (2018) Bio-nanotechnology and its role in agriculture and food industry. J Mol Genet Med 12:324

Thomas CE, Ehrhardt A, Kay MA (2003) Progress and problems with the use of viral vectors for gene therapy. Nat Rev Genet 4:346–358

Torney F (2009) Nanoparticle mediated plant transformation. Emerging technologies in plant science research. Interdepartmental Plant Physiology Major Fall Seminar Series. Phys:696

Torney F, Trewyn BG, Lin VSY et al (2007) Mesoporous silica nanoparticles deliver DNA and chemicals into plants. Nat Nanotechnol 2(5):295

Tripathi DK, Singh S, Singh S et al (2017) An overview on manufactured nanoparticles in plants: uptake, translocation, accumulation and phytotoxicity. Plant Physiol Biochem 110:2–12

Trivedi R, Kompella UB (2010) Nanomicellar formulations for sustained drug delivery: strategies and underlying principles. Nanomed-UK 5:485–505

Tsuji K (2001) Microencapsulation of pesticides and their improved handling safety. J Microencapsul 18(2):137–147

Uzu G, Sobanska S, Sarret G et al (2010) Foliar lead uptake by lettuce exposed to atmospheric fallouts. Environ Sci Technol 44:1036–1042

Van Vu T, Roy Choudhury N, Mukherjee SK (2013) Transgenic tomato plants expressing artifcial microRNAs for silencing the pre-coat and coat proteins of a begomovirus, tomato leaf curl New Delhi virus, show tolerance to virus infection. Virus Res 172:35–45

Verma SK, Das AK, Patel MK et al (2018) Engineered nanomaterials for plant growth and development: a perspective analysis. Sci Total Environ 630:1413–1435

Versiani AF, Astigarraga RG, Rocha ESO et al (2017) Multi-walled carbon nanotubes functionalized with recombinant Dengue virus 3 envelope proteins induce significant and specific immune responses in mice. J Nanobiotechnol 15:26
Wagaba H, Patil BL, Mukasa S et al (2016) Artifcial microRNA-derived resistance to Cassava brown streak disease. J Virol Methods 231:38–43
Wang Z, Qian L, Wang X et al (2009) Hollow DNA/ PLL microcapsules with tunable degradation property as efficient dual drug delivery vehicles by α-chymotrypsin degradation. Colloids Surf A Physicochem Eng Asp 332:164–171
Wang P, Zhao FJ, Kopittke PM (2019) Engineering crops without genome integration using nanotechnology. Trends Plant Sci 24(7):574–577. https://doi.org/10.1016/j.tplants.2019.05.004. Epub 2019 May 30. PMID: 31155336
Wang TT, Ying GG, Shi WJ et al (2020a) Uptake and translocation of perfluorooctanoic acid (pfoa) and perfluorooctanesulfonic acid (pfos) by wetland plants: tissue- and cell-level distribution visualization with desorption electrospray ionization mass spectrometry (desi-ms) and transmission electron microscopy equipped with energy-dispersive spectroscopy (tem-eds). Environ Sci Technol 54:6009–6020
Wang W, Rhodes G, Ge J et al (2020b) Uptake and accumulation of per- and polyfluoroalkyl substances in plants. Chemosphere 261:127584
Wu W, Ma B (2015) Integrated nutrient management (INM) for sustaining crop productivity and reducing environmental impact: a review. Sci Total Environ 512-513:415–427
Xia T, Kovochich M, Liong M et al (2009) Polyethyleneimine coating enhances the cellular uptake of mesoporous silica nanoparticles and allows safe delivery of siRNA and DNA constructs. ACS Nano 3:3273–3286
Yang FL, Li XG, Zhu F et al (2009) Structural characterization of nanoparticles loaded with garlic essential oil and their insecticidal activity against Tribolium castaneum (Herbst) (Coleoptera: Tenebrionidae). J Agric Food Chem 57:0156–10162
Yang H, Zheng K, Zhang Z et al (2012) Adsorption and protection of plasmid DNA on mesoporous silica nanoparticles modified with various amounts of organosilane. J Colloid Interface Sci 369:317–322
Yiú HH, McBain SC, Lethbridge ZA et al (2010) Preparation and characterization of polyethylenimine-coated Fe_3O_4-MCM-48 nanocomposite particles as a novel agent for magnet-assisted transfection. J Biomed Mater Res A 92A:386–392
Yuan G, Wu L (2007) Allophanenano clay for the removal of phosphorus in water and wastewater. Sci Tech Adv Mater 8:60–62
Zhang X, Li H, Zhang J et al (2011) Expression of artificial microRNAs in tomato confers efficient and stable virus resistance in a cell-autonomous manner. Transgenic Res 20:569–581
Zhang J, Sun W, Bergman L et al (2012) Magnetic mesoporous silica nanospheres as DNA/drug carrier. Mater Lett 67:379–382
Zhang J, Li M, Fan T et al (2013) Construction of novel amphiphilic chitosan copolymer nanoparticles for chlorpyrifos delivery. J Polym Res 20:1–11
Zhang M, Kim YK, Cui P et al (2016) Folate conjugated polyspermine for lung cancer-targeted gene therapy. Acta Pharm Sin B 6:336–343
Zhang M, Yang M, Morimoto T et al (2018) Size-dependent cell uptake of carbon nanotubes by macrophages: a comparative and quantitative study. Carbon 127:93–101
Zhang H, Demirer GS, Zhang H et al (2019) DNA nanostructures coordinate gene silencing in mature plants. Proc Natl Acad Sci 116(15):7543–7548
Zheng L, Hong F, Lu S et al (2005) Effect of nano-TiO_2 on strength of naturally aged seeds and growth of spinach. Biol Trace Elem Res 104:83–91
Zheng K, Yang H, Wang L et al (2013) Amino functionalized mesoporous silica nanoparticles: adsorption and protection for pcDNA3.1(þ)-PKB-HA. J Porous Mater 20:1003–1008
Zhou Y, Quan G, Wu Q et al (2018) Mesoporous silica nanoparticles for drug and gene delivery. Acta Pharmaceutica Sinica B 8(2):65–177

Zhu Y, Meng W, Gao H et al (2011) Hollow mesoporous silica/poly (L-lysine) particles for codelivery of drug and gene with enzymetriggered release property. J Phys Chem C 20 115:13630–13636

Ziaee M, Moharramipour S, Mohsenifar et al (2014) Toxicity of Carumcopticum essential oil-loaded nanogel against Sitophilus granarius and Triboliumconfusum. J Appl Entomol 138, 763-771

Zulfiqar F, Navarro M, Ashraf M et al (2019) Nanofertilizer use for sustainable agriculture: Advantages and limitations. Plant science: an international journal of experimental plant biology 289:110270

Chapter 12
Enhancement of Stress Tolerance of Crop Plants by ZnO Nanoparticles

Martin Šebesta, Sindy Kurtinová, Marek Kolenčík, and Ramakanth Illa

Abstract Zinc oxide (ZnO) nanoparticles are among the most promising nanoparticles used in precision agriculture. Since they are made of an essential element Zn, their potential applications are most vital in agricultural regions where the natural concentrations of bioavailable Zn are low, and crop plants suffer from the deficiency of this essential micronutrient. Also, a large number of genes requires Zn to protect cells from the detrimental effects of environmental stress to regulate and maintain their expression. ZnO nanoparticles are more tuneable in their properties, such as size, shape, dissolution rate, and surface properties, compared to conventional ionic formulas. Thus, they pose an effective way to supplement plants with Zn. Their nanoparticulate characteristics, such as photocatalysis, may provide additional beneficial effects for crop plants. Precise application of nanoparticles may reduce chemical inputs to agricultural fields, helping with long-term agricultural sustainability, environmental protection, and higher nutritional value of crops. In this chapter, we describe the influence of ZnO nanoparticles on the stress tolerance of crop plants. Since their effect is dependent on their properties, we elaborated on the importance of size, shape, crystal structure, and nanoparticle surface coating. Different modes of application show varying effects on crops, and foliar application may have direct positive effects connected to the photocatalytic properties of ZnO nanoparticles. We provide a summary of the positive impacts of ZnO NP on crop

M. Šebesta (✉) · S. Kurtinová
Institute of Laboratory Research on Geomaterials, Faculty of Natural Sciences, Comenius University in Bratislava, Bratislava, Slovakia
e-mail: martin.sebesta@uniba.sk

M. Kolenčík
Department of Soil Science and Geology, Faculty of Agrobiology and Food Resources, Slovak University of Agriculture in Nitra, Nitra, Slovakia

Nanotechnology Centre, VŠB Technical University of Ostrava, Ostrava, Czech Republic

R. Illa
Department of Chemistry, Rajiv Gandhi University of Knowledge Technologies, Nuzvid, India

M. Faizan et al. (eds.), *Sustainable Agriculture Reviews 53*, Sustainable Agriculture Reviews 53, https://doi.org/10.1007/978-3-030-86876-5_12

plants associated with the alleviation of biotic stresses (herbivores, pathogens) and abiotic stresses (heavy metals, heat, cold, drought, and salt). In mainly laboratory or pot studies, several beneficial effects of ZnO nanoparticles were identified. Future field experiments are needed to further our knowledge and adapt agricultural techniques to changing local and global climatic conditions.

Keywords Zinc oxide · Agriculture · Nanoparticle · Stress alleviation · Nanofertilizer

12.1 Introduction

Engineered nanoparticles are increasingly used in many diverse human activities. The ability to adjust their properties through a change in their size, shape, crystallinity, and surface properties is highly beneficial for new technological applications. The small size of nanoparticles means that large fractions of atoms that build up the nanoparticles are on the surface. Nanoparticle's surface to volume ratio dramatically changes with their size, which affects their properties and smaller nanoparticles exhibit forces typical for the atomic and molecular world, e.g. various quantum effects (Nel et al. 2006; González-Melendi et al. 2008; Nair et al. 2010; Gogos et al. 2012; Strambeanu et al. 2015; Mallakpour and Madani 2015; Rasmussen et al. 2018; Faraz et al. 2020). Nanomaterials are usually defined as materials with at least one dimension between 1 and 100 nm. They are subdivided into three types (1) nanosheets with one dimension below 100 nm, (2) nanofibres and nanotubes with two dimensions lower than 100 nm, and (3) nanoparticles with three dimensions between 1 and 100 nm (Wang et al. 2015). Due to their highly tunable properties, engineered nanoparticles are of high interest in various technological applications, including precision agriculture, where they may improve the health and yields of crops, even under suboptimal conditions of environmental stress.

Biotic and abiotic stress has a considerable influence on the growth of plants and agricultural production. Through millions of years of evolution, plants have developed various physiological responses that improve their ability to tolerate biotic and abiotic stresses (Almutairi 2019). It is, thus, essential to understand plants' stress tolerance mechanisms. The response mechanisms start with fast recognition of the nature of the stress. Afterwards, a complex cascade of signals triggers the plant's defences. Accumulation of reactive oxygen species (ROS), activation of gene networks, specific ion channels, and kinase cascades are involved in early stress response (Rejeb et al. 2014; Czarnocka and Karpiński 2018; Almutairi 2019). Also, the response leads to an elevated release of hormones such as abscisic acid (ABA), ethylene (Et), jasmonic acid (JA), and salicylic acid (SA) (Bari and Jones 2009; Davies 2010).

During the normal metabolism of oxygen, ROS are produced as a by-product of aerobic metabolism and are kept at low levels by cells' antioxidant chemicals. Furthermore, ROS are used as signalling molecules. Although ROS are necessary for basic biological processes such as cellular proliferation and differentiation, their elevated levels may have undesired effects. When under stress, ROS levels increase and impose oxidative stress in the cells. To cope with the increased levels of ROS, plants elevate the production of antioxidants through various defence systems (Allen 1995; Apel and Hirt 2004; Mittler 2017).

Plants activate multiple complex signalling pathways during different types of abiotic and biotic stresses. Signalling cascades induced by the sensing of stress trigger and change expressions of specific genes used in stress defence. Proteins and enzymes participating in these pathways function as agents for ROS detoxification; they induce signalling cascades such as nitrogen-activated kinase and salt overly sensitive kinase; they play a role in transcriptional control, and alter water and ion uptake and transport (Blumwald 2000; Scandalios 2005; Saibo et al. 2008; Choudhury et al. 2013; Flowers and Colmer 2015; Jain et al. 2018; Kosová et al. 2018).

In recent days, enhancing plant production through nanotechnology has shown promise as a new emerging strategy. Nanotechnology can be supplemental in alleviation of nutrition deficiencies, improvement of resistance to diseases and tolerance to hostile environments. Nanoparticle interaction with plants has been studied on various levels, including genetics, physiology, plant development, and changes in the morphology of plant organs, etc. Literature reports effects ranging from negative through neutral to positive, with some higher concentrations toxic to plants and lower concentrations having positives effects. Recently, nanoparticles have also been applied to mitigate the adverse effects of both biotic and abiotic stresses. Unlike conventional agricultural chemicals, the effectiveness of nanoparticles depends not only on their chemistry, dosage, repetition and time of application but also on nanoparticle size, shape, crystallinity, and surface properties (Misra et al. 2016; Wang et al. 2016a; Faraz et al. 2020; Landa 2021). There is a trend in agriculture to reduce bulk fertilizers in particle size to nanometer sizes to increase their efficiency. Liu and Lal (2015) proposed to categorise these nanofertilizers according to plant nutrition into (1) macronutrient-nanofertilizers that incorporate elements such as P, K, N, Mg, and Ca; (2) micronutrient fertilizers incorporating Zn, Mn, Fe, Cu, Mo, etc.; (3) nanomaterials-enhanced fertilizers, such as nutrient-loaded zeolites with silica, carbon-coated Fe, polymers etc.; and (4) plant-growth enhancers with unclear mechanisms of action, e.g. TiO_2 or carbon nanotubes.

Zinc is an element, a micronutrient, essential for most organisms living on the planet, including plants, fungi, animals, and humans. There is a wide range of studies concerned with the effects of zinc on the growth and proliferation of plants since it is an essential part of more than 300 enzymes in organisms. It is the only metal that is included in all six groups of enzymes, and as such, it is present in oxidoreductases, transferases, hydrolases, lyases, isomerases, and ligases (Alloway 2008, 2009). Most of the processes in plants, such as photosynthesis and production of DNA and RNA, depend on zinc being present in some form. Therefore, amendment with zinc is essential, and it is used to support cereal, vegetable, and forage production (Alloway 2008; Faraz et al. 2020). In plants, zinc is required to metabolise carbohydrates, synthesise enzymes, maintain the integrity of cell membranes, regulate auxin synthesis, and create pollen (Alloway 2008). It also regulates the expression of genes important in tolerance toward environmental stresses, such as high intensity of light or high temperature. Zinc deficiency in plants is expressed by abnormal growth of plant structures. During acute deficiency, visible signs include slowed growth, chlorosis of leaves, reduction of leaf area, and sterility. Quality of crops is threatened, including protein content, appearance, and size of the fruit or seeds. Tolerance towards heightened intensity of light, and some fungal infections is diminished. Under slight deficiency of zinc in soils, quality and quantity of produced crops is diminished without visible hinderance in growth of the plants (Alloway 2008, 2009). For humans and higher organisms, zinc is referred to as a "type 2" nutrient, meaning that its concentration in the blood does not decrease in proportion to its deficiency in the organism. Zinc deficiency results in slowed physical growth, and its secretion is lowered since the organism strives to retain it. Many children with this deficiency have stunted growth. The recommended daily intake of Zn is between 3 and 16 mg•day^{-1}, and this value depends on age, sex, diet and several other factors. Roughly, one-third of the human population suffers from zinc deficiency in their diet. The relative size of the population afflicted is highly specific for each country and ranges from 4% to 73% (Alloway 2009).

Zinc oxide (ZnO) is an amphoteric oxide with low solubility in pure water. However, it is easily soluble in many acids. It crystallizes almost exclusively in the hexagonal wurtzite type structure known in the mineral classification system as zincite (Fig. 12.1). On rare occasions, it crystalises in cubic structure (Borysiewicz 2019). In lattice, Zn^{2+} cation is coordinated four O^{2-} anions in tetrahedron arrangement (Fig. 12.1), and reciprocally, each anion of O^{2-} can be thought of as a

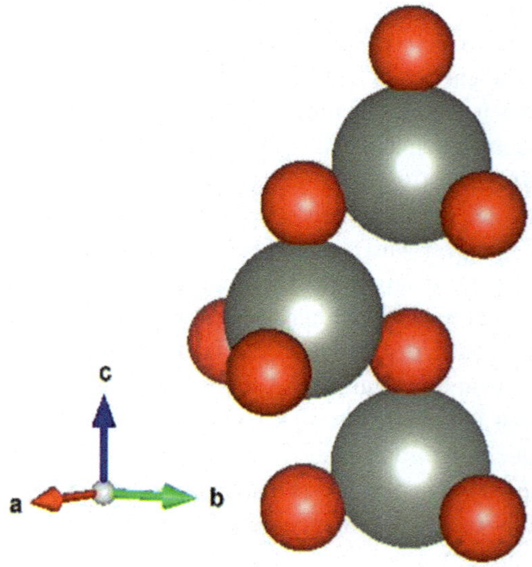

Fig. 12.1 Geometrical arrangement of atoms with tetrahedron coordination in unit cell of ZnO (Zn in grey, O in red) constructed and visualized using VESTA program in space-filling regime (Momma and Izumi 2011)

tetrahedron coordinated with four cations of Zn^{2+}. Zincite has sp^3 hybridized bonds with almost equivalent covalent and ionic nature (Borysiewicz 2019). As such, ZnO is a wide gap semiconductor (Eg = 3.37 eV) (Klingshirn et al. 2010). ZnO NPs have many useful properties, such as high binding energy, high refractive index, high thermal conductivity, piezoelectric properties, high absorbance of UV light, and antibacterial properties and it is applied in a wide variety of uses and products (Moezzi et al. 2012). As a nanomaterial, ZnO exhibits a broad range of shapes, from flowerlike structures to nanorods and nanoparticles. And since it is easy to manipulate its shape and to dope it, ZnO nanomaterials have been used in, for example, rubber production, concrete composites, electronics, solar panels, cosmetics, medicine, biosensors, food packaging and food products, and agriculture (Moezzi et al. 2012; Sabir et al. 2014). ZnO nanomaterials have been produced in a volume of approximately 30 kilotons in 2014, with a predicted rise in production (Future Markets Inc 2016). ZnO nanoparticle (ZnO NP) input in the environment was predicted to be in the range of 0.01–0.03 $\mu g \bullet kg^{-1} \bullet y^{-1}$ for soils (unintentional release), and levels of ZnO NPs of 0.05–0.29 $\mu g \bullet l^{-1}$ were predicted for fresh waters (Sun et al. 2014). Properties of ZnO NPs related to their UV protection, easily adjustable size, shape, and surface properties, and their antimicrobial and antifungal properties were shown to positively impact the growth and health of plants (Tarafdar et al. 2012; Raliya and Tarafdar 2013; Sabir et al. 2014; Raliya et al. 2015, 2016, 2018), and to help plants with coping with environmental stresses (Saxena et al. 2016; Hussain et al. 2018; Rizwan et al. 2019a, c). Under appropriate concentrations, ZnO NPs were found to increase seed germination (García-López et al. 2018), growth (Singh et al. 2019), photosynthesis (Faizan et al. 2018), activity of antioxidant enzymes (Venkatachalam et al. 2017), production of chlorophyll (Reddy Pullagurala et al. 2018a), proteins (Venkatachalam et al. 2017; Salama et al. 2019), oil, and seeds (Kolenčík et al. 2019, 2020), and they increased uptake of micronutrients (Peralta-Videa et al. 2014). They were also found to alleviate abiotic stresses, e.g. drought (Kolenčík et al. 2019; Dimkpa et al. 2020a), heavy metals (Rizwan et al. 2019c), salt (Torabian et al. 2016; Wan et al. 2020), and temperature (Fig. 12.2) (Hassan et al. 2018).

ZnO NPs play an important role in plant development, photosynthesis, and other critical physiological systems of plants. Also, ZnO NPs were applied under stress conditions and found to increase the tolerance to both biotic (herbivores and pathogens) and abiotic (heavy metals, heat, cold, drought, flooding, and salts). Up-to-date knowledge on the positive effects of ZnO NPs on (crop) plants is shortly summarized. We discuss current knowledge gaps connected to the research on environmental stress alleviation and propose research areas that may further our understanding and help with best practices in applying ZnO NPs.

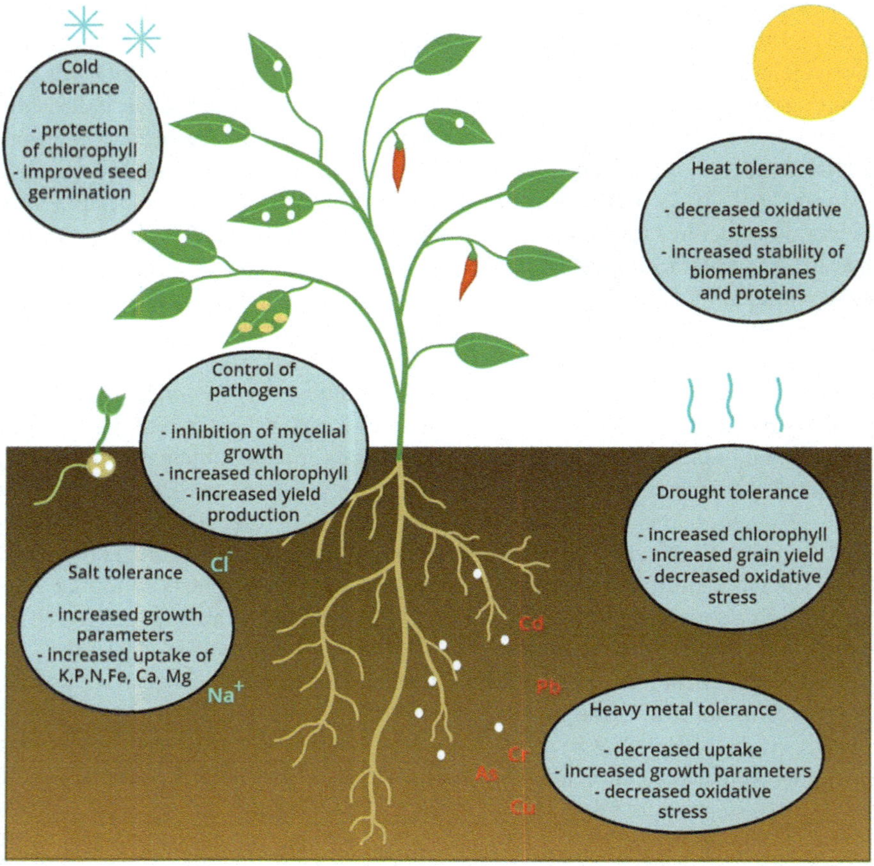

Fig. 12.2 Effects of ZnO NPs reducing abiotic and biotic stress in plants

12.2 Effect of ZnO Nanoparticles' Properties on Biological Interaction in Soils and Colloids

One of the most important properties of ZnO NPs is their ability to be transported in soils and their interaction with soil constituents. In an acidic environment, ZnO NPs can create large aggregates and dissolve easily (partially or fully), and the behaviour of released Zn ions is the most important in plant bioavailability (Bian et al. 2011; Mohd Omar et al. 2014; Sirelkhatim et al. 2015; Šebesta et al. 2020a). In more alkaline environments, the dissolution decreases dramatically, and under some circumstances, ZnO NPs can be more easily transported compared to ionic Zn (Šebesta et al. 2020a), even though the binding to various soil constituents may be similar to both forms (Šebesta et al. 2020b). The most important soil constituents that interact with ZnO NPs are living and dead organic matter, oxyhydroxides of Al, Fe and Mn, $CaCO_3$ and clay fraction (Fig. 12.3) (Bian et al. 2011; Zhao et al. 2012;

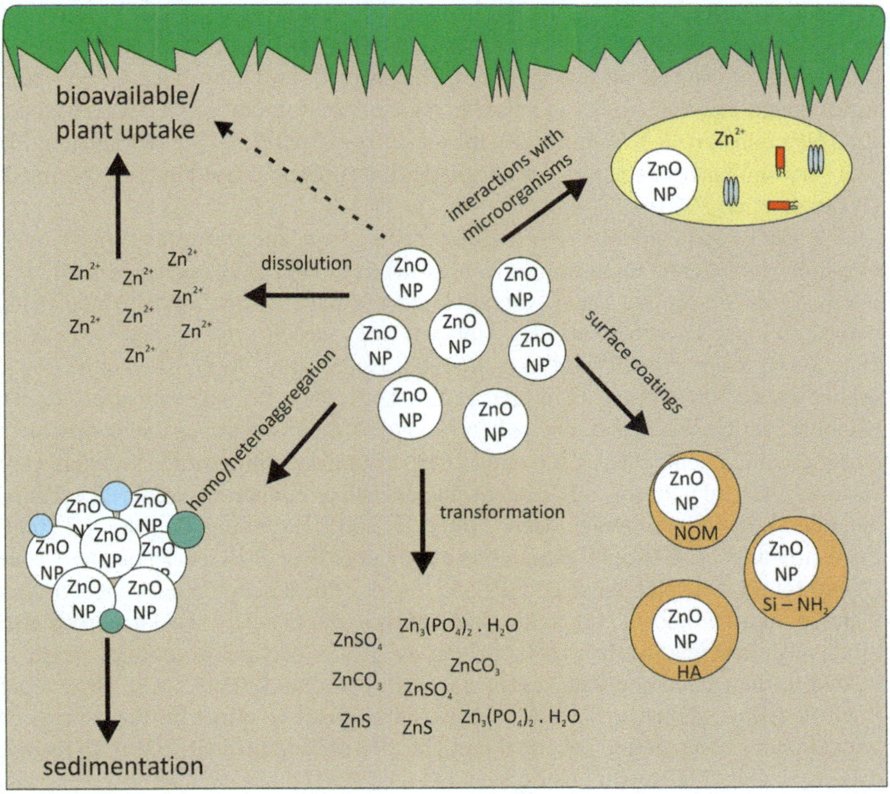

Fig. 12.3 ZnO NPs and their interaction in the soil environment

Mohd Omar et al. 2014; Han et al. 2016; Polák et al. 2019; Šebesta et al. 2019, 2020a). The soils pore water characteristics, such as ionic strength, concentration of Ca^{2+}, SO_4^{2-}, and different phosphates also play an important role, mainly in ZnO NPs' transformation and aggregation (Sivry et al. 2014; Xu et al. 2016; Peng et al. 2017). The behaviour of ZnO NPs was also dependent on the properties such as concentration (Yung et al. 2015; Šebesta et al. 2020a), particle size and surface area (Bian et al. 2011), and surface coating (Gelabert et al. 2013).

The size of ZnO NPs is important to several processes that happen in soils and during interaction with plants. The dissolution behaviour of ZnO NPs depends on their size, and smaller nanoparticles dissolve more readily (Meulenkamp 1998; Bian et al. 2011; Chang et al. 2012; Mudunkotuwa et al. 2012). The size of the nanoparticles and their aggregates is also important when absorption by plants is considered, and smaller nanoparticles can be more readily absorbed since they do not need to dissolve (Dietz and Herth 2011; Molnárová et al. 2015). Transfer of nanoparticles is hindered by the pores of plant cell walls, and they most probably only allow a passage of particles smaller than 20 nm (Wang et al. 2016a) or 5 nm (Gogarten 1988) to the apoplast. Because of the changes in dissolution and ability to penetrate deeper into plant tissues, nanoparticle size is also important in toxicity, including toxicity towards plants (Nair et al. 2009; Chang et al. 2012; Nemček et al. 2020). Size of ZnO NPs was important in the test with fava bean (*Vica fabia*) where it was found that the toxicity is linked to a greater dissolution of NPs where 25 nm ZnO NPs released ca 30% more Zn^{2+} than the 70 nm ZnO NPs. The ROS generation was very similar between the two NPs under the conditions used in the experiment (Pedruzzi et al. 2020).

The ZnO nanomaterials have highly tunable shape and form one-dimensional, two-dimensional, and three-dimensional structures (Thorny Chanu and Upadhyaya 2019). One dimensional ZnO structures are the most diverse and include belts, combs, helixes, nanorods, needles, ribbons, rings, springs, tubes, and wires (Pan et al. 2001; Kong et al. 2004; Liu et al. 2005; Huang et al. 2006; Chen et al. 2007; Wahab et al. 2007; Frade et al. 2012; Xu et al. 2012; Nikoobakht et al. 2013). Nanoplates/nanosheets and nanopellets belong to ZnO two-dimensional structures (José-Yacamán et al. 2005; Chiu et al. 2010). The three-dimensional structures are variously shaped with several different shapes such as coniferous urchin-like forms, dandelions, flowers, snowflakes, etc. (Liu et al. 2006; Bitenc and Crnjak Orel 2009; Polshettiwar et al. 2009). In a study comparing the effect of shape, Zhou and Keller (2010) found that spherical ZnO NPs had higher critical coagulation concentration than the mixture of ZnO NPs in the shape of rods and platelets. ZnO nanorods also dissolved more readily than the spherical ZnO NPs of similar volume, which is related to their uncompensated surface energies (Joo and Zhao 2017). Hexagonal ZnO NPs were shown to have a slightly higher positive effect on the growth of tomato plants compared to spherical particles (Pérez Velasco et al. 2020). In theory, the interaction of NPs with plant surfaces is related to (1) the chemistry and crystallinity of the nanoparticles; (2) nanoparticle surface in contact with the surrounding environment, and its shape, roughness, charge, and surface energy; and (3) the physicochemical properties of the environment the nanoparticles interact with, such as chemistry, input energy, e.g. sunlight or other types radiation, and changes in temperature (Konvičková et al. 2018; Holišová et al. 2019, 2021; Kolenčík et al. 2021).

Therefore, ZnO NPs are often surface modified to adjust their properties like aggregation and interaction with other constituents in the system. Nanoparticles with negative surface charge often behave more similarly to conservative tracers (such as Br^-) in model porous media, whereas positively charged nanoparticles are retained to a much higher degree (Yecheskel et al. 2016). Also, surface capping changed the toxicity of ZnO NPs toward *E. coli* bacteria and the cancer cell line, where starch-capped ZnO NPs had the lowest toxicity (Nair et al. 2009). The surface coating was found to be an important factor affecting the toxicity of ZnO NPs, and different coatings either increased or decrease cell damage or stress (Le et al. 2016). Beans (*Phaseolus vulgaris*) were exposed to bare and Z-COTE HP1® coated ZnO NPs. Coated NPs promoted more root growth and increased the concentration of nutritional elements (B, Mg, Mo, and S) compared to bare ZnO NPs (Medina-Velo et al. 2017; López-Moreno et al. 2018). Surface defects also increase the toxicity of ZnO NPs (Persaud et al. 2020).

In an experiment measuring oxidative stress response, ZnO NPs doped with Mn and Co showed increased response compared to pristine uncoated ZnO NPs. The Fe doped ZnO NPs showed a similar response when compared with pristine uncoated ZnO NPs (Le et al. 2016). Se doped ZnO NPs exhibited decreased toxicity to *Esheria coli* even though they showed higher production of reactive oxygen species due to Se leaching from NPs in culture media (Dutta et al. 2014).

To create more benign ZnO NPs for agricultural application, a biologically induced synthesis with plant, fungi or microbial extracts was used in studies. There is some evidence that there may be a synergy of nanoparticle-sized effect together with residual effects of organic extracts that are bound to nanoparticles surfaces (Gebre and Sendeku 2019). For example, Chaudhuri and Malodia (2017) biosynthesised ZnO NPs with leaf extract of *Calotropis gigantea. They* applied the created ZnO NPs on three plant species (*Azadirachta indica, Alstonia pinnata, a Pongamia scholaris*). All three species showed improvement in height after 6 months when compared to control. Biosynthesis is a new trend in the application of ZnO NPs in agriculture that may lead to better crop production with lower side effects.

12.3 Multiple Effects of Exposure Pathways

Several pathways of exposure are typical for the agricultural application of ZnO NPs. ZnO NPs were applied (1) on seeds to evaluate germination and early growth of seedling (Umavathi et al. 2020; Khan et al. 2021; Rai-Kalal and Jajoo 2021), (2) into the soil (or growth medium) to evaluate the root uptake and its toxicological and beneficial effects (Nemček et al. 2020), and lastly, (3) foliar application, that applies ZnO NPs directly on plants, mainly leaves, and may be good in reducing the number of nanoparticles needed to induce beneficial effects in plants (Kolenčík et al. 2020; Adrees et al. 2021).

12.3.1 Seed Application

Application of ZnO NPs on seeds was tested on various plants, and low concentrations had a generally positive effect on germination, seed vigour index, and the photosynthesis of seedlings (Dileep Kumar et al. 2020; Itroutwar et al. 2020b; Maslobrod et al. 2020; Rafique et al. 2020; Rani et al. 2020; Rawashdeh et al. 2020; Younes et al. 2020; Awan et al. 2021; Khan et al. 2021). Submerging seeds in ZnO NPs suspensions for 1, 2, 3, or 18 hours had a positive effect at a range of concentrations from 0.05 $mg{\bullet}l^{-1}$ to 2000 $mg{\bullet}l^{-1}$, and the concentrations that improved the germination the most were very plant species-specific (Itroutwar et al. 2020b; Kasivelu et al. 2020; Awan et al. 2021; Rai-Kalal and Jajoo 2021). In Tymoszuk and Wojnarowicz (2020), *Allium cepa* seeds were grown on the modified Murashige and Skoog (MS) medium spiked with ZnO NP at concentrations of 50, 100, 200, 400,

800, 1600, and 3200 mg•l^{-1} and the highest germination was recorded at 800 mg•l^{-1}. ZnO NP concentration of 3200 mg•l^{-1} decreased the germination. Youssef and Elamawi (2020) found that concentrations of 50 mg•l^{-1} of ZnO NPs were positively affecting germination of *Vicia faba*, while higher concentrations, higher than 100 mg•l^{-1}, had a negative effect. Corn (*Zea mays*) was primed with 2, 4, 8, and 16 mg•l^{-1} of 16 to 20 nm ZnO NPs biosynthesized with *Bacillus* sp. for 24 h. Root length, shoot length, and protein concentration was increased after growth in pots, and the maximum increase was observed at 8 mg•l^{-1} (Sabir et al. 2020). Generally, priming with higher concentrations of ZnO NPs has an inhibitory effect on germination and the early growth of plants, and lower concentrations have a positive impact. The positive effect of ZnO NPs was found to be higher than in their ionic counterparts, at least for lentil (*Lens esculentum*) and chick pea (*Cicer arietinum*) (Choudhary and Khandelwal 2020) and higher than bulk ZnO when applied on corn plants (*Zea mays*) (Esper Neto et al. 2020).

12.3.2 Soil Application

ZnO NPs can be applied to soils to ameliorate the Zn deficiency that is one of the most widespread deficiencies in plants, and it affects up to one-third of agricultural soils, mainly in tropical and subtropical regions (Alloway 2008, 2009). The germination of black mustard (*Brassica nigra*) in soils contaminated with high concentrations of ZnO NPs (200, 400, and 600 mg•kg^{-1}) caused inhibition in seed germination and had a negative effect on the root length and height of plants. Also, phenolics and flavonoids, which play a vital role in the detoxification of ROS, were increased compared to control without applied ZnO NPs. However, an increase in leaf area was observed for 200 and 600 mg•kg^{-1} (ur Rehman et al. 2020). Similarly, an increase of leaf area and also the stem height, number of leaves, number of branches, and number of nodes per black mustard plant was observed in a study by Zafar et al. (2020) when 200, 400, and 600 mg•kg^{-1} ZnO NPs were applied to soil, but a decrease in seed diameter and number of pods per plant was also observed. Application of ZnO NPs also resulted in a higher accumulation of Ca, Co, and Zn in seeds along with protein, glucosinolates and erucic acid (Zafar et al. 2020). When black mustard (*Brassica nigra)* was grown for 30 days in ZnO NPs enriched soil, concentrations below 400 mg•kg^{-1} ZnO NPs had a positive effect on root length, and higher concentrations caused increased oxidative stress (Zafar et al. 2019). In an early growth experiment with barley (*Hordeum vulgare*), only high concentrations of ZnO NPs (2000 mg Zn•kg^{-1} as ZnO NP) had a negative effect on average fresh and dry weight and root and sprout length (Nemček et al. 2020). In a 35-day growth experiment with cilantro plants (*Coriandrum sativum*), 100, 200, and 400 mg•kg^{-1} of ZnO NPs were used, and concentration of 100 and 200 mg•kg^{-1} increased the concentration of chlorophyll by at least 50% and changed the carbinolic-based compounds. The concentration of 400 mg•kg^{-1} of ZnO NPs had a negative effect and decreased lipid peroxidation by 70%. Application of ZnO NPs to soils alone or in combination with organic matter did not influence soil chemical properties, however, they may influence microbial properties and may support bacterial growth over fungal growth (Aziz et al. 2019). Nanoparticles may enhance the nutrient mobilization in soils via influence on the soil microbial population and extracellular enzymes secretion such as urease or phosphatase activity in soils which play an important role in the regulation of plant available nitrogen and phosphorus (Olander and Vitousek 2000; Raliya and Tarafdar 2013; Raliya et al. 2016, 2018). Nonetheless, there are indications that ZnO NPs may enhance microbial activity and increase their biomass that could lead to N immobilization in soils, and thus, decreasing its plant uptake (Aziz et al. 2019). Higher levels of chlorophyll, nitrogen and micronutrients such as zinc, magnesium, and potassium were observed when lettuce (*Lactuca sativa*) and carrot (*Daucus carota* subsp. *sativus*) plants were grown in commercial soil substrate spiked with 1, 5, 20, 100, and 1000 mg•kg^{-1} ZnO NPs, where the highest concentration had negative effects and the other concentrations showed mostly positive or neutral effects on both plants (Song and Kim 2020). After 120 days of growing in soil, soybean (*Glycine max* cv. Kowsar) seed yield was evaluated in experiments with different concentrations of ZnO NPs with different sizes and morphologies and Zn^{2+} to compare ionic vs nanoparticle influence. All Zn compounds (ZnO NPs, and Zn^{2+}) increased seed yield when applied at a concentration up to 160 mg Zn•kg^{-1}. At higher concentrations, ZnO NPs were toxic, with the highest toxicity elucidated by spherical 38 nm NPs and evidence suggested some nano-specific toxicological effect when compared with Zn^{2+} (Yusefi-Tanha et al. 2020). In experiments with either foliar or soil application of ZnO NPs, tomato plants (*Solanum lycopersicum*) were positively influenced by both application, i.e. their height, stem diameter and leaves, stem and root dry weight was increased, with little actual difference between the two applications (Pérez Velasco et al. 2020). Similarly, soil and foliar application led to similar Zn distribution of zinc in wheat grain (Doolette et al. 2020). Umar et al. (2020) found that foliar application of ZnO NPs on corn had the highest positive effect on the concentration of Zn in grains when compared to both soil application of ZnO NPs and more conventional Zn fertilizers and also ZnO NPs application promoted plant growth and seed yield.

12.3.3 Foliar Application

In general, uptake of nutrients via leaves, including uptake from nanoparticles and nutrient distribution in leaves, is still less well-known in comparison to nutrient translocation in root systems in the soil environment (Li et al. 2019). It depends on several factors such as concentration, particle size, the chemical composition of NPs, the timing and number of applications, plant species, etc. (Nair et al. 2010; Servin and White 2016; Wang et al. 2016a).

Currently, foliar application of ZnO NPs to plants has two distinct benefits: (1) decrease in amounts of agrochemicals used, and (2) gradual release of Zn from the

NPs (Prasad et al. 2014; Wang et al. 2016a; Li et al. 2019). In contrast, corresponding conventional ionic Zn fertilizers are absorbed faster through leaves and are more readily metabolized in plants. Additionally, our research indicates (Kolenčík et al. 2019, 2020) that in the case of ZnO NPs, similarly to TiO_2 NPs, their nano-domain effect may contribute to elevated photosynthetic activity in plants via the protection of chloroplasts or their photocatalytic properties (Siddiqui et al. 2019; Rizwan et al. 2019b). Sunlight exposition may also support photo-corrosion and dissolution of ZnO NPs (Ma et al. 2014), and the released ionic zinc is easily absorbed and utilized by the plant. However, there are still knowledge gaps when it comes to their application in field conditions. In this context, their foliar application may be an effective measure in precision agriculture that aids in adaptation to climate change (Kolenčík et al. 2019, 2020).

There are three potential pathways for foliar absorption of ZnO NPs or Zn ions released from them: cuticular, stomatal, or through trichomes (Li et al. 2019). After the application of ZnO NPs on the sunflower, there was a qualitative change in leaves in the flower bud stage of the life cycle of the plants. Trichomes diversity, ratio, width, and length were change, and a new type of trichomes was observed – capitate glandular trichomes (Kolenčík et al. 2019). Leaf part with different structures and the semi-quantitative analysis of leaf surface chemistry is shown in Fig. 12.4.

Zinc released from ZnO NPs or conventionally applied zinc, such as $ZnSO_4$, is transformed after the absorption to leaves via complexation with carboxylic groups (oxalate, pectine, citrate), phytate and cysteine, or it stays in Zn^{2+} form, with forms and their relative concentrations varying in different plants and plant stages (Li et al. 2019; Kolenčík et al. 2019).

Foliar application of 10 mg•l^{-1} ZnO NPs, 15–52 nm in size, applied twice, before and after flowering, improved grain weight, seed length, seed thickness, and seed in

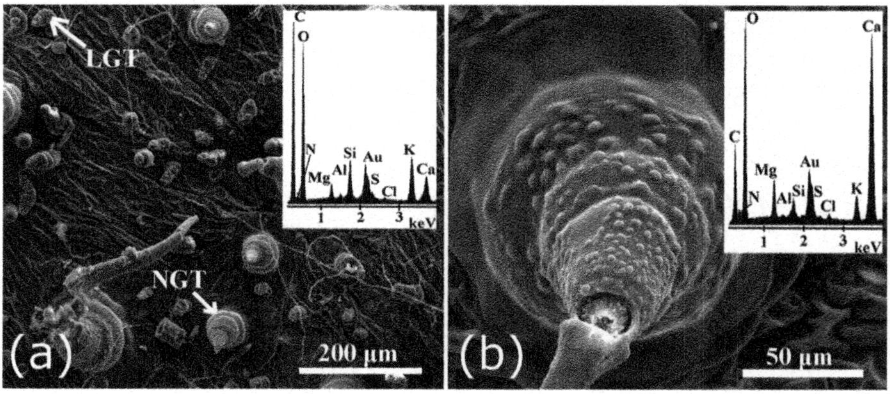

Fig. 12.4 (**a**) Sunflower leaf surface visualized by scanning electron microscopy with two types of trichomes – non-glandular trichomes (NGTs) and linear glandular trichomes (LGTs); (**b**) details of NGTs; semiquantitative analysis of element by energy-dispersive X-ray spectroscopy on surface of control (**a**) and ZnO NPs foliated leaves (**b**)

rice (*Oryza sativa*) (Itroutwar et al. 2020a). Foliar application of ZnO NPs also leads to transfer of Zn to grains, as was observed in the work of Doolette et al. (2020), where 40–50 nm ZnO NPs were applied on wheat (*Triticum aestivum*). ZnO NP application at 750 mg•l^{-1} led to a higher grain yield than in $ZnCl_2$ application. However, the authors did not have an explanation for this phenomenon, and all other measured parameters did not show a better efficacy compared to the conventional application of Zn ($ZnCl_2$ and Zn EDTA). Read et al. (2019) found that ZnEDTA had higher foliar uptake than ZnO NP in wheat (*Triticum estivum*), and already taken up ZnO NP and ZnEDTA were transported in a similar way to newly formed leaves. When applied on corn, ZnO NPs showed a higher promotion of growth and seed yield compared to more conventional fertilizers (Bala et al. 2019). Two application of low concentration of ZnO NPs at crucial points of foxtail millet (*Setaria italica*) growth positively affected several parameters necessary for crop production (Kolenčík et al. 2019). Foxtail millet plant grains had significantly higher oil and total nitrogen contents and a significantly lower crop water stress index (Kolenčík et al. 2019). Two applications of ZnO NPs at low concentrations had a positive effect on head diameter, dry-seed head weight, yield and thousand seed weight of sunflower, and also on sunflower physiological responses (*Helianthus annuus*) (Kolenčík et al. 2020). When applied on sunflowers at 60 mg•l^{-1} on day 25 and 45 after sawing, ZnO NP decreased the uptake of Cr and Pb and increased the uptake of Fe and had an even better effect when applied at 30 mg•l^{-1} in combination with rice straw biochar and cow manure biochar and additionally reduced Cu, Ni, and Cd concentrations in plants (Seleiman et al. 2020). Generally, the foliar application is preferred as less of the ZnO NP is needed overall, and the ZnO NP do not contaminate the soil to such a degree that it could be detrimental to the growth of other, more susceptible plant species grown later at the same field.

12.3.4 *Effect of Applied Nanoparticle Concentration and Soil Properties*

Despite multiple studies showing that ZnO NPs can have an influential role in the growth and development of plants, they also may have a detrimental effect on plants and the environment when applied at too high concentrations and by an inappropriate method. Release of high concentrations of Zn ions and production of ROS due to ZnO NPs higher reactivity can inflict damage to organisms in contaminated environment (Manke et al. 2013; Rajput et al. 2018). The toxic effects of ZnO NPs were studied on a large variety of organisms, including bacteria, fungi, plants, invertebrates and vertebrates, and their effect on soil health was also studied (Rajput et al. 2018). Under natural sunlight, ZnO NPs became much more toxic than during laboratory light and dark conditions to free-living nematode *Caenorhabditis elegans* and the increased toxicity was related to ROS generation (Ma et al. 2011). Seed soaking and exposure of roots to higher concentrations of ZnO NPs may lead to diminished root growth (Yang et al. 2015), reduction of plant growth, and photosynthesis (Wang et al. 2016b), reduction in catalase and ascorbate peroxidase in roots (Mukherjee et al. 2014). Tripathi et al. (2017a) found a decrease in ascorbate peroxidase, glutathione reductase, dehydroascorbate reductase, and monodehydroascorbate reductase, whereas ascorbic acid *þ* dehydroascorbic acid, and ascorbic acid increased when ZnO NPs were applied on wheat seedlings. The toxicity may come from the dissolution of Zn inside and/or outside roots and from damage caused by direct contact of roots with ZnO NPs (Lin and Xing 2008).

When applied at lower, appropriate concentrations, ZnO NPs may have a positive effect on plant growth. At higher dosages (>500 mg•kg^{-1}), ZnO NPs may often have a toxic effect on plants via the release of Zn ions and ROS production. While at lower concentrations (50 mg•kg^{-1}) or when applied on leaves or through seed coating or priming applications, they often have beneficial effects and are promising plant growth promoters, nanofertilizers, or nanopesticides. When studying impacts on plant growth, it is also advisable to ascertain that the ZnO NPs do not have an inhibitory effect on beneficial soil bacteria (Reddy Pullagurala et al. 2018b). The positive effects of ZnO NPs include an increase in shoot length, root length, fresh and dry biomass, protein content, an increase in other phytochemicals of agricultural use and an increase in photosynthetic activity. Upon application of ZnO NPs, expression of genes is altered, and various effects on biochemistry, physiology, and plant morphology have been observed (Tripathi et al. 2017b; Thorny Chanu and Upadhyaya 2019; Kolenčík et al. 2019, 2020; Faraz et al. 2020).

A different number of applications of ZnO NPs, two and four, were applied on pinto bean (*Phaseolus vulgaris*) cultivars, and the four applications (0.05%, 0.1%, or 0.15% w/v) showed an increase in plant height and internode length compared to two applications of ZnO NPs and also compared to two and four applications of $ZnSO_4$ and chelated Zn at similar concentrations. Four applications of 0.05% ZnO NPs were more effective than two applications of 0.1% or 0.15% for several measured parameters (Mahdieh et al. 2018), hinting to possibility that a higher number of applications may be more beneficial for plants, even if the total concentration of the applied ZnO NPs is similar.

Soil factors influence the response of plants to metallic nanoparticles. Both chemical factors, such as pH, organic matter, and ionic strength, and biological factors such as plant root exudates, microbes, and microbial activities, heavily influence what effects nanoparticles have on the growth and health of plants (Dimkpa 2018). One of the most important factors, soil pH, has a considerable influence on the effect of ZnO NPs on plants. More acidic pH leads to their dissolution, and their association with soil chemicals lead to similar behaviour compared to the application of ionic Zn (Wang et al. 2013). Wheat was grown in both alkaline and acidic soil, and the soil-applied ZnO NPs had a positive effect in alkaline soil but a negative effect in acidic soil (Watson et al. 2015; Anderson et al. 2017). Similar effect was observed by García-Gómez et al. (2018b) in nine plant crop species. This behaviour can be reverted by creating ZnO NP- alginate complexes that release Zn more slowly in acidic soils and thus can lower their toxicity and increase their usefulness (Martins et al. 2020). Similarly, soil enzymes produced by soil microbes were adversely affected by ZnO NPs in acidic soil more than in calcareous soil (García-Gómez et al. 2018a), which can also negatively affect the growth and health of plants.

When ZnO NPs were applied on sewage sludge-amended soil, they enhanced root growth at a concentration of 50, 250, and 500 mg•kg^{-1} compared to the soil with just sludge-amendment. The root growth enhancement at even a relatively high concentration may be due to a high organic matter content (Oleszczuk et al. 2019). Similar results were observed in collected manure-amended soil where higher concentrations of ZnO NPs (1000 mg•kg^{-1}) affected shoot weight negatively in unamended soil and positively in manure-amended soil (Moghaddasi et al. 2017).

12.4 Amelioration of Stress by ZnO NP

Plant stress is described as any unfavourable condition that affects the metabolism, growth or/and development of a plant. Plant stress can be caused by multiple factors, which are generally divided into two categories, biotic and abiotic stresses (Kranner et al. 2010). Biotic stresses are induced by living organisms like microorganisms, insects, viruses or other plant species, and abiotic stresses are initiated by environmental factors, for example, drought, salinity, and temperature (Hakim et al. 2018; Thakur et al. 2019). As a response to various types of stresses, plants have evolved immune systems and defence responses that increase their tolerance to environmental stress. Therefore, a broader study and understanding of plant tolerance mechanisms can benefit agriculture (Almutairi 2019).

The alleviation of environmental stress on crops by applying NPs has been a significant trend in the agricultural research of the last decade, as nanotechnology has been shown to be a promising tool for enhancing plant production by improving disease resistance and plant tolerance to a harsh environment. Various sources of stress, which are discussed in the following subchapters, can be mitigated by NPs (Almutairi 2019). This mitigation role of NPs depends on NPs' size, shape and dosage, as some concentrations have been toxic for plants while lower concentrations have positive effects (Jha and Pudake 2016; Siddiqi and Husen 2017).

12.4.1 Biotic Stress

Biotic stresses, like herbivore grazing and pathogen infection, are essential factors affecting crop production. The attacker has to defeat many defence strategies that plants deploy against the intruder (Thordal-Christnsen 2003; Zhao et al. 2020). In the case of pathogen infection, it requires the interactions of a susceptible host, virulent pathogen and conducive environment. Even though conventional pesticides can significantly increase agricultural production, they can also cause health and environmental risks. Therefore, the application of various engineered metal NPs, including ZnO NPs, was considered a more gentle way to protect plants from pathogen invasion or pest and insect attacks (Poschenrieder et al. 2006; Zhao et al. 2020).

12.4.1.1 Herbivores

Higher metal ion activity in the soil or on the plant surface may deter, kill or inhibit the development of herbivores. Especially for chewing herbivores, consuming polluted leaves can lead to suffering from the detrimental effects of metals. Also, herbivores eating plant tissues with high metal concentrations can be affected by the toxicity or the evocation of an aversion response. Therefore insects can learn to avoid feeding on plants with higher concentrations of metals through a post-ingestive feedback mechanism (Eeva et al. 1998; Behmer et al. 2005; Poschenrieder et al. 2006). There are many studies (Noret et al. 2005; Stolpe et al. 2017) dealing with zinc applications against stress caused by herbivores. For example, the performance of caterpillars, either chewing or sucking species, on *Arabidopsis halleri* (*Brassicaceae*) was reduced on plants grown on zinc-amended soil compared to plants grown on unamended soil (Stolpe et al. 2017). Therefore, we find it essential that ZnO-NPs, a form of particulate Zn, should also be studied. Its more gradual release in soils may be advantageous, and also foliar application may be more effective because, unlike ionic Zn that is readily absorbed by leaves, they may stay for a longer period of time on leaf surfaces and thus have a longer-lasting protective effect on the plants.

12.4.1.2 Pathogens

At least 25% of crop losses worldwide is due to plant parasites. Conventional synthetic fungicides are considered to be the most effective for plant diseases (Pandey et al. 2018; Malandrakis et al. 2019). However, pathogens can become resistant to fungicides because of long-term exposure, and residues of fungicides are also dangerous for human health and the environment (Zhang et al. 2015). NPs are promising in resolving this challenge in the future by providing a novel eco-friendly alternative to synthetic fungicides. ZnO NPs have been shown to be very effective antibacterial and antifungal agents against numerous species due to their unique physicochemical properties (Pandey et al. 2018; Sun et al. 2018; Malandrakis et al. 2019).

According to Malandrakis et al. (2019), ZnO NPs were able to inhibit *in vitro* mycelial growth of fungal strains in a dose-response manner. ZnO NPs were also more toxic at the spore germination level than at mycelial growth and more effective than the commercial fungicide containing $Cu(OH)_2$. Hafez et al. (2020) tested the application of bio-agent *Bacillus subtilis* with ZnO NPs to control powdery mildew in cucumber plants caused by *Podosphaera xanthii*. The application reduced electrolyte leakage, and the disease severity was correlated with the production of defence-related enzymes and early elevation of ROS levels. Total chlorophyll content and yield production were increased, along with most morphological and physiological characteristics and improved fruit yield. Savi et al. (2015) studied ZnO NPs treatment onto spikelets at the anthesis stage on wheat, inoculated with *Fusarium graminearum*. Results showed a reduction in the number of colonies of *Fusarium graminearum* in samples treated with ZnO NPs when compared to control. Deoxynivalenol (mycotoxin) formation in the grains was also reduced. The concentration of Zn remained within the internationally recommended levels for consumption, and the ZnO NPs treatment did not cause any damage to wheat grains. Biologically synthesized ZnO NPs using *Parthenium hysterophorus* reported maximum inhibition for *Aspergillus niger* and *A. flavus*. It was confirmed that smaller ZnO NPs have greater antifungal activity against fungal pathogens (Rajiv et al. 2013; Ingle et al. 2020). Still, more studies are needed to find the best ways of ZnO NPs application under field conditions and also to investigate their effects on a diverse range of pathogens.

12.4.2 Abiotic Stress

Abiotic stresses are estimated to be the primary factor of crop-production drops worldwide (Bajguz and Hayat 2009; Zhu 2016). ZnO NPs may enhance the defence mechanisms of plants against abiotic stresses by stimulating the activities of antioxidant enzymes and bettering the accumulation of osmolytes, free amino acids, and nutrients (Torabian et al. 2016; Hassan et al. 2018; Rizwan et al. 2019a). In Table 12.1., we show some of the known effects ZnO NPs have on the amelioration of abiotic stresses.

12.4.2.1 Heavy Metals

Heavy metal stress has become a global phenomenon causing various toxic effects at high concentrations and, thus, growth inhibition of crop plants. Although some heavy metals act as nutrients at lower concentrations, their excess in plants can lead to oxidative stress. High concentrations of some heavy metals in soil/growth medium can also increase ROS generation, denaturation of cell structures, cell membranes, and biomolecules (Sharma et al. 2012; Chibuike and Obiora 2014; Khan et al. 2017). For example, Cd can enter through roots and cause damage to the photosynthetic system, impairing plants growth and nutrient uptake and accumulation. Further, Cd affects the redox homeostasis of the plant cells and enhances ROS production. Even though plants have developed a defence system, it fails at elevated Cd stress (Bashir et al. 2018; Rizwan et al. 2019c, a). To combat the heavy metal stress, NPs have been applied to soils, and they were found to be effective in alleviating heavy metals stress in plants. They can easily penetrate into a contaminated zone due to their small size and large surface area and have a strong affinity to metals, where the metals make bonds with NPs, they are either adsorbed on their

Table 12.1 Impact of ZnO NPs on plants exposed to different abiotic stresses

Concentration of ZnO NPs (mg•l^{-1})	Plant species	Abiotic stress	Impact	References
25–100	*Triticum aestivum*	Heavy metals – Cd	Decreased concentrations of Cd in roots, shoots and gains; increased plant high, spike lenght and dry weight of shoots, roots, spikes and grains	Rizwan et al. (2019a)
75	*Gossypium hirsutum*	Heavy metals – Cd, Pb	Increased shoot, root growth and biomass under Cd, Pb stress; up-regulated chlorophyll a,b and carotenoids contents in leaves	Priyanka et al. (2021)
10–200	*Oryza sativa*	Heavy metals – As	Decreased As concentrations in roots and shoots; increased germination rate, shoot and root weight, chlorophyll content and promoted biomass	Wu et al. (2020)
60	*Helianthus annuus*	Heavy metals – Cd, Pb, Cu, Cr	Reduced availability of Cd, Pb, Cu and Cr in soil and its content in plant biomass	Seleiman et al. (2020)
10	*Triticum aestivum*	Heat	Enhanced heat tolerance by maintaining ROS production; reduced the permeability of the leaf cells' plasma wall, which decreased lipid peroxidation and protect the cellular wall	Hassan et al. (2018)
50	*Saccharum oddicinarum*	Cold	Lower reduction of chlorophyll *a,b* contents; increased carotenoids	Elsheery et al. (2020)
20–100	*Triticum aestivum*	Drought	Boosted up leaf chlorophyll contents, decreased oxidative stress and enhanced the leaf superoxide dismutase and peroxidase activities	Adrees et al. (2021)
50	*Glycine max;Sorghum bicolor*	Drought	Enhanced drought tolerance stress; improvement of shoot and root morphology	Linh et al. (2020), Dimkpa et al. (2019)
10–100	*Solanum lycopersicum*	Salts	Increased shoot lenght, root lenght, biomass, leaf area, chlorophyll content and photosynthetic attributes; ameliorate the negative effect of salt stress	Faizan et al. (2021)

(continued)

Table 12.1 (continued)

Concentration of ZnO NPs ($mg{\bullet}l^{-1}$)	Plant species	Abiotic stress	Impact	References
50; 5–10	*Linum usitatissimum; Triticum aestivum*	Salts	Improved the growth, carbon and nutrient assimilation; increased the antioxidant enzymatic system and other physiochemical reactions	Singh et al. (2021), El-Bassiouny et al. (2020)

surfaces or chemically bound in NPs and are, thus, immobilized and are less bioavailable to plants. ZnO NPs also release Zn ions that compete with Cd, Cu or other ions in soil solutions and limit their uptake (Khan et al. 2017; Tripathi et al. 2015; Worms et al. 2012).

Hussain et al. (2018), Khan et al. (2019) and Rizwan et al. (2019a, c) tested the effect of ZnO NPs on wheat under Cd stress. Both results showed increased dry weights of shoot, roots, spikes, and grains. The concentration of Cd in roots, shoots and grains were significantly reduced with ZnO NPs treatment. Rizwan et al. (2019a) further showed that ZnO NPs positively affected the photosynthesis of wheat and reduced the electrolyte leakage and superoxide dismutase and peroxidase activities in leaves of Cd-stressed wheat. Shah et al. (2021) showed that the combined application of ZnO NPs and *Bacillus fortis* IAGS 223 modulated the activity of antioxidant enzymes besides upregulation of the biochemicals and growth parameters of Cd stressed plants. They also found a decreased amount of stress markers (H_2O_2 and MDA) and a reduction of Cd content in shoots. In rice, foliar application of ZnO NPs decreased Cd uptake, and lower Cd content was found in rice roots and shoots (Ali et al. 2019). However, contrary to the studies mentioned above, Zhang et al. (2019, 2020) discovered that Cd bioavailability increased in high ZnO NPs (500 $mg{\bullet}kg^{-1}$) treatments.

Priyanka et al. (2021) tested the application of ZnO NPs (0–200 $mg{\bullet}l^{-1}$) on the development of Cd and Pb tolerance mechanism in cotton seedlings. ZnO NPs applications significantly promoted shoot and root growth as well as biomass under Cd and Pb stress. It also up-regulated chlorophyll *a,b* and carotenoids contents in leaves under Cd and Pb stress, along with the accumulation of antioxidant defence enzymes (CAT, POX, APX, SOD) and MDA contents. This indicates that the addition of ZnO NPs protects cotton seedlings by alleviating Cd and Pb stress. Sharifan et al. (2020) showed similar results after the application of ZnO NPs (100 $mg{\bullet}l^{-1}$) on different leafy greens (spinach, parsley and cilantro) under Cd and Pb stress. Results by Seleiman et al. (2020) showed positive effects after foliar applications of ZnO NPs (60 $mg{\bullet}l^{-1}$), rice straw biochar and cow-manure biochar on sunflowers under Pb, Cr, Cu and Cd stress. The application of the combination treatment reduced the availability of Pb, Cr, Cu and Cd in the soil by 78.6, 115.3, 153.3, and 178.5% in comparison to untreated plots, and it also reduced the Pb, Cr, Cu and Cd in plant biomass by 1.13, 5.19, 3.88, and 0.26 $mg{\bullet}kg^{-1}$, respectively.

Wu et al. (2020) described the role of ZnO NPs (10–200 mg•l^{-1}) in alleviating As stress in rice germination and early seedling growth. ZnO NPs increased the germination rate (2.3–8.9%), shoot weight (18.2–42.4%), root weight (5.2–23.9%), and chlorophyll content (3.5–40.1%), while elevated the SOD (2.2–22.8%) and CAT (7.2–60.7%) activities and reduced the MDA content (17.5–30.8%). The concentration of As was decreased by 8.4–72.3% in rice roots and 10.2–56.6% in rice shoots. ZnO NPs amendment increased As adsorption and promoted biomass of rice. Similarly, in the study by Wang et al. (2018), ZnO NPs reduced the accumulation of As(III) in rice roots and shoots when the As was applied as As(III) and As(V), and As(V) in rice roots. However, the concentration of As(V) in rice shoots was unaffected.

Interaction of heavy metals with ZnO NPs in plants still has gaps in knowledge, and, therefore, ZnO NPs application was studied in plants affected by Pb. Raghib et al. (2020) applied ZnO NPs, and ZnO NPs in combination with arbuscular mycorrhizal fungi. Both applications increased the growth and biochemical attributes of wheat and decreased the Pb uptake from contaminated soil. The combined formula of ZnO NPs and fungi has shown the best results, increasing growth parameters like plant height, fresh weight, dry weight, and total chlorophyll content. Also, the application of ZnO NPs with fungi had a positive effect on plant metabolism and increased proline content, H_2O_2 content, the SOD and CAT enzymes' activity, and increased lipid peroxidation content. The Pb concentration was reduced in both roots and shoots of wheat after applying ZnO NPs with fungi.

ZnO NPs have the potential to alleviate heavy metal stress in plants. However, more field studies are needed where the best mode of application is found. ZnO NPs also show promise in combined formulas with other treatments where they positively enhance the treatment effects.

12.4.2.2 Heat

Heat stress is defined as "the rise in temperature of both soil and air above the level of the threshold for a limited time such that permanent harm occurs to plants" (Lipiec et al. 2013). Generally, an impermanent phase when the temperature exceeds temperature tolerance by 10–15 °C is referred to as heat stress/shock, which reduces plant growth and crop productivity (Wahid 2007). Higher temperature also increases ROS production, and it causes oxidative stress and limits plant growth and yields. The chlorophyll content is also affected by heat stress, and low chlorophyll content is mainly present in leaves (Møller et al. 2007; Mathur et al. 2014; Faizan et al. 2020b). This could be caused by inhibited chlorophyll biosynthesis or enhanced degradation of chlorophyll pigments. The damaged chlorophyll biosynthesis under heat stress is a consequence of the presence of many heat-sensitive enzymes in the chlorophyll biosynthesis pathway (Mathur et al. 2014). An increase in leaf temperature can also lead to a deactivation of the heat-sensitive enzyme Rubisco (the enzyme responsible for CO_2 fixation during photosynthesis), initiating the photorespiratory pathway and generating H_2O_2 (a by-product of the pathway) (Sharkey 2005; Allakhverdiev et al. 2008).

According to Hassan et al. (2018), the application of ZnO NPs enhanced heat tolerance in wheat by maintaining ROS production and the stability of biomembranes and proteins. The treatment of ZnO NPs on wheat also reduced the permeability of the leaf cells' plasma wall, resulting in a decrease in lipid peroxidation and protecting the cellular wall against heat stress. The mechanism of heat stress amelioration by ZnO NPs is still poorly understood since we were able to find only one study that examined it. The heat tolerance may come from sufficient nutrition with Zn that increased levels of antioxidants, as was shown in chickpeas and winter wheat supplemented with ionic Zn (Peck and McDonald 2010; Ullah et al. 2019). However, Ag NPs also helped to increase heat tolerance in wheat (Iqbal et al. 2019) with a not well-understood process that may be linked to their nano-size. Therefore, ZnO NPs may be superior in protecting plants from heat stress compared to more conventional ionic zinc formulations.

12.4.2.3 Cold

Cold stress is abiotic stress, which can cause difficulties in plant growth and production. It is caused by temperatures cool enough (0–15 °C) to damage plants without forming ice crystals in plant tissues, whereas freezing stress (<0 °C) results in the formation of ice crystal in plant tissues (Hasanuzzaman et al. 2013). Plants exposed to cold stress suffer from loss of fluidity of membranes, leakage of solutes, poor growth and germination, and reduced crop yield. It also causes inhibition in chlorophyll levels, CO_2 assimilation, transpiration rate and degradation of Rubisco (Welti et al. 2002; Suzuki et al. 2008; Liu et al. 2012). Enhancement of carboxylation of Rubisco, the light absorption capacity of chloroplasts, electron transport rate, and inhibition of ROS generation have been described as alleviating effects of NPs on cold stress in plants (Gao et al. 2006; Giraldo et al. 2014; Khan et al. 2017; Ze et al. 2011). Foliar application of ZnO NPs may mitigate ROS generation by increasing enzymatic activities of superoxide dismutase, catalase, and peroxidase, and, more generally, they may prevent photoinhibition (Elsheery et al. 2020).

Elsheery et al. (2020) tested a foliar application of ZnO NPs ($50\ mg{\bullet}l^{-1}$) on sugarcane in an open field experiment. Results showed that during a cold front, chlorophyll *a* and *b* contents were significantly reduced, but the rate of reduction was lower in seedlings treated with ZnO NPs than that of the control group. In contrast, carotenoids were increased during the cold front. These effects demonstrate that ZnO NPs can mitigate the negative impact of cold stress in sugarcane. Maslobrod et al. (2020) treated winter wheat seeds with water dispersions of ZnO NPs and a mixture of bismuth, copper, zinc oxide NPs. Both treatments increased seed thermal stability, seed germination energy and length of coleoptiles while being exposed to low temperature (+4 °C).

12.4.2.4 Drought

Drought events are becoming more common as a result of anthropogenic influence on climate change that severely limits crop production. During these events, plants experience insufficient water uptake and, consequently, nutrient uptake that is related to a changed condition in a soil environment with limited amounts of capillary water present. During drought, plant growth and development are affected (Faizan et al. 2020a, 2021). Drought seriously disturbs plant growth, reducing the rate of cell division, leaf expansion, stem elongation, and water use efficiency. Drought stress also impairs enzyme activities, results in loss of turgor, root proliferation, plant water and nutrients. Likewise, diminished agricultural productivity and prolonged maturation of plants are caused by droughts (Poormohammad Kiani et al. 2007; Farooq et al. 2009; Faizan et al. 2020b). Plant macronutrients (N, P, K) have low uptake efficiencies (<50%) under normal soil moisture. These efficiencies are even lower during drought event, which further reduces fertilizer efficacy. Therefore, a reduction in grain yield and nutritional quality caused by drought stress can lead to food and nutrition insecurity (Baligar et al. 2001; Fischer et al. 2019). Among numerous techniques used to alleviate drought stress in crops, the application of ZnO NPs is considered to be an effective treatment. ZnO NPs help protect chlorophyll and other pigments, reduce ROS generation, and provide plants with Zn, which is important for many enzymes related to drought stress response mechanisms.

Dimkpa et al. (2020a, b) studied wheat performance after application of ZnO NPs under drought condition. Results showed that drought significantly reduced chlorophyll levels (6%), but ZnO NPs alleviated some stress by increasing chlorophyll levels (16%) compared to control. Drought delayed (3 days) panicle emergence, and ZnO NPs accelerated (5 days) panicle emergence under drought condition. Grain yield was unaffected by ZnO NPs under drought stress but increased (88%) under non-drought condition. Adrees et al. (2021) studied the foliar application of ZnO NPs (20, 50, and 100 $mg{\bullet}l^{-1}$) on wheat under drought stress (35% of water holding capacity). The foliar exposure of ZnO NPs elevated leaf chlorophyll contents and also decreased oxidative stress, and enhanced the leaf superoxide dismutase and peroxidase activities. They further showed that ZnO NPs decreased Cd concentrations in grains under water deficit conditions by 35, 66, and 81%, respectively. Furthermore, Linh et al. (2020) demonstrated that ZnO NPs treatment effectively helped soybean plants at an early vegetative stage to adapt to drought stress, and (Dimkpa et al. 2019) showed similar findings with sorghum plants.

12.4.2.5 Flooding

Similarly to drought, flooding is also a source of major abiotic stress and can have adverse effects on plant growth and development, albeit they are very different. It affects soils by altering soils structure, depleting O_2, accumulating CO_2, and inducing anaerobic decomposition of organic matter. Lack of O_2 around plant roots can cause severe damage and affect physiological processes in plants (Kozlowski 1997). Anaerobic respiration via ethanolic fermentation is considered to be an essential mechanism for plants to deal with the lack of O_2. Switching from aerobic respiration to anaerobic fermentation under flooding stress seems to be an important mechanism that helps plants survive O_2 deficiency (Drew 1997). Foliar application of Zn, especially after flooding, was shown to increase the growth of plants since Zn becomes more immobile during flood-related soil conditions (Hafeez et al. 2013). Therefore, ZnO NPs may pose an efficient way of supplying Zn to affected plants.

For example, in soybean, flooding damages plant growth mainly by damaging root length due to the loss of root tips in waterlogged soil and also reducing hypocotyl pigmentation, which leads to low intracellular O_2 levels and the synthesis of proteins related to anaerobic metabolic pathways (Russell et al. 1990; Huang et al. 2005; Hashiguchi et al. 2009). To alleviate some aspects of flood stress, Mustafa et al. (2015) exposed soybeans to various NPs, including ZnO NPs (5, 50, and 500 $mg{\bullet}l^{-1}$). After ZnO NP (50 $mg{\bullet}l^{-1}$) treatment, soybean showed a lower fresh weight of plants under flood stress compared to control. Soybeans treated with 5 and 500 $mg{\bullet}l^{-1}$ ZnO NPs experienced a further decrease in the fresh weight of plants. Also, under treatment with 50 $mg{\bullet}l^{-1}$ ZnO NPs, the length of root, including hypocotyl, was increased after 2 days of stress but then decreased during the remaining days of the treatment period. Soybean treated with 5 and 500 $mg{\bullet}l^{-1}$ of ZnO NPs showed a decreased length of root and hypocotyl compared to the flooding-stressed plants during the treatment period. The results of the experiments have shown that treating soybean with ZnO NPs before flooding did not alleviate the stress caused by flooding. However, this was a laboratory experiment, where 2 days-old soybeans were submerged in reverse osmosis water. For rice, there are known times of application in field, when Zn treatment can have the highest positive effect even before flooding of the plants, with preplant incorporated, and delayed preemergence stages having the highest positive effect (Slaton et al. 2005). Therefore, we believe that studying the application of ZnO NPs may positively affect the growth of rice and other plants that may face flooding stress (Elshayb et al. 2021).

12.4.2.6 Salts

Crop production worldwide faces an increase in land area with heightened salinity as a result of the industrialization of agriculture, incorrect agricultural practices, and changing climate. Salt affected lands increased by >100 Mha between 1986 and 2016 (Ivushkin et al. 2019). Stress from salinity affects plant growth and causes severe problems, mainly in arid and semi-arid lands (Hussain et al. 2019). The increased amount of soil salts hinders germination, morpho-physiological traits and crop yield. It can lead to extensive accumulation of ions (Na^+, Cl^-) and inhibit K^+ and Ca^{2+} uptake and result in ionic imbalance. Furthermore, salt causes the accumulation of ROS in plant cells, creating oxidative and osmotic stress (Astaneh et al. 2018; Isayenkov and Maathuis 2019). Osmotic stress causes the lower availability to take up water, which leads to dwarfed growth, while oxidative stress inhibits plant transpiration and damages cells in the transpiring leaves (Munns 2005; Amirjani 2011).

The application of ZnO NPs could be beneficial for reducing the adverse effects of salt stress. ZnO NPs positively affected the growth ratings in salt-stressed plants because ZnO NPs treatment synthesizes the indole acetic acid (IAA) and thus activates cell division and enlargement (Ali and Mahmound 2013; Latef et al. 2016; Faizan et al. 2020b). It can also be beneficial for maintaining the structural integrity of biomembranes, improving protein synthesis, increasing shoot length, chlorophyll, nutrient content, antioxidant enzyme activity, photosynthetic rate, etc. (He et al. 2015; Landa et al. 2015; Torabian et al. 2016; Hussein and Abou-Baker 2018). Moreover, Soliman et al. (2015) showed that ZnO NPs reduce Na^+ and Cl^- contents and increase N, P, K^+, Ca^{2+}, Mg^{2+}, Fe and Zn in *Moringa peregrina*.

Faizan et al. (2021) studied the foliar application of ZnO NPs (10, 50, and 100 mg•l^{-1}) in the presence/absence of NaCl (150 mM) on tomato plants. Results showed that foliar spray of ZnONPs significantly increased shoot length, root length, biomass, leaf area, chlorophyll content and photosynthetic attributes of tomato plants in the presence/absence of salt stress. The application of ZnONPs also ameliorate the negative effects of salt stress and enhanced protein content and antioxidative enzyme activity under salt stress. Alabdallah and Alzahrani (2020) showed that ZnO NPs (10 mg•l^{-1}) treatment increased the salinity tolerance in okra plants. The folial application of ZnO NPs increased the contents of photosynthetic pigments and the activity of antioxidant enzymes. ZnO NPs seed priming (5 and 10 mg•l^{-1}) of wheat increased growth parameters such as photosynthetic pigments, indole-3-acetic acid, phenol contents, and organic antioxidant enzyme activities. The treatments also significantly decreased lipid peroxidation (El-Bassiouny et al. 2020). Other studies (Gaafar et al. 2020; Singh et al. 2021) showed positive effects of ZnO NPs (50 mg•l^{-1}) under salt stress at soybean and *Linum usitassimum*, respectively. Overall, these studies suggest that the application of ZnO NPs at the appropriate dosage can be beneficial for enhancing the plant's toleration and antioxidant activity to decrease the damage caused by salt stress.

12.5 Conclusion and Future Outlook

Exposure to engineered NPs, including ZnO NPs, may have both negative or positive effects on plants and the environment. From an agricultural perspective, nanotechnology may mainly help with nutrient management that leads to more environmentally friendly agriculture, where applied nanomaterials have more targeted effects, reduce the number of chemicals used thanks to a more controlled release of nutrients and potential benefits related to the nanostructure of applied chemicals. It may help create more effective formulas with better plant responses against biotic and abiotic stresses during their life cycles. Such a limited, precise application of nanomaterials can have positive environmental impacts together with potential cost-saving measures for agriculture. However, there is also inherent risk connected to their unknown long-term application spanning several generations, detailed effects on the food chain, unclear impact on soil edaphon, and the quality of agricultural products as viewed by biogeochemical transport and transformations of NPs and their residues. Equally important is the absence of a more precise international legislative framework that determines levels of concentrations of NPs that are toxic and that determines which individual properties of NPs play the major role in their toxicity with standardised tests for not only toxicity in agricultural and model plants, but also for soil microorganisms and animals, including the impact on human health.

ZnO NPs have been shown to improve the growth of crops functioning as micronutrient nano fertilizer, nano growth promoters or nano pesticides that also protect plants from abiotic and biotic stress. A wide variety of effects, including alteration of several biochemical, metabolic, and physiological processes, e.g. production of reactive oxygen species, photosynthesis, water status, root hydraulic conductance, stress signalling and hormonal pathways, and transport and distribution of solutes in plants were found for ZnO NPs. The direct knowledge of the processes governing ZnO NPs interaction with various species of crops is growing but is not exhaustive, although much can be inferred from the action of ionic Zn in plants. ZnO NPs have the potential to alleviate different types of abiotic and biotic stresses in plants. However, more research is needed to fully understand the interaction between different species of plants and ZnO NPs with corresponding macro-sized or ionic forms, where the molecular and transcriptional alterations at the level of the plant are still not to be fully understood. Also, metabolic and proteomic changes in different plant organs need to be fully described. Mechanisms underlying the ability of ZnO NPs to alleviate abiotic stress in plants need to be characterised at the molecular and genetic levels and their actions compared to ionic forms of Zn to elucidate nano specific actions of ZnO NPs. Finally, more attention should be put on the development of strategic tools for ZnO NPs application in fields conditions in the context of climate change to further the knowledge gained from greenhouse and laboratory studies.

Acknowledgements Presented work was financially supported by Culture and Education Grant Agency of the Ministry of Education, Science, Research and Sports of the Slovak Republic and the Slovak Academy of Sciences via grant KEGA 013SPU-4/2019, by Science Grant Agency of the Ministry of Education, Science, Research and Sports of the Slovak Republic and the Slovak Academy of Sciences via grants VEGA 1/0146/18 and 1/0292/21, and by Comenius University's grant for young scientists Grant UK UK/48/2021.

References

Adrees M, Khan ZS, Hafeez M et al (2021) Foliar exposure of zinc oxide nanoparticles improved the growth of wheat (Triticum aestivum L.) and decreased cadmium concentration in grains under simultaneous Cd and water deficient stress. Ecotoxicol Environ Saf 208:111627. https://doi.org/10.1016/j.ecoenv.2020.111627

Alabdallah NM, Alzahrani HS (2020) The potential mitigation effect of ZnO nanoparticles on [Abelmoschus esculentus L. Moench] metabolism under salt stress conditions. Saudi J Biol Sci 27:3132–3137. https://doi.org/10.1016/j.sjbs.2020.08.005

Ali EA, Mahmound AM (2013) Effect of foliar spray by different salicylic acid and zinc concentrations on seed yield and yield components of mungbean in Sandy soil. Asian J Crop Sci 5:33–40

Ali S, Rizwan M, Noureen S et al (2019) Combined use of biochar and zinc oxide nanoparticle foliar spray improved the plant growth and decreased the cadmium accumulation in rice (Oryza sativa L.) plant. Environ Sci Pollut Res 26:11288–11299. https://doi.org/10.1007/s11356-019-04554-y

Allakhverdiev SI, Kreslavski VD, Klimov VV et al (2008) Heat stress: an overview of molecular responses in photosynthesis. Photosynth Res 98:541–550. https://doi.org/10.1007/s11120-008-9331-0

Allen RD (1995) Dissection of oxidative stress tolerance using transgenic plants. Plant Physiol 107:1049–1054. https://doi.org/10.1104/pp.107.4.1049

Alloway BJ (2008) Zinc in soils and crop nutrition. International Zinc Association, Brussels

Alloway BJ (2009) Soil factors associated with zinc deficiency in crops and humans. Environ Geochem Health 31:537–548. https://doi.org/10.1007/s10653-009-9255-4

Almutairi ZM (2019) Plant molecular defense mechanisms promoted by nanoparticles against environmental stresses. Int J Agric Biol 21:259–270

Amirjani MR (2011) Effect of salinity stress on growth, sugar content, pigments and enzyme activity of rice. Int J Bot 7:73–81

Anderson A, McLean J, McManus P, Britt D (2017) Soil chemistry influences the phytotoxicity of metal oxide nanoparticles. Int J Nanotechnol 14:15–21. https://doi.org/10.1504/IJNT.2017.082438

Apel K, Hirt H (2004) REACTIVE OXYGEN SPECIES: metabolism, oxidative stress, and signal transduction. Annu Rev Plant Biol 55:373–399. https://doi.org/10.1146/annurev.arplant.55.031903.141701

Astaneh RK, Bolandnazar S, Nahandi FZ, Oustan S (2018) The effects of selenium on some physiological traits and K, Na concentration of garlic (Allium sativum L.) under NaCl stress. Inf Process Agric 5:156–161. https://doi.org/10.1016/j.inpa.2017.09.003

Awan S, Shahzadi K, Javad S et al (2021) A preliminary study of influence of zinc oxide nanoparticles on growth parameters of Brassica oleracea var italic. J Saudi Soc Agric Sci 20:18–24. https://doi.org/10.1016/j.jssas.2020.10.003

Aziz Y, Shah GA, Rashid MI (2019) ZnO nanoparticles and zeolite influence soil nutrient availability but do not affect herbage nitrogen uptake from biogas slurry. Chemosphere 216:564–575. https://doi.org/10.1016/j.chemosphere.2018.10.119

Bajguz A, Hayat S (2009) Effects of brassinosteroids on the plant responses to environmental stresses. Plant Physiol Biochem 47:1–8. https://doi.org/10.1016/j.plaphy.2008.10.002

Bala R, Kalia A, Dhaliwal SS (2019) Evaluation of efficacy of ZnO nanoparticles as remedial zinc Nanofertilizer for Rice. J Soil Sci Plant Nutr 19:379–389. https://doi.org/10.1007/s42729-019-00040-z

Baligar VC, Fageria NK, He ZL (2001) Nutrient use efficiency in plants. Commun Soil Sci Plant Anal 32:921–950. https://doi.org/10.1081/CSS-200047048

Bari R, Jones JDG (2009) Role of plant hormones in plant defence responses. Plant Mol Biol 69:473–488. https://doi.org/10.1007/s11103-008-9435-0

Bashir A, Rizwan M, Ali S et al (2018) Effect of foliar-applied iron complexed with lysine on growth and cadmium (cd) uptake in rice under cd stress. Environ Sci Pollut Res 25:20691–20699. https://doi.org/10.1007/s11356-018-2042-y

Behmer ST, Lloyd CM, Raubenheimer D et al (2005) Metal hyperaccumulation in plants: mechanisms of defence against insect herbivores. Funct Ecol 19:55–66. https://doi.org/10.1146/annurev-arplant-042809-112156

Bian S-WW, Mudunkotuwa IA, Rupasinghe T, Grassian VH (2011) Aggregation and dissolution of 4 nm ZnO nanoparticles in aqueous environments: influence of pH, ionic strength, size, and adsorption of humic acid. Langmuir 27:6059–6068. https://doi.org/10.1021/la200570n

Bitenc M, Crnjak Orel Z (2009) Synthesis and characterization of crystalline hexagonal bipods of zinc oxide. Mater Res Bull 44:381–387. https://doi.org/10.1016/j.materresbull.2008.05.005

Blumwald E (2000) Sodium transport and salt tolerance in plants. Curr Opin Cell Biol 12:431–434. https://doi.org/10.1016/S0955-0674(00)00112-5

Borysiewicz MA (2019) ZnO as a functional material, a review. Crystals 9. https://doi.org/10.3390/cryst9100505

Chang Y-N, Zhang M, Xia L et al (2012) The toxic effects and mechanisms of CuO and ZnO nanoparticles. Materials (Basel) 5:2850–2871. https://doi.org/10.3390/ma5122850

Chaudhuri SK, Malodia L (2017) Biosynthesis of zinc oxide nanoparticles using leaf extract of Calotropis gigantea: characterization and its evaluation on tree seedling growth in nursery stage. Appl Nanosci 7:501–512. https://doi.org/10.1007/s13204-017-0586-7

Chen WJ, Liu WL, Hsieh SH, Tsai TK (2007) Preparation of nanosized ZnO using α brass. Appl Surf Sci 253:6749–6753. https://doi.org/10.1016/j.apsusc.2007.01.091

Chibuike GU, Obiora SC (2014, 2014) Heavy metal polluted soils : effect on plants and bioremediation methods. Appl Environ Soil Sci. https://doi.org/10.1155/2014/752708

Chiu WS, Khiew PS, Cloke M et al (2010) Photocatalytic study of two-dimensional ZnO nanopellets in the decomposition of methylene blue. Chem Eng J 158:345–352. https://doi.org/10.1016/j.cej.2010.01.052

Choudhary PK, Khandelwal V (2020) Comparative efficacy of Zn supplement and zinc oxide nanoparticles over the seed germination of lentil and Chick pea. J Pure Appl Microbiol 14:673–678. https://doi.org/10.22207/JPAM.14.1.69

Choudhury S, Panda P, Sahoo L, Panda SK (2013) Reactive oxygen species signaling in plants under abiotic stress. Plant Signal Behav 8:e23681. https://doi.org/10.4161/psb.23681

Czarnocka W, Karpiński S (2018) Friend or foe? Reactive oxygen species production, scavenging and signaling in plant response to environmental stresses. Free Radic Biol Med 122:4–20. https://doi.org/10.1016/j.freeradbiomed.2018.01.011

Davies PJ (2010) The plant hormones: their nature, occurrence, and functions. In: Davies PJ (ed) Plant hormones: biosynthesis, signal transduction, action! Springer, Dordrecht, pp 1–15

Dietz K-J, Herth S (2011) Plant nanotoxicology. Trends Plant Sci 16:582–589. https://doi.org/10.1016/j.tplants.2011.08.003

Dileep Kumar G, Raja K, Natarajan N et al (2020) Invigouration treatment of metal and metal oxide nanoparticles for improving the seed quality of aged chilli seeds (Capsicum annum L.). Mater Chem Phys 242:122492. https://doi.org/10.1016/j.matchemphys.2019.122492

Dimkpa CO (2018) Soil properties influence the response of terrestrial plants to metallic nanoparticles exposure. Curr Opin Environ Sci Heal 6:1–8. https://doi.org/10.1016/J.COESH.2018.06.007

Dimkpa CO, Singh U, Bindraban PS et al (2019) Zinc oxide nanoparticles alleviate drought-induced alterations in sorghum performance, nutrient acquisition, and grain fortification. Sci Total Environ 688:926–934. https://doi.org/10.1016/j.scitotenv.2019.06.392

Dimkpa CO, Andrews J, Fugice J et al (2020a) Facile coating of urea with low-dose ZnO nanoparticles promotes wheat performance and enhances Zn uptake under drought stress. Front Plant Sci 11:168. https://doi.org/10.3389/fpls.2020.00168

Dimkpa CO, Andrews J, Sanabria J et al (2020b) Interactive effects of drought, organic fertilizer, and zinc oxide nanoscale and bulk particles on wheat performance and grain nutrient accumulation. Sci Total Environ 722:137808. https://doi.org/10.1016/j.scitotenv.2020.137808

Doolette CL, Read TL, Howell NR et al (2020) Zinc from foliar-applied nanoparticle fertiliser is translocated to wheat grain: a 65Zn radiolabelled translocation study comparing conventional and novel foliar fertilisers. Sci Total Environ 749:142369. https://doi.org/10.1016/j.scitotenv.2020.142369

Drew MC (1997) Oxygen deficiency and root metabolism: injury and acclimation under hypoxia and anoxia. Annu Rev Plant Biol 48:223–250. https://doi.org/10.1146/annurev.arplant.48.1.223

Dutta RK, Nenavathu BP, Talukdar S (2014) Anomalous antibacterial activity and dye degradation by selenium doped ZnO nanoparticles. Colloids Surf B Biointerf 114:218–224. https://doi.org/10.1016/j.colsurfb.2013.10.007

Eeva T, Lehikoinen E, Ronka M (1998) Air pollution fades the plumage of the Great Tit. Funct Ecol 12:607–612

El-Bassiouny HMS, Abdallah MM-S, El-Enany MAM, Sadak MS (2020) Nano-zinc oxide and arbuscular mycorrhiza effects on physiological and biochemical aspects of wheat cultivars under saline conditions. Pakistan J Biol Sci 23:478–490. https://doi.org/10.3923/pjbs.2020.478.490

Elshayb OM, Farroh KY, Amin HE, Atta AM (2021) Green synthesis of zinc oxide nanoparticles: fortification for Rice grain yield and nutrients uptake enhancement. Molecules 26. https://doi.org/10.3390/molecules26030584

Elsheery NI, Sunoj VSJ, Wen Y et al (2020) Foliar application of nanoparticles mitigates the chilling effect on photosynthesis and photoprotection in sugarcane. Plant Physiol Biochem 149:50–60. https://doi.org/10.1016/j.plaphy.2020.01.035

Esper Neto M, Britt DW, Lara LM et al (2020) Initial development of corn seedlings after seed priming with nanoscale synthetic zinc oxide. Agronomy:10. https://doi.org/10.3390/agronomy10020307

Faizan M, Faraz A, Yusuf M et al (2018) Zinc oxide nanoparticle-mediated changes in photosynthetic efficiency and antioxidant system of tomato plants. Photosynthetica 56:678–686. https://doi.org/10.1007/s11099-017-0717-0

Faizan M, Hayat S, Pichtel J (2020a) Effects of zinc oxide nanoparticles on crop plants: a perspective analysis. In: sustainable agriculture reviews 41 nanotechnology for plant growth and development. Springer, Cham, pp 83–100

Faizan M, Yu F, Chen C et al (2020b) Zinc oxide nanoparticles help to enhance plant growth and alleviate abiotic stress: a review. Curr Protein Pept Sci 21:1–15. https://doi.org/10.2174/1389203721666201016144848

Faizan M, Bhat JA, Chen C et al (2021) Zinc oxide nanoparticles (ZnO-NPs) induce salt tolerance by improving the antioxidant system and photosynthetic machinery in tomato. Plant Physiol Biochem 161:122–130. https://doi.org/10.1016/j.plaphy.2021.02.002

Faraz A, Faizan M, Fariduddin Q, Hayat S (2020) Nanotechnology for plant growth and development. Springer, Cham

Farooq M, Wahid A, Basra SMA (2009) Improving water relations and gas exchange with brassinosteroids in rice under drought stress. J Agron Crop Sci 195:262–269. https://doi.org/10.1111/j.1439-037X.2009.00368.x

Fischer S, Hilger T, Piepho HP et al (2019) Do we need more drought for better nutrition? The effect of precipitation on nutrient concentration in East African food crops. Sci Total Environ 658:405–415. https://doi.org/10.1016/j.scitotenv.2018.12.181

Flowers TJ, Colmer TD (2015) Plant salt tolerance: adaptations in halophytes. Ann Bot 115:327–331. https://doi.org/10.1093/aob/mcu267

Frade T, Melo Jorge ME, Gomes A (2012) One-dimensional ZnO nanostructured films: effect of oxide nanoparticles. Mater Lett 82:13–15. https://doi.org/10.1016/j.matlet.2012.05.028

Future Markets Inc (2016) The global market for zinc oxide nanoparticles, Edinburgh

Gaafar R, Diab R, Halawa M et al (2020) Role of zinc oxide nanoparticles in ameliorating salt tolerance in soybean. Egypt J Bot 60:733–747. https://doi.org/10.21608/ejbo.2020.26415.1475

Gao F, Hong F, Liu C et al (2006) Mechanism of nano-anatase TiO2 on promoting photosynthetic carbon reaction of spinach: inducing complex of rubisco-rubisco activase. Biol Trace Elem Res 111:239–253. https://doi.org/10.1385/BTER:111:1:239
García-Gómez C, Fernández MD, García S et al (2018a) Soil pH effects on the toxicity of zinc oxide nanoparticles to soil microbial community. Environ Sci Pollut Res 25:28140–28152. https://doi.org/10.1007/s11356-018-2833-1
García-Gómez C, Obrador A, González D et al (2018b) Comparative study of the phytotoxicity of ZnO nanoparticles and Zn accumulation in nine crops grown in a calcareous soil and an acidic soil. Sci Total Environ 644:770–780. https://doi.org/10.1016/j.scitotenv.2018.06.356
García-López JI, Zavala-García F, Olivares-Sáenz E et al (2018) Zinc oxide nanoparticles boosts phenolic compounds and antioxidant activity of capsicum annuum L during germination. Agronomy:8. https://doi.org/10.3390/agronomy8100215
Gebre SH, Sendeku MG (2019) New frontiers in the biosynthesis of metal oxide nanoparticles and their environmental applications: an overview. SN Appl Sci 1:928. https://doi.org/10.1007/s42452-019-0931-4
Gelabert A, Sivry Y, Ferrari R et al (2013) Uncoated and coated ZnO nanoparticle life cycle in synthetic seawater. Environ Toxicol Chem 33:341–349. https://doi.org/10.1002/etc.2447
Giraldo JP, Landry MP, Faltermeier SM et al (2014) Plant nanobionics approach to augment photosynthesis and biochemical sensing. Nat Mater 13:400–408. https://doi.org/10.1038/nmat3890
Gogarten JP (1988) Physical properties of the cell wall of photoautotrophic suspension cells from-Chenopodium rubrum L. Planta 174:333–339. https://doi.org/10.1007/bf00959518
Gogos A, Knauer K, Bucheli TD (2012) Nanomaterials in plant protection and fertilization: current state, foreseen applications, and research priorities. J Agric Food Chem 60:9781–9792. https://doi.org/10.1021/jf302154y
González-Melendi P, Fernández-Pacheco R, Coronado MJ et al (2008) Nanoparticles as smart treatment-delivery systems in plants: assessment of different techniques of microscopy for their visualization in plant tissues. Ann Bot 101:187–195. https://doi.org/10.1093/aob/mcm283
Hafeez B, Khanif YM, Saleem M (2013) Role of zinc in plant nutrition- a review. Am J Exp Agric 3:374–391. https://doi.org/10.9734/AJEA/2013/2746
Hafez YM, Attia KA, Kamel S et al (2020) Bacillus subtilis as a bio-agent combined with nano molecules can control powdery mildew disease through histochemical and physiobiochemical changes in cucumber plants. Physiol Mol Plant Pathol 111:101489. https://doi.org/10.1016/j.pmpp.2020.101489
Hakim UA, Hussain A et al (2018) Osmotin: a plant defense tool against biotic and abiotic stresses. Plant Physiol Biochem 123:149–159. https://doi.org/10.1016/j.plaphy.2017.12.012
Han Y, Hwang G, Kim D et al (2016) Transport, retention, and long-term release behavior of ZnO nanoparticle aggregates in saturated quartz sand: role of solution pH and biofilm coating. Water Res 90:247–257. https://doi.org/10.1016/j.watres.2015.12.009
Hasanuzzaman M, Nahar K, Fujita M (2013) Extreme temperature responses, oxidative stress and antioxidant defense in plants. In: Vahdati K, Leslie C (eds) Abiotic stress – plant responses and applications in agriculture. InTech Open Access Publisher
Hashiguchi A, Sakata K, Komatsu S (2009) Proteome analysis of early-stage soybean seedlings under flooding stress. J Proteome Res 8:2058–2069. https://doi.org/10.1021/pr801051m
Hassan NS, El Din TAS, Hendawey MH, Borai IH, Mahdi AA (2018) Magnetite and zinc oxide nanoparticles alleviated heat stress in wheat plants. Curr Nanomater 3:32–43
He XC, Lin M, Li F et al (2015) Advances in studies of nanoparticle-biomembrane interactions. Nanomedicine 10:121–141. https://doi.org/10.2217/nnm.14.167
Holišová V, Konvičková Z, Kratošová G et al (2019) Phytosynthesis of Au and Au/ZrO 2 biphasic system nanoparticles with evaluation of their colloidal stability. J Nanosci Nanotechnol 19:2807–2813. https://doi.org/10.1166/jnn.2019.15851
Holišová V, Urban M, Konvičková Z et al (2021) Colloidal stability of phytosynthesised gold nanoparticles and their catalytic effects for nerve agent degradation. Sci Rep 11:4071. https://doi.org/10.1038/s41598-021-83460-1

Huang S, Greenway H, Colmer T, Millar AH (2005) Protein synthesis by Rice coleoptiles during prolonged anoxia: implications for glycolysis, growth and energy utilization. Ann Bot 96:703–715. https://doi.org/10.1093/aob/mci222

Huang Y, He J, Zhang Y et al (2006) Morphology, structures and properties of ZnO nanobelts fabricated by Zn-powder evaporation without catalyst at lower temperature. J Mater Sci 41:3057–3062. https://doi.org/10.1007/s10853-006-6978-9

Hussain A, Ali S, Rizwan M et al (2018) Zinc oxide nanoparticles alter the wheat physiological response and reduce the cadmium uptake by plants. Environ Pollut 242:1518–1526. https://doi.org/10.1016/j.envpol.2018.08.036

Hussain S, Shaukat M, Ashraf M, et al (2019) Salinity stress in arid and semi-arid climates: effects and management in field crops. In: Climate Change and Agriculture

Hussein MM, Abou-Baker NH (2018) The contribution of nano-zinc to alleviate salinity stress on cotton plants. R Soc Open Sci 5. https://doi.org/10.1098/rsos.171809

Ingle AP, Biswas A, Vanlalveni C et al (2020) Biogenic synthesis of nanoparticles and their role in the management of plant pathogenic fungi. In: Pai M, Golinsa P (eds) Microbial nanotechnology, 1st edn. CRC Press, Boca Raton, pp 135–161

Iqbal M, Raja NI, Mashwani Z-U-R et al (2019) Effect of silver nanoparticles on growth of wheat under heat stress. Iran J Sci Technol Trans A Sci 43:387–395. https://doi.org/10.1007/s40995-017-0417-4

Isayenkov SV, Maathuis FJM (2019) Plant salinity stress: many unanswered questions remain. Front Plant Sci 10:80. https://doi.org/10.3389/fpls.2019.00080

Itroutwar PD, Govindaraju K, Tamilselvan S et al (2020a) Seaweed-based biogenic ZnO nanoparticles for improving agro-morphological characteristics of rice (Oryza sativa L.). J Plant Growth Regul 39:717–728. https://doi.org/10.1007/s00344-019-10012-3

Itroutwar PD, Kasivelu G, Raguraman V et al (2020b) Effects of biogenic zinc oxide nanoparticles on seed germination and seedling vigor of maize (Zea mays). Biocatal Agric Biotechnol 29:101778. https://doi.org/10.1016/j.bcab.2020.101778

Ivushkin K, Bartholomeus H, Bregt AK et al (2019) Global mapping of soil salinity change. Remote Sens Environ 231:111260. https://doi.org/10.1016/j.rse.2019.111260

Jain S, Muneer S, Guerriero G et al (2018) Tracing the role of plant proteins in the response to metal toxicity: a comprehensive review. Plant Signal Behav 13:e1507401. https://doi.org/10.1080/15592324.2018.1507401

Jha S, Pudake R (2016) Molecular mechanism of plant–nanoparticle interactions. In: Kole C, Kumar DS, Khodakovskaya MV (eds) Plant nanotechnology: principles and practices. Springer, pp 155–181

Joo SH, Zhao D (2017) Environmental dynamics of metal oxide nanoparticles in heterogeneous systems: A review. J Hazard Mater 322(Part):29–47. https://doi.org/10.1016/j.jhazmat.2016.02.068

José-Yacamán M, Gutierrez-Wing C, Miki M et al (2005) Surface diffusion and coalescence of mobile metal nanoparticles. J Phys Chem B 109:9703–9711. https://doi.org/10.1021/jp0509459

Kasivelu G, Selvaraj T, Malaichamy K et al (2020) Nano-micronutrients [γ-Fe2O3 (iron) and ZnO (zinc)]: green preparation{,} characterization{,} agro-morphological characteristics and crop productivity studies in two crops (rice and maize). New J Chem 44:11373–11383. https://doi.org/10.1039/D0NJ02634D

Khan MN, Mobin M, Abbas ZK et al (2017) Role of nanomaterials in plants under challenging environments. Plant Physiol Biochem 110:194–209. https://doi.org/10.1016/j.plaphy.2016.05.038

Khan ZS, Rizwan M, Hafeez M et al (2019) The accumulation of cadmium in wheat (Triticum aestivum) as influenced by zinc oxide nanoparticles and soil moisture conditions. Environ Sci Pollut Res 26:19859–19870. https://doi.org/10.1007/s11356-019-05333-5

Khan MI, Fatima N, Shakil M et al (2021) Investigation of in-vitro antibacterial and seed germination properties of green synthesized pure and nickel doped ZnO nanoparticles. Phys B Condens Matter 601:412563. https://doi.org/10.1016/j.physb.2020.412563

Klingshirn C, Fallert J, Zhou H et al (2010) 65 years of ZnO research – old and very recent results. Phys Status Solidi 247:1424–1447. https://doi.org/10.1002/pssb.200983195

Kolenčík M, Ernst D, Komár M et al (2019) Effect of foliar spray application of zinc oxide nanoparticles on quantitative, nutritional, and physiological parameters of foxtail millet (Setaria italica l.) under field conditions. Nanomaterials 9:1559. https://doi.org/10.3390/nano9111559

Kolenčík M, Ernst D, Urík M et al (2020) Foliar application of low concentrations of titanium dioxide and zinc oxide nanoparticles to the common sunflower under field conditions. Nano 10. https://doi.org/10.3390/nano10081619

Kolenčík M, Nemček L, Šebesta M et al (2021) Effect of TiO2 as plant growth-stimulating nanomaterial on crop production. In: Singh VP, Singh S, Prasad SM et al (eds) Plant responses to nanomaterials. Springer, Cham, pp 129–144

Kong XY, Ding Y, Yang R, Wang ZL (2004) Single-crystal nanorings formed by epitaxial self-coiling of polar nanobelts. Science 80(303):1348–1351. https://doi.org/10.1126/science.1092356

Konvičková Z, Holišová V, Kolenčík M et al (2018) Phytosynthesis of colloidal Ag-AgCl nanoparticles mediated by Tilia sp. leachate, evaluation of their behaviour in liquid phase and catalytic properties. Colloid Polym Sci 296:677–687. https://doi.org/10.1007/s00396-018-4290-2

Kosová K, Vítámvás P, Urban MO et al (2018) Plant abiotic stress proteomics: the major factors determining alterations in cellular proteome. Front Plant Sci 9:122. https://doi.org/10.3389/fpls.2018.00122

Kozlowski TT (1997) Responses of woody plants to flooding and salinity. Tree Physiol 17:490. https://doi.org/10.1093/treephys/17.7.490

Kranner I, Minibayeva FV, Beckett RP, Seal CE (2010) What is stress? Concepts, definitions and applications in seed science. New Phytol 188:655–673. https://doi.org/10.1111/j.1469-8137.2010.03461.x

Landa P (2021) Positive effects of metallic nanoparticles on plants: overview of involved mechanisms. Plant Physiol Biochem 161:12–24. https://doi.org/10.1016/j.plaphy.2021.01.039

Landa P, Prerostova S, Petrova S et al (2015) The transcriptomic response of Arabidopsis thaliana to zinc oxide: a comparison of the impact of nanoparticle, bulk, and ionic zinc. Environ Sci Technol 49:14537–14545. https://doi.org/10.1021/acs.est.5b03330

Latef AAHA, Jan S, Abd-Allah EF et al (2016) Soybean under abiotic stress. Plant Environ Interact 2:28–42. https://doi.org/10.1002/9781119081005.ch2

Le TC, Yin H, Chen R et al (2016) An experimental and computational approach to the development of ZnO nanoparticles that are safe by design. Small 12:3568–3577. https://doi.org/10.1002/smll.201600597

Li C, Wang P, van der Ent A et al (2019) Absorption of foliar-applied Zn in sunflower (Helianthus annuus): importance of the cuticle, stomata and trichomes. Ann Bot 123:57–68. https://doi.org/10.1093/aob/mcy135

Lin D, Xing B (2008) Root uptake and Phytotoxicity of ZnO nanoparticles. Environ Sci Technol 42:5580–5585. https://doi.org/10.1021/es800422x

Linh TM, Mai NC, Hoe PT et al (2020) Metal-based nanoparticles enhance drought tolerance in soybean. J Nanomater 2020:13. https://doi.org/10.1155/2020/4056563

Lipiec J, Doussan C, Nosalewicz A, Kondracka K (2013) Effect of drought and heat stresses on plant growth and yield: a review. Int Agrophys 27:463–477. https://doi.org/10.2478/intag-2013-0017

Liu R, Lal R (2015) Potentials of engineered nanoparticles as fertilizers for increasing agronomic productions. Sci Total Environ 514:131–139. https://doi.org/10.1016/j.scitotenv.2015.01.104

Liu J, Huang X, Duan J et al (2005) A low-temperature synthesis of multiwhisker-based zinc oxide micron crystals. Mater Lett 59:3710–3714. https://doi.org/10.1016/j.matlet.2005.06.043

Liu J, Huang X, Li Y et al (2006) Selective growth and properties of zinc oxide nanostructures. Scr Mater 55:795–798. https://doi.org/10.1016/j.scriptamat.2006.07.010

Liu YF, Qi MF, Li TL (2012) Photosynthesis, photoinhibition, and antioxidant system in tomato leaves stressed by low night temperature and their subsequent recovery. Plant Sci 196:8–17. https://doi.org/10.1016/j.plantsci.2012.07.005

López-Moreno ML, Cedeño-Mattei Y, Bailón-Ruiz SJ et al (2018) Environmental behavior of coated NMs: physicochemical aspects and plant interactions. J Hazard Mater 347:196–217. https://doi.org/10.1016/j.jhazmat.2017.12.058
Ma H, Kabengi NJ, Bertsch PM et al (2011) Comparative phototoxicity of nanoparticulate and bulk ZnO to a free-living nematode Caenorhabditis elegans: the importance of illumination mode and primary particle size. Environ Pollut 159:1473–1480. https://doi.org/10.1016/j.envpol.2011.03.013
Ma H, Wallis LK, Diamond S et al (2014) Impact of solar UV radiation on toxicity of ZnO nanoparticles through photocatalytic reactive oxygen species (ROS) generation and photo-induced dissolution. Environ Pollut 193:165–172. https://doi.org/10.1016/j.envpol.2014.06.027
Mahdieh M, Sangi MR, Bamdad F, Ghanem A (2018) Effect of seed and foliar application of nano-zinc oxide, zinc chelate, and zinc sulphate rates on yield and growth of pinto bean (Phaseolus vulgaris) cultivars. J Plant Nutr 41:2401–2412. https://doi.org/10.1080/01904167.2018.1510517
Malandrakis AA, Kavroulakis N, Chrysikopoulos CV (2019) Use of copper, silver and zinc nanoparticles against foliar and soil-borne plant pathogens. 670:292–299
Mallakpour S, Madani M (2015) A review of current coupling agents for modification of metal oxide nanoparticles. Prog Org Coat 86:194–207. https://doi.org/10.1016/j.porgcoat.2015.05.023
Manke A, Wang L, Rojanasakul Y (2013) Mechanisms of nanoparticle-induced oxidative stress and toxicity. Biomed Res Int 2013:942916. https://doi.org/10.1155/2013/942916
Martins NCT, Avellan A, Rodrigues S et al (2020) Composites of biopolymers and ZnO NPs for controlled release of zinc in agricultural soils and timed delivery for maize. ACS Appl Nano Mater 3:2134–2148. https://doi.org/10.1021/acsanm.9b01492
Maslobrod SN, Lupashku GA, Gavzer SI et al (2020) Evaluation of stimulatory, antifungal and Thermo-resistant action of aqueous dispersions of nanoparticles on seeds of parental forms and reciprocal hybrids of winter wheat. In: Tiginyanu I, Sontea V, Railean S (eds) 4th international conference on nanotechnologies and biomedical engineering. Springer, Cham, pp 137–141
Mathur S, Agrawal D, Jajoo A (2014) Photosynthesis: response to high temperature stress. J Photochem Photobiol B Biol 137:116–126. https://doi.org/10.1016/j.jphotobiol.2014.01.010
Medina-Velo IA, Barrios AC, Zuverza-Mena N et al (2017) Comparison of the effects of commercial coated and uncoated ZnO nanomaterials and Zn compounds in kidney bean (Phaseolus vulgaris) plants. J Hazard Mater 332:214–222. https://doi.org/10.1016/j.jhazmat.2017.03.008
Meulenkamp EA (1998) Size dependence of the dissolution of ZnO nanoparticles. J Phys Chem B 102:7764–7769. https://doi.org/10.1021/jp982305u
Misra P, Shukla PK, Pramanik K et al (2016) Nanotechnology for crop improvement. In: Kole C, Kumar DS, Khodakovskaya MV (eds) Plant nanotechnology: principles and practices. Springer, Cham, pp 219–256
Mittler R (2017) ROS are good. Trends Plant Sci 22:11–19. https://doi.org/10.1016/j.tplants.2016.08.002
Moezzi A, McDonagh AM, Cortie MB (2012) Zinc oxide particles: synthesis, properties and applications. Chem Eng J 185–186:1–22. https://doi.org/10.1016/j.cej.2012.01.076
Moghaddasi S, Fotovat A, Khoshgoftarmanesh AH et al (2017) Bioavailability of coated and uncoated ZnO nanoparticles to cucumber in soil with or without organic matter. Ecotoxicol Environ Saf 144:543–551. https://doi.org/10.1016/j.ecoenv.2017.06.074
Mohd Omar F, Abdul Aziz H, Stoll S (2014) Aggregation and disaggregation of ZnO nanoparticles: influence of pH and adsorption of Suwannee River humic acid. Sci Total Environ 468–469:195–201. https://doi.org/10.1016/j.scitotenv.2013.08.044
Møller IM, Jensen PE, Hansson A (2007) Oxidative modifications to cellular components in plants. Annu Rev Plant Biol 58:459–481. https://doi.org/10.1146/annurev.arplant.58.032806.103946
Molnárová M, Filová A, Peško M (2015) Iónové a nanočasticové formy ťažkých kovov v prostredí a ich interakcia s fotosyntetizujúcimi organizmami

Momma K, Izumi F (2011) VESTA3 for three-dimensional visualization of crystal, volumetric and morphology data. J Appl Crystallogr 44:1272–1276. https://doi.org/10.1107/S0021889811038970

Mudunkotuwa IA, Rupasinghe T, Wu C-M, Grassian VH (2012) Dissolution of ZnO nanoparticles at Circumneutral pH: a study of size effects in the presence and absence of citric acid. Langmuir 28:396–403. https://doi.org/10.1021/la203542x

Mukherjee A, Peralta-Videa JR, Bandyopadhyay S et al (2014) Physiological effects of nanoparticulate ZnO in green peas (Pisum sativum L.) cultivated in soil. Metallomics 6:132–138. https://doi.org/10.1039/C3MT00064H

Munns R (2005) Genes and salt tolerance: bringing them together. New Phytol 167:645–663. https://doi.org/10.1111/j.1469-8137.2005.01487.x

Mustafa G, Sakata K, Komatsu S (2015) Proteomic analysis of flooded soybean root exposed to aluminum oxide nanoparticles. J Proteome 128:280–297. https://doi.org/10.1016/j.jprot.2015.08.010

Nair S, Sasidharan A, Divya Rani VV et al (2009) Role of size scale of ZnO nanoparticles and microparticles on toxicity toward bacteria and osteoblast cancer cells. J Mater Sci Mater Med 20:235–241. https://doi.org/10.1007/s10856-008-3548-5

Nair R, Varghese SH, Nair BG et al (2010) Nanoparticulate material delivery to plants. Plant Sci 179:154–163. https://doi.org/10.1016/j.plantsci.2010.04.012

Nel A, Xia T, Mädler L, Li N (2006) Toxic potential of materials at the nanolevel. Science 311(80):622–627. https://doi.org/10.1126/science.1114397

Nemček L, Šebesta M, Urík M et al (2020) Impact of bulk ZnO, ZnO nanoparticles and dissolved Zn on early growth stages of barley—a pot experiment. Plants 9. https://doi.org/10.3390/plants9101365

Nikoobakht B, Wang X, Herzing A, Shi J (2013) Scalable synthesis and device integration of self-registered one-dimensional zinc oxide nanostructures and related materials. Chem Soc Rev 42:342–365. https://doi.org/10.1039/C2CS35164A

Noret N, Meerts P, Tolrà R et al (2005) Palatability of Thlaspi caerulescens for snails: influence of zinc and glucosinolates. New Phytol 165:763–772. https://doi.org/10.1111/j.1469-8137.2004.01286.x

Olander LP, Vitousek PM (2000) Regulation of soil phosphatase and chitinase activityby N and P availability. Biogeochemistry 49:175–191. https://doi.org/10.1023/A:1006316117817

Oleszczuk P, Czech B, Kończak M et al (2019) Impact of ZnO and ZnS nanoparticles in sewage sludge-amended soil on bacteria, plant and invertebrates. Chemosphere 237:124359. https://doi.org/10.1016/j.chemosphere.2019.124359

Pan ZW, Dai ZR, Wang ZL (2001) Nanobelts of semiconducting oxides. Science 291 (80):1947–1949. https://doi.org/10.1126/science.1058120

Pandey S, Giri K, Kumar R et al (2018) Nanopesticides: opportunities in crop protection and associated environmental risks. Proc Natl Acad Sci India Sect B Biol Sci 88:1287–1308. https://doi.org/10.1007/s40011-016-0791-2

Peck AW, McDonald GK (2010) Adequate zinc nutrition alleviates the adverse effects of heat stress in bread wheat. Plant Soil 337:355–374. https://doi.org/10.1007/s11104-010-0532-x

Pedruzzi DP, Araujo LO, Falco WF et al (2020) ZnO nanoparticles impact on the photosynthetic activity of Vicia faba: effect of particle size and concentration. NanoImpact 19:100246. https://doi.org/10.1016/j.impact.2020.100246

Peng Y-H, Tsai Y-C, Hsiung C-E et al (2017) Influence of water chemistry on the environmental behaviors of commercial ZnO nanoparticles in various water and wastewater samples. J Hazard Mater 322:348–356. https://doi.org/10.1016/J.JHAZMAT.2016.10.003

Peralta-Videa JR, Hernandez-Viezcas JA, Zhao L et al (2014) Cerium dioxide and zinc oxide nanoparticles alter the nutritional value of soil cultivated soybean plants. Plant Physiol Biochem 80:128–135. https://doi.org/10.1016/j.plaphy.2014.03.028

Pérez Velasco EA, Betancourt Galindo R, Valdez Aguilar LA et al (2020) Effects of the morphology, surface modification and application methods of ZnO-NPs on the growth and biomass of tomato plants. Molecules 25. https://doi.org/10.3390/molecules25061282

Persaud I, Raghavendra AJ, Paruthi A et al (2020) Defect-induced electronic states amplify the cellular toxicity of ZnO nanoparticles. Nanotoxicology 14:145–161. https://doi.org/10.1080/17435390.2019.1668067

Polák F, Urík M, Matúš P (2019) Low molecular weight organic acids in soil environment. Chem List 113:307–314

Polshettiwar V, Baruwati B, Varma RS (2009) Self-assembly of metal oxides into three-dimensional nanostructures: synthesis and application in catalysis. ACS Nano 3:728–736. https://doi.org/10.1021/nn800903p

Poormohammad Kiani S, Grieu P, Maury P et al (2007) Genetic variability for physiological traits under drought conditions and differential expression of water stress-associated genes in sunflower (Helianthus annuus L.). Theor Appl Genet 114:193–207. https://doi.org/10.1007/s00122-006-0419-7

Poschenrieder C, Tolrà R, Barceló J (2006) Can metals defend plants against biotic stress? Trends Plant Sci 11:288–295. https://doi.org/10.1016/j.tplants.2006.04.007

Prasad R, Kumar V, Prasad KS (2014) Nanotechnology in sustainable agriculture: present concerns and future aspects. Afr J Biotechnol 13:705–713

Priyanka N, Geetha N, Manish T et al (2021) Zinc oxide nanocatalyst mediates cadmium and lead toxicity tolerance mechanism by differential regulation of photosynthetic machinery and antioxidant enzymes level in cotton seedlings. Toxicol Rep 8:295–302. https://doi.org/10.1016/j.toxrep.2021.01.016

Rafique M, Tahir R, Gillani SSA et al (2020) Plant-mediated green synthesis of zinc oxide nanoparticles from Syzygium Cumini for seed germination and wastewater purification. Int J Environ Anal Chem 0:1–16. https://doi.org/10.1080/03067319.2020.1715379

Raghib F, Naikoo MI, Khan FA et al (2020) Interaction of ZnO nanoparticle and AM fungi mitigates Pb toxicity in wheat by upregulating antioxidants and restricted uptake of Pb. J Biotechnol 323:254–263. https://doi.org/10.1016/j.jbiotec.2020.09.003

Rai-Kalal P, Jajoo A (2021) Priming with zinc oxide nanoparticles improve germination and photosynthetic performance in wheat. Plant Physiol Biochem 160:341–351. https://doi.org/10.1016/j.plaphy.2021.01.032

Rajiv P, Rajeshwari S, Venckatesh R (2013) Bio-fabrication of zinc oxide nanoparticles using leaf extract of Parthenium hysterophorus L. and its size-dependent antifungal activity against plant fungal pathogens. Spectrochim Acta Part A Mol Biomol Spectrosc 112:384–387. https://doi.org/10.1016/j.saa.2013.04.072

Rajput VD, Minkina TM, Behal A et al (2018) Effects of zinc-oxide nanoparticles on soil, plants, animals and soil organisms: a review. Environ. Nanotechnology. Monit Manag 9:76–84

Raliya R, Tarafdar JC (2013) ZnO nanoparticle biosynthesis and its effect on phosphorous-mobilizing enzyme secretion and gum contents in clusterbean (Cyamopsis tetragonoloba L.). Agric Res 2:48–57. https://doi.org/10.1007/s40003-012-0049-z

Raliya R, Nair R, Chavalmane S et al (2015) Mechanistic evaluation of translocation and physiological impact of titanium dioxide and zinc oxide nanoparticles on the tomato (Solanum lycopersicum L.) plant. Metallomics 7:1584–1594. https://doi.org/10.1039/C5MT00168D

Raliya R, Tarafdar JC, Biswas P (2016) Enhancing the mobilization of native phosphorus in the mung bean rhizosphere using ZnO nanoparticles synthesized by soil fungi. J Agric Food Chem 64:3111–3118. https://doi.org/10.1021/acs.jafc.5b05224

Raliya R, Saharan V, Dimkpa C, Biswas P (2018) Nanofertilizer for precision and sustainable agriculture: current state and future perspectives. J Agric Food Chem 66:6487–6503. https://doi.org/10.1021/acs.jafc.7b02178

Rani P, Kaur G, Rao KV et al (2020) Impact of green synthesized metal oxide nanoparticles on seed germination and seedling growth of Vigna radiata (mung bean) and Cajanus cajan (red gram). J Inorg Organomet Polym Mater 30:4053–4062. https://doi.org/10.1007/s10904-020-01551-4

Rasmussen K, Rauscher H, Mech A et al (2018) Physico-chemical properties of manufactured nanomaterials – characterisation and relevant methods. An outlook based on the OECD testing Programme. Regul Toxicol Pharmacol 92:8–28. https://doi.org/10.1016/J.YRTPH.2017.10.019
Rawashdeh RY, Harb AM, AlHasan AM (2020) Biological interaction levels of zinc oxide nanoparticles; lettuce seeds as case study. Heliyon 6:e03983. https://doi.org/10.1016/j.heliyon.2020.e03983
Read TL, Doolette CL, Cresswell T et al (2019) Investigating the foliar uptake of zinc from conventional and nano-formulations: a methodological study. Environ Chem 16:459. https://doi.org/10.1071/EN19019
Reddy Pullagurala VL, Adisa IO, Rawat S et al (2018a) ZnO nanoparticles increase photosynthetic pigments and decrease lipid peroxidation in soil grown cilantro (Coriandrum sativum). Plant Physiol Biochem 132:120–127. https://doi.org/10.1016/j.plaphy.2018.08.037
Reddy Pullagurala VL, Adisa IO, Rawat S et al (2018b) Finding the conditions for the beneficial use of ZnO nanoparticles towards plants-a review. Environ Pollut 241:1175–1181. https://doi.org/10.1016/j.envpol.2018.06.036
Rejeb IB, Pastor V, Mauch-Mani B (2014) Plant responses to simultaneous biotic and abiotic stress: molecular mechanisms. Plan Theory 3:458–475. https://doi.org/10.3390/plants3040458
Rizwan M, Ali S, Ali B et al (2019a) Zinc and iron oxide nanoparticles improved the plant growth and reduced the oxidative stress and cadmium concentration in wheat. Chemosphere 214:269–277. https://doi.org/10.1016/J.CHEMOSPHERE.2018.09.120
Rizwan M, Ali S, ur Rehman MZ, Maqbool A (2019b) A critical review on the effects of zinc at toxic levels of cadmium in plants. Environ Sci Pollut Res 26:6279–6289. https://doi.org/10.1007/s11356-019-04174-6
Rizwan M, Ali S, ur Zia Rehman M et al (2019c) Alleviation of cadmium accumulation in maize (Zea mays L.) by foliar spray of zinc oxide nanoparticles and biochar to contaminated soil. Environ Pollut 248:358–367. https://doi.org/10.1016/j.envpol.2019.02.031
Russell DA, Wong DML, Sachs MM (1990) The anaerobic response of soybean. Plant Physiol 92:401–407. https://doi.org/10.1104/pp.92.2.401
Sabir S, Arshad M, Chaudhari SK (2014) Zinc oxide nanoparticles for revolutionizing agriculture: synthesis and applications. Sci World J 2014:8. https://doi.org/10.1155/2014/925494
Sabir S, Zahoor MA, Waseem M et al (2020) Biosynthesis of ZnO nanoparticles using bacillus subtilis: characterization and nutritive significance for promoting plant growth in Zea mays L. Dose-Response 18:1559325820958911. https://doi.org/10.1177/1559325820958911
Saibo NJM, Lourenço T, Oliveira MM (2008) Transcription factors and regulation of photosynthetic and related metabolism under environmental stresses. Ann Bot 103:609–623. https://doi.org/10.1093/aob/mcn227
Salama DM, Osman SA, Abd El-Aziz ME et al (2019) Effect of zinc oxide nanoparticles on the growth, genomic DNA, production and the quality of common dry bean (Phaseolus vulgaris). Biocatal Agric Biotechnol 18:101083. https://doi.org/10.1016/j.bcab.2019.101083
Savi GD, Piacentini KC, de Souza SR et al (2015) Efficacy of zinc compounds in controlling fusarium head blight and deoxynivalenol formation in wheat (Triticum aestivum L.). Int J Food Microbiol 205:98–104. https://doi.org/10.1016/j.ijfoodmicro.2015.04.001
Saxena R, Tomar RS, Kumar M (2016) Exploring Nanobiotechnology to mitigate abiotic stress in crop plants. J Pharm Sci Res 8:974–980
Scandalios JG (2005) Oxidative stress: molecular perception and transduction of signals triggering antioxidant gene defenses. Braz J Med Biol Res 38:995–1014. https://doi.org/10.1590/S0100-879X2005000700003
Šebesta M, Kolenčík M, Urík M et al (2019) Increased colloidal stability and decreased solubility – sol-gel synthesis of zinc oxide nanoparticles with humic acids. J Nanosci Nanotechnol 19:3024–3030. https://doi.org/10.1166/jnn.2019.15868
Šebesta M, Nemček L, Urík M et al (2020a) Partitioning and stability of ionic, nano- and microsized zinc in natural soil suspensions. Sci Total Environ 700:134445. https://doi.org/10.1016/j.scitotenv.2019.134445

Šebesta M, Urík M, Kolenčík M et al (2020b) Sequential extraction resulted in similar fractionation of ionic Zn, nano- and microparticles of ZnO in acidic and alkaline soil. Forests 11. https://doi.org/10.3390/f11101077
Seleiman MF, Alotaibi MA, Alhammad BA et al (2020) Effects of ZnO nanoparticles and biochar of Rice straw and cow manure on characteristics of contaminated soil and sunflower productivity, oil quality, and heavy metals uptake. Agronomy 10. https://doi.org/10.3390/agronomy10060790
Servin AD, White JC (2016) Nanotechnology in agriculture: next steps for understanding engineered nanoparticle exposure and risk. NanoImpact 1:9–12. https://doi.org/10.1016/j.impact.2015.12.002
Shah AA, Aslam S, Akbar M et al (2021) Combined effect of Bacillus fortis IAGS 223 and zinc oxide nanoparticles to alleviate cadmium phytotoxicity in Cucumis melo. Plant Physiol Biochem 158:1–12. https://doi.org/10.1016/j.plaphy.2020.11.011
Sharifan H, Moore J, Ma X (2020) Zinc oxide (ZnO) nanoparticles elevated iron and copper contents and mitigated the bioavailability of lead and cadmium in different leafy greens. Ecotoxicol Environ Saf 191:110177. https://doi.org/10.1016/j.ecoenv.2020.110177
Sharkey TD (2005) Effects of moderate heat stress on photosynthesis: importance of thylakoid reactions, rubisco deactivation, reactive oxygen species, and thermotolerance provided by isoprene. Plant Cell Environ 28:269–277. https://doi.org/10.1111/j.1365-3040.2005.01324.x
Sharma P, Jha AB, Dubey RS, Pessarakli M (2012) Reactive oxygen species, oxidative damage, and Antioxidative defense mechanism in plants under stressful conditions. J Bot 2012:1–26. https://doi.org/10.1155/2012/217037
Siddiqi KS, Husen A (2017) Plant response to engineered metal oxide nanoparticles nanoscale. Res Lett:12. https://doi.org/10.1186/s11671-017-1861-y
Siddiqui ZA, Parveen A, Ahmad L, Hashem A (2019) Effects of graphene oxide and zinc oxide nanoparticles on growth, chlorophyll, carotenoids, proline contents and diseases of carrot. Sci Hortic (Amsterdam) 249:374–382. https://doi.org/10.1016/j.scienta.2019.01.054
Singh J, Kumar S, Alok A et al (2019) The potential of green synthesized zinc oxide nanoparticles as nutrient source for plant growth. J Clean Prod 214:1061–1070. https://doi.org/10.1016/j.jclepro.2019.01.018
Singh P, Arif Y, Siddiqui H et al (2021) Nanoparticles enhances the salinity toxicity tolerance in Linum usitatissimum L. by modulating the antioxidative enzymes, photosynthetic efficiency, redox status and cellular damage. Ecotoxicol Environ Saf 213:112020
Sirelkhatim A, Mahmud S, Seeni A et al (2015) Review on zinc oxide nanoparticles: antibacterial activity and toxicity mechanism. Nano-Micro Lett 7:219–242. https://doi.org/10.1007/s40820-015-0040-x
Sivry Y, Gelabert A, Cordier L et al (2014) Behavior and fate of industrial zinc oxide nanoparticles in a carbonate-rich river water. Chemosphere 95:519–526. https://doi.org/10.1016/j.chemosphere.2013.09.110
Slaton NA, Norman RJ, Wilson CE Jr (2005) Effect of zinc source and application time on zinc uptake and grain yield of flood-irrigated rice. Agron J 97:272–278. https://doi.org/10.2134/agronj2005.0272
Soliman AS, El-feky SA, Darwish E (2015) Alleviation of salt stress on Moringa peregrina using foliar application of nanofertilizers. J Hortic For 7:36–47. https://doi.org/10.5897/JHF2014.0379
Song U, Kim J (2020) Zinc oxide nanoparticles: a potential micronutrient fertilizer for horticultural crops with little toxicity. Hortic Environ Biotechnol 61:625–631. https://doi.org/10.1007/s13580-020-00244-8
Stolpe C, Krämer U, Müller C (2017) Heavy metal (hyper)accumulation in leaves of Arabidopsis halleri is accompanied by a reduced performance of herbivores and shifts in leaf glucosinolate and element concentrations. Environ Exp Bot 133:78–86. https://doi.org/10.1016/j.envexpbot.2016.10.003

Strambeanu N, Demetrovici L, Dragos D, Lungu M (2015) Nanoparticles: definition, classification and general physical properties. In: Lungu M, Neculae A, Bunoiu M, Biris C (eds) Nanoparticles' promises and risks: characterization, manipulation, and potential hazards to humanity and the environment. Springer, Cham, pp 3–8

Sun TY, Gottschalk F, Hungerbühler K, Nowack B (2014) Comprehensive probabilistic modelling of environmental emissions of engineered nanomaterials. Environ Pollut 185:69–76. https://doi.org/10.1016/j.envpol.2013.10.004

Sun Q, Li J, Le T (2018) Zinc oxide nanoparticle as a novel class of antifungal agents: current advances and future perspectives. J Agric Food Chem 66:11209–11220. https://doi.org/10.1021/acs.jafc.8b03210

Suzuki K, Nagasuga K, Okada M (2008) The chilling injury induced by high root temperature in the leaves of rice seedlings. Plant Cell Physiol 49:433–442. https://doi.org/10.1093/pcp/pcn020

Tarafdar JC, Agrawal A, Raliya R et al (2012) ZnO nanoparticles induced synthesis of polysaccharides and phosphatases by *aspergillus* fungi. Adv Sci Eng Med 4:324–328. https://doi.org/10.1166/asem.2012.1160

Thakur M, Bhattacharya S, Khosla PK, Puri S (2019) Improving production of plant secondary metabolites through biotic and abiotic elicitation. J Appl Res Med Aromat Plants 12:1–12. https://doi.org/10.1016/j.jarmap.2018.11.004

Thordal-Christnsen H (2003) Fresh insight into processes of nonhost resistance. Curr Opin Plant Biol 6:351–357

Thorny Chanu T, Upadhyaya H (2019) Chapter 3: Zinc oxide nanoparticle-induced responses on plants: a physiological perspective. In: Tripathi DK, Ahmad P, Sharma S et al (eds) Nanomaterials in plants, algae and microorganisms. Academic, pp 43–64

Torabian S, Zahedi M, Khoshgoftarmanesh A (2016) Effect of foliar spray of zinc oxide on some antioxidant enzymes activity of sunflower under salt stress. J Agric Sci Technol 18

Tripathi DK, Singh VP, Prasad SM et al (2015) Silicon nanoparticles (SiNp) alleviate chromium (VI) phytotoxicity in Pisum sativum (L.) seedlings. Plant Physiol Biochem 96:189–198. https://doi.org/10.1016/j.plaphy.2015.07.026

Tripathi D, Mishra R, Singh S et al (2017a) Nitric oxide ameliorates zinc oxide nanoparticles phytotoxicity in wheat seedlings: implication of the ascorbate-glutathione cycle. Front Plant Sci 8:1. https://doi.org/10.3389/fpls.2017.00001

Tripathi DK, Shweta SS et al (2017b) An overview on manufactured nanoparticles in plants: uptake, translocation, accumulation and phytotoxicity. Plant Physiol Biochem 110:2–12. https://doi.org/10.1016/j.plaphy.2016.07.030

Tymoszuk A, Wojnarowicz J (2020) Zinc oxide and zinc oxide nanoparticles impact on in vitro germination and seedling growth in Allium cepa L. Materials (Basel) 13. https://doi.org/10.3390/ma13122784

Ullah A, Romdhane L, Rehman A, Farooq M (2019) Adequate zinc nutrition improves the tolerance against drought and heat stresses in chickpea. Plant Physiol Biochem 143:11–18. https://doi.org/10.1016/j.plaphy.2019.08.020

Umar W, Hameed MK, Aziz T et al (2020) Synthesis, characterization and application of ZnO nanoparticles for improved growth and Zn biofortification in maize. Arch Agron Soil Sci 0:1–13. https://doi.org/10.1080/03650340.2020.1782893

Umavathi S, Mahboob S, Govindarajan M et al (2020) Green synthesis of ZnO nanoparticles for antimicrobial and vegetative growth applications: a novel approach for advancing efficient high quality health care to human wellbeing. Saudi J Biol Sci. https://doi.org/10.1016/j.sjbs.2020.12.025

ur Rehman R, Khan B, Aziz T et al (2020) Postponement growth and antioxidative response of Brassica nigra on CuO and ZnO nanoparticles exposure under soil conditions. IET Nanobiotechnol 14:423–427. https://doi.org/10.1049/iet-nbt.2019.0357

Venkatachalam P, Priyanka N, Manikandan K et al (2017) Enhanced plant growth promoting role of phycomolecules coated zinc oxide nanoparticles with P supplementation in cotton

(Gossypium hirsutum L.). Plant Physiol Biochem 110:118–127. https://doi.org/10.1016/j.plaphy.2016.09.004

Wahab R, Ansari SG, Kim Y-S et al (2007) Room temperature synthesis of needle-shaped ZnO nanorods via sonochemical method. Appl Surf Sci 253:7622–7626. https://doi.org/10.1016/j.apsusc.2007.03.060

Wahid A (2007) Physiological implications of metabolite biosynthesis for net assimilation and heat-stress tolerance of sugarcane (Saccharum officinarum) sprouts. J Plant Res 120:219–228. https://doi.org/10.1007/s10265-006-0040-5

Wan J, Wang R, Bai H et al (2020) Comparative physiological and metabolomics analysis reveals that single-walled carbon nanohorns and ZnO nanoparticles affect salt tolerance in Sophora alopecuroides. Environ Sci Nano 7:2968–2981. https://doi.org/10.1039/D0EN00582G

Wang P, Menzies NW, Lombi E et al (2013) Fate of ZnO nanoparticles in soils and cowpea (Vigna unguiculata). Environ Sci Technol 47:13822–13830. https://doi.org/10.1021/es403466p

Wang H, Adeleye AS, Huang Y et al (2015) Heteroaggregation of nanoparticles with biocolloids and geocolloids. Adv Colloid Interf Sci 226:24–36. https://doi.org/10.1016/j.cis.2015.07.002

Wang P, Lombi E, Zhao F-J, Kopittke PM (2016a) Nanotechnology: a new opportunity in plant sciences. Trends Plant Sci 21:699–712. https://doi.org/10.1016/j.tplants.2016.04.005

Wang X, Yang X, Chen S et al (2016b) Zinc oxide nanoparticles affect biomass accumulation and photosynthesis in arabidopsis. Front Plant Sci 6:1243. https://doi.org/10.3389/fpls.2015.01243

Wang X, Sun W, Zhang S et al (2018) Elucidating the effects of cerium oxide nanoparticles and zinc oxide nanoparticles on arsenic uptake and speciation in Rice (Oryza sativa) in a hydroponic system. Environ Sci Technol 52:10040–10047. https://doi.org/10.1021/acs.est.8b01664

Watson J-L, Fang T, Dimkpa CO et al (2015) The phytotoxicity of ZnO nanoparticles on wheat varies with soil properties. Biometals 28:101–112. https://doi.org/10.1007/s10534-014-9806-8

Welti R, Li W, Li M et al (2002) Profiling membrane lipids in plant stress responses: role of phospholipase Dα in freezing-induced lipid changes in arabidopsis. J Biol Chem 277:31994–32002. https://doi.org/10.1074/jbc.M205375200

Worms IAM, Boltzman J, Garcia M, Slaveykova VI (2012) Cell-wall-dependent effect of carboxyl-CdSe/ZnS quantum dots on lead and copper availability to green microalgae. Environ Pollut 167:27–33. https://doi.org/10.1016/j.envpol.2012.03.030

Wu F, Fang Q, Yan S et al (2020) Effects of zinc oxide nanoparticles on arsenic stress in rice (Oryza sativa L.): germination, early growth, and arsenic uptake. Environ Sci Pollut Res 27:26974–26981. https://doi.org/10.1007/s11356-020-08965-0

Xu T, Ji P, He M, Li J (2012) Growth and structure of pure ZnO micro/nanocombs. J Nanomater 2012:797935. https://doi.org/10.1155/2012/797935

Xu H, Li L, Lv H et al (2016) pH-dependent phosphatization of ZnO nanoparticles and its influence on subsequent lead sorption. Environ Pollut 208:723–731. https://doi.org/10.1016/J.ENVPOL.2015.10.052

Yang Z, Chen J, Dou R et al (2015) Assessment of the phytotoxicity of metal oxide nanoparticles on two crop plants, maize (Zea mays L.) and rice (Oryza sativa L.). Int J Environ Res Public Health 12:15100–15109. https://doi.org/10.3390/ijerph121214963

Yecheskel Y, Dror I, Berkowitz B (2016) Transport of engineered nanoparticles in partially saturated sand columns. J Hazard Mater 311:254–262. https://doi.org/10.1016/j.jhazmat.2016.03.027

Younes NA, Hassan HS, Elkady MF et al (2020) Impact of synthesized metal oxide nanomaterials on seedlings production of three Solanaceae crops. Heliyon 6:e03188. https://doi.org/10.1016/j.heliyon.2020.e03188

Youssef MS, Elamawi RM (2020) Evaluation of phytotoxicity, cytotoxicity, and genotoxicity of ZnO nanoparticles in Vicia faba. Environ Sci Pollut Res 27:18972–18984. https://doi.org/10.1007/s11356-018-3250-1

Yung MMN, Wong SWY, Kwok KWH et al (2015) Salinity-dependent toxicities of zinc oxide nanoparticles to the marine diatom Thalassiosira pseudonana. Aquat Toxicol 165:31–40. https://doi.org/10.1016/J.AQUATOX.2015.05.015

Yusefi-Tanha E, Fallah S, Rostamnejadi A, Pokhrel LR (2020) Zinc oxide nanoparticles (ZnONPs) as a novel nanofertilizer: influence on seed yield and antioxidant defense system in soil grown soybean (Glycine max cv. Kowsar). Sci Total Environ 738:140240. https://doi.org/10.1016/j.scitotenv.2020.140240

Zafar H, Abbasi BH, Zia M (2019) Physiological and antioxidative response of Brassica nigra (L.) to ZnO nanoparticles grown in culture media and soil. Toxicol Environ Chem 101:281–299. https://doi.org/10.1080/02772248.2019.1691555

Zafar H, Aziz T, Khan B et al (2020) CuO and ZnO nanoparticle application in synthetic soil modulates morphology, nutritional contents, and metal analysis of Brassica nigra. ACS Omega 5:13566–13577. https://doi.org/10.1021/acsomega.0c00030

Ze Y, Liu C, Wang L et al (2011) The regulation of TiO2 nanoparticles on the expression of light-harvesting complex II and photosynthesis of chloroplasts of arabidopsis thaliana. Biol Trace Elem Res 143:1131–1141. https://doi.org/10.1007/s12011-010-8901-0

Zhang Z, Jiang W, Jian Q et al (2015) Residues and dissipation kinetics of triazole fungicides difenoconazole and propiconazole in wheat and soil in Chinese fields. Food Chem 168:396–403. https://doi.org/10.1016/j.foodchem.2014.07.087

Zhang W, Long J, Li J et al (2019) Impact of ZnO nanoparticles on Cd toxicity and bioaccumulation in rice (Oryza sativa L.). Environ Sci Pollut Res 26:23119–23128. https://doi.org/10.1007/s11356-019-05551-x

Zhang W, Long J, Li J et al (2020) Effect of metal oxide nanoparticles on the chemical speciation of heavy metals and micronutrient bioavailability in paddy soil. Int J Environ Res Public Health 17. https://doi.org/10.3390/ijerph17072482

Zhao L, Peralta-Videa JR, Ren M et al (2012) Transport of Zn in a sandy loam soil treated with ZnO NPs and uptake by corn plants: electron microprobe and confocal microscopy studies. Chem Eng J 184:1–8. https://doi.org/10.1016/j.cej.2012.01.041

Zhao L, Lu L, Wang A et al (2020) Nano-biotechnology in agriculture: use of nanomaterials to promote plant growth and stress tolerance. J Agric Food Chem 68:1935–1947. https://doi.org/10.1021/acs.jafc.9b06615

Zhou D, Keller AA (2010) Role of morphology in the aggregation kinetics of ZnO nanoparticles. Water Res 44:2948–2956. https://doi.org/10.1016/j.watres.2010.02.025

Zhu J (2016) Review abiotic stress signaling and responses in plants. Cell 167:313–324. https://doi.org/10.1016/j.cell.2016.08.029

Chapter 13
Effects of Nanoparticles on Alleviating Phytotoxicity of Soil Heavy Metals: Potential for Enhancing Phytoremediation

Aurang Zeb, Weitao Liu, and Yinlong Zhang

Abstract Soil pollution by heavy metals is a severe global environmental problem, posing severe threats to environmental safety and the well-being of humans, since they are highly toxic and nonbiodegradable. Phytoremediation is a process using plants to remove pollutants from contaminated substrates, which has been recognized as an environmentally friendly and inexpensive technology for cleaning up soil pollution. However, phytoremediation is hindered by low remediation efficiency and long remediation cycle. Recently, the development of nanotechnology offers an efficient substitute method for improving the phytoremediation process. This chapter provides an overview of recent developments in the use of nanoparticles (NPs) to enhance phytoremediation of heavy metals (HMs) contaminated soils. NPs can act in a phytoremediation system either by directly removing contaminants and improving the physiological functions of the plant, or by intensifying the phytoavailability of contaminants. Nano-Zerovalent Iron (nZVI) is a majorly studied NP for enhancing phytoremediation as it has been used successfully for the remediation of polluted soil and groundwater. Similarly, fullerene NPs can increase the phyto availability of the contaminant. Generally, using NPs for enhancing the phytoremediation of HM contaminated soils can be a valuable approach, even though it is still in the experimental phase. Therefore, experience in multiple applications and further exploration is required on the long-term performance of NPs in phytoremediation systems.

A. Zeb · W. Liu (✉)
MOE Key Laboratory of Pollution Processes and Environmental Criteria, Tianjin Engineering Center of Environmental Diagnosis and Contamination Remediation, College of Environmental Science and Engineering, Nankai University, Tianjin, China
e-mail: lwt@nankai.edu.cn

Y. Zhang (✉)
Co-Innovation Center for Sustainable Forest in Southern China, Nanjing Forestry University, Nanjing, China
e-mail: ylzhang@njfu.edu.cn

M. Faizan et al. (eds.), *Sustainable Agriculture Reviews 53*, Sustainable Agriculture Reviews 53, https://doi.org/10.1007/978-3-030-86876-5_13

Keywords Nanoparticles · Heavy metals · Phytoremediation · Nanotechnology · Stress

13.1 Introduction

Heavy metal (HM) contamination of the soil is a major problem as the soil absorbs a hefty amount of HM emissions through human activities (Wu et al. 2021). Heavy metal is a persistent toxic pollutant with poor transferability and long retention time. They severely damage plants, animals and humans through the food chain (Hu et al. 2020; Qin et al. 2021). The key causes of HMs are traffic emissions, and discharge of industrial and municipal wastes (Zeb et al. 2020). HMs, such as copper (Cu), zinc (Zn), and manganese (Mn) and iron (Fe) are vital for life processes, while other metals, such as cadmium (Cd), nickel (Ni), and mercury (Hg), do not have a physiological function but often leads to harmful effects at higher concentrations (Xiang et al. 2021). Augmented intake of HMs such as Hg, Cr, Pb, and Cd can cause lung and renal dysfunction, liver damage, and bone degeneration in human and animal health (Liu et al. 2021a; Yin et al. 2021). Even though soil has physical (due to its sieving effect), chemical (through the adsorption, precipitation and transformation of chemical substances) and biological filter (through the decomposition of organic matter), it does not have great capacity to perform these functions (Singh et al. 2008). Unlike air pollution which directly influence human life, soil pollution indirectly ravages human and animal health.

Nanotechnology is one of the fast-growing technologies exhibiting an ability for sustainable growth in agriculture. Materials having a size range within 1–100 nm are considered as NPs. NPs exhibit several distinct physicochemical properties, as compared to bulk parent material comprising higher mechanical strength, and lower surface area and melting point (Liu et al. 2020). Due to these inimitable properties; NPs have grasped substantial attention from researchers working in different scientific fields. Different applications of nanomaterials are presented in Fig. 13.1. NPs are mainly classified into (i) metal-based nanoparticles (nano zero-valent iron (nZVI), aluminum, gold, and silver), metal-oxide NPs (ZnO, Al_2O_3, TiO_2, Fe_3O_4, NiO, CuO, CeO_2 etc., and metal-salt nanoparticles (nano-ceramics, nano-silicates etc); (ii) carbon-based nanoparticles, (e.g., single and multi-walled carbon nanotubes); (iii) quantum-dots (CdTe, CdSe); (iv) composites (compounds of different nanomaterials) and dendrimers (nano-sized polymers) (Zeb et al. 2021).

Physicochemical remediation technologies influence soil fertility, biodiversity, and other soil properties. Likewise, various washing solutions (sulfuric acid, polyglutamic acid, hydrochloric acid, phosphoric acid, phosphoric acid, etc.) are used to dissolve and mobilize HMs, eventually affecting the nature and biodiversity of soil (Zeb et al. 2020). As an alternative, the phytoremediation technique for soil remediation is cost-effective, environment friendly, and a sustainable way to refurbish the contaminated soil (Liang et al. 2017b). Plants have the

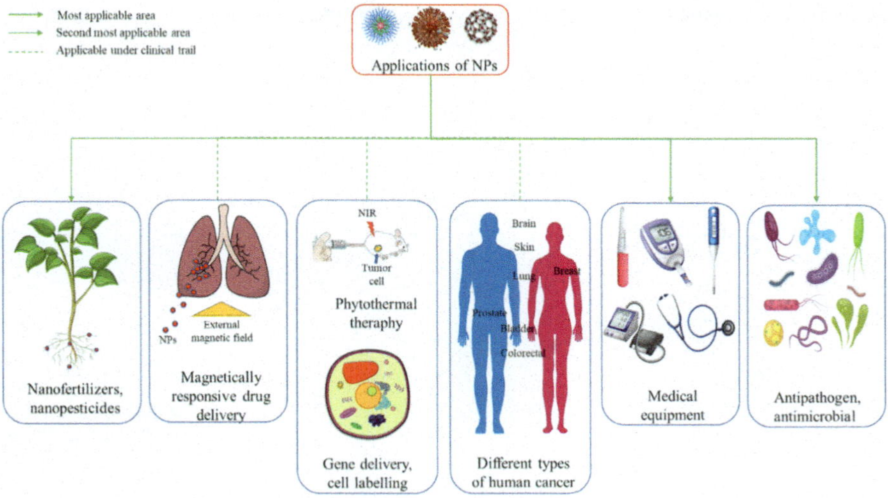

Fig. 13.1 Applications of nanomaterials in different fields

ability to remove pollutants or convert pollutants into harmless products by breaking down, bioaccumulating, fixing or extracting pollutants (Shah and Daverey 2020). Even though phytoremediation is a promising remediating technology, its effectiveness can be influenced by various factors, including soil quality, pollutant levels and bioavailability (Abdelkrim et al. 2020). Thus, new strategies are needed to be applied for enhancing the practice of phytoremediation for decontaminating polluted soils.

Nanoparticles enhanced phytoremediation (nano-enhanced phytoremediation) is an innovative technology that may reduce the retention time of phytoremediation and the expense of nanotechnology (Song et al. 2019). Nano-enhanced phytoremediation is more effective in reducing and removing 2,4,6-trinitrotoluene (TNT) from polluted soil than phytoremediation or nano-remediation alone (Sozer and Kokini 2009). Most of the plant species used for phytoextraction are hyperaccumulators, but such plant species are not fit for biomass production and their growth rate is slower than most species. Because of their sluggish growth, an alternative method is needed to improve the phytoextraction capacity of non-accumulating plants. Therefore, modern chemical methods need to be used to modify this method. The use of phytoextraction, chemically enhanced with NPs has been proposed as an alternative method for purifying heavy metal-contaminated soils (Gong et al. 2018). NPs have the ability to adsorb/absorb or reduce pollutants from the environment, while research aiming at the use of NPs to improve phytoremediation efficiency is still in its initial stage. Thus, this chapter focuses on (1) the recent advances in using NPs to alleviate the HM toxicity, (2) the use of NPs for abetting the phytoremediation of HM-polluted soils, and (3) the possible mechanisms involved in the nano-enhanced PR. Finally, the challenges of using NPs in phytoremediation are considered to identify future research needs.

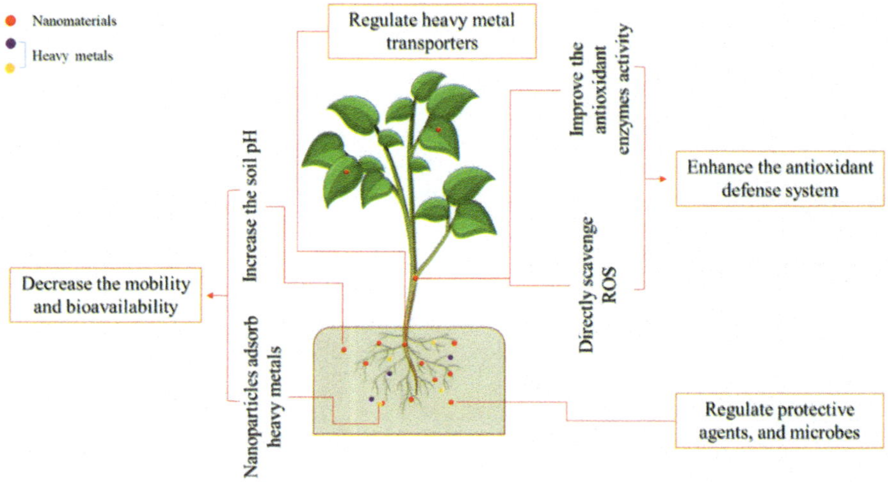

Fig. 13.2 Mechanisms involved in the alleviation of heavy metal stress by nanoparticles

13.2 Effect of NPs on Alleviating the Toxicity of HMs

It has been proved that plant physiology, biochemistry, and morphology are affected by the stress from HMs. Several strategies could be employed for reducing the HM stress in plants (Fig. 13.2). Such as, (1) decreasing the bioavailability of HMs in the soil, (2) adjusting the HMs transporting genes, (3) Increase the effectiveness of antioxidant system and improve the physiological function of plants and (4) promote the production of protective agents (e.g., organic acids, phytochelatin, and root exudates) (Zhou et al. 2021). Table 13.1 summarize the effect of different NPs on plant species under heavy metal stress.

13.2.1 Decreasing the Metal Bioavailability

HMs can be transformed by nanomaterials in the soil, thereby reducing their movement and bioavailability. For instance, green synthesized Fe NPs reduced the Na and Cd stress in rice plants by increasing the chlorophyll, carotenoids, and plant biomass (Sebastian et al. 2019). Similarly, Si NPs converted Cd into a more stable component (after 3 years) under field conditions (Wang et al. 2020). Besides, some NPs can improve soil properties, such as increasing soil pH, thus reducing the bioavailability of HMs in the soil.

Table 13.1 Influence of nanoparticles on the physicochemical properties of plants in heavy metal contaminated soil

NPs	HMs	Plant species	Effects	References
Ag	Cd, Pb, Zn, Ni	Yellow-lupin (*Lupinus luteus* L.)	A significant increase in the expression level of metallothionein was detected in plants grown under HM-stress, but no significant increase in the expression level of metallothionein was observed in plants grown on 25 mg kg^{-1} Ag NPs. In addition, the activity of guaiacol peroxidase (GPX) showed significant changes between different conditions and doses.	Jaskulak et al. (2019)
Biochar	Cd^{2+}	Rice (*Oryza sativa*)	High temperature nano-BC significantly reduced Cd^{+} phytotoxicity, MDA content, and increased root vitality, biomass, chlorophyll content	Yue et al. (2019)
Ce	Cd	Soy bean (*Glycine max*)	Improved Fv/Fm ratio and Cd concentrations in leaves; the accumulation of Ce in the roots and old leaves was markedly increased.	Rossi et al. (2017b)
Ce	Cu, Mn, Zn, Fe	Sugar pea (*Pisum sativum* L.)	Considerably decreased Mn, Cu, Fe, and Zn concentrations in roots and above ground parts of the pea plants	Skiba and Wolf (2019)
Si	Cu and Mn	Bamboo (*Arundinaria pygmaea*)	NPs increased plant growth and attenuated Mn and Cu toxicity, resulting in a significant increase in protective enzymes, chlorophyll content and fluorescence, as well as plant biomass.	Emamverdian et al. (2020)
Biosyn-Si	Pb, Ni, Cd	Common bean (*Phaseolus vulgaris* L., cv. Bronco)	Bio-Si-NPs and potassium silicate significantly improved plant growth, chlorophylls, net rate of photosynthesis, and activity of antioxidant enzymes.	El-Saadony et al. (2021)
Carbon black	Cd, Ni	*Suaeda salsa*	MNCB dramatically reduced HM availability in soil and Cd and Ni uptake by Suaeda salsa by 18% and 10%, respectively, and improved plant growth by reducing growth inhibition caused by HMs.	Cheng et al. (2019)
Si	Cr	Chickpea (*Cicer arietinum*)	Improved plant growth, protein, nitrogen and chlorophyll content, reduced ROS production and Cr accumulation in plants	Tripathi et al. (2015)
Fe	Cd	Wheat (*Triticum aestivum*)	Enhanced the plant growth indicators, photosynthesis and the antioxidant activities; decreased MDA content, EL, and the Cd accumulation in grains	Hussain et al. (2019)

(continued)

Table 13.1 (continued)

NPs	HMs	Plant species	Effects	References
FeO	Cd	Wheat (*Triticum aestivum*)	NPs in Cd-contaminated soils ultimately reduced Cd uptake by wheat plants by 72.5%, possibly due to adsorption of Cd to the large area of NPs.	Manzoor et al. (2021)
Fe	Cd and Pb	Castor (*Ricinus communis* L.)	NPs influenced the synthesis of starch granules in response to HM-stress	Zhang and Zhang (2020)
Si	Cd	Wheat (*Triticum aestivum*)	Improved indicators of plant growth, photosynthesis and POD and SOD activities; diminished the content of Cd, H_2O_2, MDA, and EL	Khan et al. (2020)
Si	Cd	Rice (*Oryza sativa*)	Augmented mineral elements content (Fe, Zn, and Mg), GSH content and POD, SOD activity in shoots; abridged Cd contents and CAT activity in shoots	Wang et al. (2015)
TiO_2	Cd	Maize (*Zea mays*)	Foliar application decreased Cd concentration in shoots	Lian et al. (2020)
TiO_2	Cd	Rice (*Oryza sativa*)	Considerably increased POD, and SOD activities, chlorophyll, and the net photosynthetic rate, and diminished MDA and Cd accumulation in leaves and roots.	Ji et al. (2017)
ZnO	Cd, Pb	Jumbay (*Leucaena leucocephala*)	Enhanced the plant growth	Venkatachalam et al. (2017)
ZnO	Cd, Pb	Jumbay (*Leucaena leucocephala*)	NPs enhanced seedling growth the level of antioxidative defense enzymes and related metabolites	Venkatachalam et al. (2017)
ZnO	Cd, Pb	Lettuce (*Lactuca sativa* L)	Markedly reduced the Pb and Cd accumulation in roots by 81%, and 49%, respectively.	Sharifan et al. (2019)
nZVI + Biochar	Cd	White Clover (*Trifolium repens*)	The joint use of nZVI and biochar favored the germination of plants and biomass in contaminated soils	Zand et al. (2020)
nZVI	As	Barley (*Hordeum vulgare* L.)	Reduced the availability of arsenic and uptake in barley, increased plant growth and biomass	Gil-Díaz et al. (2016)

NPs Nanoparticles, *HMs* Heavy metals, *EL* Electrolyte leakage

13.2.2 Influencing the Formation of Apoplastic Barriers

The apoplastic barrier of the plant roots acts physically to protect the plant and control the flow of water, oxygen, and ions (Liu et al. 2021b). NPs can affect the development of apoplastic barriers, which may reduce the HM concentration in the roots (Rossi et al. 2017a). Since the plasma membrane of plant root cells contains a variety of protein and ion transporters simultaneously transporting HMs, therefore, the apoplast barrier single-handedly may not be an efficient approach for reducing the HMs. In addition, metal transport genes in plants can be controlled by some NPs that may help seize the HM by enhancing the extracellular barrier of plants (Fox et al. 2020).

13.2.3 Production of Protective Agents

Most NPs accrue in the cell wall, bind to HMs, and form complexes, thus rendering them unavailable. These complexes adsorb onto the cell surface thus preventing the HMs' relocation in plants and plummeting its biological activity (Cui et al. 2017).

Additionally, HMs can be chelated with organic acids that accrue in the cell walls of plant roots and shoot to diminish the damage caused by HMs stress in the plant. It has been shown that the production of protective agents is promoted by NPs. For example, the use of Si NP promoted the production of organic acids and reduced the damage caused by Cd to the plant (Hussain et al. 2021).

13.2.4 Activation of the Antioxidant Defense System

Activation of the oxidation defense system is another strategy for decreasing HM stress. Generally, plants generate ROS through certain biochemical reactions. ROS functions as cell signal molecules controlling the defense, and growth of plants. However, extreme ROS accumulation (due to stress) is detrimental to cellular components such as cell membranes, and proteins, etc. (Wu et al. 2017). Antioxidant enzymes (such as peroxidase (POD), catalase (CAT), glutathione reductase (GR), ascorbate peroxidase (APX), glutathione peroxidase (GPX), and superoxide dismutase (SOD) mainly scavenge the plant ROS. A study reported that the addition of ZnO NPs with Pb and Cd exposure alleviated the HM induced phytotoxicity by promoting the antioxidative defense mechanisms in cotton seedlings (Priyanka et al. 2021). The use of NPs can improve the plants' ability to diminish ROS, eventually reducing the effects on plant growth (Guo et al. 2019).

13.3 Concept of Nano-enhanced Phytoremediation

Nano-assisted phytoremediation involves both phyto- and nanotechnology to clean up contaminated soils. Currently, NPs are widely used in numerous fields including paint, textiles, cosmetics, and medicines etc. Field applications of NP for groundwater and soil remediation have been successfully accomplished in the US and EU (Mueller and Nowack 2010). In recent times, several studies have described the use of NPs for the phytoremediation of contaminated soil. For example, Ghormade et al. (2011) examined the use of NPs in the nutrition and protection of plants and stated that NPs may help distribute fertilizers and pesticides, detect pollutants and diseases of plants, as well as protect the soil structure. The incorporation of NPs into traditional plant-based processing systems is promising.

13.3.1 Nanoparticles-Enhanced Phytoremediation of Heavy Metals

Soil pollution by heavy metals is a critical global problem, posing a major threat to human health and food safety. Phytoremediation is extensively used in the field treatment of soil contaminated by HMs. It is evident that the use of NPs can improve the phytoremediation efficiency of soil contaminated with different HMs (Table 13.2) (Liang et al. 2017c; Vítková et al. 2018). Cd and Pb are two of the most studied HMs because they are more commonly found in polluted sites (Liu et al. 2015). Due to its excessive tolerance, speedy growth, and low cost, ryegrass (*Lolium perenne*) is widely used for the phytoextraction of Pb in polluted soils. Reports suggest that, the use of NPs can effectively improve the ryegrass phytoextraction efficiency. For example, Liang et al. (2017a) explored the effect of using nano-hydroxyapatite for the phytoextraction of Pb by using ryegrass after 1.5, 3, and 12 months. The results after 1.5 months disclosed that adding 0.2% nano-hydroxyapatite markedly increased the Pb accumulation in shoots. As compared to control treatment, the addition of nano-hydroxyapatite with ryegrass removed 30%, 44.3%, and 46.5% of Pb in 1.5, 3, and 12 months, correspondingly. Likewise, Huang et al. (2018), added many doses (0, 100, 1000 and 2000 mg kg^{-1}) of nZVI particles to promote lead extraction by using ryegrass. After 45 days of treatment, the authors discovered that the accumulation of Pb in ryegrass is augmented in low concentrations of nZVI. While the higher concentration of nZVI increased oxidative stress in plants and thereby decreased the accumulation of lead.

Cadmium (Cd) is a harmful metal that is normally released to soil from various industrial products and processes (e.g., sulfur, minerals, phosphate fertilizers, electroplating, and batteries) (Liu et al. 2010, 2011). The use of hyperaccumulating plants for extracting Cd from polluted soil is the main strategy of phytoremediation, but the types and quantities of Cd hyperaccumulators available are limited (Liu et al. 2009). Some NPs have been shown to increase the efficiency of plants to extract Cd from soils. For example, the application of TiO_2 NPs to soybean plants has been reported to increase the bioaccumulation of Cd (Singh and Lee 2016). TiO_2-NP in concentrations of 100 and 300 mg kg^{-1} was added to the soil and the accumulation and distribution of Cd (after 60 days) in plants was analyzed. With the application of TiO_2 NPs, the Cd concentration augmented by about 1.9, and 2.6 folds, in shoots whereas 2.5, and 3.3 times in roots, respectively. In another study, nZVI were reported to improve the phytoextraction of Cd by ramie (*Boehmeria*

Table 13.2 Studies related to the use of nanoparticles for enhancing the phytoremediation of heavy metal contaminated soil

NPs	HMs	Plant species	Results	Role of NPs	References
Salicylic acid	As	*Isatis cappadocica*	The concentrations of arsenic accumulation in the shoot and in root reached 705 and 1188 mg/kg, respectively.	Increased rate of absorption and utilization of nutrients for plant growth	Souri et al. (2017)
nZVI	Cd	Ramie (*Boehmeria nivea*)	Cd concentrations in root and shoots increased by 29–52%, 31–73%, respectively.	Promoted plant growth at low nZVI concentration	Gong et al. (2017)
TiO_2	Cd	Soybean (*Glycine max*)	The absorption of Cd increased from 128.5 to 507.6 g per plant with an increasing concentration of TiO_2 NPs from 100 to 300 mg/kg.	Improved rate of germination, development and photosynthesis of plants	Singh and Lee (2016)
Ag	Cd, Pb, Ni	Maize (*Zea mays* L.)	The accumulation concentrations of Pb, Cd and Ni in the shoot increased from 129.1 to 232.7 mg/kg 0.65 to 0.73 mg/kg, and from 0 to 12.4 mg/kg, respectively	Enhanced root area and root length	Khan and Bano (2016)
ZnO	Cd, Pb	White popinac (*Leucaena leucocephala*)	Pb and Cd accumulation the plant increased from 1026.8 to 1343.4 mg/kg, 1253.1 to 1863.5 mg/kg, respectively.	Promoted plant growth via alleviating phytotoxicity	Venkatachalam et al. (2017)
nZVI	Pb	Ryegrass (*Lolium perenne* L.)	With 100 mg/kg nZVI, the Pb accumulation in roots and shoots reached 1175.4 lg per pot	The plant growth was promoted at lower nZVI dose	Huang et al. (2018)

(continued)

Table 13.2 (continued)

NPs	HMs	Plant species	Results	Role of NPs	References
Hydroxyapatite and carbon black	Pb	Ryegrass (*Lolium perenne* L.)	After 12 months, the Pb removal rates from the soil increased from 31.76% to 45.53%, and 46.55% with nano-carbon black and nano-hydroxyapatite, respectively	Direct stabilization of Pb by NPs, thus attenuating toxicity and promoting plant growth	Liang et al. (2017c)

NPs Nanoparticles, *HMs* Heavy metals

nivea) (Gong et al. 2017). In this study, starch-stabilized nZVI particles of different concentrations (100,.500,.and 1000.mg/kg) were added to HM polluted sediment prior to cultivating ramie seeds. The results disclosed that the addition of nZVI particles augmented the accumulation of Cd in leaves, and roots by 52%, and 33%, respectively.

Arsenic (As) is a noxious HM found commonly in the environment. Owing to its elevated toxicity, concerns have been raising regarding the soil As-contamination, due to its widespread use in industrial activities, pesticides, and phosphate fertilizers (Singh et al. 2015). Phytoextraction preferentially has been reported to be a successful remediation technique for removing As from soil (Lei et al. 2018). Souriet al. (2017) described that salicylic acid nanoparticles could be used to improve the removal efficiency of As by *Isatis cappadocica* plants. The authors integrated salicylic acid NPs in the arsenic phytoextraction system based on the idea that salicylic acid plays an indispensable role in plant growth and tolerance to As. The seedlings were pretreated with 250.mM NP of salicylic acid for 10 days for the phyto-extraction of Arsenic. Application of salicylic acid NP not only enhanced the growth of *Isatis cappadocica* plant, but also significantly increased the efficiency of remediation. Arsenic accumulation in the root and shoot reached 1188 and 705 mg kg^{-1}, respectively. Likewise, Vítková et al. (2018) presented that the application of nZVI particles has a positive effect on the arsenic stabilization in the rhizosphere of sunflower.

13.3.2 Mechanisms Involved in the Nano-enhanced Phytoremediation

The system of nano-enhanced phytoremediation comprises three key parts: plants, NPs, and pollutants. On one hand, NPs can increase the efficiency of phytoremediation by acting directly on plants and pollutants. On the other hand, the NPs used participate in the interaction between plants and pollutants and can indirectly influence the efficiency of final treatment. The next section clarifies how the application

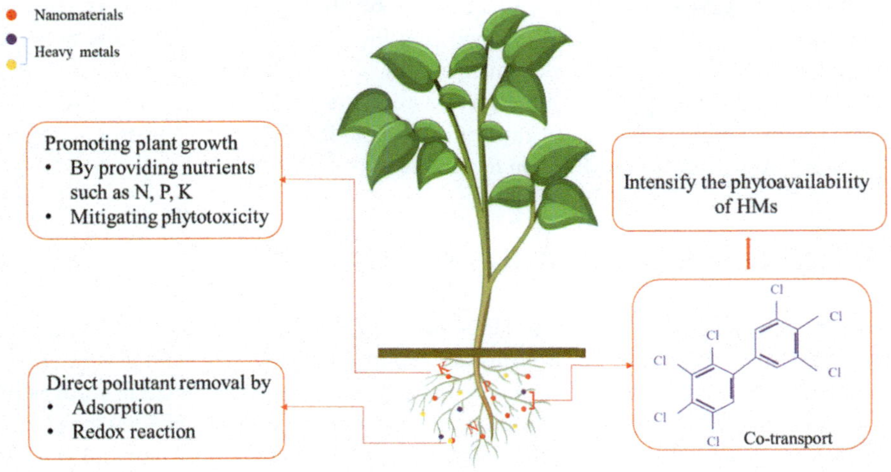

Fig. 13.3 Mechanisms involved in the nano-enhanced phytoremediation

of NPs in phytoremediation works. Figure 13.3. presents a graphical illustration of the mechanisms involved in the nano-enhanced phytoremediation.

13.3.2.1 Improvement of the Plant Physiological Functions

Plant growth and biomass are two vital considerations when selecting species of plant for phytoremediation. Owing to their low biomass, limited resistance to pollutants, and slow growth rates, several plants are generally not suitable for phytoremediation. For that reason, several strategies are employed to accelerate plant growth, including inoculation of rhizobacteria (PGPR), the utilization of transgenic plants, and application of growth regulators (Aderholt et al. 2017; Yadu et al. 2018). Research related to the interaction between plants and NPs has revealed that the application of NPs could boost plant growth, such as ZnO, nZVI, Ag NPs, carbon nanotubes and graphene quantum dots (Table 13.1). The mechanisms involved in promoting plant growth through these nanomaterials vary. For instance, graphene quantum dots may act as nano-pesticides and nano-fertilizers to increase the growth rate of *Allium sativum* and *Coriandrum sativum* (Chakravarty et al. 2015), and carbon nanotubes activate the reproductive system of plants and promote the growth of tomatoes (Khodakovskaya et al. 2013). Furthermore, NPs may increase the tolerance of plants to HMs thus increasing the plant growth (as discussed in Sect. 13.2). Apart from alleviating the toxicity of HMs, NPs may increase plant growth by increasing photosynthetic rates, promoting water and nutrient absorption, and regulating soil microbial communities. For example, the addition of nano-hydroxyapatite increased plant growth and improved the removal of Pb by ryegrass. The author justified that the increase in plant growth might be due to the higher concentration of phosphorus in the soil as a result of hydroxyapatite NPs (Ding et al. 2017).

Similarly, TiO_2 NPs are stated to be increasing the plant growth in a Cd phytoextraction study using soybean plants, by boosting the photosynthetic rate. The authors explained that the possible mechanism might be the entry of TiO_2 NPs into the chloroplasts and hastening the electron transfer and light adaptation (Singh and Lee 2016). The experiences of these cases are valuable for the use of NPs to accelerate plant growth in the phytoextraction technique.

13.3.2.2 Direct Removal of HMs by NPs

Several nanomaterials can remove HMs directly from the soil in the phytoremediation system, reducing the HM removal burden on plants. According to a study related to the influence of carbon nanotubes on the accumulation of Cd in cordgrass, it is stated that at a higher concentration of Cd, carbon nanotubes shielded the plants from growth inhibition. By further examining the ion content of calcium (Ca^{2+}), sodium (Na^+), and potassium (K^+), the instigators found that CNTs increase Ca^{2+} and K^+ for osmoregulation thereby reducing the phytotoxicity of Cd (Chai et al. 2013). In a study hydroxyapatite and carbon black nanomaterials promoted the plant extraction of Pb by ryegrass, and reduced the Pb toxicity in plants by adsorbing and stabilizing HMs (Liang et al. 2017c). The concentration of Pb accumulated in the roots of the plant decreased during the first month, but the effectiveness of phytoremediation increased after 12 months due to the reduced toxicity to the plants. Additionally, the effectiveness of phytoremediation is limited in a single growing season, and it can take years (or decades) for plants to completely eliminate large amounts of HM on their own. The utilization of nanomaterials for direct removal of some pollutants may diminish the load of contaminant removal by plants and curtail the remediation time.

13.3.2.3 Intensification in the Phytoavailability of HMs

The phytoavailability of HMs is an important factor affecting the efficiency of phytoremediation. Plants absorb only available forms of HMs. HMs phytoavailability is highly dependent on its distribution in soil and chemical speciation. For instance, a study concerning the phyto-availability of Cd in various binding forms, Cd adsorbed by gibbsite relative to other oxidized minerals in soil (manganese, aluminum oxide, magnetite and goethite) was widely available for reeds (Wang et al. 2009). In general, the phytoavailability of HMs is higher in the modifiable form (dissolved in soil solution), as compared to the combined form (with organic matter, oxides, and minerals) and is the lowest in the crystalline form (Sheoran et al. 2016). Moreover, the physicochemical properties of soil and the physiological characters of a plant also influence the HMs phytoavailability (Ren et al. 2018). Lower phytoavailability restrains the process of phytoremediation. For example, Pb in soil usually occurs in an insoluble form due to adsorption, precipitation, and complexion, making its phytoextraction difficult (Zaier et al. 2014). Therefore, several approaches have been proposed to increase the phytoavailability of HMs, such as agricultural management, utilizing genetic engineering, inoculation of microorganisms, application of chemical additives, etc. (Song et al. 2019).

13.3.3 *Ideal Characteristics of Plant for Nano-enhanced Phytoremediation*

A suitable plant to remove pollutants from the soil should have the following characteristics:

- Rapid growth and high productivity
- Well-developed root system, and higher root surface area
- Hyperaccumulators and can tolerate the contaminant stress
- Easy to harvest
- Non-consumable by human beings and animals
- Liable for genetic engineering.

13.3.4 *Suitable NPs for Enhancing Phytoremediation*

NPs suitable for enhancing the phytoremediation process should have the characteristics mentioned below:

- NPs must be harmless to the plant.
- NPs should enhance the germination, plant growth, biomass, root and shoot elongation.
- NPs should have the ability to bind contaminants and increase the phytoavailability of contaminant.

13.4 Challenges and Recommendations

The use of NPs for enhancing the process of phytoremediation is an emerging concept that came into view with the progress in nano and bioremediation technology. The practical application of nano-enhanced phytoremediation is facing several challenges. The environmental risk posed by NPs to the global ecosystem is a major concern. Some NPs are highly noxious to plants, animals, and human beings (Lian et al. 2020). The phytotoxicity of NPs is of particular concern in the phytoremediation system. Therefore, more studies need to be carried out on the environmental risk of NPs for understanding its toxicity completely. In addition, the use of NP in the phytoremediation process should be regulated in order to maximize its use but reduce the risk. Presently, the use of NPs to enhance phytoremediation is in the early stages of testing, but numerous constructive results have been obtained. Experience from many studies is required and the long-term efficacy of NPs requires further investigation. On the basis of known cases regarding nano-enhanced phytoremediation, nano-ZVI is being examined in detail. In comparison with other NPs, the use of nano-ZVI for enhancing phytoremediation has several advantages but it has been reported that the application of nZVI particles may suffer from interference of soil components, oxidation corrosion, and particle aggregation (Crane and Scott 2012). Therefore, it is crucial to design the structure of NPs and recognize the causes and conditions that inhibit the use of NPs in the phytoremediation system. Furthermore, reactions of NPs to different species of plant, pollutants, climatic conditions and soil types in the phytoremediation processes have to be explored in order to achieve broad applicability. Other methods, including agricultural management, genetic engineering, inoculating rhizosphere microorganisms, etc. may be involved in the nano-enhanced phytoremediation system to control the NP performance and improve the remediation efficiency further.

13.5 Summery

Phytoremediation is an environmentally friendly technology that offers many advantages in the treatment of polluted soils, but drawbacks such as time-consuming, change with climatic conditions, and not applicable to severely polluted soils, restricts its use. Extensive efforts have been made to increase the effectiveness of phytoremediation. In recent years, the use of NPs in phytoremediation has shown great potential for increasing its performance. This chapter provides an overview of the current research and information on the use of NPs to aid phytoremediation in HM polluted soils. NPs are able to endorse plant growth, remove HMs and increase the HM phtytoavailability, making it easier to boost the phytoremediation of HM contaminated soils. At present, the use of NPs in phytoremediation is in the experimental phase, but it provides a substitute means to improve the phytoremediation performance. However, much exploration still needs to be done for advancing the knowledge and in making sure that these findings are accurate.

Acknowledgments This work was supported by the National Natural Science Foundation of China (32171614), the National Key Research and Development Program of China (2020YFC180800), and the National Natural Science Foundation of China as a key project in collaboration with the provincial government of Shandong (U1806216).

Declaration of Competing Interest The authors declare that they have no competing interests.

References

Abdelkrim S, Jebara SH, Saadani O, Abid G, Taamalli W, Zemni H, Mannai K, Louati F, Jebara M (2020) In situ effects of *Lathyrus sativus*- PGPR to remediate and restore quality and fertility of Pb and cd polluted soils. Ecotoxicol Environ Saf 192:110260

Aderholt M, Vogelien DL, Koether M, Greipsson S (2017) Phytoextraction of contaminated urban soils by *Panicum virgatum* L. enhanced with application of a plant growth regulator (BAP) and citric acid. Chemosphere 175:85–96

Chai M, Shi F, Li R, Liu L, Liu Y, Liu F (2013) Interactive effects of cadmium and carbon nanotubes on the growth and metal accumulation in a halophyte *Spartina alterniflora* (Poaceae). Plant Growth Regul 71:171–179

Chakravarty D, Erande MB, Late DJ (2015) Graphene quantum dots as enhanced plant growth regulators: effects on coriander and garlic plants. J Sci Food Agric 95:2772–2778

Cheng J, Sun Z, Yu Y, Li X, Li T (2019) Effects of modified carbon black nanoparticles on plant-microbe remediation of petroleum and heavy metal co-contaminated soils. Int J Phytoremediat 21:634–642

Crane RA, Scott TB (2012) Nanoscale zero-valent iron: future prospects for an emerging water treatment technology. J Hazard Mater 211–212:112–125

Cui J, Liu T, Li F, Yi J, Liu C, Yu H (2017) Silica nanoparticles alleviate cadmium toxicity in rice cells: mechanisms and size effects. Environ Pollut 228:363–369

Ding L, Li J, Liu W, Zuo Q, Liang S-X (2017) Influence of Nano-hydroxyapatite on the metal bioavailability, plant metal accumulation and root exudates of ryegrass for phytoremediation in lead-polluted soil. Int J Environ Res Pub Health 14: 532

El-Saadony MT, Desoky E-SM, Saad AM, Eid RSM, Selem E, Elrys AS (2021) Biological silicon nanoparticles improve *Phaseolus vulgaris* L. yield and minimize its contaminant contents on a heavy metals-contaminated saline soil. J Environ Sci 106:1–14

Emamverdian A, Ding Y, Mokhberdoran F, Ahmad Z, Xie Y (2020) Determination of heavy metal tolerance threshold in a bamboo species (*Arundinaria pygmaea*) as treated with silicon dioxide nanoparticles. Glob Ecol Conserv 24:e01306

Fox J-P, Capen JD, Zhang W, Ma X, Rossi L (2020) Effects of cerium oxide nanoparticles and cadmium on corn (*Zea mays* L.) seedlings physiology and root anatomy. NanoImpact 20:100264

Ghormade V, Deshpande MV, Paknikar KM (2011) Perspectives for nano-biotechnology enabled protection and nutrition of plants. Biotechnol Adv 29:792–803

Gil-Díaz M, Diez-Pascual S, González A, Alonso J, Rodríguez-Valdés E, Gallego JR, Lobo MC (2016) A nanoremediation strategy for the recovery of an as-polluted soil. Chemosphere 149:137–145

Gong X, Huang D, Liu Y, Zeng G, Wang R, Wan J, Zhang C, Cheng M, Qin X, Xue W (2017) Stabilized nanoscale Zerovalent iron mediated cadmium accumulation and oxidative damage of *Boehmeria nivea* (L.) Gaudich cultivated in cadmium contaminated sediments. Environ Sci Technol 51:11308–11316

Gong X, Huang D, Liu Y, Zeng G, Wang R, Wei J, Huang C, Xu P, Wan J, Zhang C (2018) Pyrolysis and reutilization of plant residues after phytoremediation of heavy metals contaminated sediments: For heavy metals stabilization and dye adsorption. Bioresour Technol 253:64–71

Guo K, Hu A, Wang K, Wang L, Fu D, Hao Y, Wang Y, Ali A, Adeel M, Rui Y, Tan W (2019) Effects of spraying nano-materials on the absorption of metal(loid)s in cucumber. IET Nanobiotechnol 13:712–719

Hu B, Shao S, Ni H, Fu Z, Hu L, Zhou Y, Min X, She S, Chen S, Huang M, Zhou L, Li Y, Shi Z (2020) Current status, spatial features, health risks, and potential driving factors of soil heavy metal pollution in China at province level. Environ Pollut 266:114961

Huang D, Qin X, Peng Z, Liu Y, Gong X, Zeng G, Huang C, Cheng M, Xue W, Wang X, Hu Z (2018) Nanoscale zero-valent iron assisted phytoremediation of Pb in sediment: impacts on metal accumulation and antioxidative system of *Lolium perenne*. Ecotoxicol Environ Saf 153:229–237

Hussain A, Ali S, Rizwan M, Rehman MZU, Qayyum MF, Wang H, Rinklebe J (2019) Responses of wheat (*Triticum aestivum*) plants grown in a Cd contaminated soil to the application of iron oxide nanoparticles. Ecotoxicol Environ Saf 173:156–164

Hussain A, Rizwan M, Ali S, Rehman MZU, Qayyum MF, Nawaz R, Ahmad A, Asrar M, Ahmad SR, Alsahli AA, Alyemeni MN (2021) Combined use of different nanoparticles effectively decreased cadmium (Cd) concentration in grains of wheat grown in a field contaminated with Cd. Ecotoxicol Environ Saf 215:112139

Jaskulak M, Rorat A, Grobelak A, Chaabene Z, Kacprzak M, Vandenbulcke F (2019) Bioaccumulation, antioxidative response, and metallothionein expression in *Lupinus luteus* L. exposed to heavy metals and silver nanoparticles. Environ Sci Pollut Res 26:16040–16052

Ji Y, Zhou Y, Ma C, Feng Y, Hao Y, Rui Y, Wu W, Gui X, Le VN, Han Y, Wang Y, Xing B, Liu L, Cao W (2017) Jointed toxicity of TiO_2 NPs and Cd to rice seedlings: NPs alleviated Cd toxicity and Cd promoted NPs uptake. Plant Physiol Biochem 110:82–93

Khan N, Bano A (2016) Role of plant growth promoting rhizobacteria and Ag-nano particle in the bioremediation of heavy metals and maize growth under municipal wastewater irrigation. Int J Phytoremediat 18:211–221

Khan ZS, Rizwan M, Hafeez M, Ali S, Adrees M, Qayyum MF, Khalid S, Rehman MZU, Sarwar MA (2020) Effects of silicon nanoparticles on growth and physiology of wheat in cadmium contaminated soil under different soil moisture levels. Environ Sci Pollut Res 27:4958–4968

Khodakovskaya MV, Kim B-S, Kim JN, Alimohammadi M, Dervishi E, Mustafa T, Cernigla CE (2013) Carbon nanotubes as plant growth regulators: effects on tomato growth, reproductive system, and soil microbial community. Small 9:115–123

Lei M, Wan X, Guo G, Yang J, Chen T (2018) Phytoextraction of arsenic-contaminated soil with Pteris vittata in Henan Province, China: comprehensive evaluation of remediation efficiency correcting for atmospheric depositions. Environ Sci Pollut Res 25:124–131

Lian J, Zhao L, Wu J, Xiong H, Bao Y, Zeb A, Tang J, Liu W (2020) Foliar spray of TiO_2 nanoparticles prevails over root application in reducing Cd accumulation and mitigating Cd-induced phytotoxicity in maize (*Zea mays* L.). Chemosphere 239:124794

Liang J, Yang Z, Tang L, Zeng G, Yu M, Li X, Wu H, Qian Y, Li X, Luo Y (2017a) Changes in heavy metal mobility and availability from contaminated wetland soil remediated with combined biochar-compost. Chemosphere 181:281–288

Liang LC, Liu WT, Sun YB, Huo XH, Li S, Zhou QX (2017b) Phytoremediation of heavy metal contaminated saline soils using halophytes: current progress and future perspectives. Environ Rev 25:269–281

Liang S-X, Jin Y, Liu W, Li X, Shen S-G, Ding L (2017c) Feasibility of Pb phytoextraction using nano-materials assisted ryegrass: results of a one-year field-scale experiment. J Environ Manag 190:170–175

Liu W, Zhou Q, Sun Y, Liu R (2009) Identification of Chinese cabbage genotypes with low cadmium accumulation for food safety. Environ Pollut 157:1961–1967

Liu W, Zhou Q, An J, Sun Y, Liu R (2010) Variations in cadmium accumulation among Chinese cabbage cultivars and screening for Cd-safe cultivars. J Hazard Mater 173:737–743

Liu W, Zhou Q, Zhang Z, Hua T, Cai Z (2011) Evaluation of cadmium phytoremediation potential in Chinese cabbage cultivars. J Agric Food Chem 59:8324–8330

Liu W, Liang L, Zhang X, Zhou Q (2015) Cultivar variations in cadmium and lead accumulation and distribution among 30 wheat (*Triticum aestivum* L.) cultivars. Environ Sci Pollut Res 22:8432–8441

Liu W, Zeb A, Lian J, Wu J, Xiong H, Tang J, Zheng S (2020) Interactions of metal-based nanoparticles (MBNPs) and metal-oxide nanoparticles (MONPs) with crop plants: a critical review of research progress and prospects. Environ Rev 28:294–310

Liu J, Liu R, Yang Z, Kuikka S (2021a) Quantifying and predicting ecological and human health risks for binary heavy metal pollution accidents at the watershed scale using Bayesian Networks. Environ Pollut 269:116125

Liu Y, Tao Q, Li J, Guo X, Luo J, Jupa R, Liang Y, Li T (2021b) Ethylene-mediated apoplastic barriers development involved in cadmium accumulation in root of hyperaccumulator *Sedum alfredii*. J Hazard Mater 403:123729

Manzoor N, Ahmed T, Noman M, Shahid M, Nazir MM, Ali L, Alnusaire TS, Li B, Schulin R, Wang G (2021) Iron oxide nanoparticles ameliorated the cadmium and salinity stresses in wheat plants, facilitating photosynthetic pigments and restricting cadmium uptake. Sci Total Environ 769:145221

Mueller NC, Nowack B (2010) Nanoparticles for remediation: solving big problems with little particles. Elements 6:395–400

Priyanka N, Geetha N, Manish T, Sahi SV, Venkatachalam P (2021) Zinc oxide nanocatalyst mediates cadmium and lead toxicity tolerance mechanism by differential regulation of photosynthetic machinery and antioxidant enzymes level in cotton seedlings. Toxicol Rep 8:295–302

Qin G, Niu Z, Yu J, Li Z, Ma J, Xiang P (2021) Soil heavy metal pollution and food safety in China: Effects, sources and removing technology. Chemosphere 267:129205

Ren X, Zeng G, Tang L, Wang J, Wan J, Liu Y, Yu J, Yi H, Ye S, Deng R (2018) Sorption, transport and biodegradation – An insight into bioavailability of persistent organic pollutants in soil. Sci Total Environ 610-611:1154–1163

Rossi L, Zhang W, Ma X (2017a) Cerium oxide nanoparticles alter the salt stress tolerance of *Brassica napus* L. by modifying the formation of root apoplastic barriers. Environ Pollut 229:132–138

Rossi L, Zhang W, Schwab AP, Ma X (2017b) Uptake, accumulation, and in planta distribution of coexisting cerium oxide nanoparticles and cadmium in *Glycine max* (L.) Merr. Environ Sci Technol 51:12815–12824

Sebastian A, Nangia A, Prasad MNV (2019) Cadmium and sodium adsorption properties of magnetite nanoparticles synthesized from Hevea brasiliensis Muell. Arg. bark: Relevance in amelioration of metal stress in rice. J Hazard Mater 371:261–272

Shah V, Daverey A (2020) Phytoremediation: a multidisciplinary approach to clean up heavy metal contaminated soil. Environ Technol Innov 18:100774

Sharifan H, Ma X, Moore JM, Habib MR, Evans C (2019) Zinc oxide nanoparticles alleviated the bioavailability of cadmium and Lead and changed the uptake of iron in hydroponically grown lettuce (*Lactuca sativa* L. var. Longifolia). ACS Sustain Chem Eng 7:16401–16409

Sheoran V, Sheoran AS, Poonia P (2016) Factors affecting phytoextraction: A review. Pedosphere 26:148–166

Singh J, Lee BK (2016) Influence of nano-TiO_2 particles on the bioaccumulation of Cd in soybean plants (*Glycine max*): A possible mechanism for the removal of Cd from the contaminated soil. J Environ Manag 170:88–96

Singh B, Shan YH, Johnson-Beebout SE, Singh Y, Buresh RJ (2008) Crop residue management for lowland rice-based cropping systems in Asia. In: Sparks DL (ed) Advances in agronomy, vol 98, pp 117–199

Singh R, Singh S, Parihar P, Singh VP, Prasad SM (2015) Arsenic contamination, consequences and remediation techniques: a review. Ecotoxicol Environ Saf 112:247–270

Skiba E, Wolf WM (2019) Cerium oxide nanoparticles affect heavy metals uptake by pea in a divergent way than their ionic and bulk counterparts. Water Air Soil Pollut 230:248

Song B, Xu P, Chen M, Tang W, Zeng G, Gong J, Zhang P, Ye S (2019) Using nanomaterials to facilitate the phytoremediation of contaminated soil. Crit Rev Environ Sci Technol 49:791–824

Souri Z, Karimi N, Sarmadi M, Rostami E (2017) Salicylic acid nanoparticles (SANPs) improve growth and phytoremediation efficiency of *Isatis cappadocica* Desv., under As stress. IET Nanobiotechnol 11:650–655

Sozer N, Kokini JL (2009) Nanotechnology and its applications in the food sector. Trends Biotechnol 27:82–89

Venkatachalam P, Jayaraj M, Manikandan R, Geetha N, Rene ER, Sharma NC, Sahi SV (2017) Zinc oxide nanoparticles (ZnONPs) alleviate heavy metal-induced toxicity in *Leucaena leucocephala* seedlings: A physiochemical analysis. Plant Physiol Biochem 110:59–69

Vítková M, Puschenreiter M, Komárek M (2018) Effect of nano zero-valent iron application on as, cd, Pb, and Zn availability in the rhizosphere of metal(loid) contaminated soils. Chemosphere 200:217–226

Wang H, Jia Y, Wang S, Zhu H, Wu X (2009) Bioavailability of cadmium adsorbed on various oxides minerals to wetland plant species Phragmites australis. J Hazard Mater 167:641–646

Wang S, Wang F, Gao S (2015) Foliar application with nano-silicon alleviates cd toxicity in rice seedlings. Environ Sci Pollut Res 22:2837–2845

Wang Y, Liu Y, Zhan W, Zheng K, Lian M, Zhang C, Ruan X, Li T (2020) Long-term stabilization of Cd in agricultural soil using mercapto-functionalized nano-silica (MPTS/nano-silica): A three-year field study. Ecotoxicol Environ Saf 197:110600

Wu H, Tito N, Giraldo JP (2017) Anionic cerium oxide nanoparticles protect plant photosynthesis from abiotic stress by scavenging reactive oxygen species. ACS Nano 11:11283–11297

Wu D, Yu X, Lai M, Feng J, Dong X, Peng W, Su S, Zhang X, Wan L, Jacobs DF, Zeng S (2021) Diversified effects of co-planting landscape plants on heavy metals pollution remediation in urban soil amended with sewage sludge. J Hazard Mater 403:123855

Xiang M, Li Y, Yang J, Lei K, Li Y, Li F, Zheng D, Fang X, Cao Y (2021) Heavy metal contamination risk assessment and correlation analysis of heavy metal contents in soil and crops. Environ Pollut 278:116911

Yadu B, Chandrakar V, Korram J, Satnami ML, Kumar M (2018) Silver nanoparticle modulates gene expressions, glyoxalase system and oxidative stress markers in fluoride stressed *Cajanus cajan* L. J Hazard Mater 353:44–52

Yin N, Zhao Y, Wang P, Du H, Yang M, Han Z, Chen X, Sun G, Cui Y (2021) Effect of gut microbiota on in vitro bioaccessibility of heavy metals and human health risk assessment from ingestion of contaminated soils. Environ Pollut 279:116943

Yue L, Lian F, Han Y, Bao Q, Wang Z, Xing B (2019) The effect of biochar nanoparticles on rice plant growth and the uptake of heavy metals: Implications for agronomic benefits and potential risk. Sci Total Environ 656:9–18

Zaier H, Ghnaya T, Ghabriche R, Chmingui W, Lakhdar A, Lutts S, Abdelly C (2014) EDTA-enhanced phytoremediation of lead-contaminated soil by the halophyte *Sesuvium portulacastrum*. Environ Sci Pollut Res 21:7607–7615

Zand AD, Tabrizi AM, Heir AV (2020) Incorporation of biochar and nanomaterials to assist remediation of heavy metals in soil using plant species. Environ Technol Innov 20:101134

Zeb A, Li S, Wu J, Lian J, Liu W, Sun Y (2020) Insights into the mechanisms underlying the remediation potential of earthworms in contaminated soil: A critical review of research progress and prospects. Sci Total Environ 740:140145

Zeb A, Liu W, Wu J, Lian J, Lian Y (2021) Knowledge domain and emerging trends in nanoparticles and plants interaction research: a scientometric analysis. NanoImpact 21:100278

Zhang H, Zhang Y (2020) Effects of iron oxide nanoparticles on Fe and heavy metal accumulation in castor (*Ricinus communis* L.) plants and the soil aggregate. Ecotoxicol Environ Saf 200:110728

Zhou P, Adeel M, Shakoor N, Guo M, Hao Y, Azeem I, Li M, Liu M, Rui Y (2021) Application of nanoparticles alleviates heavy metals stress and promotes plant growth: An overview. Nanomaterials 11:26

Chapter 14
Bio-Fabricated Silver Nanoparticles: A Sustainable Approach for Augmentation of Plant Growth and Pathogen Control

Amna, Baber Ali, Muhammad Atif Azeem, Ayesha Qayyum, Ghazala Mustafa, Muhammad Ansar Ahmad, Muhammad Tariq Javed, and Hassan Javed Chaudhary

Abstract Sustainability in agricultural development has emerged as one of the most important concerns of present era because significant reductions in crop yield by various phytopathogens have raised the global food security issues. Among other alternatives of disease control methods, bio-control of phytopathogens by biologically produced silver nanoparticles offers a good choice for the development of eco-friendly, cost effective, and sustainable approaches. Considering the above-mentioned scenario, present study is a proposed solution to address the problem. Moreover, the mode of action of nanoparticles on plant growth and development is also a least explored aspect. Silver nanoparticles (AgNPs) are currently the widely produced nanomaterials and have been implicated in agriculture for improving crop growth and yield. Several reports indicated that appropriate concentrations of AgNPs play an important role in enhanced seed germination, improved plant growth, augmented photosynthetic quantum efficiency and chlorophyll content, increased water and fertilizer utilization efficiency. They also revealed the exclusive biological and physicochemical properties and the potential to boost up plant metabolism and antioxidant enzyme activities to control various pathogens. Therefore, this chapter will figure out the role of bio fabricated AgNPs in disease control, growth enhancement, some biochemical attributes like antioxidant enzyme activities, proline content, and yield of plant.

Keywords Antioxidant enzymes · Disease control · Phytopathogens · Plant growth · Silver nanoparticles

Amna · B. Ali · M. A. Azeem · A. Qayyum · G. Mustafa · Muhammad Ansar Ahmad
H. J. Chaudhary (✉)
Department of Plant Sciences, Quaid-i-Azam University, Islamabad, Pakistan
e-mail: hassaan@qau.edu.pk

M. T. Javed
Department of Botany, Government College University, Faisalabad, Pakistan

M. Faizan et al. (eds.), *Sustainable Agriculture Reviews 53*, Sustainable Agriculture Reviews 53, https://doi.org/10.1007/978-3-030-86876-5_14

14.1 Introduction

An American physicist Richard Feynman has given a famous visionary lecture in 1959, "There's plenty of room at the bottom," is thought to have given a conceptual birth to the field of nanotechnology. In 1974, the word 'nanotechnology' was coined for the first time by Norio Taniguichi, a professor at Tokyo University of Science (Khan et al. Khan and Rizvi 2014). Presently nanotechnology is inevitable in every field of sciences, including, agriculture, engineering, environment and medicine. Both organic, inorganic and hybrid materials are used in nanotechnology.

Keeping in consideration, nanotechnology is an emerging field in the scientific discoveries and inventions. However, it has been witnessed in last 30 years that nanotechnology made new inventions. In early 2000s, commercial applications of nanoparticles is increased in different fields. Materials having distinctive characteristics and ranging from 1 to 100 nm in size are called as nanoparticles (EU 2011; Adlakha-Hutcheon et al. 2009). Furthermore, a new branch of nanotechnology exists, that is bio-nanotechnology which generates nano-sized particles having specific functions by integrating principles of biology with different physical and chemical procedures (Kathiresan et al. 2009; Qi and Wang 2004; Roduner 2006). These nanoparticles can potentially be used for commercial, industrial, agricultural and medicinal purposes.

Furthermore, as agricultural field have several trials, including low crop yield, nutrient deficiency and environmental pollution caused by various biotic or abiotic factors; nanotechnology has emerged as a promising application for precision agriculture (Fig. 14.1). In short nanotechnology provides the best solutions for various ecological encounters i.e. the discovery of Nano sensors extensively enhances the observations of different environmental stresses and it improves the potential of plant in contradiction of diseases (Afsharinejad et al. 2016; Kwak et al. 2017). As a result of continuous improvements in the field of nanotechnology, sustainable agricultural provides significant communal and reasonable benefits.

Silver nanoparticles have size of 1–100 nm and are used in medical field mostly as tissue scaffolding, wound dressings, and protective coating applications. Furthermore, surface of silver nanoparticles coordinates various ligands providing remarkable applications regarding the surface functionalization of silver nanoparticles.

Silver is normally utilized as silver nitrate (NO_3^-) in different anti-microbial activities. As compared to free silver, silver nanoparticles are more beneficial and enhance microbial acquaintance due to greater surface area. Researchers also greatly emphasized on silver nanoparticles due to exclusive properties against many microbes and because of higher resistance to many antibiotics (Hutchison et al. 2008). So far, in this field of food processing, agro-based industries and agriculture, numerous studies have been reported (Chen et al. 2016).

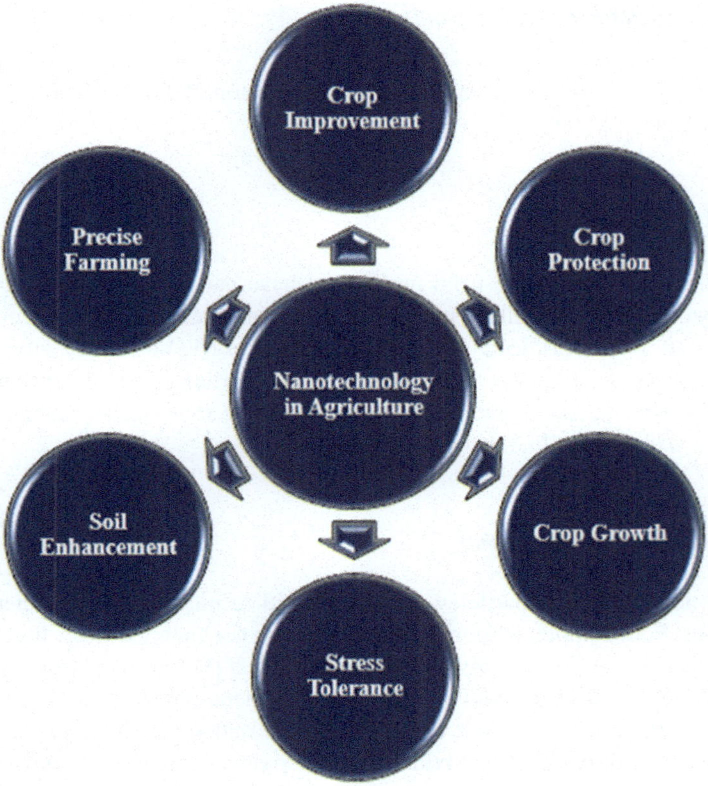

Fig. 14.1 Implementation of nanotechnology in agricultural field

14.2 Classification of NPs

Generally, NPs are categorized into three types.

- Organic nanoparticles
- Inorganic nanoparticles
- Carbon based

14.2.1 Organic Nanoparticles

Organic nanoparticles or polymers includes dendrimers, liposomes, micelles, ferritin, etc. Organic NPs are less-toxic, biodegradable and possess a hollow core and they are thermal and radiation sensitive (Tiwari et al. 2008). They are commonly used in the field of biomedicine.

14.2.2 Inorganic Nanoparticles

They are not composed of carbon. Inorganic nanoparticles include metal oxide based and metal NPs.

14.2.2.1 Metal Based

Metal based NPs are made from metals and have nano metric sizes. NPs can be synthesized from almost all the metals (Salavati-niasari et al. 2008). Cadmium (Cd), copper (Cu), gold (Au), lead (Pb), cobalt (Co), silver (Ag) iron (Fe), zinc (Zn) and aluminum (Al) are used for the synthesis of nanoparticles. These particles having size of 10–100 nm and characteristics of high surface area, surface charge and density, pore sizes, crystalline and amorphous structures.

14.2.2.2 Metal oxides Based

Metal oxide-based nanoparticles are synthesized to modify the characteristics of metal NPs. For example, at room temperature, iron (Fe) nanoparticles instantly oxidizes to iron oxide in the presence of oxygen (Fe2O3) that increases its reaction rate. Due to increased reactivity and efficiency, metal-based NPs are synthesized (Tai et al. 2007). The common example of these nanoparticles includes Titanium oxide, Silicon dioxide, Zinc oxide, Cerium oxide, Iron oxide, Magnetite and Aluminum oxide.

14.2.3 Carbon Based

They are composed of purely carbon (Bhaviripudi et al. 2007). Carbon-based NPs are characterized into graphene, carbon black, carbon nanotubes (CNT), carbon nanofibers, and fullerenes.

Crop productiveness is additionally generally influenced via biotic factors such as pests and diseases (Oerke et al. 2006). Farmers have been more dependent on pesticides to decrease the crop losses which effects the human health and climate sustainability. However, by the use of nanomaterial, pests and diseases are decreased efficiently, crop yield and environmental sustainability is increased (Table 14.1). AgNPs synthesized from cotton have a sturdy antibacterial activity aganist *Xanthomonas campestris* PV. and *Xanthomonas axonopodis* PV. *malvacearum campestris*, main pathogens of Brassicaceae and Malvaceae family crops, respectively (Vanti et al. 2019).

Metal oxide NPs, like MgO, ZnO, and CuO are effectively reduce many soil borne and palnt disease triggered via, *Alternaria alternate, Botrytis cinerea, Radicis*

Table 14.1 Nanoparticles, their mode of application and responses on plant growth and yield

Nanomaterial	Crop species	Application mode	Used Conc.	Treatment duration	Response	References
ZnO	*Coffea arabica*	Foliar application	10 mg/L	45 days	Enhanced growth, biomass accumulation and photosynthesis	Rossi et al. (2019)
FeS_2	*Cicer arietinum; pinacia oleracea; Daucus carota, Brassica juncea*	Seed priming	80–100 μg/mL	12–14 h	Increased germination and crop yield	Srivastava et al. (2014a, b); Das et al. (2016)
Fe/SiO_2	*Zea mays; Arachis hypogaea,*	Nano fertilizer	15 mg/kg	3 days	Improved plants growth and biomass	Disfani et al. (2017)
AgNPs	*Triticum aestivum*	Augmentationof pot soils	50 mg/L and 75 mg/L	Trifoliate stage	Enhanced growth and heat stress tolerance	Iqbal et al. (2019)
TiO_2	*Spinacia oleracea*	Seed priming and foliar application	0.25% suspension	48 h and 35 days	Increased biomass accumulation, chlorophyll, nitrogen and protein content.	Yang et al. (2007)
CuO	*Spinacia oleracea*	Mixed with soils	200 mg/kg	60 days	Improved photosynthesis and biomass production	Wang et al. (2019)
Ag NPs	*Vigna sinensis*	Spray on leaves	50 mg/L	40 days	Boosted plant growth and biomass by enhanced root nodulation and bacterial diversity in soils	Pallavi et al. (2016)
ZnO	*Cyamopsis tetragonoloba*	Foliar application	10 mg/L	6 weeks	More plant growth, biomass and nutrient content	Raliya et al. (2013)
CuO	*Solanum lycopersicum*	Foliar application	150–340 μg/mL	11 days	Efficient disease control (late blight) triggered by *Phytophthora infestans*	Giannousi et al. (2013)
Ag NPs	*Vigna unguiculata*	Foliar application	50–100 μg/mL	7 days	No phytotoxicity symptoms, in vitro inhibition of *Xanthomonas axonopodis* pv. *Malvacearum* and *Xanthomonas campestris* pv. *Campestris*	Vanti et al. (2019)

(continued)

Table 14.1 (continued)

Nanomaterial	Crop species	Application mode	Used Conc.	Treatment duration	Response	References
MgO	*Solanum lycopersicum*	Drenching	7–10 μg/mL	7 days	Suppress bacterial wilt disease caused by pathogen *Ralstonia solanacearum*	Imada et al. (2016)
Al_2O_3 NPs	*Solanum lycopersicum*	Foliar application	400 mg/L	20 days	Comparative control of phytopathogen root rot (*Fusarium*) in infected tomato plants	Shenashen et al. (2017)
SiO_2 NPs	*Oryza sativa*	Foliar application	2.5 mM/L	70 days	Assuaged heavy metal contamination and improved growth by reducing bio-concentration and translocation of toxic metals in plants	Wang et al. (2016)
TiO_2 and SiO_2	*Oryza sativa*	Foliar application	20 and 30 mg/L	55 days	Mitigation of Cd stress; improved growth by stimulation of antioxidant enzymes and limited Cd translocation	Rizwan et al. (2019)
ZnO	*Nicotiana tabacum*	Hydroponics	0.2 μM and 1 μM	21 days	Elevate plant growth and physiology, increased metabolites, enzyme activity and anatomical characters of plants	Tirani et al. (2019)
ZnO, CuO and ag NPs	*Prunus domestica* f	Fruit spray	100 and 1000 μg/mL	4 days	Suppressed soil borne diseases and symptoms of grey mold caused by *B. cinerea*	Malandrakis et al. (2019)

Lycopersici, Ralstonia solanacearum, Fusarium oxysporum fsp, *Verticillium Dahliae, Colletotrichum gloeosporioides, Monilinia fructicola, Phytophthora infestans, Fusarium solani,* and in other plants (Malandrakis et al. 2019; Shenashen et al. 2017; Imada et al. 2016). So, without disturbing the environment, the appropriate use of nanomaterials can extend crop productiveness. In current years, different studies proved that nanocomposites are eco-friendly and highly effective against plant diseases and plant protection. (Gomathi et al. 2019). For example, antifungal activity is increased against antracol fungicide by the usage of Ag-incorporated chitosan nanocomposites (Ag@CS) (Le et al. 2019). The efficiency and shelf life of pesticides is increased by applying NPs of *Bacillus thuringiensis* (Bt) containing lively Bt of and decreased the required dosage earlier. However, mechanisms of Bt-based nanocomposites are unknown (Devi et al. 2019).

The mechanisms through which nanoparticles increases plant growth and reduces the stresses are still unknown. Nanoparticles enhanced plant yield and growth by increasing the enzymatic activities (Shojaei et al. 2019). For example, NPs like nano-ZnO and nano-SiO2 or application increases water uptake, nutrients acquisition and the accumulation of free proline and amino acids, and activity of antioxidant enzymes which boost plant resistance to several native weather fluctuations (Shalaby et al. 2016; Wang et al. 2015).

Nanomaterials can additionally alter expression of stress genes e.g. a microarray evaluation exhibited a range of gene that is regulated by AgNPs in Arabidopsis (Banerjee et al. 2016). Plants are protected immediately from stresses by applying these nanoparticles (Onaga et al. 2016). However, plant responses are varied with species of plants, their developmental stages and applied NPs (Abdel-Aziz et al. 2016).

14.3 Characterization Techniques of Nanoparticles

For synthesized nanoparticles, characterization is an important to define the surface area, morphology, surface chemistry and nature inconsistency AgNPs. Characterization of silver nanoparticles can be done by various techniques (Fig. 14.2).

14.3.1 Transmission Electron Microscopy (TEM)

For characterization of nanoparticles TEM is very useful technique that provides information about morphology and the size of NPs (Rajeshkumar et al. 2017; Ghosh et al. 2012). Compared to SEM, TEM has a 1000-fold higher resolution (Eppler et al. 2000) and the images formed by TEM give precise statistics related to shapes, crystallographics and sizes (Vijayaraghavan et al. 2017).

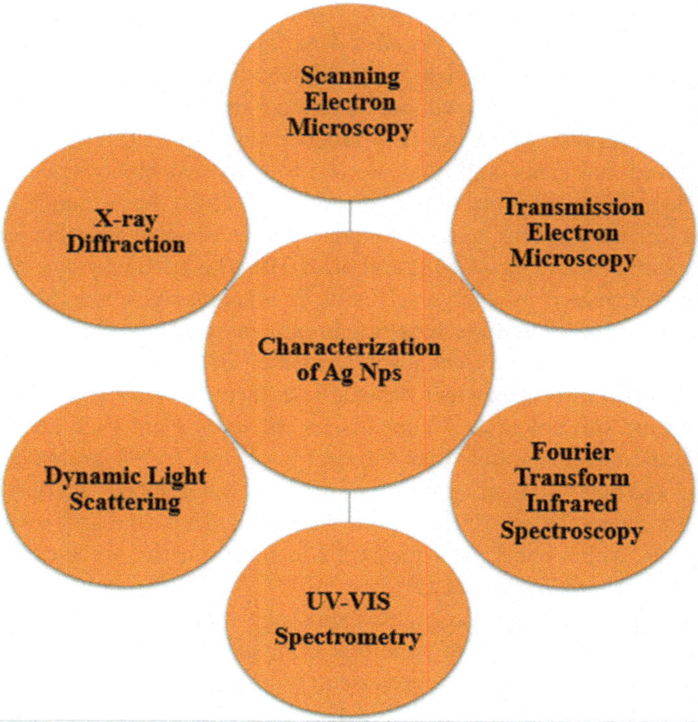

Fig. 14.2 Characterization Techniques for silver nanoparticles

14.3.2 Scanning Electron Microscopy (SEM)

SEM is useful to observe topography and morphology of nanoparticles that can also be used to measure the nanoparticles size 10^{-6} to 10^{-9} (Noruzi et al. 2011; Sundrarajan et al. 2011). A high energy beam of electrons is directed at surface of nanoparticles and it results characteristic features of samples (Hudlikar et al. 2012a, b). Electron microscopy analysis examines the morphological changes of the cell before and after the treatment of nanoparticle. It has been reported several times that the visible variations in cell profile and gaps in the cell wall, nanoparticles have been used as indicators in antimicrobial activity (Rahimi-Nasrabadi et al. 1969a, b; Zhang et al. 2014). SEM showed smooth and undamaged structures of the control bacterial cells, while AgNPs treated cells exhibit clear morphological variations in cell membrane and loss of membrane integrity and damaged cells (Roy et al. 2019).

14.3.3 Fourier Transform Infrared Spectroscopy (FTIR)

FTIR is used to analyze different capping agents, surface chemistry of NPs and involved biomolecules in the synthesis of nanoparticles (Rajeshkumar et al. 2017). In FTIR a spectrum is made when some infrared rays are absorbed, and some are passed from sample. These spectra indicate characteristic of the sample material through transmission and absorption (Rohman et al. 2010). FTIR is a less-costly, economical, suitable, and modest technique to examine the function of biomolecules to reduce silver nitrate to silver nanoparticles (Zhang et al. 2016).

14.3.4 X-Ray Diffraction Analysis (XRD)

To observe the crystalline structure of metallic nanoparticles, XRD is used. (Rajeshkumar et al. 2017; Vijayaraghavan et al. 2017). The formation of nanoparticles with crystalline structure confirms the resulting diffraction pattern (Alexander et al. 1950). The Debye–Scherrer equation is applied to calculate size of particle by determine the width through Bragg reflection law (Prathna et al. 2011). Hence, structural features of different materials, such as biomolecules, polymers, glasses and superconductors, can be observed by XRD silver (Zhang et al. 2016). Furthermore, XRD is an important method for studying nanomaterial's (Sharma et al. 2012).

14.3.5 UV-Vis Spectrophotometry

The stability and synthesis of metallic nanoparticles are monitored by UV-Vis spectrophotometry (Sastry et al. 1998). In the visible region, particular salts of metallic nanoparticles give specific peaks with absorptions (Li et al. 2012).

In general, it has been revealed by various studies that for the characterization of particles, absorption bands ranging 200 nm to 800 nm are best in size ranging 2–100 nm (Luyen et al. 2011). In silver nanoparticles the valence and conduction bands are quite closer with each other's. Freely moving electrons made a band of surface plasmon resonance absorption. Absorption of NPs depend upon size of particles, chemical surrounding and dielectric medium. Biological synthesized AgNPs were observed in order to check their stability for about 12 months and at the same wavelength a surface plasmon resonance peak was found using UV-Vis spectrophotometry (Zhang et al. 2016).

14.4 Synthesis of AgNPs

Silver nanoparticles are manufactured through different techniques. They can be synthesized by chemically, physically and biologically ways. These techniques have their advantages and disadvantages as well. The organisms play role of capping agents in biological synthesis and reduced Ag + to produce Ag^0 (Zewde et al. 2016). Biologically synthesized nanoparticles have improved reputation in recent years because of economical, less toxicity and greater yielding capacities (Shanmuganathan et al. 2019). Chemically and physically manufacturing is costly and hazardous to environment (Hulkoti et al. 2017).

14.4.1 Physical Approaches

Laser ablation and Evaporation-condensation techniques are included in AgNPs synthesis. A lot of energy is required, and long time period of completion are two main drawbacks in these approaches. Monodispersed AgNPs were synthesized by thermal decomposition of Ag^+ oleate complexes (Lee et al. 2004). Through evaporation-condensation technique, steel NPs were combined by using ceramic heater. AgNPs were sphere-shaped (Jung et al. 2006).

14.4.2 Chemical Approaches

AgNPs are manufactured by many chemical methods. These approaches are easy to handle ass compared to biological approaches. Previously reported that reducing agents donate electrons to Ag^+ ions and transformed into metallic AgNPs. Ag^+ salts are used in chemical manufacturing of AgNPs and silver nitrate is mostly used (Table 14.2) (Ge et al. 2014; Calderón et al. 2017).

Monodispersed AgNPs were synthesized via reducing NO_3^- (Sun et al. 2002). Sodium borohydride and trisodium citrate were stabilizing agents in synthesis of AgNPs. An effective reducing agent for the synthesis of silver nanoparticles is trisodium citrate ranging 60 nm to 100 nm (Agnihotri et al. 2013).

14.4.3 Biological Approaches

Physical and chemical approaches are costly, time-taking and eco-unfriendly for synthesis of AgNPs. So, an eco-friendly, economically, less hazardous and less toxic approach is required (Iravani et al. 2014) which is more advantageous as compared to physical and chemical approaches. Biological manufacturing approaches

Table 14.2 Physical and chemical syntheses of silver nanoparticles

Type	Reducing agent	Biological activity	Characterization	References
PVP-coated silver nanoparticles	Sodium borohydride	–	UV-Vis, TEM, EDS, DLS, FIFFF	Tejamaya et al. (2012)
Silver nanoparticles	Hydrazine, D-glucose	Antimicrobial	UV-Vis, TEM	Shrivastava et al. (2011)
Chitosan-loaded silver nanoparticles	Polysaccharide chitosan	Antimicrobial	TEM, FTIR, XRD, DSC, TGA	Ali et al. (2011)
Silver nanoparticles	Ascorbic acid	Antimicrobial	UV-Vis, EFTEM	Pal et al. (2011)
Polydiallyldimethylammonium chloride and polymethacrylic acid capped silver nanoparticles	Methacrylic acid polymers	Antimicrobial	UV-Vis, reflectance spectrophotometry	Dubas et al. (2006)

includes the usage of bacteria, fungi, yeast, algae and plant sources. These sources are very much important in medicinal uses of NPs.

Synthesis of NPs from microbes and plants is cost-effective, economic, safe and less harmful to ecosystem and environment (Makarov et al. 2014; Gowramma et al. 2015). Furthermore, microbes and plants can accumulate different metallic ions from ecosystem (Shah et al. 2015). Biological approaches for the amalgamation of AgNPs includes microbes and plants (Fig. 14.3) (Ahmad et al. 2019). (Table 14.3)

14.5 Factors Influencing Silver nanoparticle Synthesis

The structure, size, and form of silver nanoparticles are decided by a combination of different physical and chemical factors. The following are the basic parameters that influence the synthesis of silver nanoparticles in general:

14.5.1 Production Approaches of Silver Nanoparticles

Nanoparticles can be made using a variety of techniques, including physical and chemical techniques as well as biological protocols. In these processes, numerous inorganic or organic chemicals, and organisms as well, are used to synthesize nanoparticles (Patra and Baek. 2014). Green synthesis has already been debated as being superior approache compare to the rest, since it is both environmentally sustainable and economical. Green synthesis often avoids the practice of high temperatures, electricity, and toxic chemicals (Dhuper et al. 2012).

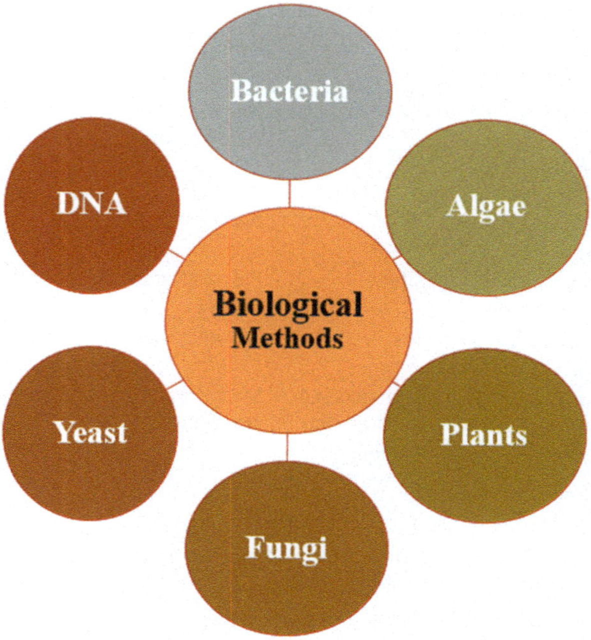

Fig. 14.3 Biological approaches for the amalgamation of AgNPs

14.5.2 Temperature

The temperature has been discovered to be a significant factor in nanoparticle production. In the presence of elevated temperature, spherical nanoparticles are synthesized. Nanotriangle formation, on the other hand, occurs mostly at lower temperatures (Rahimi-Nasrabadi et al. 1969a,b). Increases in temperature between 30 and 90 °C have been shown to increase the frequency of synthesis (Hudlikar et al. 2012a, b; Dankovich et al. 2011) and, in some cases, promote the formation of silver nanoparticles of smaller sizes (Mohammed et al. 2009). Multiple studies have suggested that the ideal temperature range for the biogenic production of metal nanoparticles is 25–37 °C (room temperature).

14.5.3 pH

According to several studies, stability of nanoparticle is better in normal media than in acidic media (Roopan et al. 2013; Sadeghi and Gholamhoseinpoor 2015). Though, a very high pH i.e., pH > 11) has some disadvantages, including the development of agglomerated and brittle silver nanoparticles (Tagad et al. 2013). As a result, it can be inferred that the pH governs the form and size of nanoparticles.

Table 14.3 Some important examples of organisms used for synthesizing silver nanoparticles

Biological synthesis of Silver nanoparticles				
Bacteria	Fungi	Plants	Algae	Yeast
Enterobacter cloacae	*Fusarium solani*	*Pinus eldarica*	*Spirogyra varians*	*S. cerevisiae.*
Escherichia coli	*Humicola sp.*	*Pelargonium graveolens*	*Padina pavonia*	
Lactobacillus casei	*Pleurotus cornucopiae*	*Emblica officinalis*	*Spirulina platensis*	
Lactobacillus strains	*Arthroderma fulvum*	*Cinnamomum camphora*	*Oscillitoria wellie*	
Klebsiella pneumonia	*Aspergillus fumigatus*	*Azadiracta indica*	*Gelidiella acerosa*	
Aeromonas sp. SH10	*Aspergillus flavus*	*Aleo vera*	*Chaetoceros calcitrans*	
Bacillus megaterium	*Fusarium oxysporum*	*Tamarix gallica*	*Chlorella salina*	
Bacillus strain CS 11	*Verticillium sp.*	*Bauhinia purpurea*	*Isochrysis galbana*	
Bacillus licheniformis	*Aspergillus fumigatus*	*Origanum vulgare* L.	*Tetraselmis gracilis*	
P. stutzeri AG259	*Fusarium semitactum*	*Moringa oleifera*		
Corynebacterium sp. SH09	*Fusarium acuminatum Ell*	*Ficus benghalensis*		
	Penicillium fellutanum	*Cleome viscosa*		
	Penicillium sp. J3	*Plasmodium falciparum*		
	Coriolus versicolor	*Vitex negundo*		
	Phanerochaete chrysosporium	*Catharanthus roseus*		

14.5.4 Time

The decrease in the reaction time influence reduction of ions to augmented metal with various shapes. Greater concentrations of nanoparticle in the medium are suggested by absorbance of high peaks during the optimal time. The optical properties, shape, and size, of anisotropic nanoparticles can be modified by varying temperatures, according to Rai and colleagues. It was determined by changing growth conditions and the creation of various nanoparticle sizes, including rectangular, hexagonal, triangular, and spherical nanoparticles (Rai et al. 2006).

14.5.5 *Size and Shape*

Nanoparticle properties are largely determined by their shape and size. It has been determined that the size and shape of the nanoparticle determine optimal activities and that most nanoparticle properties are dependent on size (Akbari et al. 2011).

14.6 Use of Silver Nanoparticles in Agriculture

The synthesis of nanoparticles (NPs) have grabbed particular attention for having peculiar properties, which can be used in cryogenic superconducting materials, biosensor materials, composite fibers, electronic components, and cosmetics (Zhang et al. 2020). The production of AgNPs and AuNPs from plant extracts, and even more so from agricultural wastes, is a major topic for encouraging sustainable growth in agro-industrial labors, in an account of natural resources depletion and climate variation. Plants-based synthesized NPs are applications in the soil to the food cycle and can be used in processes in agroindustry due to their low level of toxcicity(Awad et al. 2019). In June 2009, the Food and Agricultural Organization (FAO) and the World Health Organization (WHO) jointly declared nanotechnological food and agricultural applications, which covered a broad range of topics including nanofiltration, nanocoating, nanosized biofortification, food processing, and nanostructured foods (Takeuchi et al. 2014). NPs are known as "magic bullets" for imparting beneficial substances like nutrients, advantageous genes, and other organic compounds to specifically targeted plant parts to boost productivity. As a result, agricultural administration, with a focus on crop nutrition is insisted through a nano-delivery system represented by NPs (Marchiol et al. 2014). Many studies on the direct application of AgNPs in agriculture have concentrated on responses of plants to metal NPs, such as cytotoxicity, seed germination, and root elongation (Cox et al. 2017; Ribeiro et al. 2019). Metal NPs can also be used in the manufacture of nano-pesticides and nano-fertilizers (Vijayaraghavan and Ashokkumar 2017). The antimicrobial activity of nanoparticle has the most popular applications in food packaging (Marchiol et al. 2014). The above applications have been covered extensively in the agro-industries in a broad range of products containing NPs of these metals with particle sizes of range 100–250 nm, increasing their activity and making them water soluble (Prasad et al. 2017).

14.6.1 *Nano-Fertilizers*

A small quantity of nano-fertilizer is used to aid plants with nutrients or to improve the effectiveness of fertilizer (Rameshaiah et al. 2015). Encapsulating fertilizers in nano form can increase nutrient uptake, which discourages loss of nutrients, reduces the risk of environmental degradation, and improves crop yield and quality. These nano-fertilizers has also been shown to reduce plant stress upon foliar application (Tarafdar et al. 2012). To sync up the nutrient release with plant uptake, AgNPs-based nano fertilizers have been developed. This method helps to preserve soil fertility by reducing loss of nutrients, soil, groundwater pollution and chemical reactions between soil, water and microorganisms that turn them into unusable or hazardous nature for plants (Panpatte et al. 2016). Kang et al. 2016 added a 5 mg/L AgNPs fertilizer suspension three times per day at 14-day intervals to red ginseng shoots. After harvesting, they noticed that the nano fertilizer had increased the content of ginsenoside.

14.6.2 Nano-Pesticides

Every year, agriculture production declines due to the advent of new plant diseases and emerging resistant pathogens to pesticide composites, and each year, millions of dollars are spent on pest control. Pesticides, both natural and synthetic, used in agriculture are either ineffective or cause environmental hazards and long-term problems. Greenly produced AgNPs is a viable alternative to chemical pesticides, and AgNPs have been used against microbes in the past (Chowdappa and Gowda 2013). *Sphaerotheca pannasa, Rhizoctonia solani, Phythium ultimum, Fusarium culmorum, Colletotrichum gloeosporioides* etc. are among the phytopathogens that are suppressed by silver nanoparticles. (Gopal et al. 2011). Furthermore, Si–Ag NP (a silica nanoparticle with Ag) is absolutely effective in *cucurbits* for cure of powdery mildew (Park et al. 2006). Validamycin was also delivered using a porous silica nanomaterial, which showed specific targeting of the pesticide (Liu et al. 2006).

14.6.3 Disease Control and Pest-Management

Nematodes, fungi, bacteria, and viruses are also known as phytopathogens are important limiting factors in the processing of food. Pathogens are controlled using a variety of techniques, but there is no flawless formula for disease prevention. As a result, manipulating nanotechnology for the treatment of plant pathogens has a bright future. Because of its extensive efficacy, little toxicity, high surface-to-volume ratios, crystallographic composition, ease of use, charge capacity, and adaptability to a variety of substrates, silver is considered the most capable nanomaterial with bactericidal, viricidal, and fungicidal abilities (Nangmenyi and Economy 2009). Silver nanoparticles (AgNPs) have a powerful inhibitory effect on a variety of microorganisms, making them an effective antimicrobial agent (Clement and Jarrett 1994). It has a high degree of toxicity in microorganisms but a lower level of toxicity in mammals. The killing effects of AgNPs on microbes were found to be dependent on size (Raza et al. 2016). Various molds and yeasts such as *Candida* *krusei, Candida albicans,*and *Aspergillus brasiliensis* were treated with AgNPs/ PVP to display fungicidal activity. The hybrid materials had potent antifungal properties against the microbes that were tested. (Bryaskova et al. 2011). Antifungal behaviors of AgNPs were studied using Raman spectroscopy, scanning electron microscopy, traditional microbiological plating, as well as alterations in cellular structures and morphology of the hypha. Aziz et al. (2016) studied the antimicrobial properties of biogenic silver nanoparticles on disease causing fungi such as, *Aspergillus flavus*, *Fusarium oxysporum,* and *Candida albicans* and found that they are similar to well-known fungicides like ketoconazole, fluconazole, and amphotericin B. Significantly, when mixed with fungicides and antibiotics, these nanoparticles exhibit important synergistic properties, providing significantly increased resistance to microbial development. Nanocomposite DNA-mediated AgNPs developed on graphene oxide were created by Ocsoy et al. (2013). In culture and on plants, these complexes efficiently reduce cell viability of *Xanthomonas perforans*. Composites exhibit exceptional antibacterial potential at very low concentrations of value16 ppm, with major benefits in antibacterial activity and increased stability. As AgNPs concentration rose, inhibition improved as well in frequent cases. The reason might be attributed to the solution's ability to homogenize and cohere to hyphae of fungus and inhabit disease causing fungi at such a high density. (Kim et al. 2012). Nano Silver is a well-known potent bacteriostat with antimicrobial activity over a wide spectrum. A nanosilver colloid in well-dispersed state is more adhesive to fungi and bacteria, resulting improvement in activity against bacteria (Kim et al. 2008). Numerous forestry plants and food crops are threatened by a variety of pathogens in nature, destroying agricultural products and the loss of tree species. By disrupting hyphae of fungus, interfering with nutrient uptake, and enhancing the inhibition of fungal development (germination and growth), AgNPs have emerged as a new hope for disease control. The effect of nanoparticles carrying silver ions on spore development and disease emergence in disease causing fungi may be the mechanism at work. As a result, AgNPs demonstrate a high potential for use as nano pesticides in the regulation of phytopathogens (Alghuthaymi et al. 2015). In agricultural soil, silver nanoparticles affect a variety of bacterial communities, which may be beneficial or detrimental to environment including plants (Panyala et al. 2008). Silver nanoparticles encounter strong inhibitory effects on different bacteria species due to which it is considered a strong antibacterial agent (Joshi et al. 2018). Kamran et al. (2011) described that nano-TiO2 and nanosilver with maximum efficacy may be used to free tobacco plants from toxic bacteria. AgNP inhibits protein synthesis, replication, and toxicity in bacteria upon exposure (Chaloupka et al. 2010). In studies, various inorganic and organic (Lamsal et al. 2011; Jo et al. 2009) compounds, along with effective carriers (Ouda 2014), have been revealed significant effects on antibacterial function and lessen nanoparticles biological toxicity. While silver nanoparticles are hybridized with other oxides or metal nanoparticles serving like a shield or center to form bimetallic nanoparticles, a synergistic antimicrobial effect is achieved. (Chou and Chen 2007).

14.7 Antibacterial Mechanism of AgNP's

The deposition of silver nanoparticles in the cell membrane increased the permeability of the membrane, resulting in bacterial cell death. They even tried to figure out how their actions worked. There are five key reasons for antibacterial behavior that have been suggested so far (Lemire et al. 2013):

1. Nutrient absorption is disrupted (Pal et al. 2007a, b).
2. Their toxic ions may damage DNA (genotoxic), resulting in cell death.
3. Protein mechanism, Electron transport chain, and membrane are all disrupted.
4. Toxic ions are released and attaches to proteins containing sulfur-group, preventing proteins in the membrane from working properly and interfering with permeability of cell (Sondi and Salopek-Sondi 2004).
5. Different forms of DNA, protein, and membrane damage may be caused by the reactive oxygen species and various oxidation reactions catalyzed by metals (Aziz et al. 2015).

This mechanism cannot function independently, implying that several mechanisms are active at the same time. These nano particles having multiple action targets could enable NPs to effectively combat a variety of plant pathogens.

14.8 Abiotic Stress Reduction

Nanoparticles (NPs) help plants grow while still protecting them from abiotic stress. Because of its wide surface area and small size, poisonous metal attaches to the nanoparticle surface, limiting its supply. Drought, salinity, alkalinity, temperature variations, and mineral and metal toxicity are all examples of abiotic stress. Nano particles can follow the action of antioxidant enzymes for scavenging reactive oxygen species from oxidative stress in the form of nano-enzyme (Sharifi et al. 2020). Photosynthesis is a crucial biochemical mechanism in plants and one of the most vulnerable; therefore, its normal functioning can be sustained by reducing oxidative and osmotic stress, plants are provided with enzymatic apparatuses that helps them cope with oxidative stress. Plants, on the other hand, suffer the repercussions of such a scenario when the defensive mechanism fails. By causing improvements in antioxidant enzyme activities that are dependent on the dosage, crop type, and application time, AgNPs significantly reduced the negative effects of salt stress in wheat (Mohamed et al. 2017). Furthermore, NPs effectively contribute to the enhancement of crop plant growth, quality, and productivity in salinity induced by excessive chemical fertilizer usage (Taran et al. 2017).

14.9 Mode of Action of AgNPs for Plant Growth and Disease Control

Pathogen growth is monitored by silver nanoparticles (AgNPs) through several mechanisms. AgNPs' detailed mechanism of antimicrobial action is unclear, although it is one of the most highly discussed subjects. According to studies revealed through electron spin resonance spectroscopy, AgNPs are thought to work in a variety of ways of causing antimicrobial effects. When AgNPs come into contact with bacteria, they produce free radicals, which cause membranes to become porous, resulting in cell expiry (Kim et al. 2007). It has also been reported that nanoparticles generate silver ions which inactivate several important enzymes by interacting with the -SH (thiol) groups of enzymes (Zhang et al. 2016). The silver ions inhibit many functions and damage the bacterial cells which results in the generation of reactive oxygen species. The ROS may possibly be produced through the inhibition of a respiratory enzyme by silver ions. These nanoparticles in the cell cause the reaction to take place and ultimately lead to cell death. The DNA contains phosphorus and sulphur as their chief components, is another important fact. Nanoparticles act on phosphorus and sulphur and damage the DNA which ultimately leads to cell death (Hatchett and Henry 1996). AgNPs interact with sulphur and phosphorus in DNA, thus disrupt DNA replication, halting the bacterial growth. Silver nanoparticles have been discovered to control signal transduction in bacteria by phosphorylation of protein substrates. The nanoparticles are involed in dephosphorylation of peptide substrates on tyrosine residues, subsequent in inhibition of signal transduction due to which growth arrest. Though, it's important to keep in mind that further investigation on the subject is needed to completely support the opinions (Shrivastava et al. 2007a, b).

Silver nanoparticles are considered as most stimulating metal nanoparticles that also possess strong biological activity. They are reported to enhance seed germination, boosted plant growth and vigor, improved biomass, better proliferation and shoot induction, or boosted photosynthetic activity. Elongated seedling and increased biomass were noted in *Oryza sativa* and *Trigonella foenumgraceum* when treated with bio fabricated AgNPs. Optimum application of AgNPs in *Arabidopsis* plants directly correlates with the production and accretion of cell cycle related proteins, expression of genes involved in cell proliferation, photosynthesis and signaling pathways of hormones i.e. auxins, abscisic acid and ethylene production (Salachna et al. 2019) (Fig. 14.4).

14.10 AgNPs Uptake and Translocation in Plants

Previous investigations reveled noticeable positive and negative impacts of AgNPs on plants that depends upon various factors that regulates its uptake and translocation in plants to act accordingly. Uptake of AgNPs by plants depends upon the

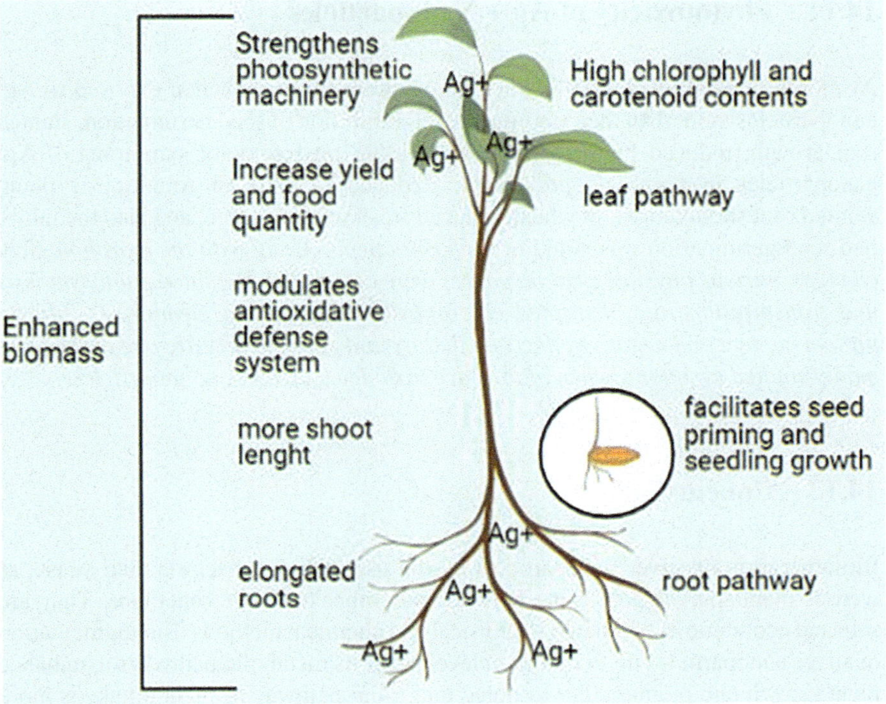

Fig. 14.4 Positive effects of Ag^+ nanoparticles on plant growth and vigor

various size and shape of AgNPs and also the cell permeability of the plant (Tripathi et al. 2017). Semi permeability of plant's cell permits the entrance of small particles. After the entrance of silver nanoparticles in plant cell wall and plasma membrane, they get translocated to stem and leaves via intercellular spaces and vascular tissues as observed in *Arabidopsis thalianas*. However, interesting point is that AgNPs can induce the formation of new and large sized pores in cell wall and membranes that documents the entrance of larger AgNPs. Two pathways were noted for the entrance of nanoparticles in plants i. root pathway; progressive uptake of AgNPs by boarder cells then successive translocation to root cap, epidermis, columella, root meristem and eventually the vascular tissues to whole plants. As observed in *Arabidopsis* plants after 14 days of exposure. ii Leaves pathway; AgNPs can be taken up by guard cells, penetrate through stomata and/or trapped by leaf cuticle when directly immersed in AgNPs containing medium or via foliar application. Accumulation of AgNP's in stomatal guard cells of *Arabidopsis* plant and cuticle of lettuce leaves was observed, furthermore, 17–200 times more Ag bioaccumulation in leaves of soybean and rice on foliar exposure was noted as compared to root application. (40, 48, 49). Vascular tissues are noted to be the marked transporter of Ag^+ nanoparticles to fruits, seeds, and other edible parts of plants.

14.11 Phytotoxicity of Ag + Nanoparticles

At morphological level, substantial changes were observed when exposed to Ag^+ nanoparticles more than their optimum level. Inhibition of seed germination, stunted root growth, reduced biomass and leaf area are phytotoxicity symptoms of Ag^+ nanoparticles in *Spirodela polyrrhiza.* Reduced biomass in Arabidopsis plant, stunted root shoot length, in wheat, reduced fresh weight in rice, and also the inhibited seed germination was noted in *Cucurbita pepo. Along with the morphological changes, various physiological responses were also noted. Declined photosynthesis and transpiration rate, disrupted chlorophyll and thylakoid membranes, altered protein content and negative effects on fluidity and permeability are one of the commonly studied symptoms when plants face toxic level of silver nanoparticles.*

14.12 Conclusion

Biofabrication of silver nanoparticles by the use of fungi, bacteria, and yeasts as well as plant sources pose more benefits than other traditional methods. They are safe and economic as compared to physical and chemical methods. Bioaugmentation of silver nanoparticles up to optimum level boost its metabolic pathways to enhance plant growth and biomass. It was noted that foliar pathway of plant uptake is more effective than root pathway. Manipulation of nanotechnology to treat phytopathogens has a bright future. Because of its extensive efficacy, little toxicity, high surface-to-volume ratios, crystallographic composition, ease of use, charge capacity, and adaptability to a variety of substrates, AgNP's are considered the most capable nanomateriasl with bactericidal, viricidal, and fungicidal abilities. Thus, the application of AgNP's in agriculture is consider as a sustainable approach for the augmentation of plant growth and pathogen control.

References

Abdel-Aziz HMM, Hasaneen MNA, Omer AM (2016) Nano chitosan-NPK fertilizer enhances the growth and productivity of wheat plants grown in sandy soil. Span J Agric Res 14:17

Adlakha-Hutcheon G, Khaydarov R, Korenstein R, Varma R, Vaseashta A, Stamm H, Abdel-Mottaleb M (2009) Nanomaterials, nanotechnology. In: Linkov I, Steevens J (eds) Nanomaterials: risks and benefits. NATO science for peace and security series C: environmental security. Springer, Dordrecht, pp 195–207

Afsharinejad A, Davy A, Jennings B, Brennan C (2016) Performance analysis of plant monitoring nanosensor networks at THz frequencies. IEEE Internet Things J 3:59–69

Agnihotri S, Mukherji S (2013) Size-controlled silver nanoparticles synthesized over the range 5–100 nm using the same protocol and their antibacterial efficacy. RSC Adv 4:3974–3983

Ahmad S, Munir S, Zeb N, Ullah A, Khan B, Ali J et al (2019) Green nanotechnology: a review on green synthesis of silver nanoparticles – an ecofriendly approach. Int J Nanomedicine 14:5087–5107
Akbari B, Pirhadi Tavandashti M, Zandrahimi M (2011) Particle size characterization of nanoparticles – a practical approach. Iran J Mater Sci Eng 8(2):48–56
Alexander L, Klug HP (1950) Determination of crystallite size with the X-ray spectrometer. J Appl Phys 21(2):137–142
Alghuthaymi MA, Almoammar H, Rai M, Said-Galiev E, Abd-Elsalam KA (2015) Myconanoparticles: synthesis and their role in phytopathogens management. Biotechnol & Biotechnol Equipment 29(2):221–236
Ali W, Rajendran S, Joshi M (2011) Synthesis and characterization of chitosan and silver loaded chitosan nanoparticles for bioactive polyester. Carbohydr Polym 83:438–446
Awad MA, Eisa NE, Virk P, Hendi AA, Ortashi KMOO, Mahgoub ASA, Elobeid MA, Eissa FZ (2019) Green synthesis of gold nanoparticles: preparation, characterization, cytotoxicity, and anti-bacterial activities. Mater Lett 256:126608
Aziz N, Faraz M, Pandey R, Shakir M, Fatma T, Varma A et al (2015) Facile algae-derived route to biogenic silver nanoparticles: synthesis, antibacterial, and photocatalytic properties. Langmuir 31(42):11605–11612
Aziz N, Pandey R, Barman I, Prasad R (2016) Leveraging the attributes of Mucor hiemalis-derived silver nanoparticles for a synergistic broad-spectrum antimicrobial platform. Front Microbiol 7:1984
Banerjee J, Kole C (2016) Plant nanotechnology: an overview on concepts, strategies, and tools. In: Kole C, Kumar D, Khodakovskaya M (eds) Plant nanotechnology. Springer, Cham, pp 1–14
Bhaviripudi S, Mile E, Iii SAS, Zare AT, Dresselhaus MS, Belcher AM, Kong J (2007) CVD synthesis of single-walled carbon nanotubes from gold nanoparticle catalysts. J Am Chem Soc 129(6):1516–1517
Bryaskova R, Pencheva D, Nikolov S, Kantardjiev T (2011) Synthesis and comparative study on the antimicrobial activity of hybrid materials based on silver nanoparticles (AgNps) stabilized by polyvinylpyrrolidone (PVP). J Chem Biol 4(4):185–191
Calderón-Jiménez B, Johnson ME, Montoro Bustos AR, Murphy KE, Winchester MR, Vega Baudrit JR (2017) Silver nanoparticles: technological advances, societal impacts, and metrological challenges. Front Chem 5:6
Chaloupka K, Malam Y, Seifalian AM (2010) Nanosilver as a new generation of nanoproduct in biomedical applications. Trends Biotechnol 28(11):580–588
Chen YW, Lee HV, Juan JC, Phang SM (2016 Oct 20) Production of new cellulose nanomaterial from red algae marine biomass Gelidium elegans. Carbohydr Polym 151:1210–1219
Chou KS, Chen CC (2007) Fabrication and characterization of silver core and porous silica shell nanocomposite particles. Microporous Mater 98:208–213
Chowdappa P, Gowda S (2013) Nanotechnology in crop protection: status and scope. Pest Manag Hortic Ecosyst 19(2):131–151
Clement JL, Jarrett PS (1994) Antibacterial silver. Met Based Drugs 1(5–6):467–82
Cox A, Venkatachalam P, Sahi S, Sharma N (2017) Reprint of: silver and titanium dioxide nanoparticle toxicity in plants: a review of current research. Plant Physiol Biochem PPB 110:33–49
Dankovich TA, Gray DG (2011) Bactericidal paper impregnated with silver nanoparticles for point-of-use water treatment. Environ Sci Technol 45(5):1992–1998
Das CK, Srivastava G, Dubey A, Roy M, Jain S, Sethy NK, Saxena M, Harke S, Sarkar S, Misra K (2016) Nano-iron pyrite seed dressing: a sustainable intervention to reduce fertilizer consumption in vegetable (beetroot, carrot), spice (fenugreek), fodder (alfalfa), and oilseed (mustard, sesamum) crops. Nanotechnol Environ Eng 1:2
Devi PV, Duraimurugan P, Chandrika K (2019) Nano-biopesticides today and future perspectives. Academic Press; Cambridge, MA, USA. Bacillus thuringiensis-based nanopesticides for crop protection; pp. 249–260

Dhuper S, Panda D, Nayak PL (2012) Green synthesis and characterization of zero valent iron nanoparticles from the leaf extract of Mangifera indica. Nano Trends J Nanotechnol App 13(2):16–22

Disfani MN, Mikhak A, Kassaee MZ, Maghari A (2017) Effects of nano Fe/SiO_2 fertilizers on germination and growth of barley and maize. Arch Agron Soil Sci 63:817–826

Dubas ST, Kumlangdudsana P, Potiyaraj P (2006) Layer-by-layer deposition of antimicrobial silver nanoparticles on textile fibers. Colloids Surf A Physicochem Eng Asp 289:105–109

Eppler AS, Rupprechter G, Anderson EA, Somorjai GA (2000) Thermal and chemical stability and adhesion strength of Pt nanoparticle arrays supported on silica studied by transmission electron microscopy and atomic force microscopy. J Phys Chem B 104(31):7286–7292

EU (2011) Commission recommendation of 18 October 2011 on the definition of nanomaterial (2011/696/EU). Off J Eur Union 2011:L275/38

Ge L, Li Q, Wang M, Ouyang J, Li X, Xing MM (2014) Nanosilver particles in medical applications: synthesis, performance, and toxicity. Int J Nanomedicine 9:2399–2407

Ghosh S, Patil S, Ahire M, Kitture R, Kale S, Pardesi K et al (2012) Synthesis of silver nanoparticles using Dioscorea bulbifera tuber extract and evaluation of its synergistic potential in combination with antimicrobial agents. Int J Nanomedicine 7:483–496

Giannousi K, Avramidis I, Dendrinou-Samara C (2013) Synthesis, characterization and evaluation of copper based nanoparticles as agrochemicals against Phytophthora infestans. RCS Adv 3:21743–21752

Gomathi T, Rajeshwari K, Kanchana V, Sudha PN, Parthasarathy K (2019) Impact of nanoparticle shape, size, and properties of the sustainable nanocomposites. In: Inamuddin TS, Kumar Mishra R, Asiri AM (eds) Sustainable polymer composites and nanocomposites. Springer International Publishing, Cham, pp 313–336

Gopal MADHUBAN, Gogoi ROBIN, Srivastava CHITRA, Kumar RAJESH, Singh PK, Nair KK et al (2011) Nanotechnology and its application in plant protection. Plant Pathol India: Vision 2030:224–232

Gowramma B, Keerthi U, Rafi M, Rao DM (2015) Biogenic silver nanoparticles production and characterization from native stain of Corynebacterium species and its antimicrobial activity. 3. Biotech 5:195–201

Hatchett DW, White HS (1996) Electrochemistry of sulfur adlayers on the low-index faces of silver. J Phys Chem 100(23):9854–9859

Hudlikar M, Joglekar S, Dhaygude M, Kodam K (2012a) Green synthesis of TiO2 nanoparticles by using aqueous extract of Jatropha curcas L. latex. Mater Lett 75:196–199

Hudlikar M, Joglekar S, Dhaygude M, Kodam K (2012b) Green synthesis of TiO2 nanoparticles by using aqueous extract of Jatropha curcas L. latex. Mater Lett 75:196–199

Hulkoti NI, Taranath TC (2017) Influence of physico-chemical parameters on the fabrication of silver nanoparticles using Petrea volubilis L. stem broth and its anti-microbial efficacy. Int J Pharm Sci Drug Res 9:72–78

Hutchison JE, Greener (2008) Nanoscience: a proactive approach to advancing applications and reducing implications of nanotechnology. ACS Nano 2:395–402

Imada K, Sakai S, Kajihara H, Tanaka S, Ito S (2016) Magnesium oxide nanoparticles induce systemic resistance in tomato against bacterial wilt disease. Plant Pathol 65:551–560

Iqbal M, Raja NI, Hussain M, Ejaz M, Yasmeen F (2019) Effect of silver nanoparticles on growth of wheat under heat stress. IJST A Sci 43:387–395

Iravani S (2014) Bacteria in nanoparticle synthesis: current status and future prospects. Int Scholar Res Not 2014:359316

Jo YK, Kim BH, Jung G (2009) Antifungal activity of silver ions and nanoparticles on phytopathogenic fungi. Plant Dis 93(10):1037–1043

Joshi N, Jain N, Pathak A, Singh J, Prasad R, Upadhyaya CP (2018) Biosynthesis of silver nanoparticles using Carissa carandas berries and its potential antibacterial activities. J Sol-Gel Sci Technol 86(3):682–689

Jung JH, Cheol OH, Soo Noh H, Ji JH, Kim SS (2006) Metal nanoparticle generation using a small ceramic heater with a local heating area. J Aerosol Sci 37:1662–1670

Kang H, Hwang Y-G, Lee T-G, Jin C-R, Cho CH, Jeong H-Y, Kim D-O (2016) Use of gold nanoparticle fertilizer enhances the Ginsenoside contents and anti-inflammatory effects of red ginseng. J Microbiol Biotechnol 26:1668–1674

Kathiresan K, Manivannan S, Nabeel MA, Dhivya B (2009) Studies on silver nanoparticles synthesized by a marine fungus, Penicillium fellutanum isolated from coastal mangrove sediment. Colloids Surf B Biointerfaces 71(1):133–137

Khan MR, Rizvi TF (2014) Nanotechnology: scope and application in plant disease management. Plant Pathol J 13:214–231

Kim JS, Kuk E, Yu KN, Kim JH, Park SJ, Lee HJ et al (2007) Antimicrobial effects of silver nanoparticles. Nanomedicine 3(1):95–101

Kim HS, Kang HS, Chu GJ, Byun HS (2008) Antifungal effectiveness of nanosilver colloid against rose powdery mildew in greenhouses. In Solid state phenomena (Vol. 135, pp. 15–18). Trans Tech Publications Ltd.

Kim SW, Jung JH, Lamsal K, Kim YS, Min JS, Lee YS (2012) Antifungal effects of silver nanoparticles (AgNPs) against various plant pathogenic fungi. Mycobiology 40(1):53–58

Kwak S-Y, Wong MH, Lew TTS, Bisker G, Lee MA, Kaplan A, Dong J, Liu AT, Koman VB, Sinclair R (2017) Nanosensor technology applied to living plant systems. Annu Rev Anal Chem 10:113–140

Lamsa K, Kim SW, Jung JH, Kim YS, Kim KS, Lee YS (2011) Inhibition effects of silver nanoparticles against powdery mildews on cucumber and pumpkin. Mycobiology 39(1):26–32

Le VT, Bach LG, Pham TT, Le NTT, Ngoc UTP, Tran D-HN, Nguyen DH (2019) Synthesis and antifungal activity of chitosan-silver nanocomposite synergize fungicide against Phytophthora capsici. J Macromol Sci Part A 56:522–528

Lee DK, Kang YS (2004) Synthesis of silver nanocrystallites by a new thermal decomposition method and their characterization. ETRI J 26:252–256

Lemire JA, Harrison JJ, Turner RJ (2013) Antimicrobial activity of metals: mechanisms, molecular targets and applications. Nat Rev Microbiol 11(6):371–384

Li Y, Chen S-M (2012) The electrochemical properties of acetaminophen on bare glassy carbon electrode. Int J Electrochem Sci 7:13

Liu F, Wen LX, Li ZZ, Yu W, Sun HY, Chen JF (2006) Porous hollow silica nanoparticles as controlled delivery system for water-soluble pesticide, Materials Research Bulletin 41(12): 2268–2275

Luyen NT, Hai LB, Nghia NX, Nga PT, Lieu NTT (2011) Effect of reaction temperature and ligand concentration on the shape of CdSe nanocrystals. Int J Nanotechnol 8(3–5):214–226

Makarov VV, Love AJ, Sinitsyna OV, Makarova SS, Yaminsky IV, Taliansky ME et al (2014) "Green" nanotechnologies: synthesis of metal nanoparticles using plants. Acta Nat 6(20):35–44

Malandrakis AA, Kavroulakis N, Chrysikopoulos CV (2019) Use of copper, silver and zinc nanoparticles against foliar and soil-borne plant pathogens. Sci Total Environ 670:292–299

Marchiol L, Mattiello A, Poš'ci'c F, Giordano C, Musetti R (2014) In vivo synthesis of nanomaterials in plants: Location of silver nanoparticles and plant metabolism. Nanoscale Res Lett 9:101

Mohamed AKS, Qayyum MF, Abdel-Hadi AM, Rehman RA, Ali S, Rizwan M (2017) Interactive effect of salinity and silver nanoparticles on photosynthetic and biochemical parameters of wheat. Arch Agron Soil Sci 63(12):1736–1747

Mohammed Fayaz A, Balaji K, Kalaichelvan PT, Venkatesan R (2009) Fungal based synthesis of silver nanoparticles – an effect of temperature on the size of particles. Colloids Surf B Biointerfaces 74:123–126

Nangmenyi G, Economy J (2009) Nanometallic particles for oligodynamic microbial disinfection. In nanotechnology applications for clean water (pp. 3–15). William Andrew Publishing

Noruzi M, Zare D, Khoshnevisan K, Davoodi D (2011) Rapid green synthesis of gold nanoparticles using Rosa hybrida petal extract at room temperature. Spectrochim Acta A Mol Biomol Spectrosc 79(5):1461–1465

Ocsoy I, Paret ML, Ocsoy MA, Kunwar S, Chen T, You M, Tan W (2013) Nanotechnology in plant disease management: DNA-directed silver nanoparticles on graphene oxide as an antibacterial against Xanthomonas perforans. ACS Nano 7(10):8972–8980

Oerke EC (2006) Crop losses to pests. J Agric Sci 144:31–43

Onaga G, Kerstin W (2016) Advances in plant tolerance to biotic stresses. In: Abdurakhmonov IY, editor. Plant genomics. InTech; Rijeka, Croatia. pp. 167–228

Ouda SM (2014) Antifungal activity of silver and copper nanoparticles on two plant pathogens, Alternaria alternata and Botrytis cinerea. Res J Microbiol 9(1):34

Pal S, Tak YK, Song JM (2007a) Does the antibacterial activity of silver nanoparticles depend on the shape of the nanoparticle? A study of the gram-negative bacterium Escherichia coli. Appl Environ Microbiol 73:1712–1720

Pal S, Tak YK, Song JM (2007b) Does the antibacterial activity of silver nanoparticles depend on the shape of the nanoparticle? A study of the gram-negative bacterium Escherichia coli. Appl Environ Microbiol 73(6):1712–1720

Pallavi, Mehta CM, Srivastava R, Arora S, Sharma AK (2016) Impact assessment of silver nanoparticles on plant growth and soil bacterial diversity. 3 Biotech 6:254

Panpatte DG, Jhala YK, Shelat HN, Vyas RV (2016) Nanoparticles: the next generation technology for sustainable agriculture. In Microbial inoculants in sustainable agricultural productivity: Vol. 2: Functional Applications; Singh, D.P., Singh, H.B., Prabha, R., Eds.; Springer India: New Delhi, India, pp. 289–300

Panyala NR, Peña-Méndez EM, Havel J (2008) Silver or silver nanoparticles: a hazardous threat to the environment and human health? J Appl Biomed 6(3)

Park HJ, Kim SH, Kim HJ, Choi SH (2006) A new composition of nanosized silica-silver for control of various plant diseases. Plant Pathol J 22(3):295–302

Patra JK, Baek KH (2014) Green nanobiotechnology: factors affecting synthesis and characterization techniques. J Nanomater 2014

Prasad R, Bhattacharyya A, Nguyen QD (2017) Nanotechnology in sustainable agriculture: recent developments, challenges, and perspectives. Front Microbiol 8

Prathna TC, Chandrasekaran N, Raichur AM, Mukherjee A (2011) Biomimetic synthesis of silver nanoparticles by Citrus Limon (lemon) aqueous extract and theoretical prediction of particle size. Colloids Surf B Biointerfaces 82(1):152–159

Qi WH, Wang MP (2004) Size and shape dependent melting temperature of metallic nanoparticles. Mater Chem Phys 88:280–284

Rahimi-Nasrabadi M, Pourmortazavi SM, Shandiz SAS, Ahmadi F, Batooli H (1969a) Green synthesis of silver nanoparticles using Eucalyptus leucoxylon leaves extract and evaluating the antioxidant activities of extract. Nat Prod Res 28(22):1964–1969

Rahimi-Nasrabadi M, Pourmortazavi SM, Shandiz SAS, Ahmadi F, Batooli H (1969b) Green synthesis of silver nanoparticles using Eucalyptus leucoxylon leaves extract and evaluating the antioxidant activities of extract. Nat Prod Res 28(22):1964–1969

Rai A, Singh A, Ahmad A, Sastry M (2006) Role of halide ions and temperature on the morphology of biologically synthesized gold nanotriangles. Langmuir 22:736–741

Rajeshkumar S, Bharath LV (2017) Mechanism of plant-mediated synthesis of silver nanoparticles – a review on biomolecules involved, characterisation and antibacterial activity. Chem Biol Interact 273:219–227

Raliya R, Tarafdar JC (2013) ZnO nanoparticle biosynthesis and its effect on phosphorous-mobilizing enzyme secretion and gum contents in Clusterbean (Cyamopsis tetragonoloba L.). Agric Res 2:48–57

Rameshaiah GN, Pallavi J, Shabnam S (2015) Nano fertilizers and nano sensors–an attempt for developing smart agriculture. Int J Engineer Res Gen Sci 3:314–320

Raza MA, Kanwal Z, Rauf A, Sabri AN, Riaz S, Naseem S (2016) Size-and shape-dependent antibacterial studies of silver nanoparticles synthesized by wet chemical routes. Nanomaterials 6(4):74

Ribeiro CAS, Albuquerque LJC, de Castro CE, Batista BL, de Souza ALM, Albuquerque BL, Zilse MS, Bellettini IC, Giacomelli FC (2019) One-pot synthesis of sugar-decorated gold nanoparticles with reduced cytotoxicity and enhanced cellular uptake. Colloids Surf A Physicochem Eng Asp 580:123690

Rizwan M, Ali S, ur Rehman MZ, Malik S, Adrees M, Qayyum MF, Alamri SA, Alyemeni MN, Ahmad P (2019) Effect of foliar applications of silicon and titanium dioxide nanoparticles on growth, oxidative stress, and cadmium accumulation by rice (Oryza sativa). Acta Physiol Plant 41:35

Roduner E (2006) Size matters: why nanomaterials are different. Chem Soc Rev 35(7):583–592

Rohman A, Man YBC (2010) Fourier transform infrared (FTIR) spectroscopy for analysis of extra virgin olive oil adulterated with palm oil. Food Res Int 43(3):886–892

Roopan SM, Bharathi A, Abdul Rahuman A, Kamaraj C, Madhumita G, Rohit et al (2013) Low-cost and eco-friendly phyto-synthesis of silver nanoparticles using Cocos nucifera coir extract and its larvicidal activity. Ind Crop Prod 43:631–635

Rossi L, Fedenia LN, Sharifan H, Ma X, Lombardini L (2019) Effects of foliar application of zinc sulfate and zinc nanoparticles in coffee (Coffea arabica L.) plants. Plant Physiol. Biochem 135:160–166

Roy A, Bulut O, Some S, Kumar Mandal A, Deniz YM (2019) Green synthesis of silver nanoparticles: biomolecule–nanoparticle organizations targeting antimicrobial activity. RSC Adv 9(5):2673–2702

Sadeghi B, Gholamhoseinpoor F (2015) A study on the stability and green synthesis of silver nanoparticles using Ziziphora tenuior (Zt) extract at room temperature. Spectrochim Acta A Mol Biomol Spectrosc 134:310–315

Salachna P, Byczyńska A, Zawadzińska A, Piechocki R, Mizielińska M (2019) Stimulatory effect of silver nanoparticles on the growth and flowering of potted oriental lilies. Agronomy 9(10):610

Salavati-niasari M, Davar F, Mir N (2008) Synthesis and characterization of metallic copper nanoparticles via thermal decomposition. Polyhedron 27:3514–3518

Sastry M, Patil V, Sainkar SR (1998) Electrostatically controlled diffusion of carboxylic acid derivatized silver colloidal particles in thermally evaporated fatty amine films. J Phys Chem B 102(8):1404–1410

Shah M, Fawcett D, Sharma S, Tripathy SK, Poinern GEJ (2015) Green synthesis of metallic nanoparticles via biological entities. Dent Mater 8:7278–7308

Shalaby TA, Bayoumi Y, Abdalla N, Taha H, Alshaal T, Shehata S, Amer M, Domokos-Szabolcsy É, El-Ramady H (2016) Nanoparticles, soils, plants and sustainable agriculture. In: Shivendu R, Nandita D, Eric L (eds) Nanoscience in food and agriculture 1. Springer, Cham, pp 283–312

Shanmuganathan R, Karuppusamy I, Saravanan M, Muthukumar H, Ponnuchamy K, Ramkumar VS et al (2019) Synthesis of silver nanoparticles and their biomedical applications – a comprehensive review. Curr Pharm Des 25:2650–2660

Sharifi M, Faryabi K, Talaei AJ, Shekha MS, Ale-Ebrahim M, Salihi A et al (2020) Antioxidant properties of gold nanozyme: a review. J Mol Liq 297:112004

Sharma R, Bisen DP, Shukla U, Sharma BG (2012) X-ray diffraction: a powerful method of characterizing nanomaterials. Recent Res Sci Technol 4:77–79

Shenashen M, Derbalah A, Hamza A, Mohamed A, El Safty S (2017) Antifungal activity of fabricated mesoporous alumina nanoparticles against root rot disease of tomato caused by Fusarium oxysporium. Pest Manag Sci 73:1121–1126

Shojaei TR, Salleh MAM, Tabatabaei M, Mobli H, Aghbashlo M, Rashid SA, Tan T (2019) Applications of nanotechnology and carbon nanoparticles in agriculture. In: Suraya AR, Raja NIRO, Mohd ZH (eds) Synthesis, technology and applications of carbon nanomaterials. Elsevier, Amsterdam, pp 247–277

Shrivastava S, Bera T, Roy A, Singh G, Ramachandrarao P, Dash D (2007a) Characterization of enhanced antibacterial effects of novel silver nanoparticles. Nanotechnology 18:1–9

Shrivastava S, Bera T, Roy A, Singh G, Ramachandrarao P, Dash D (2007b) Characterization of enhanced antibacterial effects of novel silver nanoparticles. Nanotechnology 18:225103

Sondi I, Salopek-Sondi B (2004) Silver nanoparticles as antimicrobial agent: a case study on E.coli as a model for gram-negative bacteria. J Colloid Interface Sci 275:177–182

Srivastava G, Das A, Kusurkar TS, Roy M, Airan S, Sharma RK, Singh SK, Sarkar S, Das M (2014a) Iron pyrite, a potential photovoltaic material, increases plant biomass upon seed pre-treatment. Mater Express 4:23–31

Srivastava G, Das CK, Das A, Singh SK, Roy M, Kim H, Sethy N, Kumar A, Sharma RK, Singh SK (2014b) Seed treatment with iron pyrite (FeS_2) nanoparticles increases the production of spinach. RSC Adv 4:58495–58504

Sun Y, Xia Y (2002) Shape-controlled synthesis of gold and silver nanoparticles. Science 298:2176–2179

Sundrarajan M, Gowri S (2011) Green synthesis of titanium dioxide nanoparticles by Nyctanthes arbor tristis leaves extract. Chalcogenide Lett 8:447–451

Tagad CK, Dugasani SR, Aiyer R, Park S, Kulkarni A, Sabharwal S (2013) Green synthesis of silver nanoparticles and their application for the development of optical fiber based hydrogen peroxide sensor. Sens Actuators B 183:144–149

Tai CY, Tai C, Chang M and Liu H (2007) Synthesis of magnesium hydroxide and oxide nanoparticles using a spinning disk reactor 5536–41

Takeuchi MT, Kojima M, Luetzow M (2014) State of the art on the initiatives and activities relevant to risk assessment and risk management of nanotechnologies in the food and agriculture sectors. Food Res Int 64:976–981

Tarafdar JC, Xiong Y, Wang WN, Quinl D, Biswas P (2012) Standardization of size, shape and concentration of nanoparticle for plant application. Appl Biol Res 14:138–144

Taran N, Storozhenko V, Svietlova N, Batsmanova L, Shvartau V, Kovalenko M (2017) Effect of zinc and copper nanoparticles on drought resistance of wheat seedlings. Nanoscale Res Lett 12(1):1–6

Tejamaya M, Römer I, Merrifield RC, Lead JR (2012) Stability of citrate, PVP, and PEG coated silver nanoparticles in ecotoxicology media. Environ Sci Technol 46:7011–7017

Tirani MM, Haghjou MM, Ismaili A (2019) Hydroponic grown tobacco plants respond to zinc oxide nanoparticles and bulk exposures by morphological, physiological and anatomical adjustments. Funct Plant Biol 46:360–375

Tiwari DK, Behari J, Sen P (2008) Application of nanoparticles in waste water treatment. World Appl Sci J 3:417–433

Tripathi DK, Tripathi A, Shweta, Singh S, Singh Y, Vishwakarma K, Yadav G, Sharma S, Singh VK, Mishra RK, Upadhyay RG, Dubey NK, Lee Y, Chauhan DK (2017 Jan 26) Uptake, accumulation and toxicity of silver nanoparticle in autotrophic plants, and heterotrophic microbes: a concentric review. Front Microbiol 8:07

Vanti GL, Nargund VB, Basavesha KN, Vanarchi R, Kurjogi M, Mulla SI, Tubaki S, Patil RR (2019) Synthesis of Gossypium hirsutum-derived silver nanoparticles and their antibacterial efficacy against plant pathogens. Appl Organomet Chem 33:e4630

Vijayaraghavan K, Ashok Kumar T (2017) Plant-mediated biosynthesis of metallic nanoparticles: a review of literature, factors affecting synthesis, characterization techniques and applications. J Environ Chem Eng 5(5):4866–4883

Vijayaraghavan K, Ashokkumar T (2017) Plant-mediated biosynthesis of metallic nanoparticles: a review of literature, factors a ecting synthesis, characterization techniques and applications. J Environ Chem Eng 5:4866–4883

Wang S, Wang F, Gao S (2015) Foliar application with nano-silicon alleviates cd toxicity in rice seedlings. Environ Sci Pollut Res Int 22(4):2837–2845

Wang SH, Wang FY, Gao SC, Wang XG (2016) Heavy metal accumulation in different rice cultivars as influenced by foliar application of nano-silicon. Water Air Soil Pollut 227:228

Wang Y, Lin Y, Xu Y, Yin Y, Guo H, Du W (2019) Divergence in response of lettuce (var. ramosa Hort.) to copper oxide nanoparticles/microparticles as potential agricultural fertilizer. Environ Pollut Bioavailab 31:80–84

Yang F, Liu C, Gao F, Su M, Wu X, Zheng L, Hong F, Yang P (2007) The improvement of spinach growth by nano-anatase TiO_2 treatment is related to nitrogen photoreduction. Biol Trace Elem Res 119:77–88

Zewde B, Ambaye A, Stubbs Iii J, Raghavan D (2016) A review of stabilized silver nanoparticles–synthesis, biological properties, characterization, and potential areas of applications. Nanomedicine 4(1043):1–4

Zhang M, Zhang K, Gusseme BD, Verstraete W, Field R (2014) The antibacterial and anti-biofouling performance of biogenic silver nanoparticles by Lactobacillus fermentum. Biofouling 30(3):347–357

Zhang XF, Liu ZG, Shen W, Gurunathan S (2016) Silver nanoparticles: synthesis, characterization, properties, applications, and therapeutic approaches. Int J Mol Sci 17(9):1534. https://doi.org/10.3390/ijms17091534. PMID: 27649147; PMCID: PMC5037809.

Zhang L, Mazouzi Y, Salmain M, Liedberg B, Boujday S (2020) Antibody-gold nanoparticle bioconjugates for biosensors: synthesis, Characterization and Selected Applications. Biosens Bioelectron 112370

Chapter 15
Nano-Proteomics of Stress Tolerance in Crop Plants

Ghazala Mustafa, Atikah Farooq, Zarqa Riaz, Murtaza Hasan, and Amna

Abstract Nano-proteomics is one of the emerging fields of science being widely explored for crop betterment and rapidly making progress in almost all aspects of life. Several chemical and biological approaches for synthesis of nanoparticles (NPs) have been introduced which differentially contribute towards unique properties of NPs. There is a continuous introduction of nanomaterials in environment because of their excessive use in commercial products, fertilizers, and daily life products. Understanding the nature and physiochemistry of NPs is thus a crucial step towards assessing their possible interactions with biological system while evaluating their ecotoxicity. Although wide range of studies have been performed to know the exact mechanisms of NPs internalization and distribution within the animals, humans and microorganisms, still knowledge about plant and NPs interactions at molecular level is in infancy. Use of nanomaterials in agriculture and food industry is gaining momentum day by day. Mitigation of environmental stresses by application of potential NPs has been a significant topic of research in agriculture since last decade. Modern proteomic technologies including gel-based and or gel free approaches combined with high throughput analysis are being widely employed for unraveling molecular mechanism of stress responses in plants. Examining NPs responsive changes in proteome profiles and gene expression in crop plants, particularly under stress conditions, constitute a robust approach towards producing stress tolerant crops. Keeping in view the potential role of NPs in stress mitigation and importance of proteomics as a significant methodology towards understanding underlying molecular mechanisms, the current chapter extensively reviews the current knowledge on NPs and plant interactions at protein level. Furthermore, the future expectations about role of nano-proteomics in agriculture are also discussed.

G. Mustafa (✉) · A. Farooq · Z. Riaz · Amna
Department of Plant Sciences, Quaid-i-Azam University, Islamabad, Pakistan
e-mail: mghazala@qau.edu.pk

M. Hasan
Department of Biotechnology, The Institute of Biochemistry, Biotechnology and Bioinformatics, The Islamia University of Bahawalpur, Bahawalpur, Pakistan

M. Faizan et al. (eds.), *Sustainable Agriculture Reviews 53*, Sustainable Agriculture Reviews 53, https://doi.org/10.1007/978-3-030-86876-5_15

Keywords Nano-proteomics · Nanoparticles · Proteomics · Stress tolerance

15.1 Introduction

NPs (NPs) constitute the building blocks of field of nanotechnology that deals with synthesis and applications of nanomaterials. Recent advancements in nanotechnology have revolutionized multiple fields of science including electronics, biomedical industry, theragnostic, life sciences and many others. Increasing innovations in nanotechnology has led to the identification of new methods for synthesis of different kinds of NPs that could be divided into four major categories as metal based or metal oxides metal oxides, carbon-based NPs (like fullerene and carbon nanotubes), dendrimers and bio-inorganic complexes (Jha and Pudake 2016). Besides chemical synthesis, biological (green synthesis) approaches of NPs preparations are also being concerned. Anthropogenic activities as well as natural processes can produce nanomaterials however, production of manufactured nanomaterials also known as engineered NPs or ENPs are also being introduced in the environment at larger scale.

Due to its increasing applications in diverse fields of life, almost all living forms are in continuous exposure which often poses biological risks to them. Therefore, the thirst to understand their toxicity towards different environmental components has attracted recent attention. To deal with this environmental problem, green synthesis of nanomaterials has been the focus of research for NP synthesis, besides conventional methods, using some biological material such as microorganisms or plant extracts which produce substances or itself act as stabilizing agents during synthesis procedure ultimately producing eco-friendly and sustainable NPs (Saxena et al. 2016).

NPs are materials with exclusive nano size range between 1 and 100 nm and are known since many years for their role in sustainable agriculture (Khan and Upadhyay 2019) and have successfully paved pathways towards plant biotechnology and genetic engineering permitting extensive developments in crop sciences (Wang et al. 2019; Kwak et al. 2019). NPs are being used since decades not only to suppress plant diseases but also to introduce abiotic stress resistance (Khan and Upadhyaya 2019). Their higher reactive properties and range of biochemical activities specifically depends upon their high surface to volume ratio, making them very suitable entities of modulating multiple biological activities (Dubchak et al. 2010). Large number of NPs have now been produced and successfully applied on plants for growth improvement and stress tolerance (Saxena et al. 2016). Different plant species exposed to specific kinds of NPs show improved growth and better metabolism even under stressed conditions (Giraldo et al. 2014). Either chemically synthesized or biologically prepared, NPs usage in different products leads to their direct release into the environment that might cause risks of different nature (Singh et al. 2015; Nair et al. 2010). Despite of these concerns, vast array of research has revealed their particular interaction patterns and transport among living systems. Number of metallic and metallic oxide NPs have been known so far including silver (Ag) (Mustafa et al. 2016), Al_2O_3 (Hossain et al. 2016), CeO (Salehi et al. 2018), TiO (Pošćić et al. 2016; Hu et al. 2020), ZnO (Taran et al. 2017; Faizan et al. 2020), FeO (Adrees et al. 2020; Manzoor et al. 2021) and many others. All these metal-based NPs potentially enhanced plant growth and metabolism and protected them against different abiotic stresses with further future expectations.

15.1.1 Environmental Exposure of NPs and Interactions with Plants

NPs present in the environment closely interact with the other living organisms and this interaction specifically depends upon unique properties of NPs and their metallic or nonmetallic nature. In NPs contaminated environment, these interactions are more unintentional as they are unable to control ultimately posing adverse impacts on biota. Constant applications of nanomaterials in commercial products, fertilizers, and nanomedicine have significantly increased the animal, plant and human exposure to them (Klaine et al. 2008). Unique features of NPs are the main reason behind morphophysiological changes induced in plants upon NP exposure. As population is increasing day by day, the constantly increasing application of conventional fertilizers on crops has been observed for expanding worldwide grain production so that food demands of growing population could be fulfilled. But due to several physiological processes taking place in the environment, most of the fertilizer remains unreachable to plants, therefore, as an innovative technology, application of nano-fertilizers is increasing in their demands (Singh et al. 2018). In this way NPs are introduced into the environmental matrices in one way or another.

NPs, once become part of food chain, further get accumulated in the food webs and eventually bioaccumulated into higher organisms (Zhang et al. 2012). Thus, study of NPs and plant interaction and their ultimate fate, is one of the crucial factors towards understanding NPs impacts on environment. As plants are in direct contact with the environment, their interactions with NPs, including uptake, transformation and accumulation ultimately effects plant growth as well as yield of many economically important crops (Miralles et al. 2012). Studies have indicated that these interactions could be of both positive and negative nature. Large number of publications have reported the potential phytotoxicity of NPs in plants (Ma et al. 2010). In plants nanoparticle phytotoxicity is mainly attributed to its surface modifications that causes pore clogging and induce mechanical damage (Dietz and Herth 2011; Jha and Pudake, 2016). Further studies by Miralles et al. (2012) described ROS mediated oxidative damage as the most common mechanisms of NPs phytotoxicity. Although investigations for NPs and plant interactions have been exclusively conducted at morphophysiological level but still the knowledge of underlying molecular mechanisms is in infancy. Studies have also investigated the mechanism of buildup and translocation mode of metal-based NPs in plants. It has been made clear that physiochemistry of NPs is the key property of NPs that controls its interaction with living system (Spielman-Sun et al. 2017). NPs promptly adhere to plant body and once attached they are either translocated, transformed or accumulated inside the plant body through foliar or root pathway (Zhang et al. 2012). Their adherence particularly depends upon surface charge (Schwabe et al. 2014). It is known that NPs have different effects on various growth stages of plants hance differently interact with plants. Some NPs are efficiently up taken and absorbed when exposed to seeds where they modulate germination process of plants. One such report by Vishwakarma et al. (2017) indicated that Ag NPs of different concentrations significantly enhance germination rate of plants which might be attributed to changes in cell division and defensive mechanism. while others are easily up taken and produce effects at seedling stage of plant growth.

Seed priming with Ag NPs at 20 mM induced changes in metabolite contents such as proline and also improved plant growth parameters by enhancing defensive mechanisms in pearl millet (Khan et al. 2020). Ag NPs induced seed priming in high specialty value crops such as watermelon enhances germination rate without having any negative effects on fruit quality (Acharya et al. 2020). Another study determined the impact of silicon NPs on seed priming in wheat and showed that 24 h exposure of Si-NPs to seeds at varying levels may enhance growth and yield of plants along with betterment in biochemical processes (Hussain et al. 2019). Seed exposure with ZnO-NPs mediated salt stress resistance in wheat plants indicating that NPs modulate germination rates and also induce growth related changes in seedlings (Latef et al. 2017). It is reported that NPs modify seed germination by interacting with cell elongation or division processes and bringing changes in cell membrane structure (Vishwakarma et al. 2017). Besides direct seed exposure to NPs in priming experiments, foliar contact of NPs has been found as more practical method for NP application as leaves can more readily absorb NPs and other essential elements compared to roots (Dhoke et al. 2013). Some studies indicated that dose dependent increase in metal within the plant organs, especially roots, restricting the translocation of metal towards shoots. Hernandez-Viezcas et al. (2013) grew soybean in soil emended with 1000 mg/L cerium oxide NPs and analyzed the plant tissues through XANES (X-ray absorption near edge spectroscopy). Results indicated that most of the metallic cerium remained untransformed inside the plant with very little amount found to be accumulated within the pod. An equivalent trend has been reported by Zhu et al. (2012) in which the uptake mechanisms of gold NPs having specific charges on their surfaces were compared in different plant species including crops confirming that negatively charged NPs can efficiently translocated inside the roots while those with positive charge only getting adhered with the roots that might be due to negatively charged root surface producing electrostatic interactions.

The spatial distribution, transformation and translocation of NPs also sometimes depends upon surface area rather than increasing concentration dose (Van Hoecke et al. 2008; Ma et al. 2010). Besides the phytotoxic effects of NPs in plants, some positive interactions have also been reported in terms of sustainable agriculture. Some metallic NPs are actively transported inside the plants and are less detrimental towards plant growth and metabolism such as CeO (Wang et al. 2012; Majumdar et al. 2016) ZnO (Jayarambabu et al. 2015), copper (Yasmeen et al. 2017), Ag (Mustafa et al. 2016), iron (Manzoor et al. 2021), titanium oxide (Hu et al. 2020), silicon oxide and many others (Jalil et al. 2019). NPs have been observed to enhance plant growth parameters thus ameliorating abiotic stress effects on crops (Jalil et al. 2019). These potential plant and NPs interactions mostly depend upon dose dependent applications of NPs on plants but still our knowledge of NPs interaction and potential impacts in entire food chains is in infancy. Furthermore, majority of the studies to evaluate nanotoxicology of NPs in crop plants have been performed in controlled conditions such as hydroponic media rather than at mature growth stages of plants in field conditions which might be needed to explore the remining knowledge gap (Hernandez-Viezcas et al. 2013). A few reports have also documented the growth improving potential of NPs at mature growth stages such as the one reported by Yasmeen et al. (2017). Thus, it becomes evident that NPs interact with other environmental component with potential impacts.

15.2 Proteomic Technology Adapted by Plant Sciences

Biological research is gaining advances day by day. Innovative approaches have led to identifications of potential biomolecules and their dynamic roles are being documented since many years. Among various biomolecules, with life supporting activities, proteins are the direct effectors of a biological response. Every living organism possess unique proteome profile which may get altered in response to external and internal cellular fluctuations and comprise of overall protein content of cell characterized in terms of their unique structure, post translational modifications (PTMs), interactions with other biomolecules and subcellular locations (Aslam et al. 2017). Wilkins (1996) first time used the term of "proteomics" which represent the protein complement of genome. Since then, scientists have made significant discoveries in the field of proteomics which has now become modern promising approach to assess molecular basis of biological responses to environment. On the other hand, abiotic stresses faced by plants are one of the majors constrains in sustainable and profitable agriculture. Among various such external factors, temperature fluctuations, water scarcity, salinity, heavy metals, nutrient deprivation, water flooding and UV exposure are the major reasons of declined crop productivity (Khan and Upadhyaya 2019; Tripathi et al. 2017; Yasmeen et al. 2018, Hashimoto et al. 2020; Adrees et al. 2020). All of these abiotic stresses drastically reduce crop growth all over the world threatening food security at global level (Khan et al. 2021). In past few decades studies have been dealing with harmful effects of NPs on plants but more recently application of NPs is emerging as mitigative approach for stress amelioration in agriculture sector. Unique features of NPs such as their shape, size and dose concentration represent the mitigative role of NPs (Almutairi 2019). However, evidence about toxic effects of higher concentrations of NPs on plants systems also exist (Jalil et al. 2019).

Use of NPs for stress amelioration in crops not only protects plants against stress but also boost growth and metabolism. Furthermore, under stress conditions NPs regulate plant molecular mechanisms by enhancing ROS scavenging protein production. Recent studies implies that metallic NPs help mitigate abiotic stresses in plants at higher efficiency when applied at varying concentrations in the form of nano-fertilizers, nano-herbicides or nano-pesticides (Jalil et al. 2019). In this system nanomaterials can easily be internalized by the plant cell where they induce changes at molecular level. In this regard, field of biotechnology is gaining momentum where nanocarriers made from potential NPs are being employed for targeted delivery of essential nutrients and molecules such as DNA discovering important applications for plant genetic engineering of plants (Singh et al. 2015). Vast range of studies have been performed to understand the molecular mechanisms plant show in response to NPs application which might help in identification of tolerance mechanisms of plants against biotic and abiotic stresses. In this scenario, proteomics is one of the promising and modern approaches for gaining insights into subcellular responses of plants against stress in combination with NP application (Tanaka et al. 2004). Identification of novel stress responsive proteins in plants can be a powerful approach towards understanding underlying molecular responses of plants to abiotic stresses (Barkla et al. 2013). Exploring alterations in whole proteome profiles help in understanding protein-protein interactions and functions since proteins act as direct effector molecules in abiotic stress responses. Recent proteomic approaches have led us to deep insights into subcellular proteomes, post translational modifications, and protein interactions. Understanding of such molecular processes could be an important step towards food security by producing stress tolerant crops. Following is a detailed explanation of recent proteomic approaches (gel based and gel free) employed for proteome assessment in plants in response to abiotic stresses.

Western blotting and ELISA (enzyme linked immunosorbent essay) has been used for selective proteins identification and analysis rather than entire proteome profiles. ELISA technology has been successfully employed for assessment of Cry1Ie protein in transgenic plants (Zhang et al. 2016). Li et al. (2011) performed western blotting and identified rice proteins. Results of this study indicated that EF1-α was most expressed along with other HSPs (heat shock proteins). Western blotting has found to be powerful tool for identification of proteins by enzyme conjugation enabling the targeted determination of specific proteins (Aslam et al. 2017). Other scientists earlier used western blotting for identification of proteins in peanut allergic patients where unique IgE was determined against *Ara h1, 2 and 3* against (Koppelman et al. 2004).

Other proteomic approaches employed for quantitatively assessing whole plant proteomes is isotope coded affinity tag labeling and SILAC (stable isotope labeling with amino acids in cell culture) that was used by for quantitatively assess proteome of *A. thaliana* which indicated expression of glutathione *S-transferase* under abiotic stress (Aslam et al. 2017). Wiese et al. (2007) reported iTRAQ (isobaric tag for relative and absolute quantitation) as more recent proteomic technique being used for protein profiling. Ge et al. (2013) used iTRAQ for assessment of hydrogen peroxide stresses wheat proteins. Results identified 44 novel proteins having potential role in

H_2O_2 detoxification and stress signal transduction. According to reports by Neilson et al. (2010) and Hashimoto et al. (2020) proteomic studies have facilitated in identification of proteins involved in stress responses such as those involved in various biosynthetic pathways, energy metabolism and defense. Furthermore, field of nano proteomic studies is gaining advances day by day. Modern proteomic approaches are being continuously introduced which have been successfully applied for crop improvement. Among these approaches gel-based and gel free techniques are most classic strategies that are now commonly being employed for identification of differentially expressed/regulated proteins under stress conditions (Timabud et al. 2016; Han et al. 2014). Proteomics of plants in response to NP exposure is gaining momentum but still remains elusive. There is very scarce knowledge about nanotechnology based proteomic studies on plants and mostly deal with NPs phytotoxicity. However, this approach has been widely employed for humans, fungi, bacteria and mice (Abdelhamid and Wu 2015).

Most of the proteomic analysis rely on gel-based studies for resolving proteins from crude mixture (Matros et al. 2011) followed by mass spectrometry (MS) for identification of specific stress responsive proteins (Bantscheff et al. 2007) thus representing the core proteomic technology for differential analysis of protein profiles. Another robust tool for understanding of protein biomolecule interactions is microarray or protein chip. These proteomic approaches could be helpful in understanding the pathways involved and identification of key proteins that are expressed in response to nanoparticle exposure. This approach could further pave pathways for applications of NPs as potential biomarkers under stress conditions (Jha and Pudake 2016). 2D SDS-PAGE in combination with isoelectric focusing (IEF) and LC/MS works as a powerful system for protein separation, purification and characterization (according to mass and charge) further exploring whole proteome maps using databases (Aslam et al. 2017; Lee et al. 2020). 2D-DIGE is a variant of simple 2DE approach that enables easy visualization of proteins labelled with dye. In 2008, Komatsu identified plasma membrane proteins from model plant *A. thaliana* and rice in response to salinity stress. Analysis revealed that identified proteins were those involved in protein degradation, REDOX processes, and CHO metabolism. Although gel based proteomic approaches (simple 2DE and 2D-DIGE) have their own advantages, still there are limitation that debilitate its potential compared to other latest proteomic techniques. Therefore, focus has been placed more towards identifying more reliable approaches that comprise of gel free analysis.

Not long ago, LC/MS based gel free approaches have made their place in the field of modern proteomics due to their higher and precise resolution power and higher scanning rates with precise alignment of chromatograms (Aslam et al. 2017). Recently, proteomic techniques have successfully aided in understanding the cellular responses of plants to systematically define protein and NPs interactions under stress and non-stress conditions. For example, gel free nano-LC/MS analysis was performed to understand proteomic responses of soybean in response to Ag NPs application that improved plant growth by producing defensive enzymes against flooding stress (Mustafa et al. 2016). Yasmeen et al. (2016a, b) performed gel free proteomics on wheat and soybean exposed to iron and aluminum NPs respectively under salt stress. Hossain et al. (2016) and Salehi et al. (2018) conducted gel free proteomics for identification of AlNPs ans CeNPs responsive proteins in crop plants. Besides this, gel based proteomic studies are also conducted for NPs exposed plants under abiotic stresses such as those employed for recognition of differentially changed proteins in soybean in response to Ag NPs (Galazzi et al. 2019). Some other studies have been reported by Marmiroli et al. (2015) for CdS quantum dots and Mustafa et al. (2015b) for aluminum oxide NPs. Zhao et al. (2012) performed proteomic studies on *Zea maize* seeds exposed to cerium dioxide (CeO_2) NPs using gel free approach (SDS-PAGE) and western blotting performed for assessment of heat shock protein 70. In another 2-DE gel based nanoLC–ESI–MS/MS analysis by Vannini et al. (2013) reported the comparison of proteomic profiles of *Eruca sativa* upon 5-days exposure to Ag NPs (AgNPs) and Ag nitrate (AgNO3). Their results identified changed in proteins involved in sulphur metabolism and redox processes. V-ATPase subunits were downregulated which is attributed towards potential phytotoxicity of Ag NPs. Thus, it becomes clear that nanotechnology definitely holds the promise of crop betterment by improving tolerance mechanisms in plants against challenging environment.

Other proteomic studies performed for crop plants for nanoparticle mediated abiotic stress mitigation have been summarized in Table 15.1. All these proteomic studies clearly reflect the effects of different kinds of NPs on economically important crops and pave pathways for additional studies involving NP and protein interactions along with environmental concerns of nanomaterial deposition.

15.3 Plant Proteomics Under Nanoparticle Stress

15.3.1 Ag NPs

Over the years silver is being used in jewelry making as it is one of the precious and priceless elements with multiple industrial applications. Owing to it distinctive properties, silver also acts as potential antimicrobial agent as it triggers ROS generation bacterial cells and deactivate microbial enzymes (Matsumura et al. 2003). Oxidized silver produces toxic Ag ions (McShan et al. 2014) that are recognized as bioactive molecules (Santoro et al. 2007). Silver ion toxicity stimulates ROS generation leading to oxidative damage to DNA further activating antioxidant enzymes and deplete antioxidant molecules (McShan et al. 2014). In earth' crust silver could be found as rare element. Increasing production of nanoparticle-based products in different fields leads to toxicological effects which has been the topic of concern since many years. Silver nanoparticles has been employed in many ways in the field of agriculture. Rezvani et al. (2012) has reported that Ag NPs could promote *Crocus sativus* roots, exposed to flooding stress, by blocking ethylene signaling. Syu et al. (2014) conducted a study on *Arabidopsis* and found that silver nanoparticles are responsible for root growth promotion and ROS accumulation in plants also function as ethylene perception inhibitors. Molecular studies indicated that Ag NPs also

Table 15.1 Nano-proteomics of Crops reported in last 5 years

Plant specie	Exposed plant organ	Growth stage	Nanoparticles (size/dose)	Treatment time period	Proteomic Technique	Major proteins found/associated pathways	References
Eruca sativa	Seedlings	Roots	AgNPs (10 nm/10 mg/l)	5 days	2DE and nano-LC–ESI–MS/MS	Proteins involved in sulphur metabolism significantly changed and V-ATPase subunits downregulated	Vannini et al. (2013)
Rice *Oryza sativa* L. cv. IR651	Root/soil irrigation	10 days old	AgNPs 18.34 nm/30 and 60 μg/mL	20 days	Gel-based (2-DE, nano LC/FT-ICR MS)	Proton motive force, oxidative stress tolerance, Ca^{2+} regulation and signaling, cell wall and DNA/RNA/protein direct damage, cell division and apoptosis	Mirzajani et al. (2014)
Wheat	Seeds and 5 days old seedlings	Shoot, root	AgNPs (1, 10 mg/L)	4 h seed soaking	2-DE IEF/SDS-PAGE, LC–ESI-MS/MS	Proteins related to cell defense and primary metabolism and were altered	Vannini et al. (2014)
Soybean (*Glycine max* L. Merrill cultivar Enrei)	Root	2 days old seedling	Al_2O_3 NPs 30–60 nm/ 5, 50 and 500 ppm (50 ppm for proteomics)	10 days	Gel free nanospray LTQ XL Orbitrap MS	Proteins related to protein synthesis, stress, cell wall, and signaling changed in abundance. S-adenosyl-l-methionine dependent methyltransferases and enolase might be involved in mediating recovery responses by Al2O3 NPs	Yasmeen et al. (2016b)
Soy bean (*Glycine max* L. cultivar Enrei)	Root/ suspension hydroponic media	7 days old seedlings	Al_2O_3 NPs 30–60 nm/500 ppm, 30–60 nm; ZnO <50 nm/500 ppm; Ag 15 nm/50 ppm	3 days	Gel free	GDSL motif lipase 5, galactose oxidase, quinone reductase were upregulated	Hossain et al. (2016)

(continued)

Table 15.1 (continued)

Plant specie	Exposed plant organ	Growth stage	Nanoparticles (size/dose)	Treatment time period	Proteomic Technique	Major proteins found/associated pathways	References
Bean (*Phaseolus vulgaris* L. (var. pinto bean))	Foliar and root	3 weeks old	Cerium oxide 10–30 nm/0–2000 mg/L	14 days	Gel free Q-TOF tendem MS/MS	Enzymes involved in proteins biosynthesis or proteases, lysine biosynthetic intermediates, and glutamine were altered	Salehi et al. (2018)
Tobacco (*Nicotiana tabacum* L. cv burley)	Roots/water supplement	2 months old plants	Silver oxide 50 nm/100 μM	7-days	Gel based (2-DE), MALDI/TOF MS	CAP, CAP-binding protein 20 (CBP20), β-1,3-glucanase, and Mn-SOD in roots. Glutathione S-transferase, glycine-rich RNA-binding protein, and mRNA binding protein, nucleotide metabolism in leaf	Peharec Štefanić et al. (2019)
Wheat (*Triticum aestivum* L.)	Root exposure	6-days old seedlings	AgNP 15–20 nm /5 ppm	3-days, 5 days and 7 days	Gel free	Redox and mitochondrial electron transport chain proteins decreased. Glyceraldehyde-3-phosphate dehydrogenase, Glucose-6-phosphate-1-epimerase increased /decreased, while phosphor-enol pyruvate carboxylase was decreased	Jhanzeb et al. (2019)
Soybean (transgenic and non-transgenic)	21 days old plants	Leaves	AgNPs 60 nm/50 mg/kg Ag	7 days	Gel based (2D-DIGE), nanoEI-LC-MS/MS	Nucleoside diphosphate kinase (NDK), chlorophyll a-b binding protein 6A, gamma-glutamyl hydrolase	Galazzi et al. (2019)

Tobacco (*Nicotiana tabacum* L. bright yellow-2 cell)	Tissue callus	From day one	Carbon nanoparticles 28–77 nm/0–100 mg/L	25 days total incubation period and 8 days exposure for protein analysis	Gel based (2D-LC MS/MS, iTRAQ)	Majority proteins involved in mitochondrial functions and ca signaling. Calmodulin (CaM), protein expression level significantly enhanced. Furthermore, among the top 20 up-regulated proteins, cytochrome c oxidase and cytochrome c are two other increased expressed proteins	Zhenjie et al. (2020)
Soybean (*Glycine max* L. cv. Enrei)	2-days old seedlings	Roots	Biologically synthesized AgNP (16 nm) Chemically synthesized AgNP (15 nm) 10 ppm	4 days	Immunoblot Label-free nanoLC/MS	Differentially-expressed proteins were mainly involved in protein-degradation and stress	Mustafa et al. (2020)

act as gene regulators in plants. Gene expression profiling in silver nanoparticles exposed *Arabidopsis* has showed that Ag NPs upregulated the thalianol biosynthetic pathway and metal and oxidative stress related genes whereas downregulated ethylene signaling pathway associated genes (Kaveh et al. 2013). It has been reported by Salama (2012) that increasing silver nanoparticles exposure also enhances corresponding carbohydrate, chlorophyll and protein levels in maize and *P. vulgaris.* Proteomic study of rice exposed to silver nanoparticles toxicity revealed that Ag NPs induce accumulation of those protein precursor in plants which are indicative of proton motive force dispersal. Further studies by Mirzajani et al. (2014) have reported that Ag NPs application in rice not only alters levels of proteins associated with oxidative stress tolerance but also differentially changes calcium regulation/signaling, cell division, transcription/degradation, apoptosis related proteins. Ag NPs exposure to *E. sativa* induced changes in proteins related to sulfur metabolism and redox regulation. Both of these processes have been known to play crucial roles in maintenance of cellular homeostasis. Furthermore, experimental work by Vannini et al. (2013) indicated that silver nanoparticles regulate the proteins of endoplasmic reticulum and cell vacuole, suggesting these two organelles as potential target for NPs. Although, number of experimental works have been performed till now to assess the effects of Ag NPs on plants exposed to different stresses but still the role of silver nanoparticles on flooding exposed soybean proteomic profiles has not been elaborated extensively. Tremendously increasing applications of Ag NPs in agricultural sector demands the extensive investigations on biological impacts of these particles particularly on plants which are flooding susceptible such as soybean. Very few studies have explored Ag NPs effects on biological materials including bacteria (Choi and Hu 2008), algae (Miao et al. 2009) and animals (van der Zande et al. 2012). According to report by Rezvani et al. (2012), Ag NPs treatment on C. sativus stress, positively protect plants by preventing ethylene action under flooding stress. However, there is very limited information available about the underlying molecular mechanisms altered by silver nanoparticles.

15.3.2 Aluminum

Although aluminum is not essential for crop development and growth still it occurs as most abundant metal in earth surface. Due to its toxic properties, it has been regarded as one of the major limiting factors for crop growth. Specifically, Al limits plant growth in soils that are acidic and have pH lower than five (von Uexküll and Mutert 1995). Kochian et al. (2004) has observed Al ions, formed from aluminosilicates, as the cause of its toxicity. Physiological studies have indicated that plants can acquire Al tolerance by two ways. One potential mechanism in plants against Al toxicity is the blockage of Al ion uptake. Secondly to detoxify cellular Al, plant cells form harmless complexes with organic ligands followed by their sequestration towards specific organelles (Ma 2000; Ma and Furukawa 2003). Valle et al. (2009) has reported that in Al sensitive plants, toxic amounts of aluminum restrict plant growth leading to crop yield reductions. Aluminum binds with sulfate, phosphate and carbonyl groups functional groups thereby damaging cellular components (Poschenrieder et al. 2008). Despite of having many toxicity effects, several reports have also indicated some positive effects of aluminum on plants.

Al_2O_3 NPs application on radish, rape, ryegrass and lettuce enhanced root elongation (Lin and Xing 2007) while it was negatively affected in corn, cucumber, lettuce and rye-grass (Yang and Watts 2005). It has been reported that miRNA expression levels, known to play crucial role in facilitating stress response, was significantly altered by Al_2O_3 NPs application in tobacco (Burklew et al. 2012). To cope up with deleterious effects of aluminum oxide nanoparticles, the antioxidant enzymes activities, responsible for removing free radicles, play important role in wheat roots (Riahi-Madvar et al. 2012). Poborilova et al. (2013) indicated that aluminum oxide nanoparticles exposure to BY-2 plant cell suspension drastically diminished mitochondrial activity and enhanced caspase like activity indicative of executing programmed cell death. However, further knowledge about underlying mechanisms altered by aluminum oxide nanoparticles is still unclear.

15.3.3 Iron NPs

Iron is one of the important micronutrients for plants and its deficiency leads to severe growth retardation and chlorosis. It is being widely used for nanoparticle preparation. Iron NPs are among many of the metallic NPs that are widely used for commercial and biomedical purposes (Yasmeen et al. 2016a, b). As far as its agricultural applications are concerned, iron NPs have been proved to be beneficial for crop betterment as it is being used for crop fortification. Furthermore, several studies have revealed their potential role in stress mitigation via interactions with plant protein biosynthesis (Yasmeen et al. 2017). Iron oxide NPs have been reported to enhance opening of stomata by activating proton ATPase protein in model plant *Arabidopsis thaliana* (Kim et al. 2015). FeNPs are also used as environmental bioremediation (Yan et al. 2013). It is evident from studies that Fe NPs has inhibitory effects at higher concentration while at lower amounts it enhanced growth and germination (Mushtaq 2011). Several proteomic techniques have been introduced to evaluate the role of Fe NPs in abiotic stress mitigation. Yasmeen et al. (2016a) performed gel free proteomic studies and identified stress responsive mechanisms related protein in *T. aestivum* (wheat). Nano LC/MS used in this research work provided comprehensive findings for evaluation of molecular mechanisms of iron NPs application at proteomic level. Furthermore, LC/MS data assessed through databases provided deep insights into protein functional assessment that help to identify the exact signal transduction pathways involved in stress responses of wheat. Another proteomic study by Yasmeen et al. (2017) revealed that Fe NPs enhanced antioxidant status in wheat by upregulating redox related enzymes. Major proteins successfully identified through this gel free/label free proteomic approach included those of starch degradation and TCA cycle. Jalil et al. (2019) reported the mitigative effects of Fe NPs on plants against drought stress. Other studies, such as those conducted by Manzoor et al. (2021) indicated that FeO NPs have potential ability to co-ameliorate heavy metal stress. Although Fe NPs have been assessed for their stress mitigative roles in plants, proteomic studies are still prerequisite.

15.3.4 Zinc and ZnO NPs

Zinc is a vital micronutrients found in plants having pivotal role in regulation of various cellular process. Zinc sulfate is one of the most ordinary source of Zn, exclusively due to its greater solubility related to carbonates and oxides, and reduced cost differing to synthetic chelates and complexes. Interpretation of mechanisms of Zn uptake and transfer is an essential step towards understanding plant and ZnNP interactions. Nano ZnO are extensively used in industry for numerous decades. A smaller number of studies have been performed on its advantageous effect on plant growth. However, its phytotoxicity has been described in several reports (Mahajan et al. 2011) which suggest that mechanism on zinc phytotoxicity in plants in highly specific. Studies have reported that approximately 0.05 mg/L of Zn in soil solution is required by optimum plant growth and higher dosage concentrations induces phytotoxicity. ZnO NPs phytotoxicity at cellular level is interrelated with oxidative stress and lipid peroxidation which might be due to particle dissolution and adhesion followed by internalization through plant roots (Singh et al. 2018). In another study by López-Moreno et al. (2010) has shown that ZnO NPs induce cytotoxicity at 4000 mg/L in soybean seedlings. Large number of studies have identified the potential of Zn NPs in the agriculture sector and are trying to explore it further at molecular level through research. As far as nano proteomics of Zn NPs cytotoxicity is concerned, limited set of data has been reported yet. Just one available study discusses the molecular mechanisms associated with Zn NPs-related mitigation of nanoparticle stress in *Glycine max* exposed to three different metal-based NPs (Hossain et al. 2016). Differentially abundant proteins identified through this approach were involved in redox reactions, lipid metabolism, cell organization, stress associated signaling and growth-related hormonal pathways thought to be the crucial for optimum plant growth. These results revealed that nano-Zn can promote plant defense against stress conditions opening new pathways for deep understanding of plant NP interactions and proteomic responses.

15.3.5 Other NPs

Current studies have shown that nanotechnology is gaining momentum day by day. Scientists are trying to reveal potential of nano-system in agriculture sector. Variety of NPs have been prepared and successfully used for crop betterment by enhancing growth parameters of several economically important crops (Table 15.2). Despite

Table 15.2 Proteomic technologies adapted by Plant Sciences for abiotic stress tolerance using Nanoparticles

Plant specie	Growth stage	Exposed organ	Nanoparticles (size/dose)	Treatment time period	Abiotic stress	Proteomic Technique	Major proteins found/ associated pathways	References
Soybean (*Glycine max* L. cv. Enrei)	2 days old seedlings	Root	Al_2O_3 (30–60 nm/50 ppm)	3 days	Flooding	Gel-free nanoLTQ XL Orbitrap MS/MS	Abundance of cell wall-related proteins was significantly increased. Most interactive protein was glyceraldehyde-3-phosphate dehydrogenase C subunit 1.	Mustafa et al. (2015a)
Soybean (*Glycine max* L.) cv. Enrei	Seedlings	2 and 4-days old roots and cotyledons	AgNPs 15 nm /2 ppm	4 days	Flooding	Gel free nanoLC-ESI-MS/MS	Commonly regulated protein glyoxalase II 3. Aspartyl protease family protein and expansin-like B1 increased in response to AgNPs in roots. While beta-ketoacyl reductase increased in cotyledons	Mustafa et al. (2015b)
Soybean (*Glycine max* L. cv. Enrei)	2 days old seedlings	Roots	Al_2O_3 (30–60 nm/50 ppm)	3 days exposure	Flooding	Immunoblot nanoLC/MS	Proteins of ascorbate glutathione pathway and ribosomal proteins were significantly enhanced	Mustafa and Komatsu (2016)
Soybean (*Glycine max* L. cv. Enrei)	Root	2-days old seedlings	Ag-NPs 2, 15 & 50–80 nm/ 5 ppm)	3 days	Flooding	Gel free nanoLC/MS	Most interactive proteins (i) 2 & 15 nm NP exposure: cytochrome family, beta ketoacyl reductase 1 (ii) 50–80 nm: enolase	Mustafa et al. (2016)

(continued)

Table 15.2 (continued)

Plant specie	Growth stage	Exposed organ	Nanoparticles (size/dose)	Treatment time period	Abiotic stress	Proteomic Technique	Major proteins found/ associated pathways	References
Wheat *T. aestivum* L.	Shoot	5-days old seedlings	Fe-NPs (20 nm/1, 5, 10, 50 ppm)	2 days	Salt stress	Gel free nanoLC/MS immunoblot	Proteins related to photosynthesis, cell, and protein metabolism; RubisCO small chain abundance increased	Yasmeen et al. (2016a)
Wheat varieties drought tolerant & salt tolerant	Seeds	Mature plants	Cu (15–30 nm) and Fe (20–30 nm)/20–40 ppm		Salt and drought	Gel free Nanospray LTQ XL Orbitrap MS/MS	Proteins involved in starch degradation, glycolysis, and tricarboxylic acid cycle enhanced	Yasmeen et al. (2017)
Wheat varieties	Seedlings	5-days old seedlings	Cu (1, 5, 10, 50 ppm/<50 nm)	2 days	Salt and drought	Gel free nanoLC/MS	Protein of photosynthesis and tetrapyrrole synthesis decreased, while those of glycolysis and TCA cycle enhanced	Yasmeen et al. (2018)
Soybean (*Glycine max* L.) cv. Enrei	Seedlings	4-days old seedlings	Ag NP (15 nm/5 ppm) + organic/ inorganic chemicals	2-days	Flooding	Gel free nanoLC/MS immunoblot	Protein degradation and synthesis, calnexin/calreticulin related and glycoproteins was significantly increased	Hashimoto et al. (2020)

these studies, knowledge is still missing for safe application of NP based nano-fertilizers in crop production. Proteomic and metabolomic studies have been performed on many plant species exposed to metallic NPs. Salehi et al. (2018) reported the potential impacts of CeO NPs on *Phaseolus vulgaris*. Cerium oxide NPs are a typical member of the industrially essential class of metal oxide and is one of the highly produced synthetic NPs, which has applications in different industrial products (Keller et al. 2013; Majumdar et al. 2016; Abbas et al. 2020). Studies have revealed both positive and negative impacts of nano-CeO on plants. Cerium oxide NPs applications on *Phaseolus vulgaris* revealed that it induces differential expression of proteins involved in protein biosynthesis. Gel free proteomic approach based on Q-TOF MS/MS revealed that foliar spray nano Cerium modulate folding or turnover related protein expression in plants (Salehi et al. 2018). Titanium dioxide NPs have been known since many years for their potential roles in many biochemical processes in plants such as photosynthesis (Landa 2021). TiO2 NPs exposure to spinach leaves reduces oxidative stress in chloroplast in response to UV light and also increased enzymatic activities of catalase, SOD, guaiacol peroxidase and APX (ascorbate peroxidase). All these enzymes potentially reduce SOD radicals and H2O2 (Lei et al. 2008). Jha and Pudake et al. (2016) has reported that photosynthesis and plant growth might be enhanced by titanium dioxide exposure. Titanium dioxide NPs might exert damaging effects on plant growth in a dose dependent manner and inducing genotoxic effects (Castiglione et al. 2011).

15.4 NPs Uptake and Mode of Action Under Stressed Conditions

Plants are sensitive to environmental stresses, either biotic or abiotic, which triggered a decline in photosynthesis, nitrogen fixation, and nutrient uptake. In spite of the widespread information accessible to recognize the effects of abiotic stresses on plants, the difference in response mechanisms at proteomic level is yet to be identified. The most widely accepted concept of plant response mechanisms towards a stress is enhanced redox reaction cascades which include vast array of biological pathways controlled by enzymes and proteins leading to difference in protein changes. The disparity in number and pattern of protein alterations under stress conditions will be beneficial for more legitimate diversity of protein abundances. The aspirant proteins engaged in response under these stresses provide a key to make strategies for producing tolerant plants.

As far as nano-proteomics is concerned, exact mode of action of NPs inside plants cells is still elusive. However, several evidence have been presented that indicate most probable mode of action of NP and their uptake mechanisms by plants. As plant are unable to change their location, they cannot escape from biotic or abiotic environmental stresses which induce ROS production in plants leading to severe oxidative burst. ROS production when exceeds beyond a certain limit, causes toxicity in plants by degrading biomolecules and damage lipid bilayers thereby prompting cytotoxicity (Yadav et al. 2014; Shen et al. 2010). The ultimate result of this toxicity is reduced or retarded plant growth (Begum and Fugetsu 2012). In response to these cytotoxic mechanisms, plant show protective responses which prominently comprise of ROS scavenging processes. It has been deliberately known that NPs can mediate ROS dependent cytotoxicity by upregulating antioxidant enzyme activities in plants. One such study conducted by Laware and Raskar (2014) reported that titanium dioxide NPs exposed to onion plants help in alleviation of ROS toxicity by enhancing SOD, POD, catalase, and amylase activities at increasing concentrations.

Equivalent results have also been obtained for other metallic NPs such as SiO2 which enhanced seed germination and overall plant growth in *Glycine max* (Jalil et al. 2019). However, studies are contradictory about mitigative effects of nanoparticle against ROS as NPs can also induce ROS production rather than detoxifying them. This finding has been confirmed by Rico et al. (2013) who conducted the study on rice plants by exposing them to cerium dioxide NPs and found that nano-CeO2 modify plant antioxidant status in concentration dependent manner. This investigation might be elucidated by investigating the exact role of NPs in plant stress signal transduction pathways. Although exact mode of action of NPs in signal transduction pathways is still elusive, several "omic" approaches might be adapted for understanding the mitigative role of NPs against abiotic stresses at molecular level. Jalil et al. (2019) further reported that metallic engineered Ag NPs induce expression of stress related genes in model plant *Arabidopsis*. Oligonucleotide microarray studies in model plant *Arabidopsis thaliana* exposed to zinc oxide nanoparticle amended with titanium dioxide NPs and carbon fullerene indicated that ZnO NPs not only downregulate biogenesis and cell organization related genes but also induce stress related genes (Landa et al. 2012). Ghodake et al. (2011) assessed phytotoxicity of zinc oxide NPs in *Alium cipa* and concluded that ZnO NP exposure increase ROS production in plants.

Aluminum dioxide (Al_2O_3) NPs along with titanium dioxide (TiO_2) NPs exposed to tobacco plants induced upregulation of non-coding microRNAs in response to metal stress. miRNAs are small highly conserved regulatory molecules which are known to modulate several stress related responses in plants (Burklew et al. 2012; Frazier et al. 2014). Titanium dioxide nanoparticle were also shown to enhance RuBiSCO activity in plants in response to chilling stress (Almutairi 2019). Other than ROS signaling, stress responsive pathways in plants also involve MAPK pathway and SOS. Mitogen activated protein kinase (MAPK) is an important signal transduction pathway that comprise multiple proteins in subcellular localizations. MAPK modulate several important processes such as environmental responses, antioxidant defense and plant growth and development (Krysan and Colcombet 2018; Almutairi 2019). Furthermore, Jha and Pudake (2016) extensively reviewed plant molecular responses towards NPs. Thus, it becomes clear that NPs not only imitate ROS scavenging entities but also regulate molecular mechanisms by changing gene expression patterns in response to abiotic stresses. Understanding these

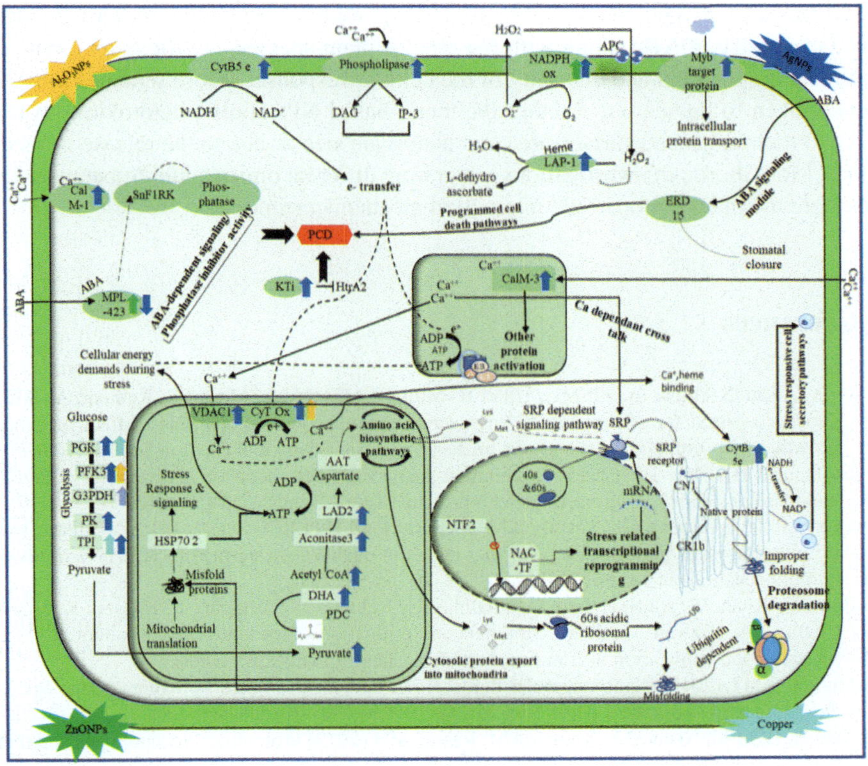

Fig. 15.1 Overall responses of subcellular proteomes to NPs stress. Differentially expressed proteins under Ag, aluminum, zinc, and copper NPs are highlighted with different colors

molecular defense mechanisms in plants might help in application of NPs in agriculture sector as a protective strategy against stressors (Fig. 15.1).

15.5 Future Perspectives

Nanotechnology has achieved enormous impetus in current times due to the broad applications of NPs in the agriculture, cellular imaging, cosmetic industry, biosensing, drug delivery, diagnosis, and cancer therapy. Nevertheless, accidental liberation of these commercially fabricated nanomaterials into environment provoked global fear. Substantial consideration is now being given to NPs synthesis methods, their release, plant-nanomaterials interactions, and their environmental destiny. As linked to conventional physical and chemical methods, green synthesis of NPs using microorganisms and plants is an environment-friendly, cost effective, safe, biocompatible, green alternative approach for giant scale production of NPs. Proteomic analyzes on NPs generated phytotoxicity disclose that concentration, particle size, and chemistry of NPs, along with the type of plant species, are the main elements influencing the kind and enormity of the cellular responses. More schemes must be undertaken to pursue out whether the metal-based NPs apply phytotoxicity solely due to their enhanced surface area and nanoscale size or due to the release of metal ions. More thorough investigations integrating different omics could be beneficial to explore the plant-nanoparticle interaction mechanisms in detail.

References

Abbas Q, Liu G, Yousaf B, Ali MU, Ullah H, Mujtaba Munir MA, Ahmed R, Rehman A (2020) Biochar-assisted transformation of engineered-cerium oxide nanoparticles: effect on wheat growth, photosynthetic traits and cerium accumulation. Ecotoxicol Environ Saf 187:109845

Abdelhamid HN, Wu HF (2015) Proteomics analysis of the mode of antibacterial action of nanoparticles and their interactions with proteins. TrAC Trends Anal Chem 65:30–46

Acharya P, Jayaprakasha GK, Crosby KM, Jifon JL, Patil BS (2020) Nanoparticle-mediated seed priming improves germination, growth, yield, and quality of watermelons (*Citrullus lanatus*) at multi-locations in Texas. Sci Rep 10(1):1–16

Adrees M, Khan ZS, Ali S, Hafeez M, Khalid S, Ur Rehman MZ, Hussain A, Hussain K, Shahid Chatha SA, Rizwan M (2020) Simultaneous mitigation of cadmium and drought stress in wheat by soil application of iron nanoparticles. Chemosphere 238:124681

Almutairi ZMT (2019) Plant molecular defense mechanisms promoted by nanoparticles against environmental stress. Int J Agric Biol 21:259

Aslam B, Basit M, Nisar MA, Khurshid M, Rasool MH (2017) Proteomics: technologies and their applications. J Chromatogr Sci 55(2):182–196

Bantscheff M, Schirle M, Sweetman G, Rick J, Kuster B (2007) Quantitative mass spectrometry in proteomics: a critical review. Anal Bioanal Chem 389(4):1017–1031

Barkla BJ, Vera-Estrella R, Pantoja O (2013) Progress and challenges for abiotic stress proteomics of crop plants. Proteomics 13(12–13):1801–1815

Begum P, Fugetsu B (2012) Phytotoxicity of multi-walled carbon nanotubes on red spinach (*Amaranthus tricolor* L) and the role of ascorbic acid as an antioxidant. J Hazard Mater 243:212–222

Burklew CE, Ashlock J, Winfrey WB, Zhang B (2012) Effects of aluminum oxide nanoparticles on the growth, development, and microRNA expression of tobacco (*Nicotiana tabacum*). PLoS One 7:e34783

Castiglione MR, Giorgetti L, Geri C, Cremonini R (2011) The effects of nano-TiO 2 on seed germination, development and mitosis of root tip cells of *Vicia narbonensis* L. and *Zea mays* L. J Nanopart Res 13(6):2443–2449

Choi O, Hu Z (2008) Size dependent and reactive oxygen species related nanosilver toxicity to nitrifying bacteria. Environ Sci Technol 42(12):4583–4588

Dhoke SK, Mahajan P, Kamble R, Khanna A (2013) Effect of nanoparticles suspension on the growth of mung (*Vigna radiata*) seedlings by foliar spray method. Nanotechnol Dev 3(1):e1

Dietz KJ, Herth S (2011) Plant nanotoxicology. Trends Plant Sci 16(11):582–589

Dubchak S, Ogar A, Mietelski JW, Turnau K (2010) Influence of silver and titanium nanoparticles on arbuscular mycorrhiza colonization and accumulation of radiocaesium in *Helianthus annuus*. Span J Agric Res 1:103–108

Faizan M, Hayat S, Pichtel J (2020) Effects of zinc oxide nanoparticles on crop plants: a perspective analysis. In: Sustainable agriculture reviews 41. Springer, Cham, pp 83–99

Frazier TP, Burklew CE, Zhang B (2014) Titanium dioxide nanoparticles affect the growth and microRNA expression of tobacco (*Nicotiana tabacum*). Funct Integr Genomics 14(1):75–83

Galazzi RM, Júnior CAL, de Lima TB, Gozzo FC, Arruda MAZ (2019) Evaluation of some effects on plant metabolism through proteins and enzymes in transgenic and non-transgenic soybeans after cultivation with silver nanoparticles. J Proteome 191:88–106

Ge P, Hao P, Cao M, Guo G, Lv D, Subburaj S, Li X, Yan X, Xiao J, Ma W, Yan Y (2013) iTRAQ-based quantitative proteomic analysis reveals new metabolic pathways of wheat seedling growth under hydrogen peroxide stress. Proteomics 13(20):3046–3058. https://doi.org/10.1002/pmic.201300042

Ghodake G, Seo YD, Lee DS (2011) Hazardous phytotoxic nature of cobalt and zinc oxide nanoparticles assessed using *Allium cepa*. J Hazard Mater 186(1):952–955

Giraldo JP, Landry MP, Faltermeier SM, McNicholas TP, Iverson NM, Boghossian AA, Reuel NF, Hilmer AJ, Sen F, Brew JA, Strano MS (2014) Plant nanobionics approach to augment photosynthesis and biochemical sensing. Nat Mater 13(4):400–408

Han C, Yang P, Sakata K, Komatsu S (2014) Quantitative proteomics reveals the role of protein phosphorylation in rice embryos during early stages of germination. J Proteome Res 13(3):1766–1782

Hashimoto T, Mustafa G, Nishiuchi T, Komatsu S (2020) Comparative analysis of the effect of inorganic and organic chemicals with silver nanoparticles on soybean under flooding stress. Int J Mol Sci 21(4):1300

Hernandez-Viezcas JA, Castillo-Michel H, Andrews JC, Cotte M, Rico C, Peralta-Videa JR, Ge Y, Priester JH, Holden PA, Gardea-Torresdey JL (2013) In situ synchrotron X-ray fluorescence mapping and speciation of CeO_2 and ZnO nanoparticles in soil cultivated soybean (*Glycine max*). ACS Nano 7(2):1415–1423

Hossain Z, Mustafa G, Sakata K, Komatsu S (2016) Insights into the proteomic response of soybean towards Al_2O^3, ZnO, and Ag nanoparticles stress. J Hazard Mater 304:291–305

Hu J, Wu X, Wu F, Chen W, White JC, Yang Y, Wang B, Xing B, Tao S, Wang X (2020) Potential application of titanium dioxide nanoparticles to improve the nutritional quality of coriander (*Coriandrum sativum* L.). J Hazard Mater 389:121837

Hussain A, Rizwan M, Ali Q, Ali S (2019) Seed priming with silicon nanoparticles improved the biomass and yield while reduced the oxidative stress and cadmium concentration in wheat grains. Environ Sci Pollut Res Int 26(8):7579–7588

Jalil SU, Zahera M, Khan MS, Ansari MI (2019) Biochemical synthesis of gold nanoparticles from leaf protein of *Nicotiana tabacum* L. *cv.* xanthi and their physiological, developmental, and ROS scavenging responses on tobacco plant under stress conditions. IET Nanobiotechnol 13(1):23–29

Jayarambabu N, Kumari BS, Rao KV, Prabhu YT (2015) Beneficial role of zinc oxide nanoparticles on green crop production. IJMART 10:273–282

Jha S, Pudake RN (2016) Molecular mechanism of plant–nanoparticle interactions. In: Plant nanotechnology. Springer, Cham, pp 155–181

Jhanzab HM, Razzaq A, Bibi Y, Yasmeen F, Yamaguchi H, Hitachi K, Tsuchida K, Komatsu S (2019) Proteomic analysis of the effect of inorganic and organic chemicals on silver nanoparticles in wheat. Int J Mol Sci 20(4):825

Kaveh R, Li YS, Ranjbar S, Tehrani R, Brueck CL, Van Aken B (2013) Changes in *Arabidopsis thaliana* gene expression in response to silver nanoparticles and silver ions. Environ Sci Technol 47(18):10637–10644

Keller AA, McFerran S, Lazareva A, Suh S (2013) Global life cycle releases of engineered nanomaterials. J Nanopart Res 15(6):1–17

Khan Z, Upadhyaya H (2019) Impact of nanoparticles on abiotic stress responses in plants: an overview. In: Nanomaterials in plants, algae and microorganisms. Elsevier, London, pp 305–322

Khan I, Raza MA, Awan SA, Shah GA, Rizwan M, Ali B, Tariq R, Hassan MJ, Alyemeni MN, Brestic M, Zhang X, Ali S, Huang L (2020) Amelioration of salt induced toxicity in pearl millet by seed priming with silver nanoparticles (AgNPs): the oxidative damage, antioxidant enzymes and ions uptake are major determinants of salt tolerant capacity. Plant Physiol Biochem 156:221–232

Khan A, Ahmad M, Ahmed M, Iftikhar Hussain M (2021) Rising atmospheric temperature impact on wheat and thermotolerance strategies. Plan Theory 10(1):43

Kim JH, Oh Y, Yoon H, Hwang I, Chang YS (2015) Iron nanoparticle-induced activation of plasma membrane H(+)-ATPase promotes stomatal opening in *Arabidopsis thaliana*. Environ Sci Technol 49(2):1113–1119

Klaine SJ, Alvarez PJ, Batley GE, Fernandes TF, Handy RD, Lyon DY, Mahendra S, McLaughlin MJ, Lead JR (2008) Nanomaterials in the environment: behavior, fate, bioavailability, and effects. Environ Toxicol Chem 27(9):1825–1851

Kochian LV, Hoekenga OA, Pineros MA (2004) How do crop plants tolerate acid soils? Mechanisms of aluminum tolerance and phosphorous efficiency. Annu Rev Plant Biol 55:459–493

Koppelman SJ, Wensing M, Ertmann M, Knulst AC, Knol EF (2004) Relevance of Ara h1, Ara h2 and Ara h3 in peanut-allergic patients, as determined by immunoglobulin E Western blotting, basophil-histamine release and intracutaneous testing: Ara h2 is the most important peanut allergen. Clin Exp Allergy 34(4):583–590

Krysan PJ, Colcombet J (2018) Cellular complexity in MAPK signaling in plants: questions and emerging tools to answer them. Front Plant Sci 9:1674

Kwak SY, Lew T, Sweeney CJ, Koman VB, Wong MH, Bohmert-Tatarev K, Snell KD, Seo JS, Chua NH, Strano MS (2019) Chloroplast-selective gene delivery and expression in planta using chitosan-complexed single-walled carbon nanotube carriers. Nat Nanotechnol 14(5):447–455

Landa P (2021) Positive effects of metallic nanoparticles on plants: overview of involved mechanisms. Plant Physiol Biochem 161:12–24

Landa P, Vankova R, Andrlova J, Hodek J, Marsik P, Storchova H, White JC, Vanek T (2012) Nanoparticle-specific changes in *Arabidopsis thaliana* gene expression after exposure to ZnO, TiO2, and fullerene soot. J Hazard Mater 241–242:55–62

Latef AAHA, Alhmad MFA, Abdelfattah KE (2017) The possible roles of priming with ZnO nanoparticles in mitigation of salinity stress in lupine (*Lupinus termis*) plants. J Plant Growth Regul 36(1):60–70

Laware SL, Raskar S (2014) Effect of titanium dioxide nanoparticles on hydrolytic and antioxidant enzymes during seed germination in onion. Int J Curr Microbiol App Sci 3(7):749–760

Lee PY, Saraygord-Afshari N, Low TY (2020) The evolution of two-dimensional gel electrophoresis – from proteomics to emerging alternative applications. J Chromatogr A 1615:460763

Lei Z, Mingyu S, Xiao W, Chao L, Chunxiang Q, Liang C, Hao H, Xiaoqing L, Fashui H (2008) Antioxidant stress is promoted by nano-anatase in spinach chloroplasts under UV-B radiation. Biol Trace Elem Res 121(1):69–79

Li X, Bai H, Wang X, Li L, Cao Y, Wei J, Liu Y, Liu L, Gong X, Wu L, Liu S, Liu G (2011) Identification and validation of rice reference proteins for western blotting. J Exp Bot 62(14):4763–4772

Lin D, Xing B (2007) Phytotoxicity of nanoparticles: inhibition of seed germination and root growth. Environ Pollut 150(2):243–250

López-Moreno ML, de la Rosa G, Hernández-Viezcas JA, Castillo-Michel H, Botez CE, Peralta-Videa JR, Gardea-Torresdey JL (2010) Evidence of the differential biotransformation and genotoxicity of ZnO and CeO2 nanoparticles on soybean (*Glycine max*) plants. Environ Sci Technol 44(19):7315–7320

Ma JF (2000) Role of organic acids in detoxification of aluminum in higher plants. Plant Cell Physiol 41(4):383–390

Ma JF, Furukawa J (2003) Recent progress in the research of external Al detoxification in higher plants: a minireview. J Inorg Biochem 97(1):46–51

Ma X, Geisler-Lee J, Deng Y, Kolmakov A (2010) Interactions between engineered nanoparticles (ENPs) and plants: phytotoxicity, uptake and accumulation. Sci Total Environ 408(16):3053–3061

Mahajan P, Dhoke SK, Khanna AS (2011) Effect of nano-ZnO particle suspension on growth of mung (*Vigna radiata*) and gram (*Cicer arietinum*) seedlings using plant agar method. J Nanotechnol 2011:1–17

Majumdar S, Peralta-Videa JR, Trujillo-Reyes J, Sun Y, Barrios AC, Niu G, Margez J, Gardea-Torresdey JL (2016) Soil organic matter influences cerium translocation and physiological processes in kidney bean plants exposed to cerium oxide nanoparticles. Sci Total Environ 569–570:201–211

Manzoor N, Ahmed T, Noman M, Shahid M, Nazir MM, Ali L, Alnusaire TS, Li B, Schulin R, Wang G (2021) Iron oxide nanoparticles ameliorated the cadmium and salinity stresses in wheat plants, facilitating photosynthetic pigments and restricting cadmium uptake. Sci Total Environ 769:145221

Marmiroli M, Imperiale D, Pagano L, Villani M, Zappettini A, Marmiroli N (2015) The proteomic response of *Arabidopsis thaliana* to cadmium sulfide quantum dots, and its correlation with the transcriptomic response. Front Plant Sci 6:1104

Matros A, Kaspar S, Witzel K, Mock HP (2011) Recent progress in liquid chromatography-based separation and label-free quantitative plant proteomics. Phytochemistry 72(10):963–974

Matsumura Y, Yoshikata K, Kunisaki S, Tsuchido T (2003) Mode of bactericidal action of silver zeolite and its comparison with that of silver nitrate. Appl Environ Microbiol 69:4278–4281

McShan D, Ray PC, Yu H (2014) Molecular toxicity mechanism of nanosilver. J Food Drug Anal 22:116–127

Miao AJ, Schwehr KA, Xu C, Zhang SJ, Luo Z, Quigg A, Santschi PH (2009) The algal toxicity of silver engineered nanoparticles and detoxification by exopolymeric substances. Environ Pollut 157:3034–3041

Miralles P, Church TL, Harris AT (2012) Toxicity, uptake, and translocation of engineered nanomaterials in vascular plants. Environ Sci Technol 46(17):9224–9239

Mirzajani F, Askari H, Hamzelou S, Schober Y, Römpp A, Ghassempour A, Spengler B (2014) Proteomics study of silver nanoparticles toxicity on *Oryza sativa* L. Ecotoxicol Environ Saf 108:335–339

Mushtaq YK (2011) Effect of nanoscale Fe(3)O(4), TiO(2) and carbon particles on cucumber seed germination. J Environ Sci Health 46(14):1732–1735

Mustafa G, Komatsu S (2016) Insights into the response of soybean mitochondrial proteins to various sizes of aluminum oxide nanoparticles under flooding stress. J Proteome Res 15(12):4464–4475

Mustafa G, Sakata K, Komatsu S (2015a) Proteomic analysis of flooded soybean root exposed to aluminum oxide nanoparticles. J Proteome 128:280–297

Mustafa G, Sakata K, Hossain Z, Komatsu S (2015b) Proteomic study on the effects of silver nanoparticles on soybean under flooding stress. J Proteomics 122:100–118

Mustafa G, Sakata K, Komatsu S (2016) Proteomic analysis of soybean root exposed to varying sizes of silver nanoparticles under flooding stress. J Proteome 148:113–125

Mustafa G, Hasan M, Yamaguchi H, Hitachi K, Tsuchida K, Komatsu S (2020) A comparative proteomic analysis of engineered and bio synthesized silver nanoparticles on soybean seedlings. J Proteomics 224:103833

Nair R, Varghese SH, Nair BG, Maekawa T, Yoshida Y, Kumar DS (2010) Nanoparticulate material delivery to plants. Plant Sci 179(3):154–163

Neilson KA, Gammulla CG, Mirzaei M, Imin N, Haynes PA (2010) Proteomic analysis of temperature stress in plants. Proteomics 10(4):828–845

Peharec Štefanić P, Jarnević M, Cvjetko P, Biba R, Šikić S, Tkalec M, Cindrić M, Letofsky-Papst I, Balen B (2019) Comparative proteomic study of phytotoxic effects of silver nanoparticles and silver ions on tobacco plants. Environ Sci Pollut Res Int 26(22):22529–22550

Poborilova Z, Opatrilova R, Babula P (2013) Toxicity of aluminium oxide nanoparticles demonstrated using a BY-2 plant cell suspension culture model. Environ Exp Bot 91:1–11

Poschenrieder C, Gunsé B, Corrales I, Barceló J (2008) A glance into aluminum toxicity and resistance in plants. Sci Total Environ 400:356–368

Pošćić F, Mattiello A, Fellet G, Miceli F, Marchiol L (2016) Effects of cerium and titanium oxide nanoparticles in soil on the nutrient composition of barley (*Hordeum vulgare L.*) kernels. Int J Environ Res Public Health 13(6):577

Rezvani N, Sorooshzadeh A, Farhadi N (2012) Effect of nano-silver on growth of saffron in flooding stress. World Acad Sci Eng Technol 6:519–524
Riahi-Madvar A, Rezaee F, Jalili V (2012) Effects of alumina nanoparticles on morphological properties and antioxidant system of *Triticum aestivum*. Iran J Plant Physiol 3:595–603
Rico CM, Hong J, Morales MI, Zhao L, Barrios AC, Zhang JY, Peralta-Videa JR, Gardea-Torresdey JL (2013) Effect of cerium oxide nanoparticles on rice: a study involving the antioxidant defense system and in vivo fluorescence imaging. Environ Sci Technol 47(11):5635–5642
Salama HMH (2012) Effects of silver nanoparticles in some crop plants, common bean (*Phaseolus vulgaris L.*) and *corn* (*Zea mays L.*). Int Res J Biotechnol 3:190–197
Salehi H, Chehregani A, Lucini L, Majd A, Gholami M (2018) Morphological, proteomic and metabolomic insight into the effect of cerium dioxide nanoparticles to *Phaseolus vulgaris* L. under soil or foliar application. Sci Total Environ 616:1540–1551
Santoro CM, Duchsherer NL, Grainger DW (2007) Antimicrobial efficacy and ocular cell toxicity from silver nanoparticles. NanoBiotechnology 3:55–65
Saxena R, Tomar RS, Kumar M (2016) Exploring nanobiotechnology to mitigate abiotic stress in crop plants. J Pharm Sci Res 8(9):974
Schwabe F, Schulin R, Rupper P, Rotzetter A, Stark W, Nowack B (2014) Dissolution and transformation of cerium oxide nanoparticles in plant growth media. J Nanopart Res 16(10):1–11
Shen CX, Zhang QF, Li J, Bi FC, Yao N (2010) Induction of programmed cell death in *Arabidopsis* and rice by single-wall carbon nanotubes. Am J Bot 97(10):1602–1609
Singh A, Singh NB, Hussain I, Singh H, Singh SC (2015) Plant-nanoparticle interaction: an approach to improve agricultural practices and plant productivity. Int J Pharm Sci Invent 4(8):25–40
Singh A, Singh NÁ, Afzal S, Singh T, Hussain I (2018) Zinc oxide nanoparticles: a review of their biological synthesis, antimicrobial activity, uptake, translocation and biotransformation in plants. J Mater Sci 53(1):185–201
Spielman-Sun E, Lombi E, Donner E, Howard D, Unrine JM, Lowry GV (2017) Impact of surface charge on cerium oxide nanoparticle uptake and translocation by wheat (*Triticum aestivum*). Environ Sci Technol 51(13):7361–7368
Syu YY, Hung JH, Chen JC, Chuang HW (2014) Impact of size and shape of silver nanoparticles on *Arabidopsis* plant growth and gene expression. Plant Physiol Biochem 83:57–64
Tanaka N, Konishi H, Khan MMK, Komatsu S (2004) Proteome analysis of rice tissues by two-dimensional electrophoresis: an approach to the investigation of gibberellin regulated proteins. Mol Gen Genomics 270(6):485–496
Taran N, Storozhenko V, Svietlova N, Batsmanova L, Shvartau V, Kovalenko M (2017) Effect of zinc and copper nanoparticles on drought resistance of wheat seedlings. Nanoscale Res Lett 12(1):1–6
Timabud T, Yin X, Pongdontri P, Komatsu S (2016) Gel-free/label-free proteomic analysis of developing rice grains under heat stress. J Proteome 133:1–19
Tripathi DK, Singh S, Singh VP, Prasad SM, Dubey NK, Chauhan DK (2017) Silicon nanoparticles more effectively alleviated UV-B stress than silicon in wheat (*Triticum aestivum*) seedlings. Plant Physiol Biochem 110:70–81
Valle SR, Carrasco J, Pinochet D, Calderini DF (2009) Grain yield, above-ground and root biomass of Al-tolerant and Al-sensitive wheat cultivars under different soil aluminum concentrations at field conditions. Plant Soil 318:299–310
van der Zande M, Vandebriel RJ, Van Doren E, Kramer E, Herrera Rivera Z, Serrano-Rojero CS, Gremmer ER, Mast J, Peters RJ, Hollman PC, Hendriksen PJ, Marvin HJ, Peijnenburg AA, Bouwmeester H (2012) Distribution, elimination, and toxicity of silver nanoparticles and ions in rats after 28-day oral exposure. ACS Nano 28:7427–7442
Van Hoecke K, De Schamphelaere KA, Van der Meeren P, Lcucas S, Janssen CR (2008) Ecotoxicity of silica nanoparticles to the green alga *Pseudokirchneriella subcapitata*: importance of surface area. Environ Toxicol Chem 27(9):1948–1957

Vannini C, Domingo G, Onelli E, Prinsi B, Marsoni M, Espen L, Bracale M (2013) Morphological and proteomic responses of *Eruca sativa* exposed to silver nanoparticles or silver nitrate. PLoS One 8(7):e68752

Vannini C, Domingo G, Onelli E, De Mattia F, Bruni I, Marsoni M, Bracale M (2014) Phytotoxic and genotoxic effects of silver nanoparticles exposure on germinating wheat seedlings. J Plant Physiol 171(13):1142–1148

Vishwakarma K, Shweta, Upadhyay N, Singh J, Liu S, Singh VP, Prasad SM, Chauhan DK, Tripathi DK, Sharma S (2017) Differential phytotoxic impact of plant mediated silver nanoparticles (AgNPs) and silver nitrate ($AgNO_3$) on *brassica* sp. Front Plant Sci 8:1501

von Uexküll HR, Mutert E (1995) Global extent, development and economic impact of acid soils. Plant Soil 171:1–15

Wang Q, Ma X, Zhang W, Pei H, Chen Y (2012) The impact of cerium oxide nanoparticles on tomato (*Solanum lycopersicum* L.) and its implications for food safety. Metallomics 4(10):1105–1112

Wang JW, Grandio EG, Newkirk GM, Demirer GS, Butrus S, Giraldo JP, Landry MP (2019) Nanoparticle-mediated genetic engineering of plants. Mol Plant 12(8):1037–1040

Wiese S, Reidegeld KA, Meyer HE, Warscheid B (2007) Protein labeling by iTRAQ: a new tool for quantitative mass spectrometry in proteome research. Proteomics 7(3):340–350

Wilkins MR, Sanchez JC, Gooley AA, Appel RD, Humphery-Smith I, Hochstrasser DF, Williams KL (1996) Progress with proteome projects: why all proteins expressed by a genome should be identified and how to do it. Biotechnol Genet Eng Rev 13(1):19–50

Yadav T, Mungray AA, Mungray AK (2014) Fabricated nanoparticles: current status and potential phytotoxic threats. Rev Environ Contam Toxicol 230:83–110

Yan W, Lien HL, Koel BE, Zhang WX (2013) Iron nanoparticles for environmental clean-up: recent developments and future outlook. *Environmental science*. Processes Impacts 15(1):63–77

Yang L, Watts DJ (2005) Particle surface characteristics may play an important role in phytotoxicity of alumina nanoparticles. Toxicol Lett 158:122–132

Yasmeen F, Raja NI, Razzaq A, Komatsu S (2016a) Gel-free/label-free proteomic analysis of wheat shoot in stress tolerant varieties under iron nanoparticles exposure. Biochim Biophys Acta 1864(11):1586–1598

Yasmeen F, Raja NI, Mustafa G, Sakata K, Komatsu S (2016b) Quantitative proteomic analysis of post-flooding recovery in soybean root exposed to aluminum oxide nanoparticles. J Proteome 143:136–150

Yasmeen F, Raja NI, Razzaq A, Komatsu S (2017) Proteomic and physiological analyses of wheat seeds exposed to copper and iron nanoparticles. Biochim Biophys Acta Proteins Proteom 1865(1):28–42

Yasmeen F, Raja NI, Ilyas N, Komatsu S (2018) Quantitative proteomic analysis of shoot in stress tolerant wheat varieties on copper nanoparticle exposure. Plant Mol Biol Report 36(2):326–340

Zhang P, Ma Y, Zhang Z, He X, Zhang J, Guo Z, Tai R, Zhao Y, Chai Z (2012) Biotransformation of ceria nanoparticles in cucumber plants. ACS Nano 6(11):9943–9950

Zhang Y, Zhang W, Liu Y, Wang J, Wang G, Liu Y (2016) Development of monoclonal antibody-based sensitive ELISA for the determination of Cry1Ie protein in transgenic plant. Anal Bioanal Chem 408(28):8231–8239

Zhao L, Peng B, Hernandez-Viezcas JA, Rico C, Sun Y, Peralta-Videa JR, Tang X, Niu G, Jin L, Varela-Ramirez A, Zhang JY, Gardea-Torresdey JL (2012) Stress response and tolerance of Zea mays to CeO2 nanoparticles: cross talk among H2O2, heat shock protein, and lipid peroxidation. ACS Nano 6(11):9615–9622

Zhenjie Z, Hu L, Chen Q, Dai H, Meng X, Yin Q, Liang T (2020) iTRAQ-based comparative proteomic analysis provides insights into tobacco callus response to carbon nanoparticles. https://doi.org/10.21203/rs.3.rs-109134/v1

Zhu ZJ, Wang H, Yan B, Zheng H, Jiang Y, Miranda OR, Rotello VM, Xing B, Vachet RW (2012) Effect of surface charge on the uptake and distribution of gold nanoparticles in four plant species. Environ Sci Technol 46(22):12391–12398

Chapter 16
Role of Chitosan Nanoparticles in Regulation of Plant Physiology Under Abiotic Stress

Yamshi Arif, Husna Siddiqui, and Shamsul Hayat

Abstract Chitosan is a biopolymer derived from chitin in crustaceans. It has emerged as bio-stimulant and elicitor in agriculture sector. It is non-toxic, biodegradable and abundant in nature with potent role in regulating plant physiological aspects. Application of chitosan in the form of nanoparticles (NPs) to promote growth and development of plant is a recent topic of interest amongst researchers. It is known to protect photosynthetic machinery during abiotic stress in plants. It mitigates toxicity symptoms in plant under abiotic stresses via induction of antioxidant defence system. Chitosan NPs are known to induce plants innate immunity responses via up-regulation of defence related genes as well as elevation of secondary metabolites. The present chapter sheds some light on recent development associated with chitosan NPs-mediated modifications of plant physiology and mainly on the abiotic stress responses in plants which could prove useful for crop improvement programs in the near future.

Keywords Antioxidants · Growth · Photosynthesis · Reactive oxygen species · Stress

16.1 Introduction

As worldwide population is rising demand for food is also rising, but ongoing environmental stress have negative impact on farmland consumption. Thus, it is increasing attention of researchers towards cheap, safe and eco-friendly product that serve as alternative to biological methods and thus enhance food demand by inducing crop productivity. Among the alternatives biopolymer-based nanoparticles shown adequate activity against several biotic and abiotic stress by inducing plant growth and development (Bandara et al. 2020). Chitosan (CHT) emerge out as the

Y. Arif · H. Siddiqui (✉) · S. Hayat
Plant Physiology Section, Department of Botany, Faculty of Life Sciences, Aligarh Muslim University, Aligarh, India

M. Faizan et al. (eds.), *Sustainable Agriculture Reviews 53*, Sustainable Agriculture Reviews 53, https://doi.org/10.1007/978-3-030-86876-5_16

biopolymer that displays effective results in increasing plant productivity under several environmental cues (Malerba and Cerana 2018).

CHT is the biopolymer made of linear unbranched β-1,4-D-glucosamine which is procured from chitin; N-acetyl-D-glucosamine and D-glucosamine are the major co-polymer constituting the exoskelton of arthropods (Kurita 2006; Fig. 1). Chitin is also found in wide group of living forms including fungi, diatoms, mollusks and sponges (Rinaudo 2006). Chitin is largely available as waste from shells of sea animals; it is also considered as the second largest renewable source of carbon following cellulose present on earth with the synthesis of about 10^{11} tons per year; this thus produce CHT on large scale (Kurita 2006). About 2000 tons of CHT is produced per year and is utilized for commercial interest (Malerba and Cerana 2018).

CHT preparation is very convenient and easy process; it majorly consist of chitin (solid form) treatment with 40–50% (*w*/*v*) NaOH at high temperature ranging between 120 and 150 °C; it removes acetyl groups and coverts N-acetyl-D-glucosamine in β-1,4-D-glucosamine (Malerba and Cerana 2018). However, chitin is distinctly insoluble in principal solvents and thus does not ensure its direct consumption; furthermore, CHT is easily solubilized in weak organic acids like acetic acid (Malerba and Cerana 2018). CHT is not easily solubilized in water which can be overcome by chemical modification "carboxymethylation" (Choi et al. 2016). Nevertheless, preparation via industrial methods implicit that the word "chitosan" does not implies a unique compound but it is several polymers having heterogeneousdeacetylation, molecular mass, viscosity, polymerization degree acid and dissociation constant (Choi et al. 2016). Thus, this heterogeneity crucially influences salient biological features of CHT (Anitha et al. 2014). Although, CHT is potent in combating stress and increasing plant metabolic responses; than other biopolymers such as starch, cellulose, galactomannans etc. (Anitha et al. 2014; Malerba and Cerana 2018). Moreover, CHT is cheap, safe and possess chemical structure which easily ensures introduction of certain molecules for designing specific polymer for selected specific function; thus these qualities make CHT of great importance in medical, biotechnological and agricultural uses (Malerba and Cerana 2016; Mutka et al. 2017; Ahmed et al. 2020a, b).

In recent year, increasing number of researches reported positive effect of CHT based compounds on plant growth and development. In this chapter, we have summarized role of CHT in increasing germination, photosynthetic traits, gas exchange parameters, antioxidant machinery and compatible osmolyte content in both healthy and stressful environmental conditions. An attempt have been made to cover contribution of CHT in augmenting enzymatic antioxidants like catalase (CAT), peroxidase (POX), superoxide dismutase (SOD) and non-enzymatic antioxidants like ascorbic acid, flavonoids, phenols, carotenoid and amino acids; and compatible osmolytes such as proline, soluble sugars, carbohydrate, gamma-aminobutyric acid (GABA) and amino acid in reducing reactive oxygen species (ROS) and MDA and maintaining osmotic balance in plants is appraised. Current review highlights role of CHT NPs in attenuating abiotic stresses like salinity, drought, heavy metal (HMs) and temperature(heat and cold) stress. Apart from this, CHT signaling and mechanism of action to enhance abiotic stress tolerance is also discussed.

16.2 Chitosan and Its Nanoparticles

Chitosan is the biodegradable, biocompatible and inexpensive polymer derived from chitin obtained by easily removing the acetate moiety from chitin (Mohammed et al. 2017). It is obtained from prawns or crabs and from cell walls of fungi; it is naturally occurring polysaccharide, bearing cationic, basic and biocompatible polymer extensively (Mohammed et al. 2017). Chitin in natural form bound to several different types of proteins and minerals which are detached from it; before CHT preparation via processes such as acidification and alkalization; purified chitin is then deacetylated to CHT; this process is modified to control end product properties like molecular weight and pKa and also control degree of deacetylation with factors like conditions of reaction such as temperature, concentration and ratio of chitin to alkali (Kaya et al. 2015). These chemical processes are carried in industrial processors; these produce deacetylated CHT which are eco-friendly and have several advantages in the future (Choi et al. 2016). CHT interacts with negatively charged compounds to for complexes via ionic or hydrogen bonding and also via hydrophobic interaction (Mohammed et al. 2017). The primary amine of CHT has pKa of ~6.5, depending on the degree of N-deacetylation; this also participates in CHT solubility in acidic pH and primary amine partial neutralization explains CHT aggregation at neutral to high pH; additionally CHT with acetylation degree in 40–60% range is easily soluble under physiological pH (Mohammed et al. 2017).

With the advancement of nanotechnology, nanosized biopolymers have regarded as the potent form in several agricultural, food, drug and biomedical industries. Nanoparticles (NPs) biosynthesized via chemical and mechanical methods displays improved properties in comparison to normal sized biopolymers because of high aspect ratio and surface area (Bandara et al. 2020). Additionally, NPs having high antimicrobial and antifungal properties have been devoted for the application in several industries; it also have thermal properties, mechanical barrier and rheological (Malerba and Cerana 2016; Mutka et al. 2017). Recent researches clearly report the nontoxic activity of chitosan NPs bearing several chemical modifications. Additionally, CHT scaffold NPsare used in tissue engineering whereas its mucoadhesive character induces its administration in poorly absorbable drugs; it is also used in delivering siRNA (Kiani et al. 2021). CHT NPs used in "green nanotechnology" that is synthesis of metal and other NPs by non-toxic and ecofriendly solvents like CHT nanomaterials; CHT serve as the immobilization of biodegradable compounds on metal oxide NPs (Kiani et al. 2021). Chitin from shells of shrimps and crabs were used for commercial production of CHT, which than undergoes several chemical processes such as demineralization, decolourization and deproteinization (Mohammed et al. 2017). Squid gladius is also the rich source of CHT and is also used in several industries; it is cheap and reduces the use of large amount of acids and alkaline compounds because of its less impurity and unavailability of coloured compounds; it also displays better solubility, reactivity and has weaker molecular hydrogen bonding (Hidangmayum et al. 2019; Bandara et al. 2020). Currently, CHT NPs were synthesized via bottom-up approach technique as the outcome of self assembling or cross-linking process (Bandara et al. 2020). CHT possess several unique properties so the NPs formulator must cautiously match the desired chemical and physical properties of the CHT, and also the predicted biological environment, via CHT processing technique (Mohammed et al. 2017).

16.3 Chitosan Mechanism of Action

CHT mechanism in plants is not yet clear; but several investigations and researches suggest that CHT elicited several defense responses in plants (Mejía-Teniente et al. 2013; Malerba and Cerana 2015). In plants, Chitin- specific receptors are localized on cell membrane which induces defense reponses in plants (Iriti and Faoro 2009). Moreover, Chitin elicitor binding proteins (CEBiP) have been isolated from several crops which induce defense machinery and influencing chitosan-responsive differential gene expression which interact with chromatin and/or it can also attach with certain receptors (Hadwiger 2015). Application of CHT and its derivative on plants induces several defense related genes which boost plant immunity and have strong antimicrobial property (Hidangmayum et al. 2019). CHT application on several plants (like peach, tomato and dragon fruit) induces activity of enzyme chitinase and glucanase; which are linked with pathogen resistance (Hidangmayum et al. 2019). Family of lectins which is a chitosan binding glycoprotein has been isolated from *Brassica campestris*; additionally isolated vesicles of *Mimosa pudica* and *Cassia fasciculate* showed increased activation of plasma membrane H^+-ATPase; which indicates presence of chitosan receptor molecule (Hidangmayum et al. 2019). It was also reported that CHT can induces several receptor like kinase genes, MAP kinase pathway and lysine motif receptor-like kinase (Iriti and Faoro 2009). However, still CHT binding receptors are undetermined and endures "a Pathogen-Associated Molecular Pattern (PAMP) in search of a Pattern Recognition Receptor (PRR)" (Iriti and Faoro 2009).

16.4 Signaling Mediated by Chitosan

CHT induced signaling cascade involves certain cellular receptors that are than transduced by secondary messengers like ROS, hydrogen peroxide (H_2O_2), calcium ions (Ca^{2+}), nitric oxide (NO) and other plant hormone (Fig. 1); which further participates in facilitating physio-biochemical responses in plants (Hidangmayum et al. 2019). CHT treatment decreases the inhibitory and toxic effect of free radicals such as superoxide anion and lipid free radical (Li et al. 2002). Furthermore, upon CHT application H_2O_2 serves as the signaling molecule which induces plant tolerance to abiotic stress by improving photosynthesis attributes and its pigment, increasing membrane stability and activating antioxidant machinery (Pongprayoon et al. 2013). In response to CHT elicitation Ca^{2+} serves as secondary messenger and regulates callose synthase activity and mediates programmed cell death in both monocot and dicot plants (Hidangmayum et al. 2019). Moreover, treatment of tobacco infected with tobacco necrosis virus with calcium channel inhibitor decreased the cell death (apoptotic) kinetics (Iriti et al. 2006). In pearl millets, NO signaling is reported when treated with CHT and NO scavenger c-PTIO (2-(4-carboxyphenyl)-4,4,5,5-tetramethylimidazoline-1-oxy-3-oxide potassium salt) or NO inhibitor LNAME (N-nitro-L-arginine methyl ester hydrochloride); which further reduces plant defense machinery against pathogen attack (Hadwiger 2013). CHT application on *Oryza sativa* and *Phaseolus vulgaris* increased the accumulation of jasmonic acid (JA) which further boost up plant immunity against several stress (Hidangmayum et al. 2019). In *Brassica napus*, cDNA microarray/ semiquantitative RT-PCR analyses displayed that JA and ethylene both serves in defense related signaling cascade mediated on CHT application (Yin et al. 2006). CHT and JA both activate genes encoding for phenylalanine ammonia lyase (PAL) and protease inhibitors (Doares et al. 1995; Khan et al. 2003). CHT facilitates synthesis of JA and ABA (Iriti and Faoro 2008). In rice, CHT application increased the signaling of octadecanoid pathway which further induced accumulation of JA and 12-oxo-phytodieonic acid; however ABA synthesis was induced by H_2O_2 signaling cascade which leads to stomatal closure (Pichyangkura and Chadchawan 2015). JA and ABA both hormones plays potent role in regulating use of water in plant under normal and stressful environment. Moreover, CHT application boosted up plant immune by increasing activation of defense related genes such as H_2O_2 via octadecanoid pathway, NO inside chloroplast, activation of MAP-kinase pathway, oxidative burst and hypersensitive reactions (Pichyangkura and Chadchawan 2015). Thus, of these signaling molecules increase plant tolerance in CHT treated plants in response to several biotic and abiotic stresses.

16.5 Chitosan in Plant Growth and Development

Several researchers have reported that CHT NPs application on plants is increasing to enhance plant productivity to cope up with the demand of food of increasing population; CHT NPs (250–500 ppm from Sigma-Aldrich (St. Louis, MO, USA)) application on strawberry prior to flowering induce plant developmental process and enhanced fruit production and improved fruit quality (Mutka et al. 2017). CHT polymers having high and low molecular weight (124 kDa and 66.4 kDa) synthesized on the basis of acetylation and deacetylation and hydrolyzed CHT derivative (13.2 kDa) when applied on *Solanumtuberosum* L. increased tuber size and yield (Falcón-Rodríguez et al. 2017). CHT application on *Capsicum annuum* L. induces fruit weight, diameter and yield traits (Mahmood et al. 2017). CHT application on *Trifoliumrepens* and *Thymus daenensis* Celakcombated drought stress and improved content of essential metabolite and oil (Li et al. 2017; Bistgani et al. 2017). Application of CHT on *Ocimumciliatum* and *O. basilicum* (sweet basil) promoted plant growth, developmental and physio-biochemical traits (Pirbalouti et al.

2017). Seed soaking of *Oryza sativa* L. with CHT reduced the deleterious effect of ozone and increased plant growth and yield traits (Phothi and Theerakarunwong 2017). CHT application improved plants morphology and chlorophyll content in *Brassica rapa* L. (Zong et al. 2017a, b). Chitooligosaccharide a hydrolysis product of CHT application on plants also increased morpho-physiological and biochemical traits which further induced growth and yield traits (Zong et al. 2017a, b). CHT application on *Stevia rebaudiana* Bertoni improved its medicinal property by increasing phenol and glycoside content (Mehregan et al. 2017). CHT application on two hairy root clones of *Gentianadinarica* Beck, increased the content of xanthoneaglyconenorswertianin (Krstić-Milošević et al. 2017). In CHT application *Phyllanthusdebilis* increased the content of tannins and other metabolite having therapeutic active constituent (Malayamana et al. 2017). CHT oligosaccharide application on *Hordeumvulgare* L. induced the accumulation of antioxidants, phenolics and anti- hyperglycemic compounds (Ramakrishna et al. 2017). In *Zea mays* L. Cu-CHT NPs application increased plant height, root length, its number and stem diameter; it also induced chlorophyll content, photosynthetic traits, ear length and weight and grain quality and quantity (Choudhary et al. 2017). CHT-polyvinyl alcohol hydrogels with absorbed copper NPs in "Jubilee" watermelon increased stem and root length and improved stomata width(González Gómez et al. 2017). Zinc complexed CHT NPs application efficiently improved micronutrient content (Deshpande et al. 2017). Application of CHT combined with silica in plants increased the growth, metabolic and cellular responses; it also improved use of NKP fertilizers (Gumilar et al. 2017). CHT NPs are also used carrier system for phytohormone which enhanced plant growth and development and also improved yield traits; such as CHT NPs combined with gibberellic acid inceased the growth, leaf area, chlorophyll and carotenoid content in *Phaseolus vulgaris* (Espirito Santo Pereira et al. 2017). However, this suggests that CHT NPs and other CHT based materials increased plant growth, development and productivity; particularly it is beneficial for the agronomic utilization in marginal lands (Hidangmayum et al. 2019).

16.6 Chitosan Nanoparticlesin Abiotic Stress

CHT and its derivatives evolve as the natural polymer which facilitates efficient responses in plants. Additionally, CHT NPs also play efficient role in mitigating abiotic stress (salinity, drought, HMs, temperature), as it enhance antioxidant activity in plants due to the existence of hydroxylated amino group which quenches ROS and induces plants morpho-physiological and biochemical traits (Hidangmayum et al. 2019; Fig. 2).

16.6.1 Salinity Stress

Salinity is the major stress factor affecting plant growth and development globally. About 20% of the agricultural land in the world faces salinity stress which hampers plant yield and productivity; moreover it is predicted by the year 2050 about 50% of the agricultural land is affected by salinity stress; thus to mitigate its toxicity on plant is major concern (Zayed et al. 2017; Arif et al. 2020). Salinity stress impairs plant growth and development by inducing ionic, osmotic and oxidative stress (Arif et al. 2020). Salinity hampers germination, growth, photosynthesis and many other important cellular and metabolic activity; however to combat it CHT is effective in many plants (Zayed et al. 2017). In *Phaseolus vulgaris* L. treatment of 0.1%, 0.2%, and 0.3% CHT NPs improved germination traits during salt stress (Zayed et al. 2017). CHT NPs releases NO which increases salinity tolerance in *Zea mays* L. (Oliveira et al. 2016). CHT NPs releases NO in plants which inducesS-nitrosothiols, chlorophyll content and PSII activity in leaf, thus it also increases bioavailability of NO in plants (Oliveira et al. 2016). In mung bean, solid matrix priming of seedlings with CHT NPs increased salinity tolerance by improving growth, protein and chlorophyll content and many important metabolic processes (Sen et al. 2019). In tomato, CHT with and without copper NPs facilitated plant growth and induced the expression of JA and antioxidant enzymes such as superoxide dismutase (SOD) which induces detoxification in plants and mitigated salt stress (Hernández-Hernández et al. 2018). Salt stress increases ROS and MDA content which hampers physio-biochemical responses in plants; however CHT treatment is efficient in alleviating deleterious effect of salt by increasing antioxidant enzyme activity (SOD, CAT, POX) in several plant species such as *Carthamustinctorius* L., *Helianthus annuus* L., *Trachyspermumammi*, *Plantago ovate*, which lower don MDA and ROS content (Hidangmayum et al. 2019). In wheat, CHT derivatives increases the photosynthetic traits such stomatal conductance, water use efficiency, stomatal conductance and chlorophyll content which reduces the toxic effect of salt and increased plant growth and yield traits (Ma et al. 2011). CHT NPs application reduced MDA content in several plant species and maintains membrane stability and integrity and reduced ion leakage during salinity (Hidangmayum et al. 2019). Proline content was increased during salt stress which is due to induction of proline biosynthesis or increase in protein turnover or reduction in protein utilization or decrease in proline oxidation to glutamate; thus CHT treatment mitigated salt stress by increasing proline level (Hidangmayum et al. 2019). CHT NPs have high surface to volume ratio which causes higher penetrability and the capacity to form more interactions in plants; thus it have remarkable effect in alleviating toxic effect of salt in several crops (Bandara et al. 2020).

16.6.2 Drought Stress

Drought or water deficit conditions impairs sustainable agriculture production leading to several toxic effects on plants health which increases ROS production to greater extent which further causes lipid peroxidation of membrane and its interaction with several other macromolecules; which hampers plant growth, developmental processes and yield traits. It affects many aspects of a plant, including anatomy, physiology, and biochemistry leading to lower yields (Hidangmayum et al. 2019; Bandara et al. 2020). However, CHT treatment enhanced plant height and induced the accessibility and water uptake and important nutrients; it reduces ROS content by increasing antioxidant enzymatic and non-enzymatic machinery (Hidangmayum et al. 2019). In white clover, CHT treatment mitigated drought stress by increasing production of osmoprotectant and metabolites like proline, valine, threonine, lysine, phenylalanine, isoleucine, serine, aspartic acid and GABA which modulated plant turgour and osmotic potential (Li et al. 2017). Thyme plant when subjected to drought stress, CHT treatment enhanced proline accumulation which helps in combating water deficit conditions (Bistgani et al. 2017). When plants are subjected to water scarcity conditions, CHT application increased the antioxidant enzymatic machinery (CAT, POX and SOD) which helps in scavenging ROS and decreasing MDA content; which helps in maintaining membrane integrity and several other metabolic responses. Additionally, CHT application also increased soluble sugar content like glucose and fructose due to the breakdown of polysaccharides, which helps in maintaining cell turgour, osmotic adjustments, cell signaling, carbohydrate transportation and maintenance of carbon balance in response to drought stress (Hidangmayum et al. 2019). Drought stress destroys the chloroplast structure, which ultimately decreases chlorophyll content, and inhibits the activity of the enzymes which participates in the Calvin cycle; it also hampers stomatal conductance, leads to stomatal closure and blocking the CO_2 intake by the leaves reducing photosynthesis and plants growth (Bandara et al. 2020). CHT application mitigated water scarcity by increasing chlorophyll content, photosynthetic attributes like net photosynthesis, stomatal conductance and water use efficiency; it also facilitates contents of nitrogen and potassium in plant shoot which promotes the synthesis of more chloroplast per cell, thus this ultimately increases synthesis of chlorophyll (Hidangmayum et al. 2019). Barley when subjected to drought stress CHT NPs application (60 and 90 ppm) via soil or foliar routes reduced the deleterious effect of water scarcity by up-regulating relative water content (RWC), plant growth and developmental processes, it also induced yield traits (Behboudi et al. 2018). CHT NPs in pearl millet and sugarcane alleviated drought stress by increasing plants water status by reducing stomatal conductance and transpiration; it also induced photosynthesis, root biomass and NO signaling (Priyaadharshini et al. 2019; Silveira et al. 2019; Bandara et al. 2020). CHT NPs application at concentration of 90 ppm on water-deficient *Triticum aestivum* plant improved physio-biochemical traits; it also improved quality and quantity of grains (Behboudi et al. 2019). In apple also CHT NPs is highly beneficial in alleviating drought stress (Avestan et al. 2017).

Foliar exogenous treatment of N-succinyl CHT and N,O-dicarboxymethylated CHT NPs enhanced drought tolerance in maize by increasing antioxidant defense machinery, induced synthesis of phenolic compound, osmoprotectant, gaseous exchange and yield traits (Rabêlo et al. 2019). Therefore, CHT NPs and their derivatives are efficient in enhancing plant tolerance to drought by displaying positive effect on growth, metabolic and cellular responses.

16.6.3 Heavy Metal Stress

Heavy metal (HMs) such as cadmium (Cd), copper (Cu), nickel (Ni), zinc (Zn), arsenic (As), lead (Pb), iron (Fe), mercury (Hg) and cobalt (Co) are accumulated in large extent in soil through industrial wastes and sewage. Plants experience oxidative stress upon exposure to HMs, they also disrupts cellular homeostasis and have detrimental effect on plant growth and development processes. However, CHT application reduces negative effect of HMs and enhances cellular and metabolic responses; as CHT has the capability to form complexes with HMs due to presence of functional amino and hydroxyl group (Hidangmayum et al. 2019). CHT and its derivatives to effectively complex metal ions (Ag(I), Pb(II), and Cu(II)) in soil that co-exist with other ionic substances like K^+, Cl^- and NO_3^- (Kamari et al. 2011; Bandara et al. 2020). Cd toxicity reduces stomatal conductance, photosynthesis and gas exchange; however CHT treatment increased photosynthetic traits and enhanced edible rape growth and development (Zong et al. 2017a, b). CHT alleviated HMs stress by increasing antioxidant machinery (CAT, POX and SOD) and by inducing ascorbic acid content (Hidangmayum et al. 2019). Hence, use of nanochitosan to overcome HMs toxicity is fields that can be explored in future research.

16.6.4 Temperature Stress

Low and high temperature stress are important factors affecting agriculture; it retards plant growth, photosynthesis and several crucial cellular responses; it also reduces membrane integrity and composition. However, use of CHT reduces heat and cold stress in bean and maize respectively; CHT increases germination traits, shoot and root length and root and shoot dry and fresh mass (Bandara et al. 2020). CHT overcome high-temperature stress by enhancing heat shock protein, ABA activity and enhance defense related ABA-responsive genes (Hidangmayum et al. 2019). Although current investigations on the application of nanochitosan to alleviate temperature stress is scarce and further needed to be investigated.

16.7 Mechanistic Approach of Chitosan Nanoparticle in Mitigating Abiotic Stress

Abiotic stress (such as salinity, drought, temperature and HMs) hampers plants developmental and metabolic processes. During signaling pathway, the initial signal start up the synthesis of secondary signaling molecules like inositol phosphate and ROS leading in the receptor-linked release of Ca^{2+} inside the cells. These processes leads to phosphorylation associated modulation of specific proteins that have stabilizing and defensive function or it may serve as transcription factor that modulates the several genes associated in abiotic stress response (Bandara et al. 2020). The primary stress signal may be occurred by various primary sensors which causes secondary signals and downstream signal transduction that takes place at different time and location from primary signaling. However, secondary signals during stress serve as "stress cross-protection" in plants (Xiong et al. 2002). Therefore, the association of signaling compound chitosan is beneficial in alleviating catastrophic effect of different abiotic stress by participating in different signaling pathways that interact with each other; stress reduces photosynthetic fixation and redistribution of carbon, and also the ETC taking place in the chloroplasts; as well as reduces activity ofribulose-1,5-bisphosphate carboxylase/oxygenase (Rubisco) (Bandara et al. 2020). Moreover, addition of CHT NPs elevated accumulation of osmoprotectants such as proline and production of solutes likes sugars, amino acid and polyols which enhance plant turgor. It also increased level of organic compounds like phenols, ketones and aldehyde which enhance plant tolerance to abiotic stress (Bandara et al. 2020). Furthermore, after CHT NPs application plant produces antioxidant machinery which balances higher levels of ROS produced during stress as a result of dysfunction ETC (Hidangmayum et al. 2019). Nanochitosan increases activity of CAT, POX and SOD; SOD forms H_2O_2 by neutralizing superoxide free radicals; and this is converted to water and oxygen (nontoxic) via CAT and POX. Thus, CHT NPs modulate antioxidant signaling and osmoprotectant production (Bandara et al. 2020). CHT based increased proline level during stress induces the activity of proteinase enzyme (Khati et al. 2017). CHT also induces production of ABA signaling during stress which regulates stomatal conductance, transpiration and induces expression of several stress-responsive genes. Thus, CHT NPs are involved in the ABA- dependent mechanism in abiotic stress tolerance in plants (Zhang et al. 2008; Iriti et al. 2009).

16.8 Chitosan Toxicity

According to researches, CHT NPs in higher concentrations cause toxicity in plants. In *Capsicum annuum* L. nan-chitosan in higher concentration (5–20 mg/L) causes phytotoxicity in plants and also elicit growth and developmental effect at low concentrations (Asgari-Targhi et al. 2018). CHT NPs in higher concentration displays toxic effect on germination and seedling growth; they also lower down defensive phenols and antioxidant content (Abdel-Aziz 2019). CHT NPs in higher concentrations have negative impact on soil microbiome (Abdel-Aziz 2019).

CHT-silver NPs increases the parasitic and predation efficiency of malaria vector and zebrafish respectively (Murugan et al. 2016). CHT NPs also have harmful effect on crab species (Bandara et al. 2020). CHT NPs when feed to rats have negative impact on liver, kidney and stomach; further also creates toxicity in blood stream (Bandara et al. 2020). However, nanochitosan-based material creates species-dependent, size and concentration- dependent toxicity; Due to small size of NPs they have potential ability to penetrate through biological membrane and have higher risk of creating toxicity (Bandara et al. 2020). Nevertheless, the concentration and size dependent properties of CHT NPs and its derivatives should be properly analyzed before the usage to avoid any toxic effects.

16.9 Conclusions and Future Prospective

CHT is biodegradable, non-toxic, biocompatibility and non-allergic in nature which offers wide potent possibility for used in agricultural sector and is beneficial for the society and agriculture. CHT NPs up-regulates many signaling molecules and participates in several signal transductions. CHT NPs proved beneficial in facilitating plant growth and development under normal and stressful environment. Furthermore, under abiotic stress (salinity, drought, temperature, HMs) it serves as potent stress reducer by reducing ROS and MDA content by up-regulating antioxidant defense machinery, also increases amino acid, phenols, sugars, osmoprotectant (proline, GABA) content which maintain cell turgor and osmoticum. Additionally, CHT NPs also increases photosynthesis, chlorophyll content; and serves as antitranspirant to combat stress. CHT have chelating properties which use to combat HMs stress. Thus, CHT NPs are efficient in improving growth, development, cellular and metabolic responses in plants, and useful in sustainable agricultural practices.

Apart from this, there are several questions and query that needs to be further explored. Such as there are not enough researches which can explain the mode of action of CHT in plant system and the transporters, sensors, receptors involved in CHT transportation and signaling inside plant cell. Furthermore, transcriptomic, proteomic, metabolomic studies, and their combined responses is useful to provide a better use of nanochitosan in the abiotic stress management.

References

Abdel-Aziz H (2019) Effect of priming with chitosan nanoparticles on germination, seedling growth and antioxidant enzymes of broad beans. Catrina Int J Environ Sci 18:81–86

Ahmed KB, Khan MM, Jahan A, Siddiqui H, Uddin M (2020a) Gamma rays induced acquisition of structural modification in chitosan boosts photosynthetic machinery, enzymatic activities and essential oil production in citronella grass (CymbopogonwinterianusJowitt). Int J Biol Mac 145:372–389

Ahmed KB, Khan MM, Siddiqui H, Jahan A (2020b) Chitosan and its oligosaccharides, a promising option for sustainable crop production-a review. Carbohydr Polymers 227:115331

Anitha A, Sowmya S, Sudheesh Kumar PT, Deepthi S, Chennazhi KP, Ehrlich H, Tsurkan M, Jayakumar R (2014) Chitin and chitosan in selected biomedical applications. Prog Polym Sci 39:1644–1667

Arif Y, Singh P, Siddiqui H, Bajguz A, Hayat S (2020) Salinity induced physiological and biochemical changes in plants: An omic approach towards salt stress tolerance. Plant Physiol Biochem 156:64–77

Asgari-Targhi G, Iranbakhsh A, Ardebili ZO (2018) Potential benefits and phytotoxicity of bulk and nano-chitosan on the growth, morphogenesis, physiology, and micropropagation of Capsicum annuum. Plant Physiol Biochem 127:393–402

Avestan S, Naseri L, Barker AV (2017) Evaluation of nanosilicon dioxide and chitosan on tissue culture of apple under agar-induced osmotic stress. J Plant Nutr 40:2797–2807

Bandara S, Du H, Carson L, Bradford D, Kommalapati R (2020) Agricultural and biomedical applications of chitosan-based nanomaterials. Nanomaterials 10(10):1903

Behboudi F, TahmasebiSarvestani Z, Kassaee MZ, ModaresSanavi SAM, Sorooshzadeh A, Ahmadi SB (2018) Evaluation of chitosan nanoparticles effects on yield and yield components of barley (Hordeum vulgare L.) under late season drought stress. Water Environ Nanotechnol 3:22–39

Behboudi F, Sarvestani ZT, Kassaee MZ, Modarres-Sanavy SAM, Sorooshzadeh A, Mokhtassi-Bidgoli A (2019) Evaluation of chitosan nanoparticles effects with two application methods on wheat under drought stress. J Plant Nutr 42:1439–1451

Bistgani ZE, Siadat SA, Bakhshandeh A, Pirbalouti AG, Hashemi M (2017) Interactive effects of drought stress and chitosan application on physiological characteristics and essential oil yield of *Thymus daenensis* Celak. Crop J 5(5):407–415

Choi C, Nam JP, Nah JW (2016) Application of chitosan and chitosan derivatives as biomaterials. J Ind Eng Chem 33:1–10

Choudhary RC, Kumaraswamy RV, Kumari S, Sharma SS, Pal A, Raliya R, Biswas P, Saharan V (2017) Cu-chitosan nanoparticle boost defense responses and plant growth in maize (Zea mays L.). Sci Rep 7:9754–9765

Deshpande P, Dapkekar A, Oak MD, Paknikar KM, Rajwade JM (2017) Zinc complexed chitosan/TPP nanoparticles: promising micronutrient nanocarrier suited for foliar application. Carbohydr Polym 165:394–401

Doares SH, Syrovets T, Wieler EW, Ryan A (1995) Oligogalacturonides and chitosan activate plant defensive gene through the octadecanoid pathway. Proc Natl Acad USA 92:4095–4098

Espirito Santo Pereira A, Mayara Silva P, Oliveira JL, Oliveira HC, FernandesFraceto L (2017) Chitosan nanoparticles as carrier systems for the plant growth hormone gibberellic acid. Colloids Surf B Biointerfaces 150:141–152

Falcón-Rodríguez AB, Costales D, Gónzalez-Peña D, Morales D, Mederos Y, Jerez E, Cabrera JC (2017) Chitosans of different molecular weight enhance potato (Solanum tuberosum L.) yield in a field trial. Span. J Agric Res 15:e0902

González Gómez H, RamírezGodina F, Ortega Ortiz H, Benavides Mendoza A, Robledo Torres V, Cabrera De la Fuente M (2017) Use of chitosan-PVA hydrogels with copper nanoparticles to improve the growth of grafted watermelon. Molecules 22:1031

Gumilar TA, Prihastanti E, Haryanti S, Subagio A, Ngadiwiyana A (2017) Utilization of waste silica and chitosan as fertilizer nanochisil to improve corn production in Indonesia. Adv Sci Lett 23:2447–2449

Hadwiger LA (2013) Multiple effects of chitosan on plant systems: Solid science or hype. Plant Sci 208:42–49

Hadwiger LA (2015) Anatomy of a nonhost disease resistance response of pea to *Fusariumsolani*: PR gene elicitation via DNase, chitosan and chromatin alterations. Front Plant Sci 6:373

Hernández-Hernández H, Juárez-Maldonado A, Benavides-Mendoza A, Ortega-Ortiz H, Cadenas-Pliego G, Sánchez-Aspeytia D, González-Morales S (2018) Chitosan-PVA and copper nanoparticles improve growth and overexpress the SOD and JA genes in tomato plants under salt stress. Agronomy 8:175

Hidangmayum A, Dwivedi P, Katiyar D, Hemantaranjan A (2019) Application of chitosan on plant responses with special reference to abiotic stress. Physiol Mol Biol Plants 25(2):313–326

Iriti M, Faoro F (2008) Abscisic acid mediates the chitosan-induced resistance in plant against viral disease. Plant Physiol Biochem 46:1106–1111

Iriti M, Faoro F (2009) Chitosan as a MAMP, searching for a PRR. Plant Signal Behav 4(1):66–68

Iriti M, Sironi M, Gomarasca S, Casazza AP, Soave C, Faoro F (2006) Cell death-mediated antiviral effect of chitosan in tabacco. Plant Physiol Biochem 44:893–900

Iriti M, Picchi V, Rossoni M, Gomarasca S, Ludwig N, Gargano M, Faoro F (2009) Chitosan antitranspirant activity is due to abscisic acid-dependent stomatal closure. Environ Exp Bot 66:493–500

Kamari A, Pulford I, Hargreaves JS (2011) Chitosan as a potential amendment to remediate metal contaminated soil – a characterisation study. Colloids Surf B Biointerfaces 82:71–80

Kaya M, Mujtaba M, Bulut E, Akyuz B, Zelencova L, Sofi K (2015) Fluctuation in physicochemical propertiesof chitins extracted from different body parts of honeybee. Carbohydr Polym 132:9–16

Khan W, Prithiviraj B, Smith DL (2003) Chitosan and chitin oligomers increase phenylalanine ammonia-lyase and tyrosine ammonia-lyase activities in soybean leaves. J Plant Physiol 160:859–863

Khati P, Chaudhary P, Gangola S, Bhatt P, Sharma A (2017) Nanochitosan supports growth of Zea mays and also maintains soil health following growth. Biotech 7:81

Kiani M, Rabiee N, Bagherzadeh M, Ghadiri AM, Fatahi Y, Dinarvand R, Webster TJ (2021) Improved green biosynthesis of chitosan decorated Ag-and Co3O4-nanoparticles: a relationship between surface morphology, photocatalytic and biomedical applications. Nanomedicine 32:102331

Krstić-Milošević D, Janković T, Uzelac B, Vinterhalter D, Vinterhalter B (2017) Effect of elicitors on xanthone accumulation and biomass production in hairy root cultures of Gentianadinarica. Plant Cell Tissue Organ Cult 130:631–640

Kurita K (2006) Chitin and chitosan: functional biopolymers from marine crustaceans. Mar Biotechnol 8:203–226

Li WJ, Jiang X, Xue PH, Chen SM (2002) Inhibitory effects of chitosan on superoxide anion radicals and lipid free radicals. Chin Sci Bull 47:887–889

Li Z, Zhang Y, Zhang X, Merewitz E, Peng Y, Ma X, Yan Y (2017) Metabolic pathways regulated by chitosan contributing to drought resistance in white clover. J Proteome Res 16(8):3039–3052

Ma L, Li Y, Yu C, Wang Y, Li X, Li N, Chen Q, Bu N (2011) Alleviation of exogenous oligochitosan on wheat seedlings growth under salt stress. Protoplasma 249:393–399

Mahmood N, Abbasi NA, Hafiz IA, Ali I, Zakia S (2017) Effect of biostimulants on growth, yield and quality of bell pepper cv. Yolo wonder. Pak J Agric Sci 54:311–317

Malayamana V, Sisubalan N, Senthilkumar RP, Sheik Mohamed S, Ranjithkumar R, GhouseBasha M (2017) Chitosan mediated enhancement of hydrolysable tannin in Phyllanthusdebilis Klein ex Willd via plant cell suspension culture. Int J Biol Macromol 104:1656–1663

Malerba M, Cerana R (2015) Reactive oxygen and nitrogen species indefense/stress responses activated by chitosan in sycamore cultured cells. Int J Mol Sci 16:3019–3034

Malerba M, Cerana R (2016) Chitosan effects on plant systems. Int J Mol Sci 17:996
Malerba M, Cerana R (2018) Recent advances of chitosan applications in plants. Polymers 10(2):118
Mehregan M, Mehrafarin A, Labbafi MR, NaghdiBadi H (2017) Effect of different concentrations of chitosan biostimulant on biochemical and morphophysiological traits of stevia plant (Stevia rebaudiana Bertoni). J Med Plants 16:169–181
Mejía-Teniente L, Duran-Flores FD, Chapa-Oliver AM, Torres-Pacheco I, Cruz-Hernández A, González-Chavira MM, Ocampo-Velázquez RV, Guevara-González RG (2013) Oxidative and molecular responses in Capsicum annuum L. after hydrogen peroxide, salicylic acid and chitosan foliar applications. Int J Mol Sci 14:10178–10196
Mohammed MA, Syeda J, Wasan KM, Wasan EK (2017) An overview of chitosan nanoparticles and its application in non-parenteral drug delivery. Pharmaceutics 9(4):53
Murugan K, Anitha J, Dinesh D, Suresh U, Rajaganesh R, Chandramohan B, Subramaniam J, Paulpandi M, Vadivalagan C, Amuthavalli P et al (2016) Fabrication of nano-mosquitocides using chitosan from crab shells: Impact on non-target organisms in the aquatic environment. Ecotoxicol Environ Saf 132:318–328
Mutka JA, Rahman M, Sabir AA, Gupta DR, Surovy MZ, Rahman M, Tofazzal Islam M (2017) Chitosan and plant probiotics application enhance growth and yield of strawberry. Biocatal Agric Biotechnol 11:9–18
Oliveira HC, Gomes BC, Pelegrino MT, Seabra AB (2016) Nitric oxide-releasing chitosan nanoparticles alleviate the effects of salt stress in maize plants. Nitric Oxide 61:10–19
Phothi R, Theerakarunwong CD (2017) Effect of chitosan on physiology, photosynthesis and biomass of rice (Oryza sativa L.) under elevated ozone. Aust J Crop Sci 11:624–630
Pichyangkura R, Chadchawan S (2015) Biostimulant activity of chitosan in horticulture. Sci Hort 196:49–65
Pirbalouti AG, Malekpoor F, Salimi A, Golparvar A (2017) Exogenous application of chitosan on biochemical and physiological characteristics, phenolic content and antioxidant activity of two species of basil (Ocimumciliatum and Ocimumbasilicum) under reduced irrigation. Sci Hortic 217:114–122
Pongprayoon W, Roytrakul S, Pichayangkura R, Chadchawan S (2013) The role of hydrogen peroxide in chitosan-induced resistance to osmotic stress in rice (*Oryza sativa* L.). Plant Growth Regul 70:159–173
Priyaadharshini M, Sritharan N, Senthil A, Marimuthu S (2019) Physiological studies on effect of chitosan nanoemulsion in pearl millet under drought condition. J Pharmacogn Phytochem 8:3304–3307
Rabêlo VM, Magalhães PC, Bressanin LA, Carvalho DT, Dos Reis CO, Karam D, Doriguetto AC, Dos Santos MH, Filho PRDSS, De Souza TC (2019) The foliar application of a mixture of semisynthetic chitosan derivatives induces tolerance to water deficit in maize, improving the antioxidant system and increasing photosynthesis and grain yield. Sci Rep 9:8164
Ramakrishna R, Sarkar D, Manduri A, Iyer SG, Shetty K (2017) Improving phenolic bioactive-linked anti-hyperglycemic functions of dark germinated barley sprouts (Hordeumvulgare L.) using seed elicitation strategy. J Food Sci Technol 54:3666–3678
Rinaudo M (2006) Chitin and chitosan: properties and applications. Prog Polym Sci 31:606–632
Sen SK, Chouhan D, Das D, Ghosh R, Mandal P (2019) Improvisation of salinity stress response in mung bean through solid matrix priming with normal and nano-sized chitosan. Int J Boil Macromol 145:108–123
Silveira NM, Seabra AB, Marcos FC, Pelegrino MT, Machado EC, Ribeiro RV (2019) Encapsulation of S-nitrosoglutathione into chitosan nanoparticles improves drought tolerance of sugarcane plants. Nitric Oxide 2019(84):38–44
Xiong L, Schumaker KS, Zhu JK (2002) Cell signaling during cold, drought, and salt stress. Plant Cell 14:S165–S183
Yin H, Li S, Zhao X, Du Y, Ma X (2006) cDNA microarray analysis of gene expression in *Brassica napus* treated with oligochitosan elicitor. Plant Physiol Biochem 44:910–916

Zayed M, ElKafafi S, Zedan A, Dawoud S (2017) Effect of Nano chitosan on growth, physiological and biochemical parameters of Phaseolus vulgaris under salt stress. J Plant Prod 8:577–585
Zhang X, Wollenweber B, Jiang N, Liu F, Zhao J (2008) Water deficits and heat shock effects on photosynthesis of a transgenic Arabidopsis thaliana constitutively expressing ABP9, a bZIP transcription factor. J Exp Bot 59:839–848
Zong H, Kecheng L, Liu S, Song L, Xing R, Chen X, Li P (2017a) Improvement in cadmium tolerance of edible rape (Brassica rapa L.) with exogenous application of chitooligosaccharide. Chemosphere 181:92–100
Zong H, Liu S, Xing R, Chen X, Li P (2017b) Protective effect of chitosan on photosynthesis and antioxidative defense system in edible rape (*Brassica rapa* L.) in the presence of cadmium. Ecotoxicol Environ Saf 138:271–278

Index

M. Faizan et al. (eds.), *Sustainable Agriculture Reviews 53*, Sustainable Agriculture Reviews 53, https://doi.org/10.1007/978-3-030-86876-5

The manufacturer's authorised representative in the EU is Springer Nature Customer Service Centre GmbH, Europaplatz 3, 69115 Heidelberg, Germany. If you have any concerns regarding our products, please contact ProductSafety@springernature.com

Printed and bound by CPI Group (UK) Ltd, Croydon, CR0 4YY
30/09/2026
02225491-0015